Molecular Diagnosis of Infectious Diseases

METHODS IN MOLECULAR MEDICINE™

John M. Walker, SERIES EDITOR

98. **Tumor Necrosis Factor:** *Methods and Protocols,* edited by *Angelo Corti and Pietro Ghezzi,* 2004

97. **Molecular Diagnosis of Cancer:** *Methods and Protocols, Second Edition,* edited by *Joseph E. Roulston and John M. S. Bartlett,* 2004

96. **Hepatitis B and D Protocols:** *Volume 2, Immunology, Model Systems, and Clinical Studies,* edited by *Robert K. Hamatake and Johnson Y. N. Lau,* 2004

95. **Hepatitis B and D Protocols:** *Volume 1, Detection, Genotypes, and Characterization,* edited by *Robert K. Hamatake and Johnson Y. N. Lau,* 2004

94. **Molecular Diagnosis of Infectious Diseases,** edited by *Jochen Decker and Udo Reischl,* 2004

93. **Anticoagulants, Antiplatelets, and Thrombolytics,** edited by *Shaker A. Mousa,* 2004

92. **Molecular Diagnosis of Genetic Diseases,** *Second Edition,* edited by *Rob Elles and Roger Mountford,* 2003

91. **Pediatric Hematology:** *Methods and Protocols,* edited by *Nicholas J. Goulden and Colin G. Steward,* 2003

90. **Suicide Gene Therapy:** *Methods and Reviews,* edited by *Caroline J. Springer,* 2003

89. **The Blood–Brain Barrier:** *Biology and Research Protocols,* edited by *Sukriti Nag,* 2003

88. **Cancer Cell Culture:** *Methods and Protocols,* edited by *Simon P. Langdon,* 2003

87. **Vaccine Protocols,** *Second Edition,* edited by *Andrew Robinson, Martin P. Cranage, and Michael Hudson,* 2003

86. **Renal Disease:** *Techniques and Protocols,* edited by *Michael S. Goligorsky,* 2003

85. **Novel Anticancer Drug Protocols,** edited by *John K. Buolamwini and Alex A. Adjei,* 2003

84. **Opioid Research:** *Methods and Protocols,* edited by *Zhizhong Z. Pan,* 2003

83. **Diabetes Mellitus:** *Methods and Protocols,* edited by *Sabire Özcan,* 2003

82. **Hemoglobin Disorders:** *Molecular Methods and Protocols,* edited by *Ronald L. Nagel,* 2003

81. **Prostate Cancer Methods and Protocols,** edited by *Pamela J. Russell, Paul Jackson, and Elizabeth A. Kingsley,* 2003

80. **Bone Research Protocols,** edited by *Miep H. Helfrich and Stuart H. Ralston,* 2003

79. **Drugs of Abuse:** *Neurological Reviews and Protocols,* edited by *John Q. Wang,* 2003

78. **Wound Healing:** *Methods and Protocols,* edited by *Luisa A. DiPietro and Aime L. Burns,* 2003

77. **Psychiatric Genetics:** *Methods and Reviews,* edited by *Marion Leboyer and Frank Bellivier,* 2003

76. **Viral Vectors for Gene Therapy:** *Methods and Protocols,* edited by *Curtis A. Machida,* 2003

75. **Lung Cancer:** *Volume 2, Diagnostic and Therapeutic Methods and Reviews,* edited by *Barbara Driscoll,* 2003

74. **Lung Cancer:** *Volume 1, Molecular Pathology Methods and Reviews,* edited by *Barbara Driscoll,* 2003

73. **E. coli:** *Shiga Toxin Methods and Protocols,* edited by *Dana Philpott and Frank Ebel,* 2003

72. **Malaria Methods and Protocols,** edited by *Denise L. Doolan,* 2002

71. *Haemophilus influenzae* **Protocols,** edited by *Mark A. Herbert, Derek Hood, and E. Richard Moxon,* 2002

70. **Cystic Fibrosis Methods and Protocols,** edited by *William R. Skach,* 2002

69. **Gene Therapy Protocols,** *Second Edition,* edited by *Jeffrey R. Morgan,* 2002

METHODS IN MOLECULAR MEDICINE™

Molecular Diagnosis of Infectious Diseases

Second Edition

Edited by

Jochen Decker

Accenture GmbH, Munich, Germany

Udo Reischl

*Institute for Medical Microbiology and Hygiene
University of Regensburg, Regensburg, Germany*

Foreword by

Christoph Schnorr and Michael Schwörer

Accenture GmbH, Munich, Germany

Humana Press   Totowa, New Jersey

© 2004 Humana Press Inc.
999 Riverview Drive, Suite 208
Totowa, New Jersey 07512

www.humanapress.com

All rights reserved.

No part of this book may be reproduced, stored in a retrieval system, or transmitted in any form or by any means, electronic, mechanical, photocopying, microfilming, recording, or otherwise without written permission from the Publisher. Methods in Molecular Medicine™ is a trademark of The Humana Press Inc.

All authored papers, comments, opinions, conclusions, or recommendations are those of the author(s), and do not necessarily reflect the views of the publisher.

This publication is printed on acid-free paper. ∞
ANSI Z39.48-1984 (American Standards Institute) Permanence of Paper for Printed Library Materials.

Cover design by Patricia F. Cleary.

Cover illustration: Figure 9 from Chapter 23, "Basic Problems of Serological Laboratory Diagnosis," by Walter Fierz.

For additional copies, pricing for bulk purchases, and/or information about other Humana titles, contact Humana at the above address or at any of the following numbers: Tel.: 973-256-1699; Fax: 973-256-8341; E-mail: humana@humanapr.com; or visit our Website: www.humanapress.com

Photocopy Authorization Policy:
Authorization to photocopy items for internal or personal use, or the internal or personal use of specific clients, is granted by Humana Press Inc., provided that the base fee of US $25.00 per copy is paid directly to the Copyright Clearance Center at 222 Rosewood Drive, Danvers, MA 01923. For those organizations that have been granted a photocopy license from the CCC, a separate system of payment has been arranged and is acceptable to Humana Press Inc. The fee code for users of the Transactional Reporting Service is: [1-58829-221-5/04 $25.00].

Printed in the United States of America. 10 9 8 7 6 5 4 3 2 1

Library of Congress Cataloging in Publication Data

Molecular diagnosis of infectious diseases / edited by Jochen Decker, Udo Reischl.— 2nd ed.
      p. ; cm. — (Methods in molecular medicine ; 94)
  Includes bibliographical references and index.
  ISBN 1-58829-221-5 (alk. paper) eISBN 1-59259-679-7
  1. Communicable diseases—Molecular diagnosis—Laboratory manuals.
  [DNLM: 1. Communicable Diseases—diagnosis. 2. Biological Assay. 3. Communicable Diseases—genetics. 4. Immunologic Tests. 5. Recombinant Proteins—diagnostic use. WC 100 M718 2004] I. Decker, Jochen. II. Reischl, Udo. III. Series.
  RC113.3.M655 2004
    616.9'047582—dc21
                                              2003012288

Foreword

Populations of the western world are now healthier and enjoying higher life expectancy than ever. They are beginning to benefit from an array of costly new therapies made possible through recent rapid advances in medical science and technology, and their demands on modern medicine are rising. Meanwhile, healthcare systems are struggling with their outdated legacy models of the mid-20th century and are experiencing ever-increasing financial pressure from governments and health insurance organizations. The equation is no longer in balance, and this predicament is forcing societies to explore new approaches to managing healthcare in the future.

Since the first edition of *Molecular Diagnosis of Infectious Diseases* was published, we have witnessed the sequencing of the (almost) complete human genome and a shift in medical research from an emphasis on genetics to the advancement and useful application of proteomics. Bioinformatics has become the key tool for managing and analyzing the upsurge of data, and faster and more effective test methods and technologies have opened up new prospects for industry and academia. The tools of modern genomics and proteomics are now being utilized to specifically guide the discovery of drugs for the prevention, diagnosis, and treatment of human disease. They may also help us to find a way out of the current healthcare calamity.

One of the aspects of new scientific research and development that could help to solve the dilemma is improved diagnostics and its integration with therapeutic approaches. Fast, sensitive, and precise diagnostic techniques will ensure early and accurate disease diagnosis, and make it possible to predict individual drug responses and to determine genetic susceptibility to specific diseases.

If information on a patient's individual response to a drug can be obtained, drug therapy can be tailored to maximize therapeutic effect and minimize side effects. Targeting patients in this way will reduce the costs of redundant and ineffective treatments and will reduce the duration of the illness and thus hospital stays and the number of physician visits. Preventive treatment may also be possible if predispositions can be identified.

Keeping the aging population healthy throughout life is the key to finding a feasible healthcare paradigm. The integration of diagnostics and individual therapies will be crucial factors in restructuring care, treatment, and payment models in the medical field. Only with the application of new scientific approaches and efficient organizational models will high quality and affordable healthcare services be available for us all in the 21st century.

Christoph Schnorr and Michael Schwörer
Accenture

Preface

Infectious diseases represent one of the main threats to public health today. Though improvements in antibiotic and antiviral therapy have been realized during the last years, pathogens were nonetheless able to develop mechanisms rendering them resistant towards specific treatments. Furthermore, a variety of recently discovered microorganisms, such as *Helicobacter pylori*, hepatitis D and E, and HIV-2, have been accepted as etiological agents of human diseases.

In order to allow a rational and specific use of antibiotic and antiviral pharmaceuticals, the detection and identification of the causative pathogen is of major importance and will require increasingly accurate diagnosis with rising demands for the specificity, sensitivity, and speed of the corresponding assay. The revolutionary progress associated with molecular biology-based technology, such as the detection of DNA or the manufacture of recombinant antigens or antibodies, has already gained major advantages in medicine and will contribute to the development of improved assay systems.

Basically, there are three different ways for the specific detection of a given pathogen: (1) the detection of the pathogen itself (e.g., by microscopy, culture or biochemical characteristics); (2) the direct detection of selected components of the pathogen (e.g., nucleic acids, antigenic proteins); and, in an indirect way, (3) the detection of specific antibodies generated by the infected organism.

Recently published books in the field of clinical microbiology are substantially focused on different aspects of PCR. As a consequence, most of them lack in highlighting the equal methodical progress of serodiagnosis by recombinant proteins and antibodies. This is of particular importance in the proteomics era. With the permanently increasing demands of both the cost per test and the expressiveness of the results, the use of recombinant antigenic proteins and recombinant antibodies for the serodiagnosis of infectious diseases has gained more and more importance. Large quantities of recombinant proteins can now be produced at comparatively low cost in suitable bacterial or cell culture systems. A variety of molecular biological techniques are available at present, and have proved to be advantageous in the preparation of recombinant proteins.

Apart from a brief description of the principles of immunological assays, the book describes both established and novel strategies that have been applied successfully in identification of valuable diagnostic markers, epitope mapping, the characterization of immunomodulatory components, the production and purification of recombinant antigens and their use as diagnostic reagents in immunological assays. Promising biosensor technology and recent developments in antibody engineering are also covered. In addition to these straightforward techniques, the book addresses basic problems of serological laboratory diagnosis as well. Some duplication of important topics has been purposely introduced to offer the reader several approaches to the same problem. Significant scientific progress has been achieved since the first edition of *Molecular Diagnosis of Infectious Diseases*. The genomic era picked its peak with the publication of the complete sequence of the human genome. Sequencing bacterial genomes has now become a routine task. Bioinformatics has been accepted as an invaluable tool for managing the large amount of data generated by high-throughput analysis, genomics and proteomic studies. Proteomics has seen a renaissance underlining the significance of proteins and protein based techniques in science and clinical diagnostics. Hence, special emphasis is placed on proteomics and the characterization and modification of proteins.

Instead of focusing on particular infectious agents, it is hoped that the collection of detailed protocols providing comprehensive and up-to-date information will be especially useful to researchers and students to get familiar with the principles of Molecular Diagnostics, and guiding them to set up test systems tailored to their specific needs.

Since molecular biology-based tools are subject to permanent improvement, this volume does not attempt the impossible task of treating all aspects of the various experimental approaches in the field. Rather, it depicts a kind of cross-section of the actual possibilities for the conception of refined assays.

Undoubtedly further progress in the rapid and specific detection of pathogenic organisms is expected with the help of modern molecular biology, and the future will show what kind of assay and what kind of diagnostic marker will prove useful for individual clinical situations.

We are especially indebted to Accenture for their support and Prof. Hans Wolf and Norbert Lehn for giving us the opportunity to gain substantial experience in the field of medical microbiology. Without their confidence and continuous support, much of our work would not have been possible. We wish to thank Prof. John Walker for his encouragement and Humana Press for their excellent assistance during the assembly of this volume. We are also thankful to our families for their patience and for placing enough time to our disposal to launch out into this promising project. Finally, we are grateful to all of the contributing authors for their constantly high level of motivation and enthusiasm and, last but not least, for providing such good manuscripts.

Jochen Decker
Udo Reischl

Contents

Foreword v
Preface vii
Contributors xiii

PART I. PROTEIN-BASED IDENTIFICATION OF ANTIGENS
1 Proteomic Approaches to Antigen Discovery
   *Karen M. Dobos, John S. Spencer, Ian M. Orme,
   and John T. Belisle* 3
2 Immunoproteomics
   Alexander Krah and Peter R. Jungblut 19
3 Immunoprecipitation and Blotting: *The Visualization
   of Small Amounts of Antigens Using Antibodies and Lectins*
   Stephen Thompson 33

PART II. DNA-BASED IDENTIFICATION OF ANTIGENS
4 Representational Difference Analysis of cDNA
   Lucas D. Bowler 49
5 Microarray Data Analysis and Mining
   *Silvia Saviozzi, Giovanni Iazzetti, Enrico Caserta,
   Alessandro Guffanti, and Raffaele A. Calogero* 67
6 Expression Cloning
   *Michael J. Lodes, Davin C. Dillon, Raymond L. Houghton,
   and Yasir A. W. Skeiky* 91

PART III. IDENTIFICATION OF EPITOPES
AND IMMUNOMODULATORY COMPONENTS
7 Determination of Epitopes by Mass Spectrometry
   Christine Hager-Braun and Kenneth B. Tomer 109
8 Identification of T Cell Epitopes Using ELISpot
   and Peptide Pool Arrays
   Timothy W. Tobery and Michael J. Caulfield 121
9 Virus-like Particles: *A Novel Tool for the Induction
   and Monitoring of Both T-Helper and Cytotoxic
   T-Lymphocyte Activity*
   Ludwig Deml, Jens Wild, and Ralf Wagner 133

10 Application of Single-Cell Cultures of Mouse Splenocytes
as an Assay System to Analyze the Immunomodulatory
Properties of Bacterial Components
*Ludwig Deml, Michael Aigner, Alexander Eckhardt,
Jochen Decker, Norbert Lehn, and Wulf Schneider-Brachert* 159

PART IV. EXPRESSION OF RECOMBINANT PROTEINS

11 High-Throughput Expression and Purification of
6xHis-Tagged Proteins in a 96-Well Format
Jutta Drees, Jason Smith, Frank Schäfer, and Kerstin Steinert 179

12 Production of Antigens in *Chlamydomonas reinhardtii*: Green
Microalgae as a Novel Source of Recombinant Proteins
Markus Fuhrmann 191

13 Codon-Optimized Genes that Enable Increased Heterologous
Expression in Mammalian Cells and Elicit Efficient Immune
Responses in Mice after Vaccination of Naked DNA
Marcus Graf, Ludwig Deml, and Ralf Wagner 197

PART V. PURIFICATION, MODIFICATION, AND RENATURATION
OF RECOMBINANT PROTEINS

14 Purification and Immunological Characterization
of Recombinant Antigens Expressed in the Form
of Insoluble Aggregates (Inclusion Bodies)
Udo Reischl 213

15 Purification of Recombinant Proteins with High Isoelectric Points
Raffaele A. Calogero and Anna Aulicino 225

16 Refolding of Inclusion Body Proteins
Marcus Mayer and Johannes Buchner 239

17 Small-Molecule–Protein Conjugation Procedures
Stephen Thompson 255

PART VI. CHARACTERIZATION OF RECOMBINANT PROTEINS

18 Structural Characterization of Proteins and Peptides
Rainer Deutzmann 269

19 Determination of Kinetic Data Using Surface Plasmon Resonance
Biosensors
*Claudia Hahnefeld, Stephan Drewianka,
and Friedrich W. Herberg* 299

20 Affinity Measurements of Biological Molecules by a Quartz Crystal
   Microbalance (QCM) Biosensor
   *Uwe Schaible, Michael Liss, Elke Prohaska, Jochen Decker,
   Karin Stadtherr, and Hans Wolf* 321

PART VII. EVALUATION OF RECOMBINANT PROTEINS
IN IMMUNOLOGICAL TEST SYSTEMS
21 Solid Supports in Enzyme-Linked Immunosorbent Assay
   and Other Solid-Phase Immunoassays
   John E. Butler 333
22 Design and Preparation of Recombinant Antigens as Diagnostic
   Reagents in Solid-Phase Immunosorbent Assays
   Alan Warnes, Anthony R. Fooks, and John R. Stephenson 373
23 Basic Problems of Serological Laboratory Diagnosis
   Walter Fierz 393
24 Molecular Diagnostics Resources on the Internet
   Larry Winger 429

PART VIII. RECOMBINANT RECEPTOR MOLECULES
25 Cloning Single-Chain Antibody Fragments (scFv)
   from Hybridoma Cells
   Lars Toleikis, Olaf Broders, and Stefan Dübel 447
Index 459

Contributors

MICHAEL AIGNER • *Institute for Medical Microbiology and Hygiene, University of Regensburg, Regensburg, Germany*
ANNA AULICINO • *Department of Biological and Clinical Sciences, University of Torino, Torino, Italy*
JOHN T. BELISLE • *Mycobacteria Research Laboratories, Department of Microbiology, Immunology, and Pathology, Colorado State University, Fort Collins, CO*
LUCAS D. BOWLER • *Trafford Centre for Medical Research, University of Sussex, Brighton, UK*
OLAF BRODERS • *University of Heidelberg, Molecular Genetics, Heidelberg, Germany*
JOHANNES BUCHNER • *Institute for Organic Chemistry and Biochemistry, Technical University Munich, Germany*
JOHN E. BUTLER • *Department of Microbiology and Interdisciplinary Immunology Training Program, The University of Iowa, Iowa City, IA*
RAFFAELE. A. CALOGERO • *Department of Biological and Clinical Sciences, University of Torino, Torino, Italy*
ENRICO CASERTA • *Department of Biology MCA, University of Camerino, Camerino (MC), Italy*
MICHAEL J. CAULFIELD • *Department of Virus and Cell Biology, Merck Research Laboratories, West Point, PA*
JOCHEN DECKER • *Accenture GmbH, Munich, Germany*
LUDWIG DEML • *Institute for Medical Microbiology and Hygiene, University of Regensburg, Regensburg, Germany*
RAINER DEUTZMANN • *Institute for Biochemistry, University of Regensburg, Regensburg, Germany*
DAVIN C. DILLON • *Corixa Corporation, Seattle, WA*
KAREN M. DOBOS • *Mycobacteria Research Laboratories, Department of Microbiology, Immunology, and Pathology, Colorado State University, Fort Collins, CO*
JUTTA DREES • *Qiagen, Hilden, Germany*
STEPHAN DREWIANKA • *Biaffin GmbH & Co KG, Kassel, Germany*
STEFAN DÜBEL • *Technical University of Braunschweig, Institute of Biochemistry and Biotechnology, Braunschweig, Germany*

ALEXANDER ECKHARDT • *Institute for Medical Microbiology and Hygiene, University of Regensburg, Regensburg, Germany*
WALTER FIERZ • *Institute for Clinical Microbiology and Immunology, St. Gallen, Switzerland*
ANTHONY R. FOOKS • *Veterinary Laboratory Agency, Addlestone, Weybridge, Surrey, UK*
MARKUS FUHRMANN • *University of Regensburg, Competence Center for Fluorescent Bioanalysis, Regensburg, Germany*
MARCUS GRAF • *Geneart, Regensburg, Germany*
ALESSANDRO GUFFANTI • *FIMO–FIRC Institute of Molecular Oncology, Milano, Italy*
CHRISTINE HAGER-BRAUN • *Laboratory of Structural Biology, National Institute of Environmental Health Sciences, National Institutes of Health, Department of Health and Human Services, Research Triangle Park, NC*
CLAUDIA HAHNEFELD • *Department of Biochemistry, University of Kassel, FB 18, Kassel, Germany*
FRIEDRICH W. HERBERG • *Department of Biochemistry, University of Kassel, FB 18, Kassel, Germany*
RAYMOND L. HOUGHTON • *Corixa Corporation, Seattle, WA*
GIOVANNI IAZZETTI • *Department of Genetics, Biological and General Biology, University "Federico II," Napoli, Italy*
PETER R. JUNGBLUT • *Core Facility Protein Analysis, Max Planck Institute for Infection Biology, Berlin, Germany*
ALEXANDER KRAH • *Department of Molecular Biology, Max Planck Institute for Infection Biology, Berlin, Germany*
NORBERT LEHN • *Institute for Medical Microbiology and Hygiene, University of Regensburg, Regensburg, Germany*
MICHAEL LISS • *Geneart, Regensburg, Germany*
MICHAEL J. LODES • *Corixa Corporation, Seattle, WA*
MARCUS MAYER • *Institute for Organic Chemistry and Biochemistry, Technical University Munich, Germany*
IAN M. ORME • *Mycobacteria Research Laboratories, Department of Microbiology, Immunology, and Pathology, Colorado State University, Fort Collins, CO*
ELKE PROHASKA • *NascaCell GmbH, Tutzing, Germany*
UDO REISCHL • *Institute for Medical Microbiology and Hygiene, University of Regensburg, Regensburg, Germany*
SILVIA SAVIOZZI • *Department of Biological and Clinical Sciences, University of Torino, Torino, Italy*
FRANK SCHÄFER • *Qiagen, Hilden, Germany*

Contributors

UWE SCHAIBLE • *Institute for Medical Microbiology and Hygiene, University of Regensburg, Regensburg, Germany*

WULF SCHNEIDER-BRACHERT • *Institute for Medical Microbiology and Hygiene, University of Regensburg, Regensburg, Germany*

YASIR A.W. SKEIKY • *Corixa Corporation, Seattle, WA*

JASON SMITH • *Qiagen, Hilden, Germany*

JOHN S. SPENCER • *Mycobacteria Research Laboratories, Department of Microbiology, Immunology, and Pathology, Colorado State University, Fort Collins, CO*

KARIN STADTHERR • *Institute for Medical Microbiology and Hygiene, University of Regensburg, Regensburg, Germany*

KERSTIN STEINERT • *Qiagen, Hilden, Germany*

JOHN R. STEPHENSON • *London School of Hygiene and Tropical Medicine, London, UK*

STEPHEN THOMPSON • *Department of Clinical Biochemistry, University of Newcastle upon Tyne, The Medical School, Newcastle upon Tyne, UK*

TIMOTHY W. TOBERY • *Department of Virus Vaccine Research, Merck Research Laboratories, West Point, PA*

LARS TOLEIKIS • *Hospital for Women, University of Heidelberg, Heidelberg, Germany*

KENNETH B. TOMER • *Laboratory of Structural Biology, National Institute of Environmental Health Sciences, National Institutes of Health, Department of Health and Human Services, Research Triangle Park, NC*

RALF WAGNER • *Institute for Medical Microbiology and Hygiene, University of Regensburg, Regensburg, Germany*

ALAN WARNES • *St. Marks Hospital, London, UK*

JENS WILD • *Institute for Medical Microbiology and Hygiene, University of Regensburg, Regensburg, Germany*

LARRY WINGER • *School of Clinical Laboratory Sciences, The Medical School, Univiversity of Newcastle upon Tyne, Newcastle upon Tyne, UK*

HANS WOLF • *Institute for Medical Microbiology and Hygiene, University of Regensburg, Regensburg, Germany*

I

PROTEIN-BASED IDENTIFICATION OF ANTIGENS

1

Proteomic Approaches to Antigen Discovery

Karen M. Dobos, John S. Spencer, Ian M. Orme, and John T. Belisle

Abstract

Proteomics has been widely applied to develop two-dimensional polyacrylamide gel electrophoresis maps and databases, evaluate gene expression profiles under different environmental conditions, assess global changes associated with specific mutations, and define drug targets of bacterial pathogens. When coupled to immunological assays, proteomics may also be used to identify B-cell and T-cell antigens within complex protein mixtures. This chapter describes the proteomic approaches developed by our laboratories to accelerate the antigen discovery program for *Mycobacterium tuberculosis*. As presented or with minor modifications, these techniques may be universally applied to other bacterial pathogens or used to identify bacterial proteins possessing other immunological properties.

Key Words: Proteomics; antigen; B-cell; T-cell; *Mycobacterium*; bacteria; pathogens.

1. Introduction

The ability to identify antigens produced by bacterial pathogens that are effective diagnostic or vaccine candidates of disease depends on multiple variables. Two of the most important variables are the source of clinical specimens (immune T-cells or sera) used to identify potential antigens and the source or nature of the crude materials containing the putative antigens. Factors that influence the usefulness of clinical specimens include whether samples were obtained from diseased individuals or experimentally infected animals and the state of disease at the time of specimen collection. Likewise, the choice between native bacterial products and recombinant products as the starting material for antigen discovery efforts may significantly influence whether a useful antigen will be identified. An equally important factor is the number of potential antigens that can be screened in a single experiment. The use of recombinant molecular biology methods and the screening of large recombinant libraries is one approach toward maximizing the number of potential antigen targets *(1–4)*. Although recombinant expression systems have been widely used for antigen identification, there are several potential drawbacks. Specifically, variability between the folding of a recombinant and native protein can complicate B-cell antigen discovery efforts *(5)*, and the contamination of recombinant proteins with other bacterial products such as endotoxin is a major obstacle for cellular assays used to identify T-cell antigens *(6)*.

The ability to sequence whole genomes rapidly and the availability of several fully annotated bacterial genomes have profoundly altered the basic experimental approach to the study of bacterial physiology and pathogenesis *(7,8)*. Previous to the sequencing of whole bacterial genomes, investigators would typically focus on a relatively small number of genes or gene products and develop specific assays to assess the activities or relevance of these gene products. In contrast, the availability of whole genome sequences has now allowed for the development of methodologies such as DNA microarrays and proteomics to identify all the genes that are potentially involved in a specific cellular process *(9–11)*. Unlike DNA microarrays, the technologies commonly used for proteomics studies [two-dimensional polyacrylamide gel electrophoresis (2D-PAGE) and mass spectrometry (MS) of peptides] have been around for decades *(12,13)*. The power of these two technologies was brought together by the need to assess rapidly all the proteins produced in a particular bacterial species, as well as the development of innovative software that allows for the interrogation of MS data against genome sequences to identify proteins of interest *(14)*. These technologies and the philosophy that we no longer need to focus on select sets of proteins, but should be evaluating the complete proteome in a single experiment can now be applied to antigen discovery efforts. Moreover, proteomics technologies allow antigen discovery programs to focus on large sets of native proteins and eliminate a reliance on recombinant technologies to expand the pool of proteins to be screened.

In our laboratory, 2D-PAGE, Western blot analysis, and liquid chromatography-mass spectrometry (LC-MS) were used to define 26 proteins of *Mycobacterium tuberculosis* that reacted with patient sera; three of these subsequently were determined to have significant potential as serodiagnostic reagents *(15,16)*. This approach is a relatively facile method to screen for B-cell antigens. The use of proteins resolved by 2D-PAGE and transferred to nitrocellulose was also applied to T-cell antigen identification *(17)*. Although this work revealed several potential T-cell antigens, there are restrictions to its use. In particular, the concentration of protein tested is unknown and the amount of protein obtained is most likely insufficient for multiple assays. Thus, to increase the protein yield, we recently applied 2D liquid-phase electrophoresis (LPE) coupled with an in vitro interferon-γ (IFN-γ) assay and LC-tandem MS to identify 30 proteins from the culture filtrate and cytosol of *M. tuberculosis* that possess a potent capacity to induce antigen-specific IFN-γ secretion from the splenocytes of *M. tuberculosis*-infected mice *(18)*. In this chapter, we detail the proteomics approach used in the identification of candidate B- and T-cell antigens from *M. tuberculosis*. However, these methods can be universally applied to the discovery of antigens from other bacterial pathogens as well as parasites.

2. Materials
2.1. Preparation of Subcellular Fractions

1. Bacterial cell cultures (400 mL or greater) (*see* **Note 1**).
2. Breaking buffer: phosphate-buffered saline (PBS; pH 7.4), 1 mM EDTA, 0.7 µg/mL pepstatin, 0.5 µg/mL leupeptin, 0.2 mM phenylmethylsulfonyl fluoride (PMSF), 0.6 µg/mL DNase, µg/mL RNase (*see* **Note 2**).
3. NaN$_3$.
4. Dialysis buffer: 10 mM ammonium bicarbonate, 1 mM dithiothreitol (DTT), 0.02% NaN$_3$.
5. 10 mM ammonium bicarbonate.
6. Vacuum pump.
7. Amicon ultrafiltration unit with a 10,000 Da MWCO membrane (Millipore, Bedford, MA; cat. no. PLGC07610).
8. 0.2 µm Zap Cap S Plus bottle filtration units (Nalgene, Rochester, NY).
9. Dialysis tubing (3500 Da MWCO).
10. French Press and French Press cell.
11. Sterile 250-mL high-speed centrifuge tubes.
12. Sterile 30-mL ultracentrifuge tubes.

2.2. Identification of B-Cell Antigens via 2D Western Blot Analysis with Protein/Antigen Double Staining

2.2.1. Optimization of Serum Titers for Detection of Antigens by Western Blot

1. Human or experimental animal sera samples (*see* **Notes 3** and **4**).
2. 15% sodium dodecyl sulfate (SDS)-PAGE gels (7 × 10 cm).
3. Protein molecular weight standards.
4. Laemmli sample buffer (5X): 0.36 g Tris-base, 5.0 mL glycerol, 1.0 g SDS, 5.0 mg bromophenol blue, 1.0 mL β-mercaptoethanol; QS to 10 mL with Milli-Q water, vortex, and store at 4°C or less *(19)*.
5. SDS-PAGE running buffer: 3.02 g Tris-base (pH 8.3), 14.42 g glycine, 1 g SDS per 1 L.
6. Nitrocellulose membrane, 0.22 µm.
7. Transfer buffer: 3.03 g Tris-base (pH 8.3), 14.4 g glycine (pH 8,3), 800 mL H$_2$O, 200 mL CH$_3$OH. Dissolve reagents in water before adding CH$_3$OH *(20)*.
8. Tris-buffered saline (TBS): 50 mM Tris-HCl (pH 7.4), 150 mM NaCl.
9. Wash buffer: TBS containing 0.5% vol/vol Tween 80.
10. Blocking buffer: wash buffer containing 3% w/v bovine serum albumin (BSA).
11. Anti-human IgG conjugated to horseradish peroxidase (HRP).
12. BM Blue POD substrate, precipitating (Roche Molecular Biochemicals, Indianapolis, IN; cat. no. 1442066).
13. Mini incubation trays for 2–4 mm nitrocellulose membrane strips (Bio-Rad, Hercules, CA; cat. no. 170-3902).

2.2.2. 2D Western Blot Analysis for Identification of B-Cell Antigens

1. Isoelectric focusing (IEF) rehydration buffer: 8 M urea, 1% 3[(3-cholamidopropyl) dimethylammonio]-1-propanesulfonate (CHAPS), 20 mM DTT, 0.5% ampholytes, 0.001% bromphenol blue (*see* **Notes 5–7**).
2. Immobilized pH gradient (IPG) strips *(21)* (*see* **Note 8**).
3. SDS-PAGE equilibration buffer: 150 mM Tris-HCl (pH 8.5), 0.2% SDS, 10% glycerol, 20 mM DTT, and 0.001% bromphenol blue.

4. 1% agarose dissolved in Milli-Q water.
5. Preparative SDS-PAGE gels (16 × 20 cm).
6. Protein molecular weight standards.
7. SDS-PAGE running buffer (see **Subheading 2.2.1.**, **item 5**).
8. Coomassie stain: 1% coomassie brilliant blue R-250 in 40% methanol, 10% acetic acid (see **Note 9**).
9. Coomassie destain 1: 40% methanol, 10% acetic acid.
10. Coomassie destain 2: 5% methanol.
11. Transfer buffer (see **Subheading 2.2.1.**, **item 7**).
12. Nitrocellulose, 0.22 μm.
13. Digoxigenin-3-O-methylcarbonyl-ε-aminocaproic acid-N-hydroxysuccinimide ester (DIG-NHS) (Roche Molecular Biochemicals; cat. no. 1333054); 0.5 mg/mL in N,N-dimethylformamide (DMF).
14. 50 mM potassium phosphate buffer (pH 8.5).
15. TBS: 50 mM Tris-HCl (pH 7.4), 150 mM NaCl.
16. Nonidet P-40, 10% solution.
17. H_2O_2, 30% solution.
18. Anti-digoxigenin-alkaline phosphatase, Fab fragments (Roche Molecular Biochemicals; cat. no. 1093274).
19. INT/BCIP stock solution (Roche Molecular Biochemicals; cat. no. 1681460).
20. INT/BCIP buffer: 100 mM Tris-HCl (pH 9.5), 50 mM $MgCl_2$, 10 mM NaCl.
21. Anti-human IgG conjugated to HRP.
22. IPGphor IEF unit (Amersham Biosciences, Piscataway, NJ) or similar system.
23. SDS-PAGE electrophoresis unit.
24. Gel documentation system.
25. PDQuest 2D-gel analysis software (Bio-Rad) or similar software.

2.2.3. Molecular Identification of Serum Reactive Proteins

1. 0.2 M ammonium bicarbonate.
2. Trifluoroacetic acid (TFA; 10% solution).
3. Destain solution: 60% acetonitrile, in 0.2 M ammonium bicarbonate.
4. Extraction solution: 60% acetonitrile, 0.1% TFA.
5. Modified trypsin, sequencing grade (Roche Molecular Biochemicals; cat. no. 1418025).
6. Washed microcentrifuge tubes (0.65 mL) (see **Note 10**).
7. Electrospray tandem mass spectrometer (such as LCQ classic, Thermo-Finnigan) coupled to a capillary high-performance liquid chromatography (HPLC) device.
8. Capillary C_{18}-reverse phase (RP)-HPLC column.
9. Sequest software *(14)* for interrogating MS and MS/MS data against genomic or protein databases, or similar software.

2.3. Identification of T-Cell Antigens

2.3.1. 2D-LPE of Subcellular Fractions

1. IEF protein solubilization buffer: 8 M urea, 1 mM DTT, 5% glycerol, 2% Nonidet P-40, and 2% ampholytes (pH 3.0–10.0 and pH 4.0–6.5 in a ratio of 1:4; see **Note 7**).
2. Preparative SDS-PAGE gels (16 × 20 cm).
3. SDS-PAGE running buffer (see **Subheading 2.2.1.**, **item 5**).
4. Laemmli sample buffer (see **Subheading 2.2.1.**, **item 4**).
5. 10 mM ammonium bicarbonate.

6. Rotofor preparative IEF unit (Bio-Rad).
7. Whole Gel Eluter (Bio-Rad).

2.3.2. Assay of IFN-γ Induction for Identification of T-Cell Antigens

1. Spleens from infected and naive mice (*see* **Note 11**).
2. Complete RPMI medium (RPMI-1640 medium with L-glutamine, supplemented with 10% bovine fetal calf serum and 50 μM β-mercaptoethanol) (*see* **Note 12**).
3. Hanks' balanced salt solution (HBSS).
4. Gey's hypotonic red blood cell lysis solution: 155 mM NH_4Cl, 10 mM $KHCO_3$; use pyrogen-free water and filter-sterilize.
5. IFN-γ enzyme-linked immunosorbent assay (ELISA) assay kit (Genzyme Diagnostics, Cambridge, MA) (see **Note 13**).
6. Concanavalin A (Con A) (*see* **Note 14**).
7. 70-μm nylon screen (Becton/Dickinson, Franklin Lakes, NJ; cat. no. 35-2350).
8. 96-well sterile tissue culture plates with lids.
9. 96-well microtiter ELISA plates, Dynex Immunlon 4 (Dynex, Chantilly, VA).
10. Conical polypropylene centrifuge tubes (15 and 50 mL).
11. Sterile tissue culture grade Petri dishes (60 mm).
12. Syringes (3 mL) with 1-inch 22-gage needles.
13. Hemocytometer.
14. Tissue culture incubator (5% CO_2, 37°C).
15. ELISA plate reader.

3. Methods

3.1. Preparation of Bacterial Subcellular Fractions (see Note 15)

1. Grow bacterial cells to mid log phase. The medium used should be devoid of exogenous proteins, such as BSA, that may interfere with 2D-PAGE analyses (*see* **Note 1**).
2. Harvest cells by centrifugation at 3500g. Decant the culture supernatant and save. Wash the cell pellet with PBS (pH 7.4) and freeze until preparation of subcellular fractions (*see* **Note 16**).
3. Add NaN_3 to the culture supernatant at a final concentration of 0.04% (w/v).
4. Filter the supernatant using a 0.2-μm filtration unit.
5. Concentrate the culture filtrate to 0.5% of its original volume using an Amicon ultrafiltration unit fitted with a 10-kDa MWCO membrane.
6. Dialyze the concentrated culture filtrate proteins (CFP) against dialysis buffer with at least two changes of this buffer, followed by a final dialysis step against 10 mM ammonium bicarbonate.
7. Filter-sterilize the dialyzed CFP using a 0.2-μm syringe filter or filtration unit.
8. Determine the protein concentration.
9. Store the final culture filtrate preparation at –80°C (*see* **Note 17**).
10. Place the frozen cell pellet at 4°C and thaw.
11. Suspend the cells in ice-cold breaking buffer to a final concentration of 2 g of cells (wet weight) per mL of buffer.
12. Place the cell suspension in a French press cell and lyse via mechanical shearing by applying 20,000 psi of pressure with a French press. Collect the lysate from the French press cell and place on ice (*see* **Note 18**).
13. To the lysate add an equal volume of breaking buffer and mix.

14. Remove unbroken cells by centrifugation of the lysate at 3500g, 4°C in the tabletop centrifuge for 15 min.
15. Collect the supernatant; this is the whole cell lysate.
16. Further separation of the whole cell lysate into cell wall, cell membrane, and cytosol is achieved by differential centrifugation *(22)*. First, centrifuge the lysate at 27,000g, 4°C for 30 min. Collect the supernatant and again centrifuge under the same conditions. Collect the supernatant and store at 4°C.
17. Suspend the 27,000g pellets in breaking buffer, combine them, and centrifuge at 27,000g, 4°C for 30 min. Discard the supernatant, and wash the pellet twice in breaking buffer. The final 27,000g pellet represents the purified cell wall. Suspend this pellet in 10 mM ammonium bicarbonate and extensively dialyze against 10 mM ammonium bicarbonate.
18. Store the dialyzed cell wall suspension at –80°C, after estimating the protein concentration (*see* **Note 17**).
19. Add the supernatant collected in **step 16** to ultracentrifuge tubes and centrifuge at 100,000g, 4°C for 4 h.
20. Collect the supernatant and centrifuge again under the same conditions.
21. The final supernatant solution represents the cytosol fraction. Dialyze the cytosol extensively against 10 mM ammonium bicarbonate, determine the protein concentration, and store at –80°C (*see* **Note 17**).
22. Suspend the 100,000g pellets obtained in **steps 19** and **20** in breaking buffer, combine, and again centrifuge at 100,000g, 4°C for 4 h. This should be repeated twice.
23. The final 100,000g pellet represents the total membrane. Suspend the membranes in 10 mM ammonium bicarbonate, dialyze extensively against 10 mM ammonium bicarbonate, determine the protein concentration, and store at –80°C (*see* **Note 17**).

3.2. Identification of B-Cell Antigens via 2D Western Blot Analysis with Protein/Antigen Double Staining

3.2.1. Optimization of Serum Titers for Detection of Antigens by Western Blot

1. Obtain and thaw a 100-µg aliquot (based on protein concentration) of the subcellular fraction to be analyzed. One 100-µg aliquot will provide enough protein to optimize one patient's serum or one sera pool and one matched control.
2. Dry the samples using a lyophilizer or speed vac.
3. Suspend the subcellular fraction in 80 µL PBS (pH 7.4).
4. Add 20 µL of 5X Laemmli sample buffer to the sample and heat at 100°C for 5 min.
5. Apply 100 µL of sample to one preparative 15% SDS-polyacrylamide gel.
6. Resolve the proteins by 1D SDS-PAGE *(19)*.
7. After the electrophoresis is completed, assemble the gel into a Western blot apparatus and electrotransfer the proteins to a nitrocellulose membrane *(20)*.
8. Remove the nitrocellulose membrane and cut 2-mm vertical strips.
9. Place individual nitrocellulose strips in the wells of a mini incubation tray. Add 500 µL of blocking buffer to each well and rock for 1 h at room temperature (RT) or overnight at 4°C.
10. Thaw the serum samples and generate dilutions (200 µL for each dilution) of the sera from 1:10 to 1:10,000 (*see* **Note 19**).
11. Incubate each strip with a single dilution of serum for 1 h at RT with gentle agitation.
12. Wash the strips repeatedly with 500 µL of wash buffer (minimally, five times).
13. Incubate the strips with 200 µL of the appropriate dilution of the anti-human IgG HRP for 1 h at RT with gentle agitation.

14. Wash the strips repeatedly with 500 μL of TBS (minimally, five times).
15. Develop the strips by addition of BM Blue POD substrate (200 μL, final volume per strip).
16. Stop the reaction by decanting the substrate and rinsing the strips with Milli-Q water.
17. Determine optimum titer based on band intensity and background staining (*see* **Note 20**).

3.2.2. 2D Western Blot Analysis for Identification of B-Cell Antigens

1. Obtain and thaw 400-μg aliquots (*see* **Note 17**) of the subcellular fractions to be analyzed. For each subcellular fraction or analysis, at least two aliquots of the selected subcellular fraction will be required, one for the 2D Western blot and one for a Coomassie-stained 2D gel. More aliquots will be required if replicate Western blots are to be performed or if comparisons between serum samples/pools are to be performed, such as a comparison between infected and healthy control serum.
2. Dry each aliquot of the subcellular fraction by lyophilization (*see* **Note 21**).
3. Add 200 μL of rehydration buffer to each dried subcellular fraction and allow to stand at RT for 4 h or at 4°C overnight. Gentle vortexing can be applied if required.
4. Centrifuge the samples at 10,000g for 30 min, and remove the supernatant without disturbing the pellet (*see* **Note 22**).
5. Transfer samples to IPG strips and allow the strips to rehydrate per manufacturer's recommendations.
6. Resolve proteins by IEF *(12)*.
7. Remove strips from the IEF apparatus, and place in 16 × 150-mm glass test tubes with the acidic end of the strip near the mouth of the tube. Apply 10 mL of SDS-PAGE equilibration buffer, and incubate at RT for 15 min.
8. Warm 1% agarose solution while IPG strips are equilibrating.
9. Assemble 16 × 20-cm preparative SDS-PAGE gels into an electrophoresis apparatus. Add running buffer to cover the lower half of the gels and inner core of the electrophoresis apparatus.
10. Remove IPG strips, clip ends of IPG strips (where no gel is present), and guide each strip into the well of the preparative SDS-PAGE gels using a flat spatula or small pipet tip. The acidic end of the IPG strip should be next to the reference well for the molecular weight standards.
11. Overlay each strip with 1% agarose. When adding agarose, move from one end of the strip to the other. This helps prevent the trapping of air bubbles between the strip and interface of the SDS-PAGE gel. Use enough agarose to cover the strip fully.
12. Add molecular weight standards to the reference well of each gel.
13. Resolve proteins in the second dimension by electrophoresis *(19)*.
14. Remove the preparative SDS-PAGE gels. Stain one gel with Coomassie (*see* **Note 9**).
15. After staining, use a gel documentation system to scan a tif image of the gel. Ensure that the gel is scanned under parameters compatible with the 2D gel analysis software that will be used.
16. Place the gel in a storage tray, cover with Milli-Q water, seal, and store at 4°C.
17. Assemble the second gel into a Western blot apparatus and electrotransfer the proteins to a nitrocellulose membrane *(20)*.
18. Wash nitrocellulose membrane five times with 20 mL of 50 m*M* potassium phosphate (pH 8.5).
19. Prepare total protein labeling solution by adding 10 μL of DIG-NHS and 20 μL of 10% Nonidet P-40 to 20 mL of 50 m*M* potassium phosphate buffer (pH 8.5).

20. Incubate membrane in labeling solution for 1 h at RT with gentle agitation.
21. Wash the membrane five times in 20 mL of TBS.
22. Incubate the labeled membrane in blocking buffer for 1 h at RT with gentle agitation.
23. Wash the membrane briefly with TBS.
24. Add serum at the titer optimized in **Subheading 3.2.1.** and incubate at RT for 1 h with gentle agitation.
25. Wash the membrane five times with 20 mL TBS.
26. Add 20 µL of anti-digoxigenin-alkaline phosphatase to 20 mL of TBS and incubate the membrane in this solution for 1 h with gentle agitation.
27. Wash the membrane five times with 20 mL of TBS.
28. Add anti-human IgG HRP diluted per manufacturer's instructions into TBS, and incubate the membrane in this solution for 1 h at RT with gentle agitation.
29. Wash the membrane five times with 20 mL of TBS.
30. Rinse the membrane briefly in Milli-Q water.
31. To visualize the serum-specific antigens, incubate the membrane in 10 mL of BM Blue POD substrate without agitation. Watch the membranes for blue/purple color development.
32. Aspirate the substrate as soon as color develops, and rinse briefly with Milli-Q water.
33. Using a gel documentation system, capture a tif image of the Western blot showing the serum reactive proteins.
34. To visualize the total protein profile on the Western blot, generate the alkaline phosphatase substrate by adding 75 µL of INT/BCIP stock solution to 10 mL of INT/BCIP buffer. Add to the membrane, incubate without agitation, and watch for color development.
35. Aspirate the substrate solution when reddish brown spots are well defined, and rinse the membrane briefly with Milli-Q water.
36. Using a gel documentation system, capture a tif image of the Western blot showing the total protein profile.
37. Transfer the images of the Coomassie-stained gel, the serum reactive proteins, and the total protein profile of the 2D Western blot to a 2D analysis program. Using this program, match the spots of the three images to allow for identification of the protein spots within the Coomassie-stained 2D gel that correspond to the serum reactive proteins.

3.2.3. Molecular Identification of Serum Reactive Proteins

1. From the Coomassie-stained 2D-gel, excise the protein spots corresponding to those reactive to serum on the 2D Western blot.
2. Cut each gel slice into small pieces (1 × 1 mm), and place the gel pieces from each spot in separate washed microcentrifuge tubes.
3. Destain by covering the gel pieces with destain solution, and incubate at 37°C for 30 min.
4. Discard the acetonitrile solution and repeat **step 3** until the gel slices are completely destained.
5. Dry the gel pieces under vacuum.
6. Dissolve 25 µg of modified trypsin in 300 µL of 0.2 M ammonium bicarbonate.
7. Add 3–5 µL of the trypsin solution to the gel slices.
8. Incubate at room temperature until the trypsin solution is completely absorbed by the gel, approx 15 min.
9. Add 0.2 M ammonium bicarbonate in 10–15 µL increments to rehydrate the gel pieces completely. Allow about 10 min for each aliquot of ammonium bicarbonate to be absorbed by the gel. Also avoid adding an excess of ammonium bicarbonate solution.
10. Incubate the gel slices for 4–12 h at 37°C.

11. Terminate the reaction by adding 0.1 vol of 10% TFA.
12. Collect the supernatant, and place it in a new washed microcentrifuge tube.
13. Add 100-µL of the extract solution to the gel slices and vortex.
14. Incubate the extract solution and gel slices at 37°C for 40 min.
15. Centrifuge the extract, collect the supernatant, and add it to the supernatant collected in **step 12**.
16. Repeat **steps 13–15**.
17. Dry the extract under vacuum.
18. Store the dried peptide extracts at –20°C until analysis by LC-MS/MS *(14)*.

3.3. Identification of T-Cell Antigens (see Note 23)

3.3.1. 2D-LPE of Subcellular Fractions

1. Obtain and thaw 250-mg aliquots of the subcellular fraction(s) to be analyzed.
2. Dry the subcellular fraction by lyophilization (*see* **Note 21**)
3. Solubilize the proteins by adding 60 mL IEF rehydration buffer, and incubate at RT for 4 h or at 4°C overnight.
4. Centrifuge the suspended material at 27,000*g* to remove particulates (*see* **Note 22**).
5. Collect the supernatant and apply this material to the Rotofor (Bio-Rad) apparatus per the manufacturer's instructions. Preparative IEF of the sample should be performed at a constant power of 12 W until the voltage stops increasing and stabilizes (*see* **Note 24**).
6. Harvest the samples from the Rotofor per manufacturer's instructions.
7. Evaluate 8–10 µL of each preparative IEF fraction by SDS-PAGE and Coomassie staining.
8. Pool those fractions that have a high degree of overlap in their protein profile as observed by SDS-PAGE (*see* **Note 25**).
9. Dialyze the IEF fractions extensively against ammonium bicarbonate. After dialysis determine the protein concentration.
10. Split the IEF fractions into 5-mg aliquots and dry by lyophilization or speed vac.
11. To separate each IEF fraction in the second dimension, solubilize 5 mg of each fraction in 1.6 mL of PBS and add 0.4 mL of 5X Laemmli sample buffer.
12. Apply each fraction to the preparative well of a 16 × 20-cm SDS-PAGE gel (*see* **Note 26**).
13. Resolve proteins in the second dimension by electrophoresis.
14. Remove polyacrylamide gels from glass plates and soak in 100 mL of 10 m*M* ammonium bicarbonate for 30 min with one change of the 10 m*M* ammonium bicarbonate.
15. While the gel is equilibrating in 10 m*M* ammonium bicarbonate, assemble the Whole Gel Eluter (Bio-Rad) per manufacturer's instructions and fill the Whole Gel Eluter wells with 10 m*M* ammonium bicarbonate (*see* **Note 27**).
16. Cut SDS-PAGE gel to the dimension of the Whole Gel Eluter.
17. Lay the SDS-PAGE gel on top of Whole Gel Eluter wells. Orientate the gel so that protein bands run parallel to the wells.
18. Complete the setup of the Whole Gel Eluter per manufacturer's instructions, and elute the proteins at 250 mA for 90 min.
19. At the end of the elution, reverse the current on the Whole Gel Eluter for 20 s.
20. Harvest the samples from the Whole Gel Eluter per manufacturer's instructions, This will yield 30 fractions of approximately 2.5 mL each.
21. Filter sterilize each fraction with a 0.2-µm PTFE syringe filter. (Use aseptic techniques for subsequent manipulation of the 2D-LPE fractions.)
22. Determine the protein concentration of each fraction.
23. Split each fraction into 10-µg aliquots, lyophilize, and store at –80°C.

3.3.2. Assay of IFN-γ Induction for Identification of T-Cell Antigens

1. Euthanize infected and naive mice, harvest spleens, and place into separate sterile petri dishes containing a minimal volume of HBSS.
2. Fill a syringe (3 mL) with 3 mL HBSS, and attach a 22-gage needle. Insert the needle into the spleen while holding it with sterile forceps. Flush the cells from the spleen, and collect the cells in the sterile Petri dish, Repeat this procedure multiple times using different injection sites. Transfer the flushed cells to a 50-mL conical tube (*see* **Note 28**).
3. Press the spleen remnant against a 70-μm nylon mesh screen. Use the black rubber end of a 1-mL syringe plunger to press the spleen remnant and to force the remaining cells through the nylon screen. Filter the cells through the screen by holding it above the Petri dish and flushing with 3 mL of HBSS.
4. Pool collected splenocytes and centrifuge at $1000g$, 4°C for 10 min.
5. Suspend the cell pellet in a minimal volume of HBSS (3 mL) using a sterile pipet. Add Gey's hypotonic solution (5 mL per spleen), and incubate for 10 min at RT to lyse RBCs. Centrifuge the cells at $1000g$, 4°C for 10 min.
6. Decant the supernatant and gently suspend cells with HBSS.
7. Centrifuge the cells at $1000g$, 4°C for 10 min.
8. Repeat **steps 6** and **7**.
9. Suspend the cell pellet in complete RPMI medium (5 mL per spleen).
10. Determine the cell concentration by counting in a hemocytometer.
11. Dilute the cell suspension to 2×10^6 cells/mL in complete RPMI medium.
12. Plate the cells into 96-well tissue culture plates at 2×10^5 cells/well (100 μL).
13. Suspend protein antigens produced by 2D-LPE in complete RPMI medium at 20 μg/mL.
14. Add protein antigens/fractions (2 μg/well) to each of the tissue culture wells. Each individual protein fraction should be added to triplicate wells of the tissue culture plates containing the splenocytes from infected or uninfected animals. (Include Con A and a known protein antigen as positive controls and complete RPMI medium alone as the negative control.)
15. Incubate in a tissue culture incubator at 37°C for 4 d.
16. Remove 150 μL of cell culture supernatant from each well and assay for IFN-γ levels using an IFN-γ ELISA kit per manufacturer's instructions.
17. Evaluate the IFN-γ production to splenocytes from naive mice. Any protein fraction that induces an IFN-γ response threefold or greater than the positive control antigen with the naive splenocytes would be considered to be producing a nonspecific response. Determine the average IFN-γ response induced by all the 2D-LPE fractions added to splenocytes of infected mice. Compare the IFN-γ response of individual 2D-LPE fractions with the above average to identify the immunodominant 2D-LPE fractions. In general, those fractions that induce an IFN-γ response threefold or greater than the average IFN-γ response are considered immunodominant fractions.
18. Resolve the protein(s) of the immunodominant 2D-LPE fractions by SDS-PAGE. Excise the protein band(s) and identify the protein(s) via proteolytic digestion and LC-MS/MS as described in **Subheading 3.2.3**.

4. Notes

1. A 400-mL culture will generate adequate quantities of material for proteomic analyses of B-cell antigens. Larger cultures (5–20 L), however, should be grown to obtain adequate amounts of starting material for the preparation of subcellular fractions to be used in the identification of T-cell antigens.

2. Fresh breaking buffer should be prepared for each use. The protease inhibitors and nucleases can be prepared as stock solutions and stored at −20°C (pepstatin, 3 mg/mL in ethanol; leupeptin, 1 mg/mL in ethanol; PMSF 100 mM in isopropanol; DNase 1 mg/mL in PBS; and RNase 1 mg/mL in PBS).
3. In this chapter, the method described for the identification of B-cell antigens utilizes human sera. However, this same methodology can be applied with sera for experimentally or naturally infected animals. Optimally, when defining B-cell antigens, the sera should be grouped according to disease state, and matched control sera should also be included.
4. In some cases of sera from naturally infected hosts, preabsorption with lysate of a heterologous pathogen may be required to remove antibodies to highly crossreactive antigens. For example, we have utilized a lysate of *E. coli* to preabsorb sera from tuberculosis patients and healthy controls *(15)*.
5. Deionize urea by stirring 5 g of washed AG 501-X8 resin per 100 mL of 8 M urea for 1 h.
6. The pH range of the ampholytes should be equal to the pH range of the IPG strips. Additionally, the final concentration of ampholytes recommended may differ between the manufacturers of IPG strips.
7. Different detergents and concentrations may be required to solubilize membrane or cell wall proteins. For instance, the IEF rehydration or protein solubilization buffers we use to solubilize cell wall proteins of *M. tuberculosis* contain 1% Nonidet P-40 and 1% ASB-14. Others have recommended using 2 M thiourea along with 7 M urea or 1% Zwittergent 3-10 for solubilization of cell wall and membrane proteins *(23,24)*.
8. The IPG strips selected should reflect the general pI range of the proteins to be separated. The IPG strips selected must be able to fit in the prep well of the SDS-PAGE gels available to the laboratory.
9. The procedures described use Coomassie R-250 for the staining of gels. However, MS-compatible silver stains or fluorescence stains such as Sypro-Ruby may be substituted *(25)*.
10. Wash microcentrifuge tubes by filling with 60% acetonitrile/0.1% TFA, followed by incubation at RT for 1 h. This process is repeated two times for each tube.
11. The procedures described in this chapter utilize spleens from mice experimentally infected with *M. tuberculosis* to obtain immune T-cells. The method for experimental infection will vary based on the pathogen. Other organs or tissues, such as the lungs or lymph nodes, may be used to obtain immune T-cells. Additionally, the T-cell assays can be performed using whole blood or peripheral blood mononuclear cells from human donors *(26)*.
12. The quality of fetal bovine serum may vary depending on the source. Individual lots should be tested in tissue culture to ensure that they support optimal growth and are free of endotoxin.
13. The kit described is for use in T-cell assays involving mouse blood or tissue homogenates. Kits compatible for use with other experimentally or naturally infected hosts are also available *(27)*.
14. Con A is a lectin with T-cell mitogenic properties and is included for use as a positive control *(28)*. The dose of Con A that induces maximum levels of IFN-γ should be tested empirically but is usually around 1–10 µg/mL.
15. The methods described in this chapter are those commonly used to prepare subcellular fractions in our laboratory. Nevertheless, other methods are available for the preparation of subcellular fractions *(29,30)*.

16. For biosafety level 3 pathogens such as *M. tuberculosis*, the cells should be inactivated with methods that do not destroy protein structure. We commonly use γ-irradiation; however, other methods, including supercritical carbon dioxide *(31)*, hydrostatic pressure *(32)*, and ultrasound treatment *(33)*, are available.
17. Subcellular fractions should be aliquoted prior to storage. The aliquots generated depend on the downstream use of the proteins. For B-cell antigen identification, aliquots of 100 and 400 μg should be generated. For T-cell antigen identification, 250-mg aliquots should be generated.
18. One pass through the French press cell may not be sufficient for complete lysis. If required, the cell suspension may be passed through the French press multiple times. Gram staining and light microscopy should be used to check the efficiency of bacterial cell lysis.
19. Serum is generally pooled according to disease state and the Western blot optimization, and antigen detection procedures are conducted with the pooled sera. This is done to conserve sera samples and to define the immunodominant B-cell antigens representative for an entire population *(34)*. Dilutions used are 1:10, 1:50, 1:100, 1:500; 1:1000, 1:5000, and 1:10,000.
20. The optimum serum titer is defined as the dilution of serum that provides a strong signal intensity with the largest number of protein bands after 2–3 min of color development. The individual reactive protein bands at the optimal titer should be clearly defined and not appear as one large undefined area of serum reactivity.
21. Lyophilization should be used for drying of protein samples that are targeted for 2D-PAGE or 2D-LPE. This method of drying is relatively gentle and allows for more efficient resolubilization of the proteins. Ammonium bicarbonate should be completely volatilized in the drying process. Incomplete removal of ammonium bicarbonate may interfere with the IEF separation of the proteins. If the odor of ammonia can be detected in the sample after lyophilization, add a small volume of Milli-Q water (200 μL) and repeat lyophilization.
22. The presence of insoluble material in the protein solution added to the IPG strip will cause horizontal streaking in the 2D gel.
23. The cellular assays used for the detection of T-cell activation can be significantly influenced by contamination of protein samples with endotoxin *(6)*. Thus, great care must be taken to avoid the introduction of endotoxin during the separation of proteins by 2D-LPE. Buffers should be made with high-quality endotoxin-free water and in endotoxin-free glassware.
24. During the preparative IEF of proteins with the Rotofor unit, the voltage will gradually increase during the course of the run and then plateau (generally around 1200–1400 volts). The voltage of the Rotofor unit should be taken every 15–30 min and graphically displayed. The IEF is complete once the voltage has stabilized. Allow the run to continue for 30 min after the voltage has stabilized and then harvest the proteins. A typical IEF run with the buffer described in this chapter will take 3–4.5 h to complete.
25. Separation of large protein pools by preparative IEF under the conditions described generally results in excellent resolution of proteins contained in the pH 4–10 fractions. However, considerable protein overlap can be observed between fractions at the extreme ends of the pH gradient (3.0–4.0 and 10–12). As a general rule, the two to three fractions at each end of the gradient should be pooled based on their protein profile by SDS-PAGE.
26. The separation of proteins by preparative SDS-PAGE can cause a bottleneck in the 2D-LPE procedure. A single SDS-PAGE run at 35 mA will take approx 5 h. Additionally, it is not recommended to run more gels than can be immediately electroeluted. The storage of unfixed gels will result in diffusion of proteins within the gel.

27. In setting up the Whole Gel Eluter, it is essential to remove any air bubbles trapped between the cellulose filter paper and the cellophane that are placed on the cathode of the apparatus, and air bubbles between the gel and the wells of the Whole Gel Eluter. The presence of bubbles will interfere with efficient recovery of sample and may distort the flow of current, resulting in the uneven movement of proteins from the gel to the eluter wells.
28. Other methods of harvesting splenocytes include mincing the spleen into small fragments, followed by passage through a nylon cell strainer *(35)* or crushing the spleen in a glass tissue homogenizer *(36)*. We have found that gentle flushing of spleen cells using a syringe, followed by passage of the remnant through a nylon strainer, produces highly viable cells that are less likely to undergo the autolysis and cell death associated with other techniques of cell isolation.

References

1. Weldon, S. K., Mosier, D. A., Simons, K. R., Craven, R. C., and Confer, A. W. (1994) Identification of a potentially important antigen of *Pasteurella haemolytica*. *Vet. Microbiol.* **40,** 283–291.
2. Skeiky, Y. A., Ovendale, P. J., Jen, S., et al. (2000) T cell expression cloning of a *Mycobacterium tuberculosis* gene encoding a protective antigen associated with the early control of infection. *J. Immunol.* **165,** 7140–7149.
3. Cortese, R., Felici, F., Galfre, G., Luzzago, A., Monaci, P., and Nicosia, A. (1994) Epitope discovery using peptide libraries displayed on phage. *Trends Biotechnol.* **12,** 262–267.
4. Amara, R. R. and Satchidanandam, V. (1996) Analysis of a genomic DNA expression library of *Mycobacterium tuberculosis* using tuberculosis patient sera: evidence for modulation of host immune response. *Infect. Immun.* **64,** 3765–3771.
5. Chen, X. G., Gong, Y., Hua, L., Lun, Z. R., and Fung, M. C. (2001) High-level expression and purification of immunogenic recombinant SAG1 (P30) of *Toxoplasma gondii* in *Escherichia coli*. *Protein Expr. Purif.* **23,** 33–37.
6. Gao, B. and Tsan, M. F. (2003) Endotoxin contamination in recombinant human heat shock protein 70 (Hsp70) preparation is responsible for the induction of tumor necrosis factor alpha release by murine macrophages. *J. Biol. Chem.* **278,** 174–179.
7. Laub, M. T., McAdams, H. H., Feldblyum, T., Fraser, C. M., and Shapiro, L. (2000) Global analysis of the genetic network controlling a bacterial cell cycle. *Science* **290,** 2144–2148.
8. Schoolnik, G. K. (2002) Microarray analysis of bacterial pathogenicity. *Adv. Microb. Physiol.* **46,** 1–45.
9. Banerjee, N. and Zhang, M. Q. (2002) Functional genomics as applied to mapping transcription regulatory networks. *Curr. Opin. Microbiol.* **5,** 313–317.
10. Conway, T. and Schoolnik, G. K. (2003) Microarray expression profiling: capturing a genome-wide portrait of the transcriptome. *Mol. Microbiol.* **47,** 879–889.
11. Yue, H., Eastman, P. S., Wang, B. B., et al. (2001) An evaluation of the performance of cDNA microarrays for detecting changes in global mRNA expression. *Nucleic Acids Res.* **29,** E41–51.
12. O'Farrell, P. H. (1975) High resolution two-dimensional electrophoresis of proteins. *J. Biol. Chem.* **250,** 4007–4021.
13. Johnson, R. S. and Biemann, K. (1987) The primary structure of thioredoxin from *Chromatium vinosum* determined by high-performance tandem mass spectrometry. *Biochemistry* **26,** 1209–1214.

14. Eng, J. K., McCormack, A. L., and Yates, J. R. (1994) An approach to correlate tandem mass-spectral data of peptides with amino-acid-sequences in a protein database. *J. Am. Soc. Mass Spectrom.* **5,** 976–989.
15. Laal, S., Samanich, K. M., Sonnenberg, M. G., Zolla-Pazner, S., Phadtare, J. M., and Belisle, J. T. (1997) Human humoral responses to antigens of *Mycobacterium tuberculosis*: immunodominance of high-molecular-mass antigens. *Clin. Diagn. Lab. Immunol.* **4,** 49–56.
16. Samanich, K. M., Belisle, J. T., Sonnenberg, M. G., Keen, M. A., Zolla-Pazner, S., and Laal, S. (1998) Delineation of human antibody responses to culture filtrate antigens of *Mycobacterium tuberculosis*. *J. Infect. Dis.* **178,** 1534–1538.
17. Gulle, H., Fray, L. M., Gormley, E. P., Murray, A., and Moriarty, K. M. (1995) Responses of bovine T cells to fractionated lysate and culture filtrate proteins of *Mycobacterium bovis* BCG. *Vet. Immunol. Immunopathol.* **48,** 183–190.
18. Covert, B. A., Spencer, J. S., Orme, I. M., and Belisle, J. T. (2001) The application of proteomics in defining the T cell antigens of *Mycobacterium tuberculosis*. *Proteomics* **1,** 574–586.
19. Laemmli, U. K. (1970) Cleavage of structural proteins during the assembly of the head of bacteriophage T4. *Nature* **227,** 680–685.
20. Towbin, H., Staehelin, T., and Gordon, J. (1979) Electrophoretic transfer of proteins from polyacrylamide gels to nitrocellulose sheets: procedure and some applications. *Proc. Natl. Acad. Sci. USA* **76,** 4350–4354.
21. Gorg, A., Obermaier, C., Boguth, G., and Weiss, W. (1999) Recent developments in two-dimensional gel electrophoresis with immobilized pH gradients: wide pH gradients up to pH 12, longer separation distances and simplified procedures. *Electrophoresis* **20,** 712–717.
22. Hirschfield, G. R., McNeil, M., and Brennan, P. J. (1990) Peptidoglycan-associated polypeptides of *Mycobacterium tuberculosis*. *J. Bacteriol.* **172,** 1005–1013.
23. Lanne, B., Potthast, F., Hoglund, A., et al. (2001) Thiourea enhances mapping of the proteome from murine white adipose tissue. *Proteomics* **1,** 819–828.
24. Henningsen, R., Gale, B. L., Straub, K. M., and DeNagel, D. C. (2002) Application of zwitterionic detergents to the solubilization of integral membrane proteins for two-dimensional gel electrophoresis and mass spectrometry. *Proteomics* **2,** 1479–1488.
25. Lauber, W. M., Carroll, J. A., Dufield, D. R., Kiesel, J. R., Radabaugh, M. R., and Malone, J. P. (2001) Mass spectrometry compatibility of two-dimensional gel protein stains. *Electrophoresis* **22,** 906–918.
26. Katial, R. K., Hershey, J., Purohit-Seth, T., et al. (2001) Cell-mediated immune response to tuberculosis antigens: comparison of skin testing and measurement of in vitro gamma interferon production in whole-blood culture. *Clin. Diagn. Lab. Immunol.* **8,** 339–345.
27. Mazurek, G. H., LoBue, P. A., Daley, C. L., et al. (2001) Comparison of a whole-blood interferon gamma assay with tuberculin skin testing for detecting latent *Mycobacterium tuberculosis* infection. *JAMA* **286,** 1740–1747.
28. Passwell, J. H., Shor, R., Gazit, E., and Shoham, J. (1986) The effects of Con A-induced lymphokines from the T-lymphocyte subpopulations on human monocyte leishmanicidal capacity and H_2O_2 production. *Immunology* **59,** 245–250.
29. Osborn, M. J. and Munson, R. (1974) Separation of the inner (cytoplasmic) and outer membranes of Gram-negative bacteria. *Methods Enzymol.* **31,** 642–653.
30. Schnaitman, C. A. (1981) Cell fractionation, in *Manual of Methods for General Bacteriology*, vol. 1 (Gerhardt, P., ed.), ASM Press, Washington, D.C., pp. 52–61.

31. Dillow, A. K., Dehghani, F., Hrkach, J. S., Foster, N. R., and Langer, R. (1999) Bacterial inactivation by using near- and supercritical carbon dioxide. *Proc. Natl. Acad. Sci. USA* **96,** 10,344–10,348.
32. Wuytack, E. Y., Diels, A. M., and Michiels, C. W. (2002) Bacterial inactivation by high-pressure homogenisation and high hydrostatic pressure. *Int. J. Food Microbiol.* **77,** 205–212.
33. Raso, J., Palop, A., Pagan, R., and Condon, S. (1998) Inactivation of *Bacillus subtilis* spores by combining ultrasonic waves under pressure and mild heat treatment. *J. Appl. Microbiol.* **85,** 849–854.
34. Samanich, K., Belisle, J. T., and Laal, S. (2001) Homogeneity of antibody responses in tuberculosis patients. *Infect. Immun.* **69,** 4600–4609.
35. Lee, N. A., McGarry, M. P., Larson, K. A., Horton, M. A., Kristensen, A. B., and Lee, J. J. (1997) Expression of IL-5 in thymocytes/T cells leads to the development of a massive eosinophilia, extramedullary eosinophilopoiesis, and unique histopathologies. *J. Immunol.* **158,** 1332–1344.
36. Campisi, J. and Fleshner, M. (2003) Role of extracellular HSP72 in acute stress-induced potentiation of innate immunity in active rats. *J. Appl. Physiol.* **94,** 43–52.

2
Immunoproteomics

Alexander Krah and Peter R. Jungblut

Abstract

Two-dimensional electrophoresis results in an adequate resolution of the proteome of microorganisms to allow the detection and identification of specific antigens after blotting on membranes and overlaying the protein pattern with patient's sera. The complement of all identified antigens presents the immunoproteome of a microorganism. All the antigens specific for a microorganism or even for a disease are identified by mass spectrometry. For identification, peptide mass fingerprinting is used, and post-translational modifications are detected by mass spectrometry MS/MS techniques. High-resolution two-dimensional electrophoresis and unambiguous identification are prerequisites for reliable results. After statistical analysis, the resulting antigens are candidates for diagnosis or vaccination and targets for therapy.

Key Words: Two-dimensional electrophoresis; immunoproteome; diagnostics; therapy; vaccination; mass spectrometry; peptide mass fingerprinting; proteome database.

1. Introduction

For about 50 yr scientists have tried to elucidate the protein composition of biological compartments. The first two-dimensional electrophoretic separation resolved 15 proteins from human serum *(1)*. Ribosomes with about 60 protein components were the first completely resolved organelles *(2)*. For more complex compartments, the resolution power had to be improved. This goal was reached by the combination of isoelectric focusing (IEF) and sodium dodecyl sulfate polyacrylamide gel electrophoresis (SDS-PAGE) for two-dimensional electrophoresis (2-DE) *(3,4)*. With this improvement, several hundred proteins were separated within one 2-DE gel. The method was further optimized for high resolution, and at present more than 10,000 protein species may be separated within large (30 × 40 cm) gels *(5)*. In 1995 the term proteome was defined: the *proteome* refers to the total protein complement of a genome *(6)*.

The identification of antigens within gels was not possible for a long time. Therefore, blotting of proteins to nitrocellulose membranes was an important milestone for the detection of antigens *(7,8)*. After transfer of proteins from gels to membranes, antibodies can interact and bind to the antigens for which they are specific. To detect the antigen–antibody complex, a second antibody is overlaid with specificity against the class of the first antibody (e.g., IgG). The secondary antibody is coupled with a detection system. This principle was used early for identification of proteins in 2-DE gels *(9)*. Overlaying blots of gels from microorganisms with sera from patients infected with these microorganisms reveals the immunoproteome of the microorganism under investigation. For many years one-dimensional (1D) gels were used to detect antigens for diagnostics. Because of the high complexity of the proteome, the assignment of a 1D band to a distinct protein is very difficult, and 2D separation is required for unambiguous identification of antigens. *Borrelia garinii* and *Helicobacter pylori* are two of the first immunoproteomes analyzed by 2-DE and mass spectrometry (MS) *(10,11)*. Direct immunodetection within the gels avoiding blotting has been reported recently (UnBlot In-Gel Chemiluminescent Detection Kit, Pierce, Rockford, IL). The future will show whether this procedure can substitute for immunoblotting.

A view of the immunoproteome may also be obtained by enzyme-linked immunosorbent assay (ELISA) tests and immunoprecipitation. ELISA tests do not differentiate between the different proteins, and immunoprecipitation requires larger amounts of antibodies and suffers from problems with removal of antibodies before identification of the antigens. Therefore, at present the 2-DE/MS approach reveals the most complete view of the immunoproteome.

The major steps of immunoproteomics are as follows:

1. Two-dimensional electrophoresis.
2. Semidry blotting.
3. Immunodetection.
4. Data analysis.
5. Antigen identification.

1.1. Two-Dimensional Electrophoresis

The smallest unit of a proteome is the protein species *(12)*, which is defined by its chemical structure. Therefore, a myosin phosphorylated at position x and a myosin phosphorylated at position y are two different protein species of one protein.

A proteome of a microorganism with 3000 genes may comprise 9000 or more protein species. Even if not all of the genes are represented by proteins in a certain biological situation, several thousands of protein species may be expected to be present. Indeed, about 1800 protein species were detected for *H. pylori*, which contains a genome of about 1600 genes *(13)*. Therefore, high-resolution 2-DE techniques are a prerequisite to resolve this complexity. Resolution may be improved by increasing the gel size or by the production of several gels with different separation ranges. The p*I* range of a gel may be modified by the use of different ampholyte or immobiline gradients. The molecular weight (M_r) range depends on the porosity of the gel matrix, which itself may be modified by different acrylamide and crosslinker concentrations. The strategy of using large gels has the advantage that the complete information is contained in one gel. High quality of 2-DE gels, as shown in **Fig. 1**, is a necessary prerequisite for successful immunoproteomics.

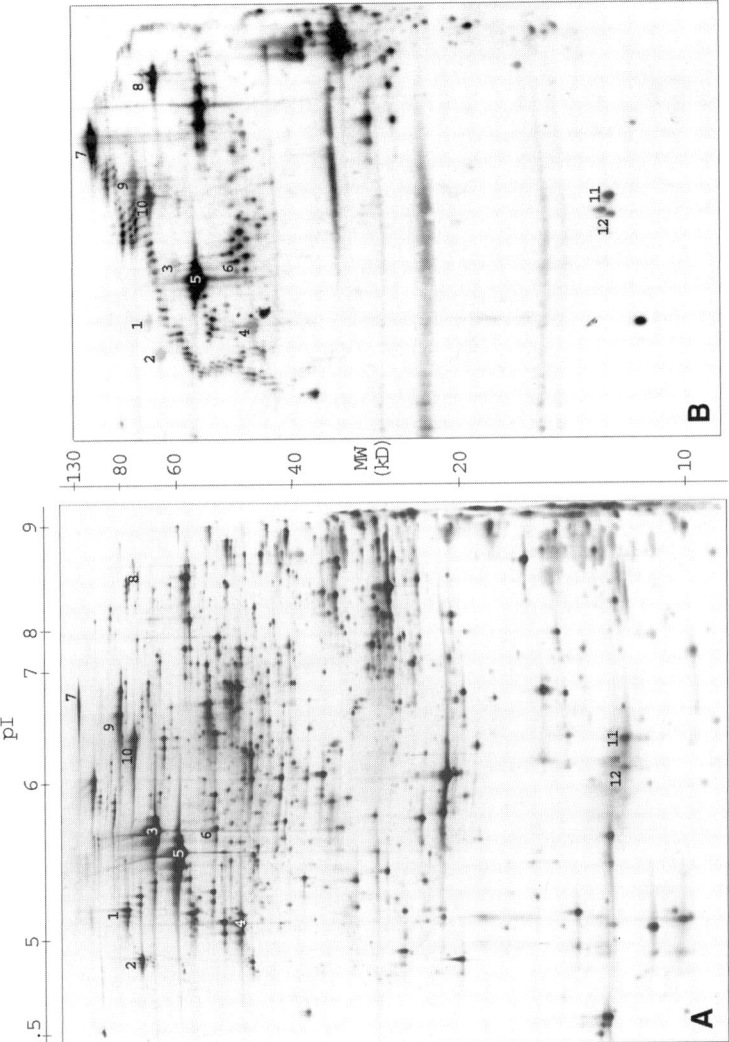

Fig. 1. Comparison of a 2-DE gel and a 2-DE gel immunoblot. (**A**) Silver-stained large 2-DE gel (23 × 30 cm) of *H. pylori* lysate containing about 1800 spots. This gel is the standard gel in the 2D-PAGE database (http://www.mpiib-berlin.mpg.de/2D-PAGE/EBP-PAGE/index.html). (**B**) Large 2-DE immunoblot (23 × 30 cm) of *H. pylori* lysate probed with a human serum. About 400 spots are recognized. Twelve of the spots are numbered in A and B. These spots contain the following proteins: 1, elongation factor G; 2, heat shock protein 70; 3, urease β-subunit; 4, elongation factor TU; 5, GroEL; 6, glutamine synthetase; 7, cag26; 8, not yet identified; 9, N-methylhydantoinase; 10, hydantoin utilization protein A; 11 and 12, GroES.

1.2. Semidry Blotting

Towbin et al. *(7)* blotted the proteins within a tank, where the blot sandwich was surrounded by large volumes of buffer. Potential impurities from the buffer are avoided by the use of semidry blotting, which is also easier to perform *(14)*. The critical point of the blotting mechanism is the time point at which the SDS is stripped off from the protein *(15)*. If SDS is removed from the protein still in the gel, the protein cannot be transferred to the membrane. If SDS is not stripped off from the protein at the moment the SDS–protein complex reaches the membrane, the protein cannot bind to the membrane and moves through the membrane to the anode. High-M_r proteins tend to lose SDS early and remain in the gel. Low-M_r proteins tend to lose SDS too late and do not bind to the membrane. Improving the SDS–protein binding by addition of SDS to the cathode buffer improves the blotting efficiency for large proteins. A better blotting efficiency for low-M_r proteins is obtained by improving the hydrophobic interaction between membrane and protein by increasing the ionic strength of the blotting buffer.

1.3. Immunodetection

After blotting, the proteins are immobilized in the membrane. Sera of patients are overlaid to detect the antigens against which the patients have produced antibodies. Here one has to be aware of two causes of variability. First, the genetic variability of the microorganism and second, the variability of the immunological response of the host. This response depends on the strain of the microorganism the host is infected with and its own genome and environment. Because individual medicine is only a vision at the moment, for the development of diagnostics, therapeutics, and vaccines at present we have to search for proteomic signatures independent of the strain and individual host. Therefore large series of patient sera with one or several common strains of the microorganisms have to be searched for antigens; only those antigens common for a majority of them are potential candidates for diagnosis, therapy, and vaccination. Attempts were also made to correlate antigen composition with certain disease manifestations *(10,11,16)*. For each microorganism, its own rules for acceptance of immunologically relevant candidates have to be delineated from the immunoproteomes obtained.

With the standard procedure, primary antibody/secondary antibody coupled with peroxidase or alkaline phosphatase, a better sensitivity than with silver staining is already mostly obtained. When chemiluminescence is applied, a further enhancement of sensitivity may be reached. For optimal sensitivity, the blocking reagents and washing procedures play an important role to avoid background staining.

1.4. Data Analysis

After blotting and immunostaining, a pattern of spots arises, which normally is completely different compared with the silver-stained 2-DE pattern. The assignment of spots between these two patterns is easy if highly intense characteristically formed spots are found in both patterns. However, assignment becomes more and more complicated the lower the number of spots in the immunostained pattern and the higher the number of antigens not stained in the silver-stained pattern. Several strategies are used for unambiguous assignment. One is replica blotting. Here, during the blotting procedure, the proteins are blotted to both sides of the gel by changing the direction of the electric field strength during the blotting procedure *(17)*. One blot is immunostained, and the other is stained by Coomassie Brilliant Blue (CBB) or by more sensitive stains like Aurodye. Another strategy is to counterstain the membrane with CBB after immunostaining *(18)*. This is possible because the surface-bound blocking proteins are removed from the membrane during the washing procedures, and the blotted proteins, which are bound within the membrane, remain in the membrane during washing.

Because of the potentially high variance, each immunostaining experiment has to be repeated at least three times. The resulting spot patterns are spot detected and matched by commercial image processing software. Within a virtual master gel, the spot intensity differences of all the tested sera can be visualized (**Fig. 2**), and nonspecific reactions may be eliminated by comparison with control sera.

1.5. Antigen Identification

If highly specific antibodies are available, antigens may be identified by them after stripping of the antibodies from the serum directly from the same membrane used for the serum tests. Protein chemical identification is more reliable and may also lead to identification at the protein species level. Here MS is the method of choice. Peptide mass fingerprinting *(19)* after tryptic digestion from spots out of preparative gels stained with CBB G-250, results in secure identification, if the genome of the microorganism is already completely sequenced (**Fig. 3**). Sequence information by MS/MS techniques gives information about post-translational modifications and also of genes not described before *(20)*. For identification of an antigen, it is important to show by MS that the immunostained spot contains only one protein. Because of the high sensitivity of the immunostaining, minor components of a spot may also be detected.

2. Materials

2.1. Two-Dimensional Electrophoresis

The 2-DE procedure is beyond the scope of this chapter; *see* **Subheading 3.1.**

2.2. Semidry Blotting

1. Polyvinylidene difluoride (PVDF) blotting membrane (Immobilon-P Transfer Membrane, Millipore, Bedford, MA).
2. Filter paper (GB003 Gel-Blotting-Papier, Schleicher & Schuell, Dassel, Germany).
3. Blotting buffers: *methanol is toxic by inhalation—prepare and use solutions under a hood!*
   a. For the high-M_r part of the gel (30–150 kDa):
   Cathode buffer: 50 mM boric acid, 10% methanol, 5% SDS; add NaOH to adjust to pH 9.0.
   Anode buffer: 50 mM boric acid, 20% methanol; add NaOH to adjust to pH 9.0.
   b. For the low-M_r part of the gel (4–30 kDa):
   Cathode and anode buffers: 100 mM boric acid, 20% methanol; add NaOH to adjust to pH 9.0.
4. Blotting chambers (Hoefer Large SemiPhor, semidry transfer unit, Amersham Pharmacia Biotech, San Francisco, CA).

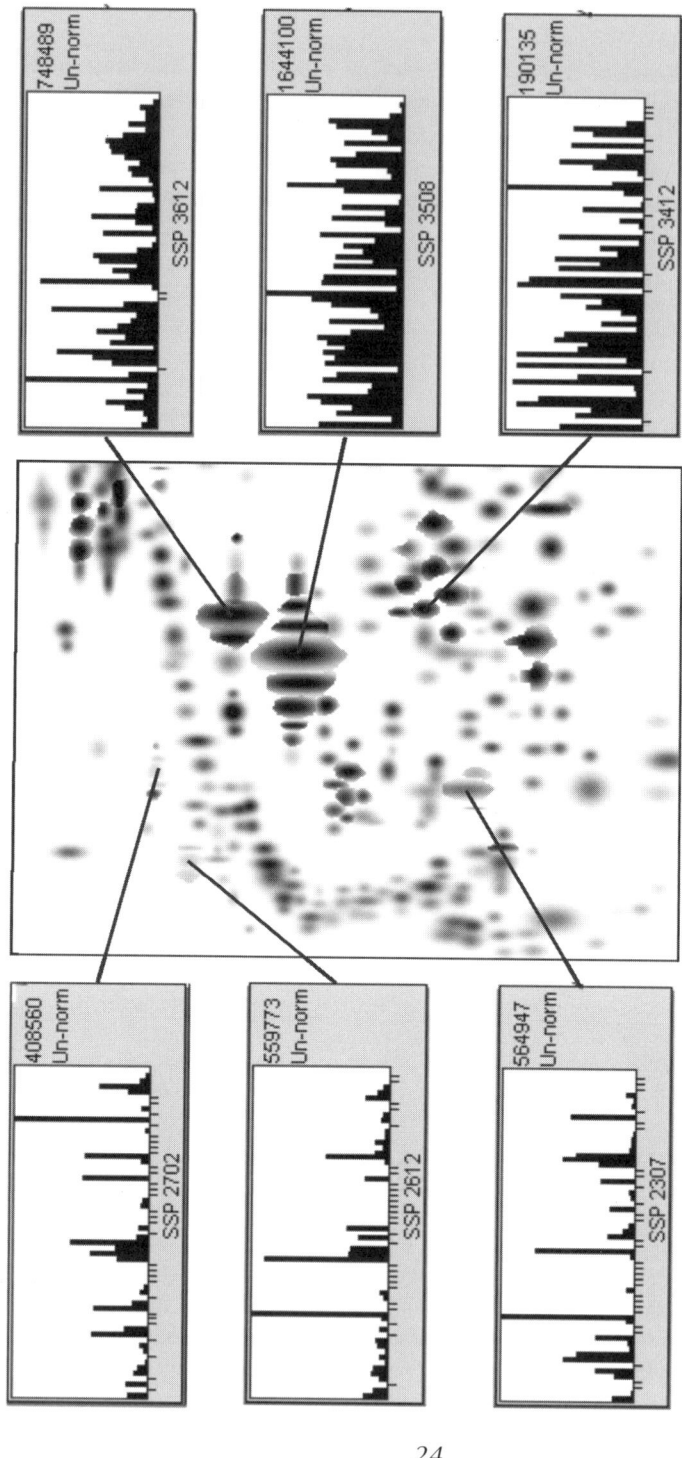

Fig. 2. Result of a data analysis of 2-DE immunoblots using PDQuest software (*see* **Subheading 2.4.**). Upper left section of the master blot, which contains the gaussian fitted spots of all immunoblots in the analysis set. Marked are spots 1–6 from **Fig. 1**. For each spot the columns in the small boxes correspond to the intensity in each immunoblot of the analysis set. The numbers in the upper right corners of the small boxes show the maximum intensity of each spot. SSP numbers are unique spot numbers that are automatically generated by the software.

Immunoproteomics 25

Mascot Search Results

```
User         : MPIIB
Email        : krah@mpiib-berlin.mpg.de
Search title :
Database     : NCBInr 20021113 (1231734 sequences; 391905809 residues)
Timestamp    : 20 Nov 2002 at 10:30:08 GMT
Top Score    : 118 for gi|15644739, chaperone and heat shock protein 70 (dnaK) [Helicobacter pylori 26695]
```

Probability Based Mowse Score

Score is -10*Log(P), where P is the probability that the observed match is a random event.
Protein scores greater than 73 are significant (p<0.05).

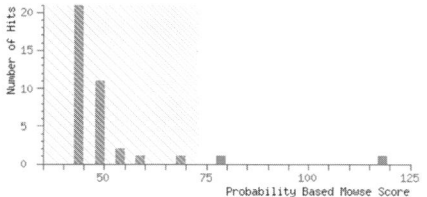

Concise Protein Summary Report

Switch to full Protein Summary Report

To create a bookmark for this report, right click this link: Concise Summary Report (../data/20021120/FtgtizTt.dat)

[Re-Search All]   [Search Unmatched]

```
1. gi|15644739        Mass: 67011    Total score: 118   Peptides matched: 14
   chaperone and heat shock protein 70 (dnaK) [Helicobacter pylori 26695]

2. gi|15611171        Mass: 67081    Total score: 80    Peptides matched: 11
   70kDa chaperone [Helicobacter pylori J99]
```

Fig. 3. Search result of a protein identification by a peptide mass fingerprint using the search machine Mascot (*see* **Subheading 2.5.**). The search result of spot 2 (*see* **Fig. 1**). The columns show the number of protein hits with a certain score value. Scores above 73 are considered to be significant (outside the hatched area). Hits that have no significant score must be considered random events. In this case, two significant hits are found. The first is the heat shock protein 70 of *H. pylori* 26695, the strain used in this experiment. The second hit is the corresponding protein in strain J99. This protein has a highly similar, but not identical, sequence in both strains. Links are given in the search result to gain more information about the proteins.

2.3. Immunodetection

Prepare stackable boxes a little larger in size than the membranes.

1. PBST buffer: add Tween-20 to phosphate-buffered saline (PBS) at pH 7.6 to obtain a concentration of 0.05%.
2. Dry milk (Blotting Grade Blocker, Non-Fat Dry Milk, Bio-Rad, Hercules, CA).
3. Primary antibodies or patient sera.
4. Secondary antibody directed against the primary antibody used (e.g., goat anti-human polyvalent IgG-Peroxidase Conjugate, Sigma; cat. no. A-8400): store at –20°C and avoid thawing–freezing cycles by freezing in aliquots.

5. Chemiluminescence reagents (Western Lightning Chemiluminescence Reagent NEL-101, NEN, Perkin Elmer, Boston, MA): store at 4°C.
6. Film (Biomax MR, Kodak, Rochester, NY).
7. Film cassette (Hypercassette, Amersham Pharmacia Biotech UK, Buckinghamshire, UK).
8. Photo machine in a dark room.
9. CBB R-250 staining solution: 50% methanol, 10% acetic acid, 0.1% CBB R-250 (Bio-Rad, Hercules, CA). *Methanol is toxic by inhalation!*
10. CBB destaining solution: 50% methanol, 10% acetic acid.

2.4. Data Analysis

1. Scanner (Umax Mirage IIse, Taiwan).
2. 2-DE analysis software (PDQuest, Version 7.1, Bio-Rad).

2.5. Antigen Identification

Avoid contamination of the buffers by dust and keratin!

1. 2-DE public database for the organism examined (in-house if available or via Internet: World 2D PAGE: http://www.expasy.org/ch2d/2d-index.html).
2. Fixing solution for preparative gels: 50% methanol, 2% phosphoric acid.
3. CBB G-250 staining solution for preparative gels: 34% methanol, 17% (w/v) ammonium sulfate, 2% phosphoric acid; 0.66 g/L CBB G-250 is added later (Bio-Rad).
4. Spot destaining solution: 200 mM NH$_4$HCO$_3$, 50% acetonitrile.
5. Digest buffer: 50 mM NH$_4$HCO$_3$, 5% acetonitrile.
6. Sequencing grade modified trypsin (Promega, Madison, WI) is dissolved to 0.2 µg/µL in resuspension buffer provided by the manufacturer. Freeze in aliquots for storage.
7. Shrink buffer: 60% acetonitrile, 0.1% trifluoroacetic acid.
8. Sample buffer: 33% acetonitrile, 0.1% trifluoroacetic acid.
9. Matrix solution: 50 mg/mL 2,5-dehydroxybenzoic acid dissolved in 33% acetonitrile, 0.33% trifluoroacetic acid.
10. Matrix-assisted laser desorption/ionization-time of flight (MALDI-TOF) mass spectrometer (Voyager Elite DE, Perseptive Biosystems or others; *see* **Note 19**).
11. Search machines for peptide mass fingerprints, e.g., Mascot (http://www.matrixscience.com), ProFound (http://129.85.19.192/profound_bin/WebPro Found.exe), or MS-Fit (http://prospector.ucsf.edu/ucsfhtml4.0/msfit.htm).

3. Methods

3.1. Two-Dimensional Electrophoresis

It would go beyond the scope of this chapter to explain sample preparation and 2-DE in detail. The procedure we are using (as shown in **Fig. 1**) is comprehensively described in refs. *5* and *21*. Briefly, for the first dimension, 150 µg of protein sample was applied to the anodic side of the IEF gel. In the second dimension, after equilibration in SDS-containing buffer, the IEF gel was placed onto a 23 × 30-cm gel and proteins were separated according to their M_r.

It is important to consider that the quality of the immunoblots strongly depends on the quality of the gels that are used (such as resolution power; *see* **Note 1**). Depending on the scientific goal, one should only use the gel system (size, sample buffer, detergents, pH gradient) that is able to resolve the proteins of interest. Because experiments often search for unknown proteins, it is better to use the gel system with the highest resolution power. We recommend the use of large gels for immunoproteomics of microorganisms. The following procedure is used for 23 × 30-cm gels.

3.2. Semidry Blotting

1. Prepare two blotting chambers and cut PVDF membrane to 23 × 15 cm (half the size of the gel). Filter papers should be at least 22 × 17 cm. Mark PVDF membrane unambiguously using a pencil (*see* **Note 2**).
2. Soak filter papers into the blotting buffers, three sheets each in anode and cathode buffers of high-M_r and six sheets in low-M_r buffer (*see* **Note 3**).
3. Cut the large 2-DE gel into two pieces (high and low M_r) right after the run has finished.
4. Build up the following sandwich in both blotting chambers, avoiding any air bubbles: anode (+), three filter papers (soaked in anode buffer), PVDF, gel, three filter papers (soaked in cathode buffer), cathode (–) (*see* **Note 4**).
5. Apply a constant current of 1 mA/cm^2 for 2 h.
6. Discard filter papers and gel (*see* **Note 5**).
7. Dry PVDF membranes at room temperature and freeze at –20°C for prolonged storage.

3.3. Immunodetection

Volumes of buffers depend on the size of the boxes. Here volumes for half an immunoblot in 20 × 20-cm boxes are given.

1. Thaw PVDF membranes and put each half into separate boxes.
2. Soak membrane in 100% methanol for about 1 min (for PVDF only; *see* **Note 6**). Do not let membrane dry out from this point until the film is exposed (*see* **Note 7**).
3. Block membrane in 100 mL 5% milk in PBST buffer for at least 1 h (or overnight at 4°C) using an appropriate shaker (*see* **Note 8**).
4. Add primary antibody or serum directly to the solution so that an appropriate dilution is reached (from 1:200 for human sera to 1:10,000 for some monoclonal antibodies; *see* **Note 9**). Shake for at least 1 h or overnight at 4°C.
5. Wash membranes four times for 15 min in 100 mL of PBST buffer.
6. Incubate in secondary antibody solution diluted in 100 mL 5% milk in PBST buffer (e.g., 1:5000) for 1 h.
7. Wash membranes four times for 15 min in 100 mL of PBST buffer.
8. Warm up chemiluminescence reagents to room temperature during this time (*see* **Note 10**).
9. Freshly mix equal amounts of both reagents and apply to the membranes (up to 50 mL per half blot). Shake for 1 min. Make sure the whole membrane is covered with liquid.
10. Drip off membrane and wrap it into a foil to keep it wet.
11. Quickly go to the dark room as chemiluminescence will drop in intensity after some minutes. (There will be some intensity until about 30 min after mixing the reagents.)
12. Place the membrane and a film on top in the cassette and expose for 1 min. Develop this film to decide on the appropriate exposure time. This can vary from a few seconds up to half an hour, depending on the strength of the signal. The exposure of several films within half an hour after mixing of reagents is possible.
13. Store developed films in a dark and dry place.
14. For quality assessment and spot assignment, stain the PVDF membranes in CBB R-250 staining solution for about 5 min.
15. Destain the background in destaining solution three times for about 2 min.
16. Dry membrane at room temperature and store in a dry place.

3.4. Data Analysis

Usually data analysis is performed by comparing spot patterns of immunoblots to search for differences according to the biological state represented by the immunoblots (*see* **Note 11**). This analysis can only be done by eye when a few immunoblots containing not too many spots are compared. For extensive studies it is necessary to apply a 2-DE analysis software, preferably with implemented statistical tools. One example of such a software is PDQuest.

To perform the computer-aided analysis, it is necessary to scan the films properly. All films from one experiment should be scanned using exactly the same parameters. These parameters strongly affect the following analysis and, therefore, they need to be optimized for resolution (for PDQuest: 150 dpi), contrast, and brightness. Files should be saved as gray-scaled images with at least 8-bit depth (256 gray values per pixel).

It is not possible to explain data analysis in great detail here as there are many different software solutions available. However, to our knowledge there is no 2-DE analysis software currently available that is able to perform the analysis automatically and achieve acceptable results without interactive corrections. We therefore strongly recommend careful review of all results manually (*see* **Note 12**).

When using PDQuest, spot detection and quantitation are performed first. Then spots in different immunoblots that represent the same protein species are matched. It is now very important to take the time to check and correct the spot detection and matching thoroughly. Depending on the number of immunoblots in the experiment, this can last for up to several weeks—but there is no alternative if reliable results are to be achieved. After this procedure, it is possible to group immunoblots according to their biological state, e.g., each group may contain three or more replicate immunoblots of the same sample. Using statistical tools like the t-test or Mann–Whitney test, it is now possible to compare the intensities of all the spots in all the immunoblots in order to find significantly regulated spots (significance level adjustable to 90, 95, or 99%). It is also possible to perform analyses like "show all spots that are up-/downregulated by factor x" or spots unique to one of the groups of immunoblots. Using these tools, one will be able to find spots that are reliable antigen candidates (*see* **Note 13**).

3.5. Antigen Identification

After detection of the spots, which are differently recognized by the antibodies or sera, the question of the identity of these spots arises. Usually one will find information about the protein contained in the spot rather than the protein species (*see* **Note 14**).

One possible way to determine the protein(s) contained in the spot is to compare the spot patterns produced by the immunoblots with a spot pattern in a database. There are 2-DE databases for many different species available for the public via the Internet (*see* **Subheading 2.5.**, **item 1**). With a little luck, one can find one's spot of interest already identified. Take care to avoid erroneous assignments, because finding the corresponding spot can be very difficult (*see* **Note 15**).

Another way to identify the proteins of interest is to run a preparative gel of the sample applying more sample protein than for analytical gels, e.g., 500 µg for 23 × 30-cm gels. These gels are stained with CBB G-250 *(22)*. CBB G-250 is more sensitive than CBB R-250. The spots of interest can be excised for MALDI-TOF analysis (*see* **Note 16**). In this case, assignment problems as described above will also occur. Take as much care as you can to avoid contamination of the gel or spots with dust or keratin (*see* **Note 17**).

3.5.1. CBB G-250 Staining of 23 × 30-cm Gels

1. Directly after the end of the gel run, move the gel to 1 L of fixing solution and shake overnight.
2. Wash three times for 30 min in 1 L distilled water.
3. Shake for 1 h in 1 L staining solution (still without CBB G-250).
4. Add 0.66 g/L CBB G-250 to the solution and shake for about 5 d.
5. Rinse with 25% methanol for 1 min. Destaining is not required.
6. The stained gel can be stored shrink-wrapped in distilled water for several weeks at 4°C.

3.5.2. Protein Digest

1. Excise spots of interest using a Pasteur pipet that was cut in length to achieve an internal diameter of about 1 mm. Place spot in a clean dust-free tube.
2. Destain spots by shaking in 500 µL spot destaining solution for 30 min at 37°C.
3. Shake spots in 500 µL digest buffer for 30 min at 37°C.
4. Dry spots in Speed Vac for about 30 min at 30°C.
5. Apply 0.25 µL trypsin solution directly to the dried spot and add 25 µL of digest buffer.
6. Incubate overnight at 37°C on a shaker.
7. Spin down and transfer supernatants into new tubes.
8. Shrink spots in 20 µL shrink buffer for 10 min (to recover all the peptides) and combine supernatants in the new tubes.
9. Dry supernatants in a Speed Vac for about 60 min at 45°C.
10. Dissolve peptide pellet in 1.3 µL sample buffer.
11. Apply 0.25 µL sample solution to the MALDI template and add 0.25 µL matrix solution and wait for the drop to be dried (*see* **Note 18**).
12. Sample solution can be stored at –20°C.

The peptide mass fingerprints can be obtained from different MALDI-TOF mass spectrometers. Therefore, the exact procedure cannot be described here. The parameter settings must be optimized according to the device used.

After measuring the peptide mass fingerprints by MALDI-TOF MS, the protein can be identified by a database search (*see* **Note 19**). In such a search the "fingerprint" of peptide masses created by the trypsin digest is compared with a list of theoretically digested proteins in a database. Search machines are available to the public via the Internet (*see* **Subheading 2.5., item 11**), and a protein sequence database can be chosen in the search masks, e.g., NCBI, SwissProt, or OWL. Make sure the protein hit meets reliable identification criteria, e.g., the hit has the highest score value when performing an "all species" search, 30% sequence coverage is achieved, and few modifications are found.

4. Notes

1. It is most important to use 2-DE gels of high resolution. Much experience is required to produce gels of high resolution and quality reproducibly. Remember, the better the quality of the gels, the better the immunoblot quality that can be achieved.
2. Keep dust or chemicals from reaching the membrane at any time because detection with chemiluminescence is extremely sensitive. Use gloves and a lab coat at all times. Never touch the membrane—use tweezers instead.
3. It is possible to use nitrocellulose instead of PVDF membranes. However, these membranes are usually more fragile. Blotting parameters must then be adapted.
4. If there are air bubbles left in between the sandwich layers, it is possible to remove them by carefully rolling a glass pipet over the filter papers. Do not let the filters dry out.
5. To assess the blotting efficiency, it is recommended to stain a test gel with CBB G-250 (as described in **Subheading 3.5.**) after blotting. If too much protein is left in the gel, SDS should be added to the cathodic blotting buffer, and the blotting time may be extended. If low-M_r proteins did not bind to the membrane, the ionic strength of the blotting buffer has to be increased.
6. It is essential to soak PVDF membranes in 100% methanol to prepare the surface for aqueous solutions.
7. Always apply enough solution for the membrane to swim freely. The protein surface should be on top.
8. The quality of the milk strongly influences the background detection. One might try one from the supermarket, but this must be tested in advance.
9. It is always necessary to optimize the dilution of primary and secondary antibodies or sera. Tests should be made using a dilution series. Sometimes milk does not work well as a blocking reagent—there are other blocking reagents available, e.g., bovine serum albumin. Some antibodies may not work properly in PBS—try TBS instead.

    If human sera or expensive antibodies are used, it is advantageous to reduce the necessary amount of antibody to a minimum. To do this, it is possible to shrink-wrap the membranes. For this purpose, cut a piece of polyethylene foil into an appropriate size, place the two membrane pieces from one gel (faces to the outsides) inside, and shrink-wrap three from four sides. Now the blocking solution can be applied. (Do not forget the methanol at first.) Take care not to catch air bubbles within the package. After the blocking (of membrane and plastic foil!), one side is carefully sliced, antibody is added, and the package is sealed again. Use an appropriate shaker and cover the package using a glass slide. Make sure the glass slide can shake freely; thus the solution can circulate properly within the package. Care must be taken to keep the glass from slipping, e.g., by the use of adhesive tape. By applying this method, the minimum amount of antibody solution can be reduced to as little as 0.1 mL/cm^2 immunoblot area (50 mL for one 23 × 30-cm blot).
10. Use film, chemiluminescence reagents, and cassette only at a certain temperature (room temperature, if not fluctuating), because the light-emitting biochemical reaction is strongly temperature-dependent. Otherwise spot intensities from different immunoblots cannot be compared.
11. Data analysis is as important as your experiments in the lab. For this reason, take a little time and think about the analysis strategy your experiment requires. According to the biological questions that are to be answered, the analysis must be performed in such a way that these are addressed properly. For instance, it is important to think about how many replicate immunoblots must be made to achieve significant results. Also, the selec-

tion criteria for spots of interest should be fixed in advance of the analysis in order to avoid "wishful analyzing." Do not forget appropriate control immunoblots.

12. Many different 2-DE analysis software packages are available. It is very time-consuming to test all the new software versions that are brought to market continuously. We have chosen PDQuest mainly for two reasons. First, statistical analysis tools are implemented. Second, PDQuest allows the user to correct spot detection and perform manual matching easily. This is a matter of particular interest for blot analysis, as immunoblots usually contain fewer spots compared with gels, which makes it harder for the automated matching procedure to find corresponding spots. Act with caution when software claims to process "every" gel automatically without giving you the chance to correct errors.

13. After exhaustive use of the 2-DE analysis software, it is still possible to export spot intensity data for even further analysis. One possibility is the application of multivariate statistics, e.g., principal component analysis or hierarchical clustering. This can give information on whether whole spot patterns cluster according to the biological state of the sample.

14. Caution should be used since spots do not represent proteins but rather protein species, i.e., proteins as they are found in vivo may have post-translational modifications, may be partly degraded, or might be splicing variants. Additionally, spots often contain several different proteins or protein species. However, when using peptide mass fingerprints for identification, little or no information about modifications or variants can be given. Search results for spots containing more than one protein can be very difficult to interpret.

15. Comparing spot patterns from immunoblots (or in fact films) with spot patterns from the gel in the database is not as easy as one might think. To find the corresponding spot unambiguously, it is important to take account of "local" spot patterns close to the spot of interest. Only by looking at these is it possible to overcome the problem of different running behaviors of proteins under different running conditions. The same assignment problem occurs between immunoblots and preparative gels.

16. The use of CBB G-250 leads to a higher sensitivity compared with CBB R-250. However, methylation of the proteins will occur that must be taken into account for the database search since methylated peptides have a 14 Dalton higher mass.

17. Contamination of spot samples with keratin is a serious problem. It is highly recommended to clean the bench, pipets, boxes, and Speed Vac prior to use. Otherwise keratin peaks will appear in the peptide mass fingerprints and may interfere with the identification.

18. The matrix α-cyano cinnamic acid can be used instead of 2,5-dehydroxybenzoic acid.

19. For identification, many other mass spectrometers can also be used, e.g., ESI-MS, MALDI-TOF/TOF, or other devices that allow sequence information to be revealed (MS/MS).

References

1. Smithies, O. and Poulik, M. D. (1956) *Nature* **177**, 1033.
2. Kaltschmidt, E. and Wittmann, H. G. (1970) Ribosomal proteins. VII. Two-dimensional polyacrylamide gel electrophoresis for fingerprinting of ribosomal proteins. *Anal. Biochem.* **36**, 401–412.
3. Klose, J. (1975) Protein mapping by combined isoelectric focusing and electrophoresis of mouse tissues. A novel approach to testing for induced point mutations in mammals. *Humangenetik* **26**, 231–243.
4. O'Farrell, P. H. (1975) High resolution two-dimensional electrophoresis of proteins. *J. Biol. Chem.* **250**, 4007–4021.
5. Klose, J. and Kobalz, U. (1995) Two-dimensional electrophoresis of proteins: an updated protocol and implications for a functional analysis of the genome. *Electrophoresis* **16**, 1034–1059.

6. Wasinger, V. C., Cordwell, S. J., Cerpa-Poljak, A., et al. (1995) Progress with gene-product mapping of the Mollicutes: *Mycoplasma genitalium*. *Electrophoresis* **16**, 1090–1094.
7. Towbin, H., Staehelin, T., and Gordon, J. (1979) Electrophoretic transfer of proteins from polyacrylamide gels to nitrocellulose sheets: procedure and some applications. *Proc. Natl. Acad. Sci. USA* **76**, 4350–4354.
8. Burnette, W. N. (1981) "Western blotting": electrophoretic transfer of proteins from sodium dodecyl sulfate—polyacrylamide gels to unmodified nitrocellulose and radiographic detection with antibody and radioiodinated protein A. *Anal. Biochem.* **112**, 195–203.
9. Celis, J. E., Ratz, G. P., Madsen, P., et al. (1989) Computerized, comprehensive databases of cellular and secreted proteins from normal human embryonic lung MRC-5 fibroblasts: identification of transformation and/or proliferation sensitive proteins. *Electrophoresis* **10**, 76–115.
10. Jungblut, P. R., Grabher, G., and Stoffler, G. (1999) Comprehensive detection of immunorelevant *Borrelia garinii* antigens by two-dimensional electrophoresis. *Electrophoresis* **20**, 3611–3622.
11. Haas, G., Karaali, G., Ebermayer, K., et al. (2002) Immunoproteomics of *Helicobacter pylori* infection and relation to gastric disease. *Proteomics* **2**, 313–324.
12. Jungblut, P., Thiede, B., Zimny-Arndt, U., et al. (1996) Resolution power of two-dimensional electrophoresis and identification of proteins from gels. *Electrophoresis* **17**, 839–847.
13. Jungblut, P. R., Bumann, D., Haas, G., et al. (2000) Comparative proteome analysis of *Helicobacter pylori*. *Mol. Microbiol.* **36**, 710–725.
14. Khyse-Andersen, J. (1984) Electroblotting of multiple gels: a simple apparatus without buffer tank for rapid transfer of proteins from polyacrylamide to nitrocellulose. *J. Biochem. Biophys. Methods* **10**, 203–209.
15. Jungblut, P., Eckerskorn, C., Lottspeich, F., and Klose, J. (1990) Blotting efficiency investigated by using two-dimensional electrophoresis, hydrophobic membranes and proteins from different sources. *Electrophoresis* **11**, 581–588.
16. Krah, A., Miehlke, S., Pleissner, K. P., et al. (2003) Identification of candidate antigens for serologic detection of *Helicobacter pylori* infected patients with gastric carcinoma. *Int. J. Cancer* (in press).
17. Johannsson, K. E. (1986) Double replica electroblotting: a method to produce two replicas from gels. *J. Biochem. Biophys. Methods* **13**, 197–203.
18. Zeindl-Eberhart, E., Jungblut, P. R., and Rabes, H. M. (1997) A new method to assign immunodetected spots in the complex two-dimensional electrophoresis pattern. *Electrophoresis* **18**, 799–801.
19. Pappin, D. J. (1997) Peptide mass fingerprinting using MALDI-TOF mass spectrometry. *Methods Mol. Biol.* **64**, 165–173.
20. Jungblut, P. R., Muller, E. C., Mattow, J., and Kaufmann, S. H. (2001) Proteomics reveals open reading frames in *Mycobacterium tuberculosis* H37Rv not predicted by genomics. *Infect. Immun.* **69**, 5905–5907.
21. Jungblut, P. R. and Seifert, R. (1990) Analysis by high-resolution two-dimensional electrophoresis of differentiation-dependent alterations in cytosolic protein pattern of HL-60 leukemic cells. *J. Biochem. Biophys. Methods* **21**, 47–58.
22. Doherty, N. S., Littman, B. H., Reilly, K., Swindell, A. C., Buss, J. M., and Anderson, N. L. (1998) Analysis of changes in acute-phase plasma proteins in an acute inflammatory response and in rheumatoid arthritis using two-dimensional gel electrophoresis. *Electrophoresis* **19**, 355–363.

3

Immunoprecipitation and Blotting

The Visualization of Small Amounts of Antigens Using Antibodies and Lectins

Stephen Thompson

Abstract

The practical problems encountered when purifying and visualizing small amounts of antigens from complex cellular and protein mixtures are explored. Practical aspects and the relative advantages and disadvantages of immunoprecipitation and blotting, the two most commonly used antibody techniques, are discussed. As glycosylation of antigens is becoming recognized as an important factor in the progress of many diseases, a short section on the use of lectins in precipitation and blotting techniques is also included. It is highly likely that a combination of precipitation followed by blotting, using either lectin followed by antibodies or antibody followed by lectins, will become a valuable tool in characterizing cellular antigens and the progression of disease.

Key Words: Cell labeling; immunoprecipitation; blotting; antibody-antigen complexes; antigens; lectins.

1. Introduction

Immunoprecipitation and blotting both use antibodies (normally, but not exclusively monoclonal antibodies) to detect and quantitate specific protein antigens in complex cellular or protein mixtures. Immunoprecipitation has an advantage in that the antigens are allowed to react with the antibodies in their native conformation prior to their subsequent separation and quantification. A further advantage is that a protein at a very low concentration can be concentrated from the relatively large volume of 1–2 mL. The major disadvantage is that the proteins normally have to be radiolabeled to facilitate their detection. In Western blotting the proteins do not have to be labeled, but they have to be separated by electrophoresis in polyacrylamide gels prior to their transfer to either nitrocellulose, polyvinyldifluoride (PVDF), or nylon membranes. This seriously restricts the size of the sample, and hence the protein antigen has to be present at higher concentrations. A further disadvantage is that the antigen is not normally in its native conformation when it reacts with the antibody, because the electrophoresis usually being carried out in the presence of sodium dodecyl sulfate (SDS) to maximize the resolution of the separated proteins. If an antibody has a lower affinity for an antigen, it may well immunoprecipitate an antigen but not react with it on a Western blot. This is the reason why some workers slot-blot their protein mixtures rather than separate them by electrophoresis. This maintains their native conformation. The main problem encountered here is that crossreactions of the primary antibody with all the other proteins in the mixture can outweigh the antigen-specific binding. An unlabeled antigen is often therefore prepurified using an immobilized lectin or even immunoprecipitation itself prior to its quantitation on a slot-blot. Such a combination of lectin and antibody techniques has tremendous potential both in more precise analysis of cellular antigens and in the characterization of disease progression.

This chapter therefore discusses the following major steps involved in immunoprecipitation:

1. The labeling and lysis of cells.
2. The formation of antibody–antigen complexes.
3. The removal and separation of the complexes.
4. The quantitation of the separated antigens.

This is followed by a shorter second section discussing Western and slot-blotting. The actual processes involved in Western blotting are not covered in great detail. This has recently been excellently reviewed (*1*). However, common practical problems are addressed.

Finally, a third and final section then discusses the use of lectins in both of the above techniques. This section is included because post-translational glycosylation of proteins is becoming increasingly recognized as an important factor in determining the course of many diseases including cancer.

2. Materials

All chemicals were of the purest grade possible (analar grade). All enzymes and second-layer antibodies were purchased from Sigma unless otherwise stated.

2.1. Immunoprecipitation

1. Radioactive amino acids (20–200 µCi), sugars or $^{32}PO_4$.
2. Minimal essential medium (MEM) depleted of the appropriate amino acid.
3. Fetal calf serum (FCS).
4. Phosphate-buffered saline (PBS).
5. Glucose: 0.5 M in PBS.
6. Lactoperoxidase: 1 mg/mL in PBS (stored in frozen aliquots).
7. KI: 100 µM in PBS.
8. Glucose oxidase: 10 µL in 10 mL PBS (make up just before use).
9. Na^{125}I: 100–200 µCi.
10. Cell lysis buffer: 25 mM Tris-HCl buffer, pH 8.0, 150 mM NaCl, 1 mM MgCl (TBS), 0.5–1.0% Nonidet P-40 (NP-40), 0.1 mM phenylmethylsulfonyl fluoride (PMSF; made up as 100 mM stock solution in ethanol).
11. CNBr-activated Sepharose beads (Pharmacia).

12. Primary antibody.
13. LP3: 3 mL round-bottomed plastic tubes.
14. Primary monoclonal or polyclonal antisera
15. Protein A-Sepharose beads: washed in TBS to remove preservatives and kept as a 50% suspension.
16. Second-layer sheep or rabbit polyclonal anti-antibody if your primary antibody is a mouse or rat antibody (*see* **Note 9**).
17. SDS: 5–10%.
18. Light-proof film cassettes.
19. X-ray film.
20. Flash gun with filters to presensitize the film.
21. Calcium tungstate scintillation screens (enhanced autoradiography).
22. 2,5-Diphenyloxazole (PPO).
23. Dimethylsulphoxide (DMSO, fluorography).
24. Gel drier.
25. Scanners or densitometers.

2.2. Blotting

1. Nitrocellulose or nylon membranes.
2. Polyoxyethylenesorbitan monolaurate (Tween-20).
3. Apparatus to suck the samples onto a membrane under vacuum (slot-blotting).

2.3. Precipitation and Blotting with Lectins

1. Lectins coupled to Sepharose beads: commercially available or made in an identical procedure to that given for antibodies in **Subheading 3.1.2.1.**
2. Labeled cellular antigens or complex protein mixtures.
3. Dioxigenin (DIG)-coupled lectins (Boehringer Mannheim, Germany).
4. Alkaline phosphate (AP)-labeled anti-DIG antibody second layer.
5. Proteins with known glycosylation structures as positive and negative controls.
6. Biotinylated lectins.
7. Anti-biotin-AP or avidin-AP second layer.

3. Methods

3.1. Immunoprecipitation

3.1.1. The Labeling and Lysis of Cells

There are two main methods of labeling cellular antigens: (1) metabolic labeling and (2) cell surface labeling. Metabolic labeling is normally carried out for at least 16 h in an attempt to label all the cellular proteins, even those with low turnover rates. Cell surface labeling allows an accurate analysis of the surface of a cell at any given time. Metabolic labeling is performed as follows:

1. Add approx 20 µCi/mL ^{35}S, ^{14}C, or ^{3}H amino acids to near confluent cell cultures for 16–18 h in MEM depleted of the appropriate amino acid *(2,3)*. A typical flask would contain between 5×10^5 and 10^7 adherent cells (*see* **Notes 1–3**).
2. Wash adherent cells once with PBS.
3. Remove cells by treatment with 0.02% EDTA or EGTA for 5–10 min followed by two further washes with PBS (*see* **Note 4**).
4. Wash nonadherent cells three times with PBS to remove surplus radioactivity.

Many techniques have been developed to label the surface of cells, but those utilizing radioactive iodine have proved to be the most popular, probably because of the ease of detection of radioactive iodine in labeled proteins after their separation in polyacrylamide gels. The most commonly used technique is the lactoperoxidase-catalyzed iodination procedure using H_2O_2 generated by the glucose–glucose oxidase system *(4)*. I have found this technique to be very reliable *(5–7)*, and it can also be used to iodinate protein mixtures with a very high efficiency (around 95%). The procedure is given below.

1. Wash adherent cells once with PBS.
2. Remove cells by treatment with 0.02% EDTA or EGTA for 5–10 min followed by two further washes with PBS.
3. Resuspend the cells immediately in 1 mL PBS.
4. Add 10 µL glucose, 10 µL glucose oxidase, 5 µL KI, 10 µL lactoperoxidase, and 1–2 µL $Na^{125}I$ (100–200 µCi) (*see* **Subheading 2.1.** and **Note 5**).
5. Leave to react for 20–30 min at 37°C.
6. Wash away any unbound ^{125}I by three further washes with PBS containing 5 µM KI (*see* **Subheading 2.1.** and **Note 5**).

Many procedures can be used to lyse labeled cells. These all utilize isotonic buffers with pH values from 7.4 to 8.0 containing 0.5–1.0% nonionic detergent to solubilize the cell membranes. Some workers include EDTA at concentrations up to 10 mM, whereas others prefer to add Ca and Mg ions. Cell lysis buffer, as described in **Subheading 2.**, works with many antibodies and will give good immunoprecipitates from cell lysates solubilized in this buffer. However, these antibodies will not give immunoprecipitates if PBS or Tris buffers at pH 7.4 with the same additives are used.

1. Collect labeled cells by centrifugation.
2. Solubilize the cell pellet in 500 µL to 1 mL of the cell lysis buffer for 30 min at 4°C with repeated vortexing (*see* **Note 6**).
3. Separate the solubilized components from residual cellular debris by microcentrifugation at 13,000g for 10 min. The solubilized extracts can be used immediately or can be stored frozen at –70°C until required.

3.1.2. Formation of Antibody–Antigen Complexes

Antigens can be immunoprecipitated by direct or indirect methods. The direct method uses the antibody directly coupled to Sepharose beads. In the indirect method the protein mixture is incubated with the antibody, and then the antigen–antibody complexes are removed using Protein A immobilized on Sepharose beads. The indirect method is more commonly used, as the antibody binding is not constrained by its immobilization to beads.

3.1.2.1. Direct Precipitation

The antibody has to be bound to activated-Sepharose beads before it can be used. This is a very simple procedure:

1. Place 0.3 g of CNBr-activated Sepharose in 100 mL of 1 mM HCl and allow to swell (to give approx 1 mL of beads).

Immunoprecipitation and Blotting

2. Decant the clear supernatant after 30 min.
3. Gently resuspend the beads in another 100 mL of 1 mM HCl to remove the preservatives.
4. After 1 h decant the supernatant again (by suction)
5. Wash the beads rapidly in 10 mL 0.1 M sodium bicarbonate and pack by minimal centrifugation (30 s at 500g).
6. Add immediately 1–2 mg (1 mg/mL in bicarbonate) of antibody to 1 mL of beads and mix gently overnight at 4°C.
7. Leave the beads to settle out. Remove the supernatant and measure its absorbance. The OD of the supernatant should be close to zero if the coupling has worked.
8. Add 5–10 mL of a 0.2 M glycine solution for 2 h to block residual active groups.
9. Wash the antibody-coated beads twice in 0.1 M bicarbonate and PBS. They are then ready for use.
10. For immunoprecipitation, add 25–50 µL of beads (50–100 µL of a 50% bead suspension) to a LP3 tube and add up to 0.5 mL of cell lysate or protein mixture.
11. Leave with gentle mixing for 1 h at room temperature for complexes to form.
12. Wash away unbound proteins (5 × 2 mL; *see* **Note 7**).
13. Dissociate the immune complexes from the beads (see below).

3.1.2.2. Indirect Precipitation

1. For indirect immunoprecipitation add the antibody to up to 1 mL of cell lysate (more often 100–200 µL) and incubate the mixture for 30–40 min at room temperature for antigen–antibody complexes to form (*see* **Note 8**).
2. Add 50 µL of a 50% suspension of Protein-A Sepharose to each sample to bind to the antigen-antibody complexes (*see* **Note 9**).
3. After a further 1 h of incubation with frequent gentle mixing, wash the beads five or six times, by gravity or very gentle centrifugation (*see* **Note 7**), with 2.5 mL TBS/NP-40 to remove unbound proteins.

3.1.3. Removal and Separation of the Complexes

1. Remove the immunoprecipitates from the small Sepharose-bead pellets by the direct addition of 50 µL of double-strength SDS-PAGE sample buffer. The ionic detergent totally disrupts the antigen–antibody complexes (*see* **Note 10**).
2. Harvest the solubilized antigen in the supernatant after microcentrifugation (*see* **Note 11**).
3. Add 5% (v/v) 2-mercaptoethanol to reduce the sample.
4. Boil for 5 min to ensure complete solubilization of the proteins.
5. Separate the immunoprecipitates by SDS-PAGE. The discontinuous system of Laemmli *(5,8)* gives the best resolution of proteins.

3.1.4. Quantitation of the Separated Antigens

Separated antigens are normally quantitated by autoradiography or fluorography of the dried polyacrylamide gels followed by densitometry of the developed X-ray film.

3.1.4.1. Autoradiography

This is the simplest procedure, as the polyacrylamide gel is simply dried onto filter paper and placed directly in contact with X-ray film. It is not very sensitive, with bands needing to contain more than 1500 dpm of ^{35}S or ^{14}C to allow their detection in 24 h. Much lower levels of ^{125}I or ^{32}P can be detected owing to their stronger γ and

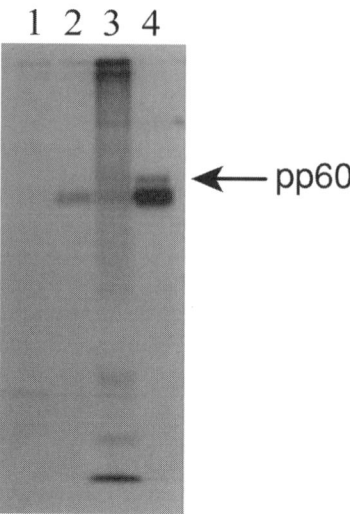

Fig. 1. A pp60^{src} immunoprecipitate of rat fibroblasts (lanes 1 and 2) and their virally transformed (A23) counterparts (lanes 3 and 4). Lanes 1 and 3 were controls using normal serum; lanes 2 and 4 were immunoprecipitated with pp60^{src} antibody.

β emissions. Their irradiation is so strong it penetrates completely though the X-ray film. This has allowed the development of an ultrasensitive sandwich detection technique in which a sensitized X-ray film is placed between the dried gel and a calcium tungstate intensifying screen *(10)*. Emissions pass through the X-ray film and hit the screen, and multiple photons of visible light are emitted that superimpose a photographic image on top of the autoradiographic image. The X-ray film is preexposed to a brief flash of light (*see* **Note 12**), and autoradiography is carried out at −70°C to maximize the detection of the emitted photons *(10)*.

An indirect autoradiograph of pp60^{src} immunoprecipitates is shown in **Fig. 1**. The pp60 kinase band is clearly visible in the positive control virally transformed cell line (lane 4) but is not visible in the untransformed parent cell line (lane 2) or the normal rabbit serum control lanes (lanes 1 and 3). The antibody in the precipitate is also phosphorylated by the kinase but at a much lower level in the parental cell line. Unlabeled lysates of cells (500 µL) were immunoprecipitated as above using 3 µL of antibody or normal rabbit serum and 25 µL of Protein A beads; 2 µCi (4 µL) of [γ-^{32}P]ATP were then added to the immunoprecipitate, and it was left to phosphorylate for 30 min. After washing to remove excess ATP, the immunoprecipitates were solubilized by SDS and separated by PAGE in a 13% polyacrylamide gel.

3.1.4.2. Fluorography

This technique also utilizes the detection of photons/light emitted by scintillators to detect the presence of low levels of ^{35}S or ^{14}C *(11)*. It can also be used to increase massively the detection limits of ^{3}H, a very weak α emitter. It is possible to measure as low an amount as 300 dpm of ^{3}H in a band in 24 h *(12)*. The principles are exactly the same as those described for enhanced autoradiography. However, here the radioactive emissions are not strong enough to pass out of the dried gel and through the film to reach a scintillation screen. The scintillant has instead to be impregnated directly into the gel, as follows:

1. Totally dehydrate fixed gels by two 30-min to 1-h immersions in a 20X excess of dimethylsulfoxide (DMSO).
2. Immerse the gel in a saturated solution of the scintillator PPO (20% w/v) in DMSO for 1 h with gentle shaking.
3. Remove and place into a large excess of water where the scintillant immediately precipitates.
4. After 1 h, dry the gel normally and expose against preexposed X-ray film at –70°C.

3.2. Blotting

As mentioned in **Subheading 1.**, blotting is described elsewhere in detail. Briefly, blotting is performed by the following steps:

1. Separate the protein mixture by SDS-PAGE and blot transversely onto a nitrocellulose or nylon membrane (*see* **Note 13**).
2. Block spare sites on the membrane (*see* **Note 14**).
3. Add the primary antibody (*see* **Note 15**).
4. After 1 h, wash the blot and add an enzyme-labeled anti-antibody (*see* **Note 16**).
5. Perform further washes and add a colored enzyme substrate. The substrate is precipitated onto the antigen by the enzyme *(1,13)*. Dried blots (or photographs) can then be scanned for quantitation purposes.

3.2.1. Slot-Dot Blotting

To avoid the problems associated with denaturing the antigen you want to quantitate, it is possible to absorb a sample directly onto nitrocellulose by vacuum in either dots or slots. As either of the primary and secondary antibodies could be crossreacting with other components in the mixture and giving a false signal, it is essential to check that your antibodies are highly specific. This is carried out by using negative and positive control slots. These should contain a mixture of proteins that are known to be negative and a highly positive protein. Even then, artifactual results can occur. A more correct control experiment is to add varying amounts of your positive control in the presence of the same large amount of your negative control. The staining of the slots should then be in a linear relationship to the amount of positive sample added.

3.3. Precipitation and Blotting with Lectins

Precipitation with lectins is performed with microbatch lectin-affinity chromatography *(14)*. This is used to take all the proteins with a given state of glycosylation out of a complex mixture prior to silver staining or blotting or ELISA quantitation. This could be used to study the glycosylated states of cellular antigens. Alternatively, it can be used to examine body fluids. I have used this procedure to study the fucosylation of serum glycoproteins in cancer progression *(15–17)*, active and inactive arthritis *(18)*, and inflammatory bowel disease *(19)* and have found similar changes in haptoglobin in the sera of "healthy" blood donors who smoke, compared with those found in can-

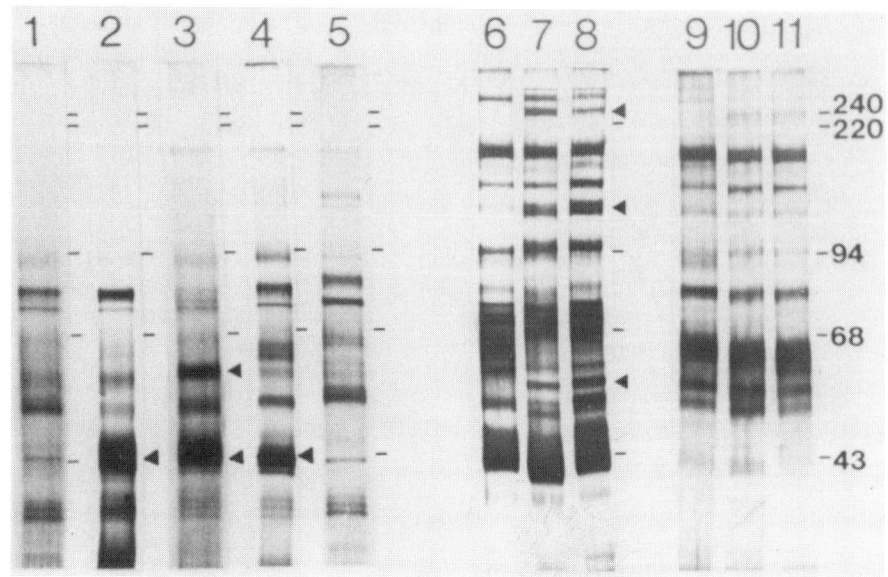

Fig. 2. Silver-stained SDS-PAGE patterns of sera extracted with *Lotus* (lanes 1–5), WGA (lanes 6–8), or lentil (lanes 9–11) lectins. Lanes 1, 6, and 9, healthy individuals; lane 2, advanced hepatoma; lane 3, recurrent ovarian cancer; lane 4, active rheumatoid arthritis; lane 5, bronchopneumonia; lanes 7 and 10, renal failure, lanes 8 and 11, liver cirrhosis.

cer patients *(20)*. This may be useful as a serum marker of risk of liver disease and/or cancer for the "healthy" population that smokes/drinks.

Figure 2 shows the serum glycoproteins precipitated from patients with five diseases by three lectins and demonstrates some of the major changes that can occur. When the fucose-specific *Lotus* lectin is used, cancer sera are characterized by strong bands around 43 kDa (lanes 2 and 3) and occasionally at 57 kDa. These are haptoglobin and α_1-antitrypsin, the former being related to tumor burden *(16)* and the latter to tumor progression *(17)*. Active rheumatoid sera (lane 4) also contain abnormal haptoglobin, but this is of a lower molecular weight than that obtained in cancer sera *(15)* and is associated with high serum haptoglobin levels (4–6 g/L). Surprisingly, fucosylated haptoglobin is not detected in untreated broncopneumonia patients even when the total haptoglobin level is highly elevated (7–8 g/L). Changes could also be seen in these specimens when the N-acetylglucosamine-specific lectin wheat germ agglutinin (WGA) was used, but they were not as marked as those found with *Lotus* lectin. WGA extracts of renal failure (lane 7) and cirrhosis patients (lane 8) did give characteristically altered patterns that were highly reproducible. Here, however, very little change was found using lentil (lanes 9–11) or *Lotus* lectins. The altered proteins were identified by blotting as described above.

Figure 3 shows an α_1-antitrypsin blot of *Lotus* extracts from serial serum samples of a cancer patient that were collected at roughly 3-mo intervals. The fucosylated

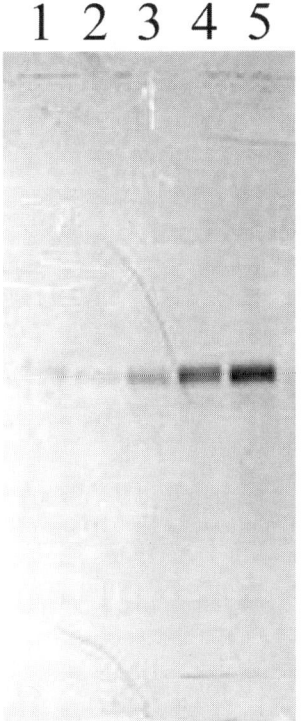

Fig. 3. An α_1-antitrypsin Western blot of fucosylated serum proteins extracted from a terminally ill cancer patient. Lanes 1–5 represent samples taken 0, 3, 6, 9, and 12 mo after the commencement of treatment, respectively. Total silver-stained components would be similar to lane 3 in **Fig. 2**.

α_1-antitrypsin levels increased as her disease progressed. A weakly stained second band can also be seen just under the antitrypsin band, owing to the heavy chain human serum immunoglobulins crossreacting weakly with the AP-labeled sheep anti-rabbit second layer. A polyclonal rabbit antiserum (5 µL) was used at 1:1000 dilution followed by 5 µL of a second-layer AP-labeled sheep-anti-rabbit antiserum (1:2000).

Blotting with lectins requires the lectin to be coupled to another molecule such as digoxigenin. Your purified protein (or mixture of proteins) is slot or Western blotted, the blot is blocked (with nonionic detergent), and the lectin is added in the same manner as a first-layer antibody (*see* **Note 17**). After incubation with gentle shaking for up to 2 h, the blot is washed and an AP- or horseradish peroxidase-labeled anti-digoxigenin antibody is added. Chemiluminescence can increase the sensitivity of this procedure and allow blots to be reprobed with different lectins *(21)*. An alternative and now more commonly used procedure is to utilize biotinylated lectins. They are relatively easy to prepare, and many commercial antibiotin or AP-avidin second layers are available. They are often used to compare the glycosylation of recombinant proteins with their natural products and have also been used to study the expression of cell surface proteins in parasites *(22)*.

4. Notes

1. $^{32}PO_4$ can also be used as a metabolic label for the specific labeling of phosphoproteins and measurement of kinase activity. However, here it is often more correct to label the cell lysates or even the specific kinase immunoprecipitates with [γ-^{32}P] ATP. This not only reduces the problems encountered because $^{32}PO_4$ also labels DNA and RNA but also minimizes the amount of ^{32}P required for protein labeling.
2. Do not label confluent cell cultures, as they will have a slower turnover rate than nonconfluent cultures. The cells may also run out of the labeling amino acid.
3. Low levels of dialyzed FCS (1–5%) are normally retained to maintain essential growth factors. Nonadherent cells are normally labeled at around 10^7 cells/mL.
4. Do not use trypsin to remove attached cells as you will alter cell surface protein expression. If labeled cells do not detach with EDTA, they should be washed *in situ* and then removed by rubbing them off the surface of the plate in the presence of lysis buffer.
5. It is essential to add nonradioactive KI; if this is not done, the radioactive ^{125}I sticks ionically to the surface of the cells and subsequently washes off in the washing steps. The addition of 0.5 µ*M* "cold" KI, *before* the ^{125}I, saturates the cell surface, and all the ^{125}I is then available for labeling. Higher levels of "cold" KI reduce the efficiency of labeling.
6. If $^{32}PO_4$ is used as a metabolic label, it is essential to include large amounts of DNAase and RNAse and possibly micrococcal nucleases to digest small labeled pieces of DNA and RNA sticking to the proteins. Otherwise numerous artifactually phosphorylated proteins will be present in your mixture.
7. In both direct and indirect precipitation, the most critical factor is how the beads are washed after antibody–antigen complexes have been formed. If the beads are spun at any speed, the complexes split off the surface of the beads and are washed away. Furthermore, proteins stick nonspecifically to the newly revealed surface. I demonstrated this in 1982. Two identical immunoprecipitates were washed, one by gravity and one using a microfuge. A very clean precipitate with one band of the correct molecular weight was obtained with the beads that were allowed to settle out under gravity. However, numerous background bands with only a faintly visible correct band were obtained with the centrifuged immunoprecipitate.
8. Enough antibody has to be added to precipitate all the antigen present. Five microliters of polyclonal antisera, 5 µL of monoclonal ascites, or 50 µL of hybridoma tissue culture supernatant should easily be sufficient. The same amount of nonimmune sera or ascites should also be incubated with a second aliquot as a negative control.
9. In indirect precipitation, occasionally your antibody will not react very well with the protein A-Sepharose beads, and hence the antigen–antibody complex will not be precipitated properly. This is especially true with some subclasses of mice antibodies. It is possible to use alternative antibody binders with different binding specificities such as protein G-Sepharose. A simpler solution is to add 2 or 3 µL of a polyclonal sheep (or rabbit) anti-mouse antibody for a further 30 min after the initial antigen–antibody binding step *(6,7)*. Sheep (and rabbit) antibodies bind very well to protein A and carry the mouse antibody-antigen complex onto the protein A. Rat monclonals may also be helped to immunoprecipitate with a sheep anti-rat second antibody.

10. If further chemical analysis of a purified immunoprecipitated antigen is required, the antigen can be released from the beads by the addition of a high (>9) or low (<4) pH buffer followed by centrifugation and immediate neutralization. Here direct immunoprecipitation is required or the purified antigen will be contaminated with the antibody *(9)*. Trifluoroacetic acid (TFA; 0.1 *M*) instantly solubilizes purified antigens from antibody-coated beads without removing any antibody *(9)* from the beads.
11. The microcentrifugation step after addition of the SDS sample buffer can be omitted. The presence of a few microliters of Sepharose beads in a sample well does not have any deleterious effect on the migration of the immunoprecipitated proteins in discontinuous SDS-PAGE.
12. In both indirect autoradiography and fluorography, preexposed X-ray film is required to permit a linear detection of photons. It is very easy to do this. A flash gun is taken, and a red filter is taped over the flash. A strip of X-ray film is flashed from between 18 and 24 in. away (from above). This ensures a fairly even spread of the flash over the film. Normally the light is still far too strong. This can be further reduced by taping a piece of Whatman No. 1 filter paper over the filter followed by several layers of colored plastic tape. After each layer of tape is added, a strip of film is exposed and the film is developed until a background film OD of 0.15 is reached *(12)*. X-ray film is two-sided; the side closest to the flash is placed against the scintillation screen or the impregnated gel.
13. There are two obvious major problems with Western blotting. The first is that a relatively high level of antigen is required in the sample, as most polyacrylamide gels have a maximum sample well size of 50 µL. Protein mixtures (serum samples, culture supernatants, and so on) are often preconcentrated by immunoprecipitation *(9)* or with lectins bound to Sepharose prior to blotting *(14,15)* to circumvent this problem.
14. I always block remaining sites on the membrane with 0.5% Tween-20 after transfer. This can have the fortuitous effect of renaturing the antigen by displacing the SDS as well as blocking remaining active sites of the membrane. I have also found that the alternative blocking reagents of 0.5% BSA or milk powder can cause very high backgrounds, especially when used in conjunction with nylon membranes, owing to nonspecific crossreactions with either or both the antibody preparations.
15. A primary antibody may sometimes crossreact with other proteins. More commonly it will not react at all due to the denatured nature of the antigen.
16. If an antibody has a high specificity but low avidity (as is the case with many antibodies to glycosylated antigens), the antibody/antibody–second antibody complexes will often fall off the blot during the washing steps, and no color will appear. This effect can be minimized or cured by two simple practical precautions:
    a. Use only an Fc-specific enzyme-labeled secondary antibody. This will then not interfere with the primary antigen–antibody reaction in the Fab region.
    b. The secondary anti-Fc antibody can be preincubated with the primary antibody for 1 h, and this primary-secondary complex can then be added in one step.

    Try to use anti-Fc-specific second-layer antibodies; this is especially important when blotting with first-layer monoclonals. Remember to preincubate the primary and secondary antibodies if the primary is suspected of having a low affinity. Not a lot else can be done about nonspecific crossreactions of either the primary or secondary antibodies. Samples should always be blotted in duplicate with one blot containing an irrelevant primary antibody raised in the same species as the correct primary. Specific staining is then guaranteed, especially if the staining is at the correct molecular weight.

17. Most if not all lectins require the presence of Ca and Mg ions. Most buffers for lectin immunoprecipitation *(16)* contain 5 mM Ca and Mg ions in isotonic Tris-HCl buffers (50 mM, pH 7.4). Blotting solutions are similar to those used with antibodies but always contain 1 mM Ca and Mg.

References

1. Gallagher, S., Winston, S. E., Fuller, S. A., and Hurrell, J. G. R. (1998) Immunoblotting and immunodetection. *Curr. Protocols Immunol.* **8,** 1–21.
2. Thompson, S., et al. (1984) Cloned human teratoma cells differentiate into neuron-like cells and other cell types in retinoic acid. *J. Cell Sci.* **72,** 37–64.
3. Stern, P. L., Beresford, N., Thompson, S., Johnson, P. M., Webb, P. D., and Hole, N. (1986) Characterisation of the human trophoblast-leukocyte antigenic molecules defined by a monoclonal antibody. *J. Immunol.* **137,** 1604–1609.
4. Hubbard, A. L. and Cohn, Z. A. (1972) The enzymic iodination of the red blood cell. *J. Cell Biol.* **55,** 390–405.
5. Thompson, S., Rennie, C. M., and Maddy, A. H. (1980) A re-evaluation of the surface complexity of the intact erythrocyte. *Biochim. Biophys. Acta* **600,** 756–768.
6. Stern, P. L., Beresford, N. A., Bell, S. M., and Thompson, S. (1986) Murine lymphocyte and embryonal carcinoma cell surface antigens recognised by rabbit anti-murine embryonal carcinoma serum. *Exp. Cell Biol.* **54,** 250–262.
7. Blackmore, M., Thompson, S., and Turner, G. A. (1990) A detailed study of the effects of in vitro interferon treatment on the growth of two variants of the B16 mouse melanoma in the lungs: evidence for non-specific effects. *Clin. Exp. Metastasis* **8,** 449–460.
8. Laemmli, U. K. (1970) Cleavage of structural proteins during the assembly of the bacteriophage T4. *Nature (Lond.)* **227,** 680–685.
9. Thompson, S., Dargan, E., and Turner, G. A. (1992) Increased fucosylation and other carbohydrate changes in haptoglobin in ovarian cancer. *Cancer Lett.* **66,** 43–48.
10. Laskey, R. A. and Mills, A. D. (1977) Enhanced autoradiographic detection of ^{32}P and ^{125}I using intensifying screens and hypersensitised film. *FEBS Lett.* **82,** 314–316.
11. Bonner, W. M. and Laskey, R. A. (1974) A film detection method for tritium-labelled proteins and nucleic acids in polyacylamide gels. *Eur. J. Biochem.* **46,** 83–88.
12. Laskey, R. A. and Mills, A. D. (1975) Quantitative film detection of ^{3}H and ^{14}C in polyacrylamide gels by fluorography. *Eur. J. Biochem.* **56,** 335–341.
13. Blake, M. S., Johnson, K. H., Russel-Jones, G. J., and Gotschilich, E. C. (1984) A rapid, sensitive method for detection of alkaline phosphatase-conjugated anti-antibody on Western blots. *Anal. Biochem.* **136,** 175–179.
14. Thompson, S., Latham, J. A. E., and Turner, G. A. (1987) A simple, reproducible and cheap batch method for the analysis of serum proteins using Sepharose coupled lectins and silver-staining. *Clin. Chim. Acta* **167,** 217–223.
15. Thompson, S. and Turner, G. A. (1987) Elevated levels of abnormally-fucosylated haptoglobins in cancer sera. *Br. J. Cancer* **56,** 605–610.
16. Thompson, S., Cantwell, B. M. J., Cornell, C., and Turner, G. A. (1991) Abnormally-fucosylated haptoglobin: a cancer marker for tumour burden but not gross liver metastasis. *Br. J. Cancer* **64,** 386–390.
17. Turner, G. A., Goodarzi, M. T., and Thompson, S. (1995) Glycosylation of alpha-1-proteinase inhibitor and haptoglobin in ovarian cancer; evidence for two different mechanisms. *Glycoconjugate J.* **12,** 211–218.

18. Thompson, S., Kelly, C. A., Griffiths, I. D., and Turner, G. A. (1989) Abnormally-fucosylated serum haptoglobins with inflammatory joint disease. *Clin. Chim. Acta* **184,** 251–258.
19. Thompson, S., Record, C. O., and Turner, G. A. (1991) Studies of lotus-extracted haptoglobin in inflammatory bowel disease. *Biochem. Soc. Trans.* **19,** 514.
20. Thompson, S., Matta, K. L., and Turner, G. A. (1991) Changes in fucose metabolism associated with heavy drinking and smoking: a preliminary report. *Clin. Chim. Acta* **201,** 59–64.
21. Jadach, J. and Turner, G. A. (1993) An ultrasensitive technique for the analysis of glycoproteins using lectin blotting with enhanced chemiluminescence. *Anal. Biochem.* **212,** 293–295.
22. Nolan, D. P., Geuskens, M., and Pays, E. (1999) N-linked glycans containing linear poly-N-acetyllactosamine as sorting signals in endocytosis in *Trypanosoma brucei. Curr. Biol.* **9,** 1169–1172.

II

DNA-BASED IDENTIFICATION OF ANTIGENS

4

Representational Difference Analysis of cDNA

Lucas D. Bowler

Abstract

In this chapter I describe the PCR-coupled subtractive hybridization technique of representational difference analysis of cDNA (cDNA RDA).

cDNA RDA is based on the representational difference analysis (RDA) method previously described by Lisitsyn et al., and can be used to identify genes whose expression is modified between two populations of cells.

cDNA RDA is relatively inexpensive to perform and requires no prior knowledge of genome sequence data. The combining of PCR with a subtractive methodology results in a highly effective and extremely sensitive technique with application to very low amounts of starting material.

The procedure can be divided into three main phases: PCR generation of amplicons representative of the starting populations of RNA molecules being compared; the two-step subtractive hybridization of these *representations*, leading to the enrichment of amplified fragments of differentially expressed genes and the sequential depletion of sequences common to both populations; and the purification, cloning, and sequencing of the resulting *difference* products.

Key Words: PCR; subtractive hybridization; cDNA RDA; differential gene expression; RNA.

1. Introduction

There is a growing realization that aspects of nearly every stage of the disease process involve a highly complex series of interactions between parasite and host. Successful pathogens have evolved a variety of specific gene products that facilitate their survival and growth within the host, as well as mechanisms to regulate expression of these virulence-associated genes in response to their environment. Not surprisingly, identification of these differences is a frequent goal in modern biomedical research, and as a result, a variety of differential screening methods have been developed over the last few years *(1,2)*.

cDNA RDA is based on the representational difference analysis (RDA) technique previously described by Lisitsyn et al. *(3)*. RDA is a method for the identification of differences between two complex genomes and was a significant advance in the field of subtractive cloning, bringing together the advantages afforded by both subtractive hybridization and the polymerase chain reaction (PCR)-based amplification used in differential display-type techniques. RDA belongs to the general class of DNA subtractive methodologies, in which one DNA population (known as the *Driver*) is hybridized in excess against a second population (the *Tester*), to remove common (hybridizing) sequences, thereby enriching for *target* sequences unique to the Tester population. This is achieved by ligation of defined oligonucleotide adaptors to the 5' end of the Tester molecules. After annealing of the Tester and Driver sequences, DNA polymerase is used to fill in the 3' ends of the double-stranded molecules. Only Tester molecules that have annealed to other Tester-originating sequences will yield molecules with double-stranded adaptor sequences at both the 5' and 3' ends of the double-stranded sequences. Accordingly, only these molecules will be amplified exponentially by PCR (using the specific oligonucleotides as primer), thus facilitating enrichment of Tester-specific sequences.

RDA successfully combines this subtractive approach with positive selection of target sequences by what is termed *kinetic enrichment (3)*. Kinetic enrichment takes advantage of the second-order kinetics of DNA reannealing, i.e., the rate of double-stranded DNA formation is higher for DNA species of higher concentration. Thus, the more abundant DNA species in a mixture of fragments can be further partitioned from less abundant species by reannealing for low C_0t values (the product of initial concentration and time), and subsequent collection of the resulting double-stranded molecules. The molar ratio of abundant to less abundant sequences in the product will then be of the order of the square of the initial ratio of the concentrations, which eventually leads to purification of the more abundant species. In RDA, this kinetic enrichment is achieved by degradation of single-stranded molecules after the initial amplification by PCR. The target sequences will be enriched exponentially by this PCR and will therefore finish at a much higher concentration than nontarget Tester sequences (therefore forming relatively more double-stranded molecules). By degrading the single-stranded DNA using mung bean nuclease, only the double-stranded sequences remain (enriched for the target sequences). Subsequently, the target sequences are further enriched for, by another round of PCR.

A modification of RDA, cDNA RDA, has been described *(4)* in which the starting material is derived from mRNA rather than DNA, and accordingly targets only genes expressed at the time the RNA is isolated. More recently, the cDNA RDA methodology has been further adapted to facilitate the identification of genes whose expression is modified between different bacterial populations *(5)*. This approach is illustrated schematically in **Fig. 1**. The method is flexible, sensitive, and relatively inexpensive to perform; in contrast to alternative methods such as differential display PCR, and the conceptually similar technique, RNA fingerprinting by arbitrarily primed PCR (AP-PCR), it has the major advantage that sequences common to both groups of cells are eliminated. This greatly simplifies the interpretation of results and identification of the differentially expressed genes. In addition, the exponential degree of enrich-

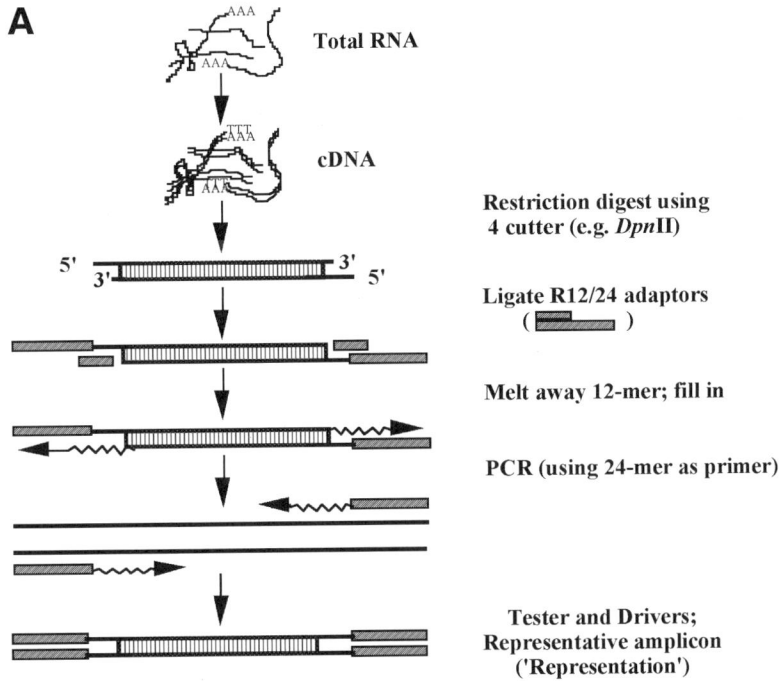

Fig. 1. **(A)** Schematic view of the procedure for generation of the initial representations in cDNA RDA. cDNA is produced by random-primed reverse transcription of total RNA. The cDNA is then digested with a restriction enzyme. To the four nucleotide 5' overhang created by the restriction enzyme, four nucleotides of the 12-mer oligonucleotide (J-Bgl-12) of the adaptor hybridize. The 24-mer oligonucleotide (J-Bgl-24) of the adaptor hybridizes to the remaining eight nucleotides of the 12-mer, and can thus be covalently joined to the 5'-phosphate group of the digested starting material. Because the 12-mer is not phosphorylated, it does not become covalently attached, and accordingly dissociates at higher temperature (72°C in the cDNA RDA protocol). The *Taq* polymerase is added, and can then "fill in" the sequence complementary to the 24-mer. This creates a binding site for the J-Bgl-24 oligonucleotide that is used as primer in subsequent PCR. **Figure 1B** continued on next page.

ment achieved by the use of PCR in cDNA RDA allows the detection of very rare transcripts *(6)*.

Examination of differential gene expression using cDNA RDA requires the sampling of a population (of cells) grown under the condition(s) of interest and a population grown under conditions that differ only by those of interest. For example, in a study to identify iron-regulated genes in *Neisseria meningitidis* *(5)*, one population of bacteria was grown under iron limitation (to provide the so-called Tester material), and a second was grown under identical conditions, except that iron was freely available (to provide the *Driver*). To identify genes whose expression is *suppressed* under iron limitation, the converse would apply. RNA is then extracted from both populations and used as a template for cDNA synthesis. With eukaryotic material, mRNA

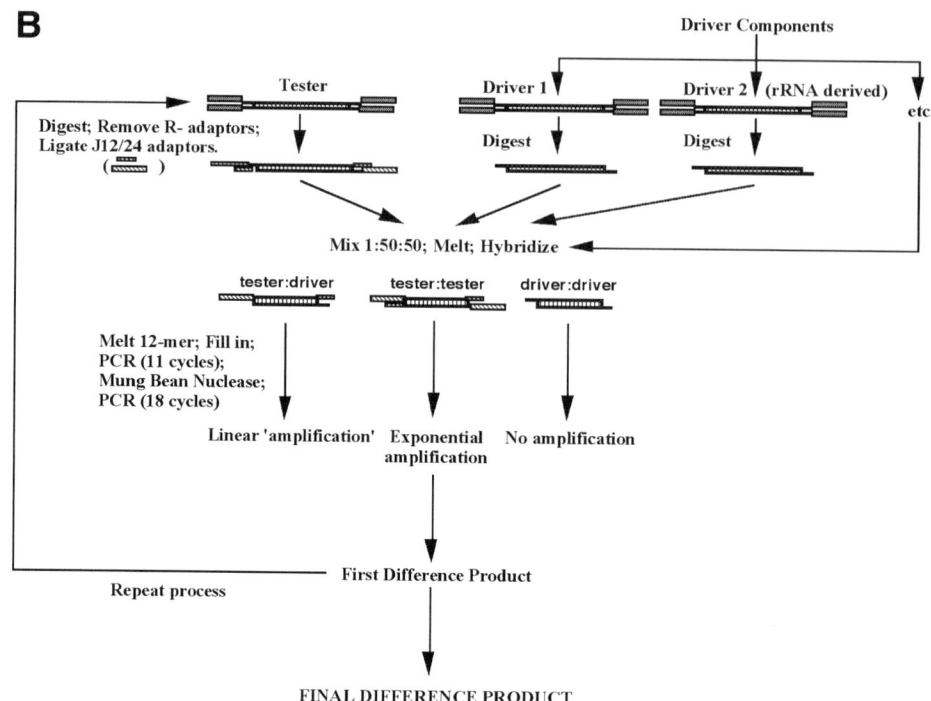

Fig. 1. **(B)** Overview of the hybridization/amplification steps. The adaptors are digested from the representative amplicons of Tester and Driver, and from the 16S and 23S rRNA-derived Driver (Driver 2). A set of adaptors is ligated to the Tester only (in a similar way to that outlined in **A**). The Tester–adaptor DNA is then mixed with excess Driver DNA, melted to obtain single-stranded DNA, and allowed to hybridize. Tester sequences present in both the Tester and Driver pools will form heteroduplexes with Driver sequences, whereas unique Tester sequences (the "targets") can only hybridize with their complementary Tester-originating sequences. The subsequent "fill in" reaction (see **A**) will create molecules with the 24-mer sequence and its complement at both the 5'- and 3'-ends of the molecule for target sequences only. In the following 11 PCR cycles, target sequences are amplified exponentially, and non-target sequences are linearly "amplified" or are not amplified at all (Driver–Driver hybrids). Further enrichment is achieved by degradation of single-stranded DNA (including single-stranded overhangs of double-stranded molecules) with mung bean nuclease, and further PCR amplification. If target sequences are not sufficiently enriched in the first difference product, the procedure can be repeated. (In cDNA RDA, proceeding to the generation of DP2 is normal and DP3 may be required in some cases.) (Adapted from ref. **5**.)

may be isolated directly or oligo (dT) may be used for priming cDNA first-strand synthesis. However, with prokaryotic RNA, given that bacterial mRNAs possess short, or no, polyA tails, random hexamers must be used for priming cDNA synthesis *(5)*. The derived cDNAs are then cut with a restriction enzyme, and oligonucleotide adaptor is ligated onto the fragments. PCR, using primers specific to these oligonucleotides, is then used to amplify each population of DNA fragments. After amplification, the adaptors are removed with the same enzyme, and the amplified cDNA fragments are purified (these preparations are referred to as *representations*).

For the hybridization and amplification stages, new oligonucleotides are ligated to the Tester DNAs only. Tester and Driver DNAs (the latter in excess) are mixed, denatured, and allowed to hybridize. PCR using primers specific to the new oligonucleotide extensions results in enrichment of sequences unique to the Tester DNA population. By repeating these steps, the degree of enrichment is increased. The result is a number of PCR products (difference products) that represent the messages unique to the Tester population. These can then be cloned and characterized. In most studies to date, the restriction endonuclease *Dpn*II has been the enzyme of choice for cutting the derived cDNAs, prior to linker addition. Mathematical modeling predicts that a *Dpn*II-derived representation will typically include at least one potentially amplifiable fragment from over 86% of expressed mammalian genes, and this can in principle be increased to 94% if the primer used for cDNA first-strand synthesis is designed to contain a *Dpn*II site *(6)*. Other enzymes may be substituted for *Dpn*II in subsequent experiments, which will maximize the number of differentially expressed genes potentially detectable.

cDNA RDA is most efficient in detecting transcripts when the differences in the level of expression between two populations are relatively high; however, the methodology can be modified to facilitate detection of mRNAs with lower levels of differential expression using a process known as melt depletion *(4)*. This technique effectively depletes representations of low copy sequences.

cDNA RDA can be divided into three main phases: the generation of PCR amplicons representative of the RNA isolated from given bacterial populations; the PCR-coupled subtractive hybridization of the different representative amplicons; and the cloning and screening of the resultant products (which represent the differences between the two populations that were compared). This chapter describes the procedures involved in carrying out these analyses.

2. Materials

Other than standard molecular biology laboratory equipment, the following items are required:

1. A high-quality Thermal Cycler.
2. A refrigerated microfuge.
3. A FastPrep FP120 instrument (Q-Biogene) or similar device; generally required for work with Gram-positive bacteria only.

2.1. RNA Isolation

RNA is generally highly sensitive to the action of ribonucleases. Accordingly, for the successful isolation of high-quality RNA, it is absolutely essential that all solutions and equipment used be RNase-free. Hands are a major source of nuclease contamination, and powder-free gloves should be worn at all times. (Indeed, it is advisable to change gloves several times during the course of an isolation, because the outsides of the gloves themselves can become contaminated through contact with items in the lab environment.) The use of sterile disposable plasticware is recommended, and when glassware is used, they should be foil sealed and baked for at least 4 h at 200°C before use. Because of the potential for residual traces of diethylpyrocarbonate (DEPC) to inhibit some enzymatic processes, we prefer to use untreated, ultra-high-purity (UHP) water to make up all solutions. These solutions are then filter-sterilized (using nuclease-free filters). UHP water from a good purification system is usually RNase-free; however, it is advisable to check, e.g., using an RNaseAlert kit (Ambion).

1. 0.1-mm-diameter zirconia/silica beads (BioSpec Products).
2. Decon 90 (Decon Laboratories, UK).
3. 0.5 M sodium acetate, pH 4.0.
4. 2-Mercaptoethanol.
5. Lysis solution: 1% decon 90, 125 mM sodium acetate, pH 4.0, 100 mM 2-mercaptoethanol. The solution should be made up fresh each time in RNase-free water. The 2-mercaptoethanol is added just before use.
6. Acid phenol:chloroform (5:1, v/v, acid-equilibrated to pH 4.7).
7. Chloroform.
8. RNase-free 1.0 mM sodium citrate pH 6.4 (Ambion).
9. DNase–free treatment and removal kit (Ambion).
10. TBE (10X): 0.9 M Tris-HCl, 0.9 M boric acid, 20 mM EDTA, pH 7.8.
11. 1% (w/v) nondenaturing agarose gel in TBE (pH 7.8). The gel should contain 0.1 mg/mL ethidium bromide.
12. Ethidium bromide (10 mg/mL). Store at 4°C in the dark. (**Caution:** Ethidium bromide is highly carcinogenic; handle under appropriate guidelines.)
13. DNA molecular size standards (e.g., GeneRuler 1-kb DNA ladder, MBI Fermentas 0.1 µg/mL).
14. Sterile 50-mL Falcon tubes.
15. Nuclease-free filter tips for micropipets.
16. Nuclease-free microfuge tubes (e.g., from Ambion).

2.2. cDNA Synthesis

1. TimeSaver cDNA synthesis kit (Amersham Pharmacia Biotech).
2. Superscript II reverse transcriptase (Invitrogen, 200 U/µL).
3. Microspin S-300HR spin columns (Amersham Pharmacia Biotech).
4. GlycoBlue glycogen co-precipitant, 15 µg/mL (Ambion).

2.3. Isolation of Chromosomal DNA

1. Genomic-tip 20/G system (Qiagen).

2.4. Isolation of Streptococcus pneumoniae 16S and 23S rRNA Genes

1. Pneumococcal chromosomal DNA (*see* **Subheading 2.3.**).
2. PCR primer pairs for amplification of pneumococcal 16S and 23S rRNA genes (*see* **Table 1**).
3. Taq DNA polymerase (e.g., TaKaRa *Ex-Taq*, Takara Shuzo, 5 U/µL).
4. PCR buffer and dNTPs (as supplied with *Ex-Taq*).
5. 5 M betaine (Sigma-Aldrich).
6. TBE (10X): 1.0 M Tris-HCl, 90 mM boric acid, 2 mM EDTA, pH 8.3.
7. 1% (w/v) nondenaturing agarose gel in TBE (pH 8.3). The gel should contain 0.1 mg/mL ethidium bromide.

Table 1
Primer Pairs for Amplification of Pneumococcal 16S and 23S Ribosomal RNA Genes

Name	Orientation	Sequence
pn16Sf	Forward primer	5'-TGGCTCAGGACGAACGCT-3'; 25 pM
pn16Sr	Reverse primer	5'-CTTGTTACGACTTCACCCCA-3'; 25 pM
pn23Sf	Forward primer	5'-CCTTGGCACTAGAAGCCGA-3'; 25 pM
pn23Sr	Reverse primer	5'-CCTGATCATCTCTCAGGGCT-3'; 25 pM

Table 2
Oligonucleotide Adaptor Pairs

Name	Sequence
R-Bgl-12	5'-GAT CTG CGG TGA-3'; 0.25 mM
R-Bgl-24	5'-AGC ACT CTC CAG CCT CTC ACC GCA-3'; 0.5 mM
J-Bgl-12	5'-GAT CTG TTC ATG-3'; 0.25 mM
J-Bgl-24	5'-ACC GAC GTC GAC TAT CCA TGA ACA-3'; 0.5 mM
N-Bgl-12	5'-GAT CTT CCC TCG-3'; 0.25 mM
N-Bgl-24	5'-AGG CAA CTG TGC TAT CCG AGG GAA-3'; 0.5 mM

8. Ethidium bromide (10 mg/mL). Store at 4°C in the dark. (**Caution:** Ethidium bromide is highly carcinogenic; handle under appropriate guidelines.)
9. DNA molecular size standards (e.g., GeneRuler 1-kb DNA ladder, MBI Fermentas, 0.1 µg/mL).
10. Razor blades or scalpels.
11. QIAquick DNA purification kit (Qiagen).

2.5. cDNA RDA

1. Double-stranded cDNA (*see* **Subheading 3.2.**).
2. 16S and 23S rRNA amplicons (*see* **Subheading 3.4.**).
3. Restriction enzyme: for example, *Dpn*II (New England Biolabs, 50 U/µL).
4. 10X *Dpn*II buffer (as supplied with enzyme).
5. T4 DNA ligase (New England Biolabs, 2000 U/µL).
6. 10X T4 DNA ligase buffer (as supplied with enzyme).
7. High-performance liquid chromatography (HPLC; or equivalent) purified oligonucleotide adaptors/primers (*see* **Table 2**).
8. 10 M ammonium acetate.
9. *Taq* DNA polymerase: *Amplitaq* DNA polymerase (Applied Biosystems, 5 U/µL).
10. PCR buffer (5X): 335 mM Tris-HCl, pH 8.9, 20 mM MgCl$_2$, 80 mM (NH$_4$)$_2$SO$_4$, 166 µg/mL bovine serum albumin (BSA).
11. 4 mM dNTP mix (4 mM each dGTP, dATP, dCTP, dTTP).
12. Phenol:chloroform:isoamyl alcohol (25:24:1, v/v, saturated with 10 mM Tris-HCl, 1 mM EDTA, pH 8.0).
13. Chloroform:isoamyl alcohol (49:1, v/v).
14. 3 M sodium acetate (pH 5.3).
15. TE: 10 mM Tris-HCl, 1 mM EDTA, pH 7.5.
16. GFX PCR DNA and Gel Band Purification kit (Amersham Pharmacia Biotech).

17. EE buffer (3X): 30 mM EPPS, 3 mM EDTA, pH 8.0.
18. 5 M NaCl.
19. Mung bean nuclease (New England Biolabs, 10 U/μL).
20. Mung bean nuclease buffer (as supplied with enzyme).
21. 50 mM Tris-HCl, pH 8.9.
22. Yeast tRNA, 10 mg/mL (Ambion).
23. GlycoBlue, 15 μg/mL (Ambion).
24. 1.5 % (w/v) nondenaturing agarose gel in TBE (pH 8.3). The gel should contain 0.1 mg/mL ethidium bromide.
25. Ethidium bromide (10 mg/mL). Store at 4°C in the dark. (**Caution:** Ethidium bromide is highly carcinogenic; handle under appropriate guidelines.)
26. DNA molecular size standards (e.g., GeneRuler 1-kb DNA ladder, MBI Fermentas 0.1 μg/mL).
27. Suitable concentration standards (*see* **Note 1**).

2.6. Cloning of Difference Products

1. Razor blades or scalpels.
2. QIAquick gel extraction kit (Qiagen).
3. Cloning system for PCR products, e.g., TA cloning kit or TOPO TA cloning kit, (Invitrogen).

3. Methods

Successful application of the cDNA RDA technique requires the reproducible isolation of high-quality representative RNA preparations from the bacteria. Accordingly, the utmost care should be taken during the RNA isolation procedure. Bacterial RNA is highly unstable, and, accordingly, isolation should carried out quickly and efficiently. A rapid-lysis technique optimized for use with streptococci is described. (Other methods that result in the production of high-quality RNA could be employed.)

Because bacterial mRNA is poorly polyadenylated, it is not possible to purify it efficiently using polyT-affinity methodologies. As a consequence, the protocol for cDNA RDA of bacteria differs from the original cDNA RDA methodology in that it uses total RNA rather than purified message as its starting point. Accordingly, given the great abundance of rRNA in total RNA preparations, it is necessary to supplement the derived Driver component with additional rRNA-derived material to increase selection against these common sequences.

3.1. RNA Isolation

Given the great instability of bacterial RNA, it is important to work as quickly and efficiently as possible during its isolation. Work should be carried out on ice where possible.

The following method works well with both *Streptococcus pneumoniae* and *S. uberis* and should be valid for other Gram-positive bacteria. RNA isolation from Gram-negatives can also be carried out in the same way. However, for Gram-negatives the use of zirconia/silica beads and the FastPrep/Ribolyser instrument is unlikely to be necessary.

Following DNase treatment of RNA preparations, it is advisable that elimination of contaminating DNA be confirmed by PCR, e.g., using 16S rRNA primers (*see* **Table 1**).

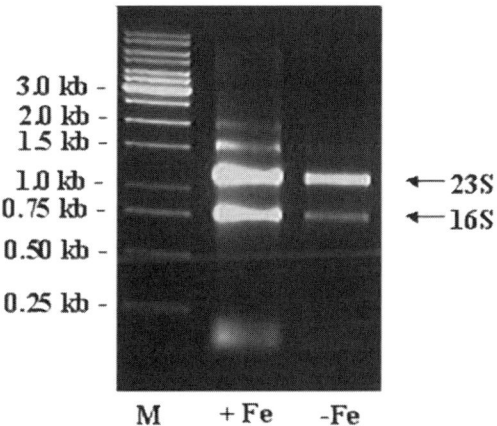

Fig. 2. Total RNA from *S. uberis* isolated using the decon90 lysis method. +Fe, total RNA isolated from *S. uberis* O140J grown under iron-sufficient conditions. –Fe, total RNA isolated from *S. uberis* O140J grown under iron-limited conditions. M, molecular weight markers, sizes are given in bp. Positions of the dominant 16S and 23S rRNA bands are indicated. Figure kindly provided by S. D. Copsey.

For approx 100 µL cell pellet volumes.

1. Harvest bacterial cultures by centrifugation of 10-mL aliquots in Falcon tubes at 7500g for 5 min at 4°C.
2. Quickly and carefully remove supernatant, and, keeping cells cool, resuspend thoroughly by vigorously vortexing cell pellets in 200 µL UHP water.
3. Add to a 2-mL screw-cap tube containing approx 250 µL zirconia/silica beads, 500 µL lysis buffer, 500 µL acid phenol:chloroform, and 100 µL chloroform.
4. Process cells in a FastPrep FP120 instrument for 3 × 20 s at a speed of 6.5 m/s.
5. Place tube(s) on ice for 10 min and then centrifuge at 16,000g for 10 min at 4°C.
6. Remove supernatant to a new tube and add equal volume of 100% ethanol.
7. Precipitate for at least 2 h at –20°C.
8. Pellet the RNA by centrifugation at 16,000g for 30 min at 4°C.
9. Wash the pellet in 70% ethanol, and air-dry. (Do *not* overdry, or the RNA may become difficult to resuspend.)
10. Resuspend in 1 mM sodium citrate, pH 6.4.
11. Treat samples to remove contaminating DNA, using DNase-free kit, according to manufacturer's instruction.
12. Aliquots of the total RNA preparations can be stored at –80°C.
13. Quantitation and crude quality assessment can be carried out by measuring optical density of the preparation at 260 and 280 nm and by examination on a 1% nondenaturing agarose gel. **Figure 2** shows a typical result obtained with a good-quality RNA preparation.

3.2. cDNA Synthesis

cDNA synthesis is performed by random priming of total RNA using the Pharmacia TimeSaver cDNA synthesis kit (*see* **Note 2**). The protocol is according to the manufacturer's instructions using 5 µg total RNA (*see* **Subheading 3.1.**) as the template, and the random hexamers (as supplied with kit) at 1:200 dilution (0.037 μg/reaction), with the following modifications:

1. After 20 min of incubation of the first-strand reaction at 37°C, add 1 μL of Superscript II (or similar) reverse transcriptase to the reaction mix and continue the incubation for a further 1 h.
2. Carry out the second-strand incubation at 12°C for 1 h and then 22°C for a further 2 h, and heat to 65°C for 10 min.
3. Extract the reactions with equal volumes of phenol:chloroform:isoamyl alcohol.
4. Purify the cDNA using a Microspin S-300 HR column, according to manufacturer's instructions.
5. Add 100 μg/mL GlycoBlue, 0.1 vol of 3 M sodium acetate, pH 5.3, 2.5 vol of cold ethanol, and precipitate at –20°C for 1 h. Resuspend in 44 μL UHP water.

3.3. Isolation of Chromosomal DNA

Chromosomal DNA isolation is carried out using the Qiagen Genomic-tip or QIAamp systems; the protocol is according to the manufacturer's instructions.

3.4. Isolation of S. pneumoniae 16S and 23S rRNA Genes

Isolation of pneumococcal 16S and 23S rRNA genes is described, although the primers used (*see* **Table 1**) also work well for *S. uberis* rRNAs. For other bacterial species/genera, different relevant primer pairs may be substituted. These can be designed using conserved regions identified by comparison of available rRNA sequences from related species.

1. Use 0.1–0.5 μg of chromosomal DNA (*see* **Subheading 3.3.**) as the template in 12X 100-μL reactions for each of the rRNA genes to be isolated.
2. To each template in the PCR tube add 10 μL 10X buffer, 8 μL dNTP mixture, and 20 μL 5 M betaine solution (*see* **Note 3**). Add 1 μL each of the two relevant primers, and make up total volume to 100 μL with sterile UHP water. Denature template DNA in a thermal cycler by heating to 96°C for 5 min. Add 0.5 μL (2.5 U) *Taq* polymerase. Cycle reactions: 96°C for 1 min, 58°C for 1 min, and 72°C for 3 min, for 25–30 cycles, with a final extension at 72°C for 10 min.
3. Examine reactions for specificity and yield by running out a 5-μL sample of each on a 1.0% nondenaturing agarose gel. The primer pairs used (pn16Sf and pn16Sr, and pn23Sf and pn23Sr) should give rise to products of approx 1450 and 2850 bp, respectively.
4. Run out the remainder (approx 95 μL) of each reaction on a 1% nondenaturing agarose gel. Excise these bands and purify, e.g., using a QIAquick DNA purification kit (Qiagen), according to the manufacturer's instructions. Dilute products to approx 0.5 mg/mL. These purified products constitute the starting material for the generation of the rRNA gene-derived representations.

3.5. PCR-Coupled Subtractive Hybridization

The following methodology is adapted from Hubank and Schatz (*6*).

cDNA RDA is a highly sensitive technique, and accordingly it is important that all possible precautions be taken to prevent cross-contamination of materials. All reagents (including enzymes) should be aliquoted before use. Micropipet tips with integral filters should be used throughout, and it is recommended that all PCR reactions be set up in an airflow cabinet.

Because the complexity of the derived cDNA populations is considerably less than that of genomic DNA, the simplification of the starting material required for RDA (the generation of a subpopulation of amplified restriction fragments) is no longer essential. However, to be able to utilize PCR to enrich for differences, it is still necessary to generate amplified populations of cDNA restriction fragments, although, because of the reduced complexity, four-cutter restriction enzymes can be used, thus increasing the proportion of amplifiable fragments generated (*see* **Note 4**). For the generation of rRNA-derived Driver, the PCR products from **Subheading 3.4.** are substituted for cDNA.

3.5.1. Ligation of R-Bgl- Adaptors

1. Add 5 μL 10X *Dpn*II buffer (as supplied with the *Dpn*II enzyme) and 1 μL *Dpn*II to 44 μL cDNA preparation (*see* **Subheading 3.2.**). Incubate for 3 h at 37°C.
2. To ligate R-Bgl- adaptors to the rRNA-derived amplicons, digest 1–2 μg of each purified PCR product (*see* **Subheading 3.4.**), by adding 1 μL of *Dpn*II, 10 μL 10X *Dpn*II buffer, and sterile UHP water to a total volume of 100 μL. Incubate for 3 h at 37°C.
3. Extract the reactions with an equal volume of phenol:chloroform:isoamyl alcohol and then with an equal volume of chloroform:isoamyl alcohol. Add 1 μL (15 μg) GlycoBlue carrier, 30 μL ammonium acetate, 600 μL cold 100% ethanol, and precipitate on ice for 30 min.
4. Spin down precipitate at 16,000g for 30 min at 4°C. Wash pellet with 70% ethanol.
5. Air-dry and resuspend in 20 μL TE; transfer to a 0.5-mL PCR tube.
6. Add 25 μL sterile UHP water, 6 μL ligase buffer (as supplied with T4 DNA ligase), and 4 μL each of 0.25 mM R-Bgl-12 and 0.5 mM R-Bgl-24 adaptors.
7. Anneal oligonucleotide adaptors in a PCR machine, by heating the reaction to 50°C for 2 min, and then cool to 10°C at no more than 1°C/min.
8. Add 1 μL T4 DNA ligase, mix well, and incubate for 18 h at 14°C.

3.5.2. Generation of Representations

Pilot reactions should be carried out for each representation to be generated to establish optimum amplification conditions (*see* **Note 5**). This is to determine the number of PCR cycles required to generate suitable ("good") representations. The criteria are that a 10-μL sample run out on a 1.5% agarose gel should give a smear ranging in size from approx 0.2 to 1.5 kb and contain approximately 0.5 μg DNA. Too few cycles will not provide sufficient material for the subsequent subtraction step, and overamplification will bias the populations and reduce average fragment size (*see* **Note 6**). For the generation of the rRNA-derived representation, however, no pilot is required; instead use the same number of cycles determined as optimum for the other Tester and Driver components.

1. Dilute the ligation (*see* **Subheading 3.5.1.**) 1:3, by adding 120 μL TE.
2. For each pilot reaction, add 3 μL diluted ligation, 139 μL sterile UHP water, 40 μL PCR buffer, 16 μL 4 mM dNTP mix, and 1 μL 0.5 mM R-Bgl-24 adaptor/primer to a 0.5-mL PCR tube.
3. Incubate at 72°C for 3 min in a Thermal Cycler. Add 1 μL (5 U) *Amplitaq* DNA polymerase, and continue incubation for a further 5 min.
4. Cycle reactions at 95°C for 1 min and then 72°C for 3 min for 25 cycles. Remove 10-μL aliquots at intervals from about cycle 16 onward. (The optimum number for generation of representations usually lies in the range of 17–24 cycles.)

5. Run out the 10-μL samples along with size and concentration standards on a 1.5% nondenaturing agarose gel.
6. From examination of this gel, select the number of cycles that generates suitable representations, and set up 9X 200-μL PCR reactions for each sample intended for use as Driver, and 3X 200-μL reactions for each sample intended for use only as Tester. (If reciprocal subtractions are to be carried out, set up 12X 200 μL reactions of each.)
7. Cycle reactions for the determined number of cycles, finishing with a 10-min extension at 72°C.
8. Extract reactions with an equal volume of phenol:chloroform:isoamyl alcohol and then with an equal volume of chloroform:isoamyl alcohol. Add 0.1 vol of 3 M sodium acetate (pH 5.3), an equal volume of 2-propanol, and precipitate the DNA on ice for 30 min.
9. Pellet the DNA by centrifugation at 16,000g for 30 min at 4°C. Wash the pellet with 70% ethanol, air-dry, and resuspend in TE to give a concentration of approx 0.5 mg/mL (a rough guideline is to use about 25 μL of TE per reaction).
10. Check quality and concentration of DNAs by running 1-μL samples on a 1.5% nondenaturing agarose gel, alongside standards.

3.5.3. Preparation of Driver and Tester Components

1. Digest 100 μg (200 μL) each of Driver and rRNA-derived representations by adding 5 μL *Dpn*II, 60 μL 10X *Dpn*II buffer, and UHP water to final volumes of 600 μL. Incubate for 3 h at 37°C.
2. Extract digests with equal volumes of phenol:chloroform:isoamyl alcohol and then chloroform:isoamyl alcohol.
3. Add 0.1 vol of 3 M sodium acetate (pH 5.3) and an equal volume of 2-propanol to each digest and incubate on ice for 30 min.
4. Precipitate restricted DNA by centrifugation at 16,000g for 30 min at 4°C. Wash the pellets with 70% ethanol, air-dry, and resuspend in 150-μL TE. Combine 16S and 23S representations.
5. Determine concentration of cut Driver and rRNA-derived representation by running 1-μL samples on a 1.5% nondenaturing gel with standards. Adjust to approx 0.5 mg/mL with TE as necessary. These represent the Driver components (Driver 1 and 2, respectively) shown in **Fig. 1B**.
6. If a representation is to be used as a Tester, digest 10 μg (20 μL) with 0.5 μL *Dpn*II, 6 μL 10X *Dpn*II buffer in a final volume of 60 μL. Incubate 3 h at 37°C.
7. Extract digests with equal volumes of phenol:chloroform:isoamyl alcohol and then chloroform:isoamyl alcohol.
8. Add 0.1 vol of 3 M sodium acetate (pH 5.3) and 3 vol of cold ethanol to each digest, and precipitate at −20°C for 30 min.
9. Collect precipitates at 16,000g for 30 min at 4°C. Wash the pellets with 70% ethanol, air-dry, and resuspend in 20 μL TE.
10. Remove digested R-Bgl-adaptors using a purification method such as the GFX system, according to the manufacturer's protocol.
11. Estimate DNA concentration by running a 1-μL sample on a 1.5% nondenaturing gel with standards.
12. Combine 1 μg of purified DNA, 3 μL 10X T4 DNA ligase buffer, 2 μL 0.5 mM J-Bgl-24 adaptor, 2 μL 0.25 mM J-Bgl-12 adaptor, and sterile UHP water to a final volume of 29 μL.
13. Anneal oligonucleotide adaptors in a PCR machine, by heating the reaction to 50°C for 2 min and then cool to 10°C at no more than 1°C/min.

14. Add 1 µL T4 DNA ligase, mix well, and incubate for 18 h at 14°C.
15. Dilute the ligation to approx 10 ng/µL by the addition of 70 µL TE. This preparation is the J-ligated Tester.

3.5.4. Subtractive Hybridization

See **Fig. 1B** for a schematic representation of the procedure.

1. Combine 5 µg (10 µL) of digested Driver (Driver 1), 5 µg (10 µL) digested rRNA-derived representation (Driver 2), and 0.1 µg (10 µL) of J-ligated Tester in a 0.5-mL microcentrifuge tube. Make up to 100 µL with sterile UHP water. This gives a Driver:Tester ratio of 100:1 (50:50:1).
2. Extract digests with equal volumes of phenol:chloroform:isoamyl alcohol and then chloroform:isoamyl alcohol.
3. Add 0.2 vol of 10 M ammonium acetate (pH 5.3), and 3 vol of cold ethanol to each digest, and precipitate at $-70°C$ for 10 min.
4. Incubate tube containing precipitate at 37°C for 1 min, and then spin at 16,000g for 20 min at 4°C to collect DNA. Very carefully wash the pellet with 70% ethanol.
5. Air-dry pellet, and resuspend very thoroughly in 4 µL 3X EE buffer by pipeting up and down for at least 3 min.
6. Incubate at 37°C for 5 min, vortex vigorously, and spin solution to the bottom of the tube.
7. Overlay solution with a few drops of mineral oil (even if PCR machine has heated lid), and denature DNA for 5 min at 98°C. Cool block to 67°C.
8. Incubate hybridization mix at 67°C for 24 h (to allow complete annealing).
9. Remove hybridization mix to a fresh tube, and dilute stepwise in 200 µL TE: add 10 µL TE and mix by pipeting, add a further 25 µL TE and mix, and then make up to 200 µL and vortex thoroughly. This diluted, hybridized DNA is then used to generate the first difference product (see **Note 7**).
10. For each subtraction, set up two PCR reactions, comprising 122 µL sterile UHP water, 40 µL 5X PCR buffer, 16 µL of 4 mM dNTP mix, and 20 µL diluted hybridization mix.
11. In a PCR machine, incubate reactions at 72°C for 3 min, add 1 µL (5 U) *Amplitaq* DNA polymerase, and continue incubation at 72°C for a further 5 min.
12. Add 1 µL of 0.5 mM J-Bgl-24 primer, and cycle at 95°C for 1 min and 70°C for 3 min for 11 cycles (see **Note 8**), with a final extension at 72°C for 10 min.
13. Combine the two reactions in a single microfuge tube. Extract with equal volumes of phenol:chloroform:isoamyl alcohol and then chloroform:isoamyl alcohol. Add 100 µg (10 µL) tRNA, 0.1 vol 3 M sodium acetate (pH 5.3), and an equal volume of 2-propanol. Precipitate on ice for 30 min.
14. Sediment the precipitate at 16,000g for 20 min at 4°C. Very carefully wash the pellet with 70% ethanol. Resuspend in 20 µL TE.
15. Add 4 µL mung bean nuclease buffer, 2 µL mung bean nuclease, and make up volume to 40 µL with sterile UHP water. Incubate at 30°C for 45 min (see **Note 9**).
16. Terminate reaction by the addition of 160 µL of 50 mM Tris-HCl (pH 8.9) and heating to 98°C for 5 min. Cool the reaction to 4°C on ice.
17. On ice, set up one PCR reaction, comprising 122 µL sterile UHP water, 40 µL 5X PCR buffer, 16 µL dNTP mix, and 1 µL 0.5 mM J-Bgl-24 oligo.
18. Add 20 µL of the MBN-treated DNA. Incubate reactions in a PCR machine at 95°C for 1 min, add 1 µL (5 U) *Amplitaq* DNA polymerase, and cycle at 95°C for 1 min and 70°C for 3 min, for 18 cycles, with a final extension at 72°C for 10 min.
19. Estimate DNA concentration by running a 10-µL sample on a 1.5% nondenaturing agarose gel, with standards.

20. Extract reactions with equal volumes of phenol:chloroform:isoamyl alcohol and then chloroform:isoamyl alcohol. Add 0.1 vol 3 M sodium acetate (pH 5.3), 1 vol 2-propanol, and precipitate on ice for 30 min.
21. Collect precipitates at 16,000g for 30 min at 4°C. Wash the pellets with 70% ethanol, air-dry, and resuspend in TE to 0.5 µg/µL. (Volume is yield-dependent but is typically approx 20–30 µL.)
22. This is the first difference product (DP1).

3.5.5. Generation of a Second Difference Product
See **Note 10** and **Fig. 3**.

1. Mix 2 µg (4 µL) of DP1 (see **Subheading 3.5.4.**, **item 1**) with 84 µL sterile UHP water, and add 10 µL 10X *Dpn*II buffer and 2 µL of *Dpn*II. Incubate for 3 h at 37°C.
2. Extract the reactions with an equal volume of phenol:chloroform:isoamyl alcohol and then with an equal volume of chloroform:isoamyl alcohol. Add 1 µL (15 µg) GlycoBlue carrier, 0.1 vol sodium acetate (pH 5.3), and 3 vol cold ethanol. Precipitate at –20°C for 30 min.
3. Collect precipitates at 16,000g for 30 min at 4°C. Wash the pellets with 70% ethanol, air-dry, and resuspend in 20 µL TE. (This gives approx 100 ng/µL.) Optional: You can remove digested J-Bgl adaptors using the purificaiton system as before for R-Bgl adaptors.
4. Take 2 µL of the restricted DP1, and add 3 µL 10X ligase buffer, 2 µL each of 0.25 mM N-Bgl-12 and 0.5 mM N-Bgl-24 adaptors, and make up to a final volume of 29 µL with sterile UHP water.
5. Anneal oligonucleotide adaptors in a PCR machine, by heating the reaction to 50°C for 2 min, and then cool to 10°C at no more than 1°C/min.
6. Add 1 µL T4 DNA ligase, mix well, and incubate for 18 h at 14°C.
7. Dilute ligation mix to approx 1.25 ng/µL by the addition of 130 µL TE.
8. Combine 5 µg (10 µL) of digested Driver (Driver 1) and 5 µg (10 µL) digested rRNA-derived representation (Driver 2) (see **Subheading 3.5.3.**) Mix 10 µL (approx 10 µg) of Driver with 10 µL (approx 12.5 ng) of N-Bgl-ligated DP1 (see **Subheading 3.5.4.**). This gives a Driver-Tester ratio of 800:1 (400:400:1).
9. Extract digests with equal volumes of phenol:chloroform:isoamyl alcohol and then chloroform:isoamyl alcohol.
10. Add 0.2 vol of 10 M ammonium acetate (pH 5.3), and 3 vol of cold ethanol to each digest, and precipitate at –70°C for 10 min.
11. Incubate tube containing precipitate at 37°C for 1 min, and then spin at 16,000g for 20 min at 4°C to collect DNA. Very carefully wash the pellet with 70% ethanol.
12. Air-dry the pellet, and resuspend very thoroughly in 4 µL 3X EE buffer by pipeting up and down for at least 3 min.
13. Incubate at 37°C for 5 min, vortex vigorously, and spin solution to the bottom of the tube.
14. Overlay solution with a few drops of mineral oil (even if PCR machine has heated lid), and denature DNA for 5 min at 98°C. Cool block to 67°C.
15. Incubate hybridization mix at 67°C for 24 h (to allow complete annealing).
16. Remove hybridization mix to a fresh tube, and dilute stepwise in 200 µL TE: add 10 µL TE and mix by pipeting, add a further 25 µL TE and mix, and then make up to 200 µL and vortex thoroughly. This diluted, hybridized DNA is then used to generate the first difference product.
17. For each subtraction, set up two PCR reactions, comprising 122 µL sterile UHP water, 40 µL 5X PCR buffer, 16 µL 4 mM dNTP mix, and 20 µL diluted hybridization mix.

18. In a PCR machine, incubate reactions at 72°C for 3 min, add 1 μL (5 U) *Amplitaq* DNA polymerase, and continue incubation at 72°C for a further 5 min.
19. Add 1 μL of 0.5 mM N-Bgl-24 primer, and cycle at 95°C for 1 min and 72°C for 3 min for 11 cycles, with a final extension at 72°C for 10 min.
20. Combine the two reactions in a single microfuge tube. Extract with equal volumes of phenol:chloroform:isoamyl alcohol and then chloroform:isoamyl alcohol. Add 100 μg (10 μL) tRNA, 0.1 vol 3 M sodium acetate (pH 5.3), and an equal volume of 2-propanol. Precipitate on ice for 30 min.
21. Spin down precipitate at 16,000g for 20 min at 4°C. Very carefully wash the pellet with 70% ethanol. Resuspend in 20 μL TE.
22. Add 4 μL 10X mung bean nuclease buffer, 2 μL mung bean nuclease, and make up volume to 40 μL with sterile UHP water. Incubate at 30°C for 45 min.
23. Terminate reaction by the addition of 160 μL of 50 mM Tris-HCl (pH 8.9) and heating to 98°C for 5 min. Cool the reaction to 4°C on ice.
24. On ice, set up two PCR reactions for each subtraction, each comprising 122 μL sterile UHP water, 40 μL 5X PCR buffer, 16 μL dNTP mix, and 1 μL 0.5 mM N-Bgl-24 oligo.
25. Add 20 μL of the mung bean nuclease-treated DNA to each. Incubate reactions in a PCR machine at 95°C for 1 min, add 1 μL (5 U) *Amplitaq* DNA polymerase, and cycle 95°C for 1 min and 72°C for 3 min, for 18 cycles, with a final extension at 72°C for 10 min. Combine each pair of reactions.
26. Estimate DNA concentration by running a 10-μL sample on a 1.5% nondenaturing agarose gel, with standards.
27. Extract reactions with equal volumes of phenol:chloroform:isoamyl alcohol, and then chloroform:isoamyl alcohol. Add 0.1 vol 3 M sodium acetate (pH 5.3), 1 vol 2-propanol, and precipitate on ice for 30 min.
28. Collect precipitates at 16,000g for 30 min at 4°C. Wash the pellets with 70% ethanol, air-dry, and resuspend in TE to 0.5 μg/μL. (Volume is yield-dependent but is typically approx 40–50 μL.) This is the second difference product (DP2).

3.5.6. Generation of Further Difference Products

There are both advantages and disadvantages in proceeding to a third difference product (DP3). It can be useful if DP2 contains many poorly defined and/or weak bands. However, a third round of PCR-coupled subtractive hybridization can result in the loss of some difference products—particularly those derived from transcripts expressed at low levels.

For the generation of DP3, the procedure is essentially as in **Subheading 3.5.5.**, except:

1. Digest DP2 (*see* **Subheading 3.5.5**) with *Dpn*II to remove the N-Bgl- adaptors, and replace with J-Bgl- adaptors.
2. Dilute the J-Bgl-ligated DP2 to 1 ng/μL with TE. Set up hybridizations (*see* **Subheading 3.5.5., steps 8–15**) using Driver-Tester ratios of between 5000 and 20,000:1 (2500:2500: 1–10,000:10,000:1). To achieve this, vary quantity of J-Bgl-ligated Tester and keep the combined Driver concentration at 10 μg/mL (i.e., 5 μg/mL each of Drivers 1 and 2; *see* **Subheading 3.5.3.**).
3. Generate DP3 according to the protocol given in **Subheading 3.5.5., steps 16–28**, setting up four PCR reactions for each subtraction.
4. Resuspend pellet from the four combined reactions (from each subtraction) to a final concentration of approx 0.5 μg/mL in TE. This is DP3.

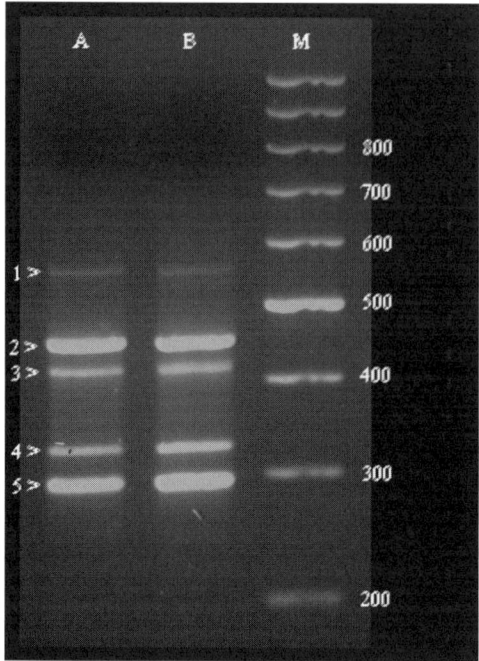

Fig. 3. Difference products (DP2) generated during an analysis of iron-regulated gene expression in *N. meningitidis* by cDNA RDA. A and B, show DNA fragments amplified in two independent experiments initiated from separate aliquots of same total RNA preparation (showing reproducibility). M, molecular mass markers. Arrows indicate bands selected for isolation, cloning and sequencing. Adapted from ref. 5.

3.5.7. Cloning of Difference Products

1. Run out each difference product on a 1.5% nondenaturing agarose gel in TBE, pH 8.3. **Figure 3** shows a typical result.
2. Carefully excise each band from the gel using a sharp razor blade. Purify amplicons using, e.g., QIAquick gel extraction kit (Qiagen), according to the manufacturer's instructions.
3. Clone amplicon in a suitable PCR cloning vector, e.g., using TA cloning/TOPO TA cloning kits (Invitrogen), according to the manufacturer's instructions. Recombinant plasmid DNA can then be isolated (using standard methods) and inserts sequenced for identification of differentially expressed genes (*see* **Note 11**).

4. Notes

1. It is important that concentrations be determined accurately. Suitable standards can be prepared by digesting genomic DNA of known concentration (sheared to an average length of about 20 kb), with *Dpn*II. Dilutions of this standard (e.g., 0.1–1.0 µg) should be loaded on the agarose gels, alongside the PCR products to be quantified.
2. The modified Amersham-Pharmacia cDNA synthesis kit and protocol were found to give very good results in our hands, although other methodologies could be used.

3. The addition of betaine to the PCR reaction mixes was found to aid amplification of rRNA sequences.
4. During generation of the representations, the resulting products are limited in their complexity by the ability of each product to be amplified within the mixture, under the conditions used. In general, template (cDNA) restriction fragments that are either too large (in excess of approx 1 kb), or too small (under approx 0.2 kb) do not amplify efficiently. The product (representation) therefore only "represents" the amplifiable proportion of the digest. Accordingly, to ensure that the greatest proportion of differentially expressed genes are identified, it is advisable to repeat the cDNA RDA analyses with a variety of different restriction enzymes (and relevant oligonucleotide adaptor/primers). The choice of alternative restriction enzymes to be used can be facilitated by examination of available sequence data.
5. Pilot reactions should also be carried out on a negative RT control (i.e., on total RNA preparations to which attempts to ligate the J-Bgl- adaptors have been made). This is an additional check for contamination with genomic DNA.
6. It is important that the PCR amplification to produce representations be kept within the linear range if the relative proportions of individual species are to be maintained with respect to the starting RNA populations. Accordingly, it is vital that care be taken over this "titration" (*see* **Subheading 3.5.2.**). This is particularly important for the detection of relative (rather than absolute) differences in expression.
7. During reannealing, three types of hybrid molecules can be formed (*see also* **Fig. 1B**). Given the abundance of material, Driver–Driver hybrids are most common, but lacking adaptors, they cannot generate primer binding sites during the T4 DNA polymerase "fill-in" reaction and are therefore not amplified. Driver–Tester hybrids are the next most common products, but because the Driver strand cannot generate a primer binding site, these molecules can only be amplified linearly (i.e., DNA synthesis can only be primed from one strand (the Tester). Tester–Tester hybrids, on the other hand (representing the differentially expressed genes), will possess primer binding sites at both ends and will thus be amplified exponentially.
8. The initial 11-cycle PCR helps prevent loss of genuine, but less abundant, difference products during the precipitation steps prior to the mung bean nuclease treatment.
9. The mung bean nuclease treatment removes all single-stranded nucleic acids, including linear products of Driver–Tester hybrids, tRNA carrier, and others.
10. Because of random annealing events, many amplified molecules present at the DP1 stage will not represent genuine differences. A second round of PCR-coupled subtractive hybridization is therefore required. However, because of the partial enrichment that has occurred, this can be carried out at higher stringencies (Driver–Tester ratios). A high degree of background smearing at this stage can indicate failure of the mung bean nuclease treatment or incomplete denaturation of Tester and Driver components prior to hybridization.
11. Identification/characterization of the difference products is initiated by sequencing of the inserts in the resulting recombinant plasmids. Multiple clones arising from the same amplicon should be examined, as, although bands appear discrete, the amplicons are heterogeneous, i.e., can comprise multiple different products. Similarity searches can then be carried out on genuine differences in DNA and protein databases. Because the identified differences should only be present or upregulated in the Tester, each putative difference should be checked for validity by Southern hybridization against the original representations, and ideally by performing RT-PCR, using different independently iso-

lated RNA preparations as template (the cloning of high numbers of false positives usually indicates that the Driver–Tester ratio used was too low). Gene libraries can be constructed and/or screened by standard techniques, or genome sequence databases can be screened to obtain full-length genes.

References

1. Sagerström, C. G., Sun, B. I., and Sive, H. L. (1997) Subtractive cloning: past, present, future. *Annu. Rev. Biochem.* **66,** 751–783.
2. Handfield, M. and Levesque, R. C. (1999) Strategies for isolation of *in vivo* expressed genes from bacteria. *FEMS Microbiol. Rev.* **23,** 69–91.
3. Lisitsyn, N., Lisitsyn, N., and Wigler, M. (1993) Cloning the differences between two complex genomes. *Science* **259,** 946–951.
4. Hubank, M. and Schatz, D. G. (1994) Identifying differences in mRNA expression by representational difference analysis of cDNA. *Nucleic Acids Res.* **22,** 5640–5648.
5. Bowler, L. D., Hubank, H., and Spratt, B. G. (1999) Representational difference analysis of cDNA for the detection of differential gene expression in bacteria: development using a model of iron-regulated gene expression in *Neisseria meningitidis*. *Microbiology* **145,** 3529–3537.
6. Hubank, M. and Schatz, D. G. (1999) cDNA representational difference analysis: a sensitive and flexible method for the identification of differentially expressed genes. *Methods Enzymol.* **303,** 325–349.

5

Microarray Data Analysis and Mining

Silvia Saviozzi, Giovanni Iazzetti, Enrico Caserta, Alessandro Guffanti, and Raffaele A. Calogero

Abstract

DNA microarray is an innovative technology for obtaining information on gene function. Because it is a high-throughput method, computational tools are essential in data analysis and mining to extract the knowledge from experimental results. Filtering procedures and statistical approaches are frequently combined to identify differentially expressed genes. However, obtaining a list of differentially expressed genes is only the starting point because an important step is the integration of differential expression profiles in a biological context, which is a hot topic in data mining. In this chapter an integrated approach of filtering and statistical validation to select trustable differentially expressed genes is described together with a brief introduction on data mining focusing on the classification of co-regulated genes on the basis of their biological function.

Key Words: GeneChip; SAM; dCHIP; MAS 5.0; geneontology.

1. Introduction

DNA microarray technology is an high-throughput method for obtaining information on gene function. Microarray technology is based on the availability of gene sequences arrayed on a solid surface (i.e., nylon filters, glass slides), and it allows parallel expression analysis of thousands of genes. Microarray can be a valuable tool to define transcriptional signatures bound to a pathological condition (*see* **Note 1**) or to rule out molecular mechanisms tightly bound to transcription (*see* **Note 2**). However, because our actual knowledge of gene function in high eukaryotes (e.g., human, mouse) is quite limited (*see* **Note 3**), microarray analysis frequently does not imply a final answer to a biological problem but allows the discovery of new research paths for exploration from a different perspective.

Computational tools are essential in microarray data analysis and mining to grasp knowledge from experimental results. In this chapter we present an integrated

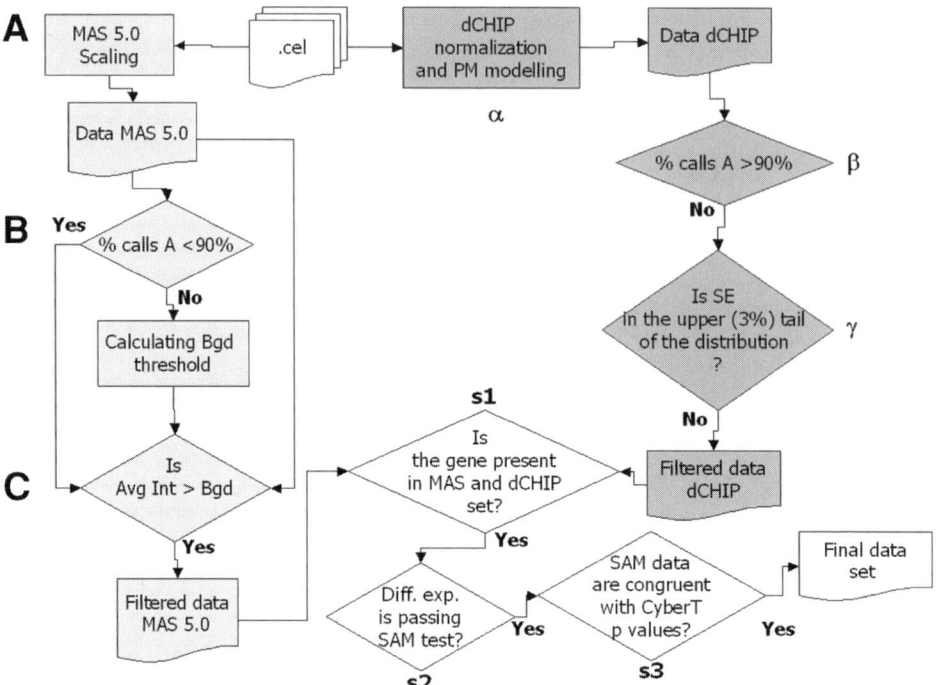

Fig. 1. Flow chart of the microarray data analysis involving the use of MAS 5.0 and dCHIP for data filtering, as well as SAM and CyberT softwares for statistical validation of differential expressions. Analysis steps are described in text. Bgd, background; SE, standard error.

filtering/statistical approach to identify differentially expressed genes (**Fig. 1**, flow chart of the procedure). Filtering and statistical validation are based on the use of robust computational tools that are relatively simple to use by scientists with limited or no computational experience. The approach is focused on obtaining a robust set of data, keeping the number of false positives low, although this might imply the loss of some correct data. Furthermore, we briefly explore the data mining problem, focusing on the classification of coregulated genes on the basis of their biological function.

1.1. Microarray Technological Platforms

1.1.1. DNA Spotted Arrays

Microarray technology was initially developed by Schena and coworkers (*1*); it is based on spotting, in an ordered manner, on a solid surface of thousands of expressed sequence tag (EST) sequences/genes (*see* **Note 4**). mRNA relative expression levels (differential expression) are measured by cohybridization of cDNAs, derived by mRNA preparations from two different samples (e.g., normal and cancer-derived cell lines), labeled with two fluorescent dyes (e.g., Cy3 → green color emission, Cy5 → red color emission). Recently gene-specific oligonucleotides ranging between 40 and

Microarray Data Analysis and Mining

70 bases have seemed to be an appropriate replacement for cDNA (*see* **Note 5**) *(2,3)*, as they are suitable for high (>10,000 genes) and low (<1000) density spotting (*see* **Note 6**). The cohybridization technology has, however, a number of problems that can strongly affect the gene expression analysis. These problems can be solved, at least to some extent, by using special reagents or computational techniques:

- The direct incorporation of the two dyes is not identical using conventional enzymes *(4)*; however, using a postlabeling technique (*see* **Note 7**), this problem can be overcome.
- The fluorescence emission efficiency of the two dyes is not identical; therefore, dye swapping as well as mathematical adjustments are needed before differential expression evaluation *(5)*.
- Background noise associated with each spot is linearly dependent on the dye signal *(5)*; therefore, background noise correction should be done using mathematical approaches considering its dependence on signal intensity.
- Array printing can be different within the various subsections of the array due to printing head consumption. This problem can be overcome by using noncontact printing technology (*see* **Note 8**) or, to some extent, by performing local mathematical normalization *(5)*.
- A mathematical model describing the hybridization of probes onto cDNA arrays is not available; therefore, it is not possible to define the hybridization specificity of each probe and thus assess the false-hybridization rate.

1.1.2. Oligonucleotide Chips

Microrrays have also been developed using photolithographic oligonucleotide synthesis (Affymetrix, Santa Clara, CA). This approach allows *in situ* synthesis of up to 300,000 (approx 25 mers) oligos/cm^2. cDNA spotted arrays are characterized by the use of one long stretch of bases (>300) for each gene; in Affymetrix GeneChip, up to 20 short oligonucleotides (probe set) are used to probe each gene/EST (**Fig. 2**), and probe sets describing the same gene are distributed in various locations on the chip. To assess the target hybridization specificity of each oligo (PM: perfect match) of the probe set, a "negative control" oligonucleotide (MM: miss match) is associated with each PM. This oligonucleotide has a sequence equal to PM but with a single central mismatch, which strongly destabilizes the hybridization of the target; the PM/MM couple is called the probe pair. Consequently, evaluation of the hybridization signals on PM and MM probes gives an indication of the aptitude of any PM to identify a specific target, as strong signal in the MM probe is a warning for the presence of crosshybridizing targets.

Target hybridization to Affymetrix GeneChips allows the generation of absolute intensity values describing the mRNA expression level (*see* **Note 9**); therefore, to generate a "virtual two-dye" experiment, two GeneChips have to be used (e.g., normal sample on chip A, pathological sample on chip B). Although Affymetrix arrays are far from the ultimate solution for the characterization of gene expression, they offer some advantages with respect to DNA spotted arrays:

- The hybridization specificity of each PM can be assessed by the level of fluorescence associated with the corresponding MM.
- Probe pairs are distributed all over a gene sequence; therefore hybridization intensity is bound to retrotranscription reaction efficiency and can give are indication of the quality

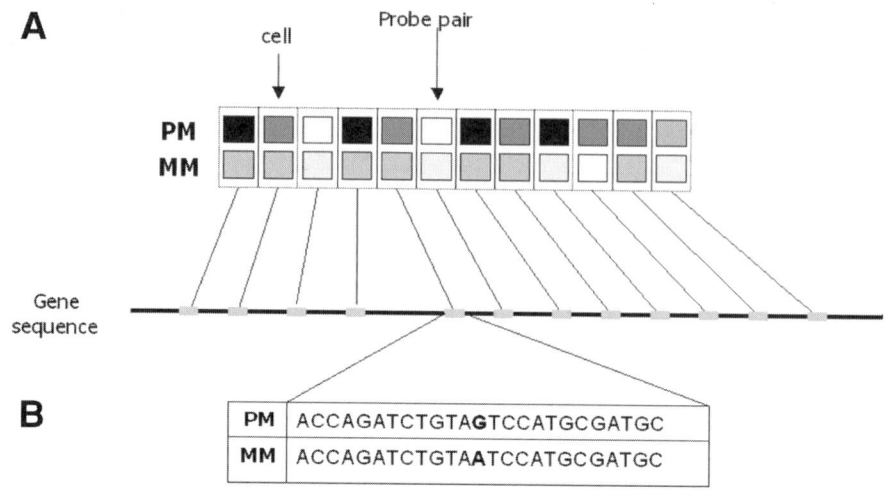

Fig. 2. Each gene/EST is represented by various probe sets scattered in the GeneChip. (**A**) Each probe set is made by up to 20 couple of oligonucleotides (probe pair) scattered over the target gene. (**B**) Each probe set is made by two oligonucleotides of the same length perfect match (PM) and miss match (MM). PM perfect match probe has a sequence perfectly matching the target sequence; MM has the same sequence as PM but with a central mismatch which radically alter its hybridization kinetic to the target gene sequence.

of the retrotranscribed RNA (degradation, presence of reverse transcriptase inhibitors, and so on).
- Recently a mathematical model assigning a degree of hybridization accuracy to each probe set *(6)* has been developed.

Microarray transcriptional profiling has a large variety of applications, but in this chapter we focus on the use of microarrays to explore molecular transcriptional mechanisms. The issue of data normalization and statistical analysis is addressed only for Affymetrix GeneChips.

2. Data Normalization

The experimental conditions (e.g., cell environment, RNA degradation, inhibitors of reverse transcriptase, yield of RNA purification steps, photomultiplier gain, amount of exposure, and so on) can strongly affect microarray hybridization intensities. It is assumed that sources of errors are multiplicative and strongly affect true expression levels *(7)*, especially if the genes are moderately expressed *(8)*. Therefore, normalization of gene expression data is a crucial preprocessing procedure that is essential for nearly all gene expression studies in which data from one array must be compared with data on another array. A number of normalization approaches may be taken into account *(9,10)*; however, so far a gold standard method has not been defined for microarray data normalization *(11)*. Thus, the chosen method should be motivated by the application at hand and the goals of the data analysis. In this section we describe a microarray analysis approach in which various methodologies are used in such a way as to emphasize their strengths. It is important to mention that microarray analysis field is evolving very fast; consequently, any microarray analysis approach can be improved on the basis of the availability of new computational tools.

In our laboratory, MAS 5.0 (www.affymetrix.com) and dCHIP *(6)* softwares are used for GeneChips signal intensity normalization. SAM *(12)* and CyberT *(13)* (*see* **Note 10**) tools are utilized for statistical validation of differential expression data. All four tools have user-friendly interfaces, and they are a good starting point for inexperienced people to analyze microarray data. In this chapter, we will not use Bioconductor (www.bioconductor.org), a microarray analysis suite based on R package (cran.r-project.org), which is an integrated suite of software facilities for data manipulation, calculation, and graphical display. Bioconductor is a powerful package offering a great deal of flexibility, more than the previously described tools possess, but it does not have a user-friendly interface and it can be tricky to use.

2.1. Affymetrix Microarray Suite 5.0 (MAS 5.0) Background Subtraction, Absolute Call, Array Scaling, and Data Filtering

MAS 5.0, produced by Affimetrix, allows reading and manipulation of the raw image file (.DAT) acquired by the GeneChip microarray scanner. The raw data file is then converted by MAS into an image file (.CEL), containing probe set's intensities and locations (*see* **Note 11**). MAS 5.0 performs a background correction across the entire array, and subsequently an expression call (i.e., call P: gene is expressed; call A: gene is not expressed; call M: gene is marginally expressed) is assigned to each probe set. A gene is call absent (A) if all probe pairs of a probe set are excluded from the analysis and a probe pair can be excluded if its MM cell is saturated. A gene is also call absent (A) if there is an insignificant different between PM and MM in all probe pairs. To determine whether the difference between PM and MM is significant, a discrimination value is calculated and its median is compared with a user-defined cutoff value (τ) (*see* **Note 12**). If all the probe pairs are not excluded (i.e., a probe pair is excluded if PM-MM is nonsignificant), a one-sided Wilcoxon's test is used to calculate a *p* value that reflects the significance of the differences between PM and MM, and it is used to assign a call (i.e., P, A, M) to the probe set. The borderline between a call P and a call M is defined by a user-definable threshold ($\alpha 1$), as the border line between M and A call is defined by an $\alpha 2$ threshold, which is also user-definable (*see* **Note 13**).

The probe set intensity signal is calculated as the one step bi-weight estimate of the combined differences of all the probe pairs in the probe set. MAS also offers the possibility of performing data scaling, which is a mathematical technique that can minimize discrepancies due to variables such as sample preparation, hybridization conditions, staining, or probe array lot (**Fig. 1**, flow chart step a). The effects of scaling can be visualized by looking at log intensity ratio versus average log intensity. In **Fig. 3A**, two biological replicates are compared before and after scaling; scaling does not affect the overall similarity between the samples, as shown by the r^2 correlation coefficient, which is 0.8995 for both raw and scaled samples. The tendency line of the scaled sample (black line) is much nearer to 0 with respect to raw data (dashed

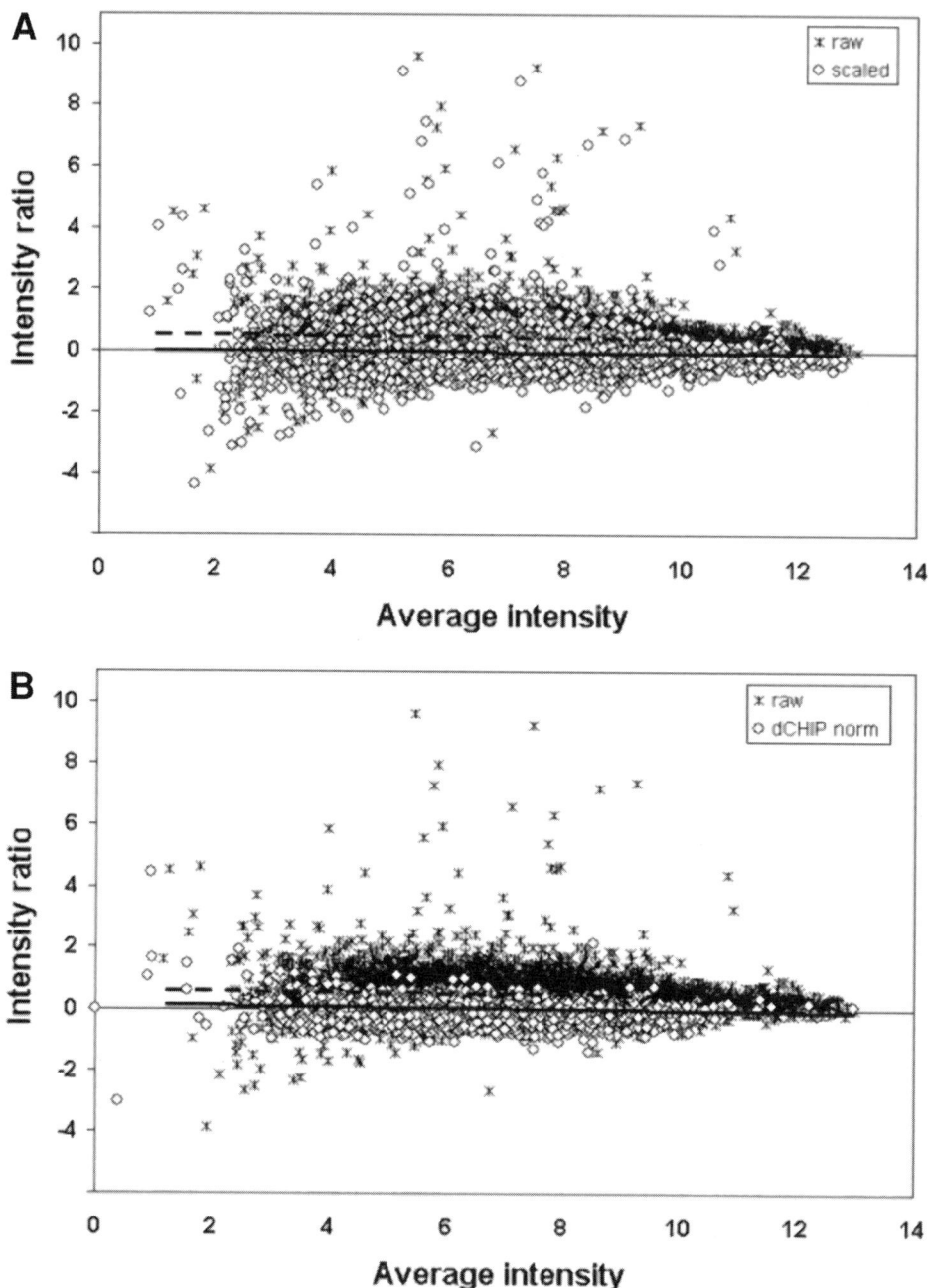

Fig. 3. (**A**) Plotting log ratio versus average log intensity for two replicates without data scaling (∗, $r^2 = 0.8995$) and after data scaling (circles, $r^2 = 0.8995$) using MAS 5.0 software. (**B**) Plotting log ratio versus average log intensity for two replicates without data scaling (∗, $r^2 = 0.8995$) and after data scaling (circles, $r^2 = 0.9716$) using dCHIP software.

line), indicating that the discrepancies between the two datasets have somehow been reduced. Scaled data can undergo filtering procedures to reduce the size of the dataset under analysis, purging those data that can induce ambiguous results (e.g., differential expression generated by genes with signal in the background range; such data are particularly dangerous, as low-intensity signals have very narrow standard deviations and therefore are likely to pass statistical validation based on a t-test although they are not informative). A first filtering step is performed by using as a threshold the number of call As detected for each probe in all the arrays under analysis (**Fig. 1**, flow chart step b), and taking out all probe sets called A in more than 90% of the analyzed arrays (not expressed gene set). The not expressed gene set is afterward used to define the threshold for an additional filtering step, based on average background intensity signal (**Fig. 1**, flow chart step c). The background intensity threshold (Bgd) is estimated by plotting the frequency distribution of intensity signals of the not expressed gene set and defining the value delimiting the 10% upper tail of the intensity distribution (**Fig. 4A**). The Bgd value is then used to filter out genes with average expression level lower than Bdg (in all experimental groups left from the first filtering step) (**Fig. 1**, flow chart step c).

2.2. dCHIP Probe Set Quality Assessment, Data Normalization, and Filtering

dCHIP, a software developed by Li and Wong *(6)*, allows analysis of GeneChip data. dCHIP calculates gene calls and performs normalization in a different way than MAS 5.0. Furthermore, if the number of arrays is at least 10, dCHIP allows the calculation of a model-based expression index in the array as well as a probe-sensitivity expression index. Fitting experimental probe set values with the calculated model, it is possible to define a standard error value (SE) that gives an indication of the hybridization quality for each probe set. SE is quite useful for filtering out probe sets that differentiate too much from the mathematical model. The Invariant Set Normalization method is used in dCHIP *(14)* to normalize arrays (**Fig. 1**, flow chart step α). In this normalization procedure, an array with median overall intensity is chosen as the baseline array against which other arrays are normalized at probe intensity level. Subsequently, a subset of PM probes, with small within-subset rank difference in the two arrays, serves as the basis for fitting a normalization curve. This normalization method produces a better fitting of the replicates, with respect to the MAS scaling procedure, as shown by the r^2 correlation value (0.9716) (**Fig. 3B**) and by the disappearing of probe sets showing strong fluctuations between replicates. Normalized dCHIP data can undergo filtering procedures to reduce the size of the dataset under analysis, purging those data that can induce ambiguous results. Also, in this case a first filtering step is done using as threshold the number of call As detected for each probe in all the arrays under analysis (**Fig. 1**, flow chart step β), taking out all probe sets called A in more than 90% of the analyzed arrays. The second filtering step is instead based on SE threshold (**Fig. 1**, flow chart step γ), in order to filter out probe sets that loosely fit the mathematical probe set profile model. Taking for granted that the vast majority of the probe sets are characterized by a good match between the experimental data and the calculated model probe set profiles (narrow SE), genes within the upper 3% tail of the SE distribution (**Fig. 4B**) are filtered out.

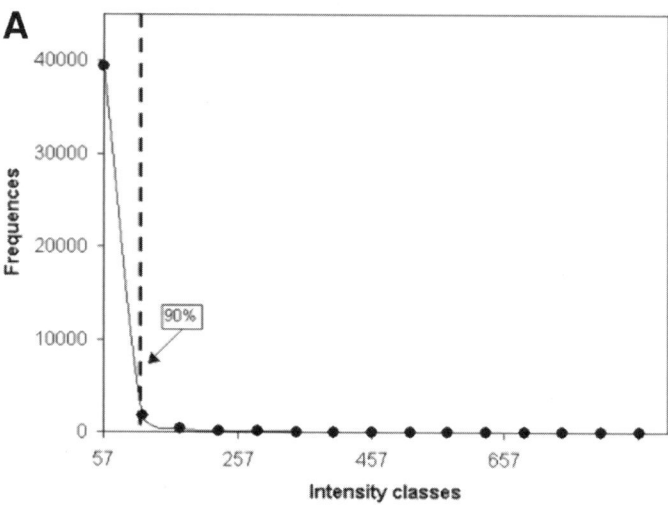
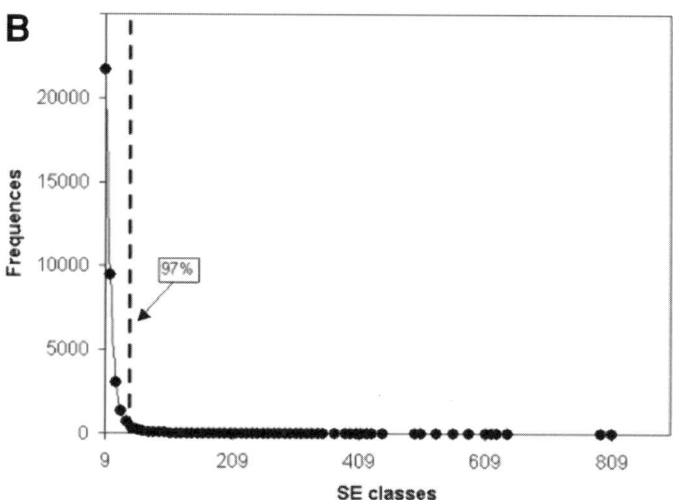

Fig. 4. (**A**) Distribution of intensity values in the not expressed genes set (genes called A in >90% of the arrays under analysis). Dashed line indicates the threshold value delimiting the 10% upper tail of the distribution. (**B**) The standard error (SE) provides a measure of probe set hybridization quality (i.e., high values indicate a low correspondence between the calculated—mathematical model—probe set hybridization profile and the experimental hybridization profile). The distribution of the SE associated to all probe sets was evaluated, and, assuming that the vast majority of the probe sets have a good hybridization quality profile, the SE value delimiting the 3% upper tail of the distribution (dashed line) is used as threshold to filter out those probe sets that could give misleading results owing to their low hybridization quality.

2.3. Combining MAS 5.0 and dCHIP Data

Datasets generated by applying the filtering procedures suggested for MAS 5.0 and for dCHIP only overlap partially, indicating that filtering can strongly affect the composition of the dataset under analysis. Because, as previously pointed out, a gold standard normalization and filtering procedure has not been defined, in our lab we prefer to perform statistical validation of differential expression only on probe sets selected by using more than one approach (**Fig. 1**, flow chart step s1, MAS 5.0 scaling, expression calls + Bgd filtering ∩ dCHIP normalization, expression calls + SE filtering). This is the final dataset generated by combing two different analysis approaches, and thus we have two sets of intensity values, those derived from MAS 5.0 and those from dCHIP. Inspecting the intensity values generated by MAS and by dCHIP for the same array, it seems that the intensity range of MAS data is wider than that of dCHIP (**Fig. 5A**). This observation indicates that the two approaches produce different intensity levels, but it does not give any indication about which of the two sets is preferred to generate differential expression values. Calculating differential expressions using MAS or dCHIP data and plotting the distribution of differential expression values (**Fig. 5B**), it seems that although the dCHIP normalization and probe set intensity calculation approach improves the quality of the replicates (**Fig. 3B**), it also strongly minimizes the differences existing between controls and treated samples (**Fig. 5B**). As shown in **Fig. 5**, MAS 5.0 data produces a broad range of differential expressions, which are instead meanly distributed in the range ±0.5-fold change for dCHIP data. The use of data within the ±0.5-fold change range can produce ambiguous results because it relies on very narrow expression changes, which can be caused by experimental fluctuations unrelated to biologically meaningful effects. Therefore, for fold change estimation, we prefer to use the intensity values derived from MAS 5.0 scaling.

3. Microarrays Differential Expression Statistical Validation

Once the final dataset is generated, robustness of differential expression can be evaluated by combining fold change with statistical validation. Because microarray results are influenced by various experimental errors *(5)*, it is important to perform replicates of the experiments to assess the variability of the gene expression levels in the treatment and control groups and to evaluate the statistical meaning of those variations (*see* **Note 14**). Statistical validation is quite important because the simple-minded fold approach, in which a gene is declared to have significantly changed if its average expression level varies by more than a constant factor, usually 2, between the treatment and the control conditions, is unlikely to yield optimal results because the fold change factor can have different significance depending on expression level *(13)*. Usually, for a limited number of replicates, parametric (e.g., t-test) or nonparametric tests (e.g., Wilcoxon's rank test) can be carried out. However, when multiple hypotheses are tested, as in the case of thousands of genes present on a microarray, the probability that at least one type I error (i.e., a gene is considered differentially expressed although it is not true) is committed can increase sharply with the number of hypotheses. For these reasons, a variety of approaches have been developed to avoid this kind of error *(5,12,13)*.

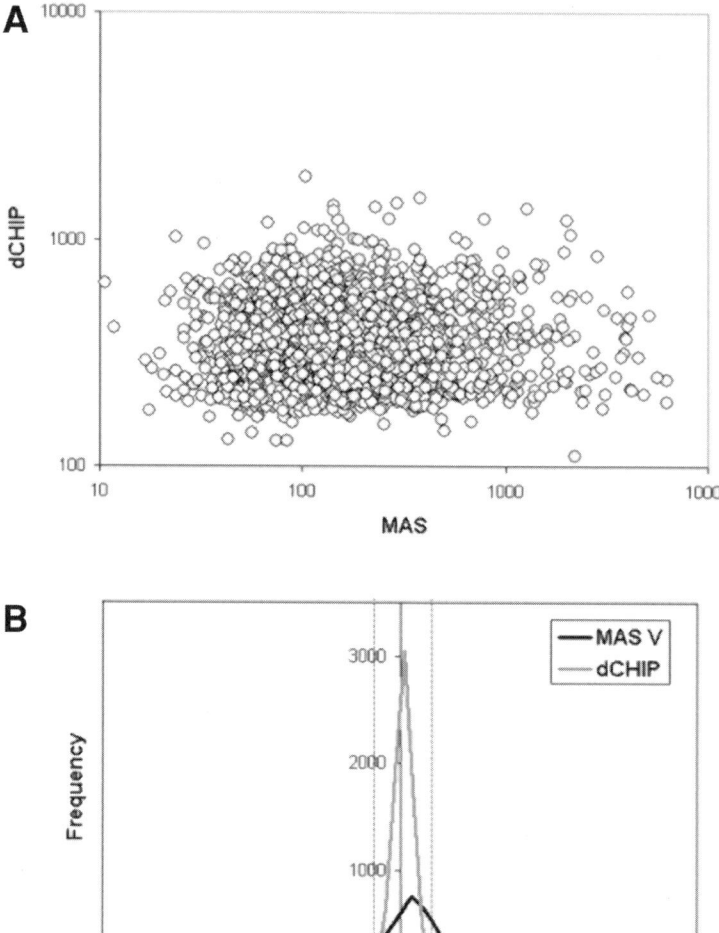

Fig. 5. (**A**) Intensity values for the same data set calculated with MAS 5.0 and with dCHIP. (**B**) Distribution of fold change generated using MAS scaled data (black line) or dCHIP normalization (gray line). Although it performs better than MAS scaling in normalizing replicates (see r^2 correlation in **Fig. 3**), in our hands dCHIP probe set intensity calculation has a strong effect on the differences existing between treatment and control samples and minimizes the fold change variations. The distribution of fold changes obtained using MAS scaled data is spread over a broad fold change range with respect to dCHIP data. dCHIP data are mainly distributed between –0.5 and 0.5 $\log_2$ fold change, which is frequently considered a sort of gray zone in which it is very hard to discriminate between real differential expressions and false positives.

SAM (Significance Analysis of Microarrays), developed by Tusher and coworkers *(12)*, is a statistical technique for finding significantly differentially expressed genes in a set of microarray experiments. The input to SAM is gene expression measurements from a set of microarray experiments, as well as a response variable from each experiment. In SAM the "unpaired response variable" that refers to grouping like "untreated" (i.e., 1) or "treated" (i.e., 2) perfectly fits the analysis of differential gene expression measured using GeneChip. SAM measures the strength of the relationship between gene expression and the response variable, and it uses repeated permutations of the data to determine whether the expression of any gene is significantly related to the response. The user can decide on the acceptable false discovery rate, setting a significance cutoff, and he or she can also set a specific fold change threshold to ensure that called genes change at least a prespecified amount. To calculate the false discovery rate, data are randomly permuted, and the user should indicate the number of permutations to be used; because the number of permutations is affected by the number of replicates up to triplicates, we suggest that the user perform the full set of permutations (>700). Concerning the definition of the threshold parameters, we usually perform SAM analysis selecting a significant cutoff that gives less than one false positive, and we combine this threshold with a fold change of at least |1.5| (**Fig. 1**, flow chart step s2).

CyberT, developed by Baldi and Long *(13)*, allows calculation of how meaningful a differential expression is using a Bayesian probabilistic framework. In particular, CyberT uses a Bayesian approach to calculate a background variance for each of the genes under analysis, using such values to balance experimental fluctuations within a limited number of replicates. In CyberT, a Bayesian version of the *t*-test can be performed as the user has defined the number of neighboring genes needed to estimate the background variance for any of the genes in the dataset and the degree of confidence in the background variance versus the empirical variance. As shown by the authors *(13)*, the Bayesian approach appears robust relative to the use of fold change alone, as large nonstatistically significant fold changes are often associated with large measurement errors. Furthermore, the Bayesian approach is about twice as consistent as a simple *t*-test for identifying differentially expressed genes over twofold expression change in independent samples of size 2 (i.e., two experiments vs two controls).

SAM and CyberT are different ways to identify differentially expressed genes; because we want to obtain a robust set of differentially expressed data, we validate SAM data by CyberT (**Fig. 1**, flow chart step s3, SAM analysis ∩ CyberT analysis). Genes found to be differentially expressed by SAM analysis are mapped on a plot in which differential expression values are plotted with respect to CyberT *p* values (*see* **Note 15**), calculated for the same dataset used in SAM analysis. We keep in consideration for further analysis only those genes that passed the SAM test and show a *p* value < 0.005 (**Fig. 6**, dashed line) in CyberT analysis. As can be observed in **Fig. 6**, all genes passing the SAM test and showing a $\log_2$ fold change = |0.5| also have *p* values <0.005.

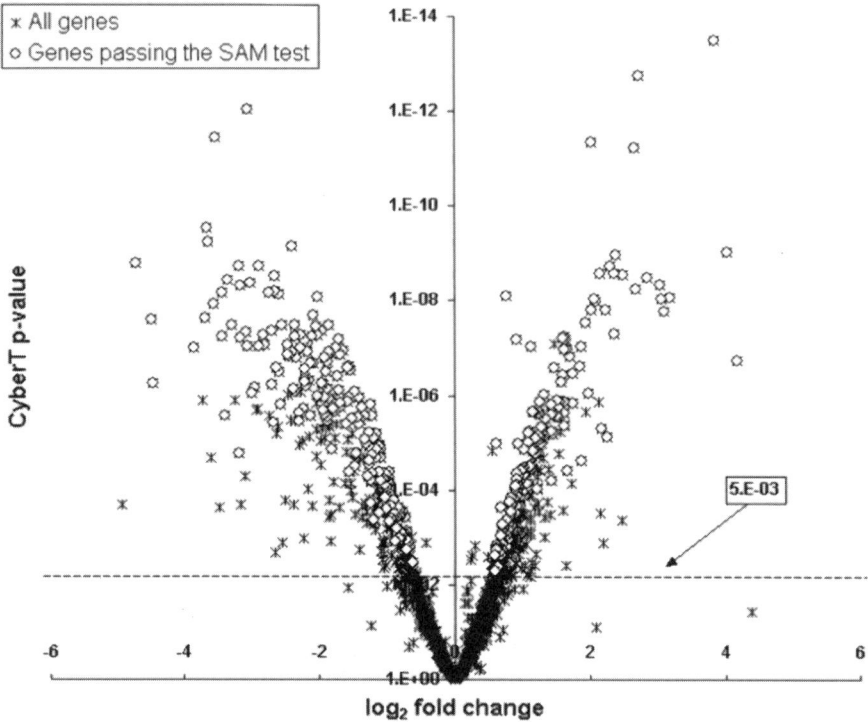

Fig. 6. *p* values generated by CyberT (which indicate the probability that an observed fold change is a casual event) for all differential expressions are plotted with respect to fold change (∗). Genes shown as differentially expressed in SAM analysis are shown as circles. Differential expressions are considered statistically validated if they pass SAM analysis and show in CyberT at $p < 0.005$ (dashed line).

4. Microarray Data Mining
4.1. Transcription Profiles Clustering

Array technology has made it straightforward to monitor simultaneously the expression patterns of thousands of genes. The challenge now is to make sense of such a massive dataset. In a simple experimental design in which a control sample is compared with a treated sample, the user gets a set of differentially expressed genes that can be ranked by their relative induction. In a more complex experimental design, involving, for example, time-course or parallel analysis of different transcription activators (e.g., different isoforms of the same transcription factor), a key goal is to extract the fundamental patterns of gene expression inherent in the data. These techniques are essentially different ways to cluster points in multidimensional space. Various clustering approaches have been applied to transcriptional expression profiles generated by microarray analysis *(15–18)*; however, in gene expression classification it is not possible to identify a universal clustering approach because the optimal clustering algo-

Microarray Data Analysis and Mining

rithm to be used depends on the nature of the dataset and what constitutes meaningful clusters in the problem under analysis. Therefore, in our lab we test various tools to generate transcription profile clustering. We usually prefer clustering approaches in which the number of clusters is not an arbitrary parameter defined by the user but is defined by the algorithm on the basis of the dataset under analysis (www.esat. kuleuven.ac.be/~thijs/Work/Clustering.html; www.stat.washington.edu/fraley/mclust).

An example of a transcription profile obtained using adaptive quality-based clustering *(17)* can be seen in **Fig. 7**. The three clusters refer to transcription profiles derived by microarray analysis of six p63 isoform-driven gene expressions (Saviozzi et al., unpublished results), and they are the clustering solution obtained using the web-based clustering tool developed by De Smet and coworkers (www.esat. kuleuven.ac.be/~thijs/Work/Clustering.html) *(17)*. This tool is based on a two-step algorithm. As the first step, transcription profiles are grouped in spheres in which the density of expression profiles is locally maximal (based on a user-defined estimate of the radius of the cluster). In the second step, the radius of each cluster is optimized so that only the significant coexpressed genes are included in the cluster. It is interesting to note that using a *k*-way partition clustering approach *(19)* (http://www-users.cs. umn.edu/~karypis/cluto/) in which we forced the program to generate three clusters, we got 95% overlaps between the *k*-way clustering and the adaptive quality-based clustering (data not shown). This result indicates that the partition in three clusters is probably a good solution for the dataset under analysis, as homogeneous results can be obtained using different clustering approaches.

4.2. Associating Functional Meaning to Transcriptional Expression Fold Change

Even after transcriptional profiles are clustered, much work must be done to extract some functional information from microarray data. Clustering, when possible, classifies genes by their transcription profile similarity, which gives only partial information about the functional correlation existing between differentially expressed genes, and belonging to the same cluster; it does not imply a functional correlation. An important topic in microarray data mining is therefore to bind transcriptionally modulated genes to functional pathways or classify them in functional classes to understand how transcriptional modulation can be associated with specific biological events (genetic disease phenotypes, molecular mechanism of drugs action, cell differentiation, development, and so on) or at least to find out whether some specific biological process is strongly affected by transcriptional modulation in the experiment under analysis. In other words, researchers need to rely on robust gene functional annotations and on tools to link functional annotations to transcriptional profiling.

Genomic sequencing has clarified that a large fraction of the genes specifying the core biological functions are shared by all eukaryotes; therefore, knowledge of the biological role of such shared proteins in one organism can often be transferred to other organisms. The Gene Ontology (GO) Consortium (www.geneontology.org) has as a main task to define a structured, precisely defined, common controlled vocabulary for describing the roles of genes and gene products in any organism. Therefore

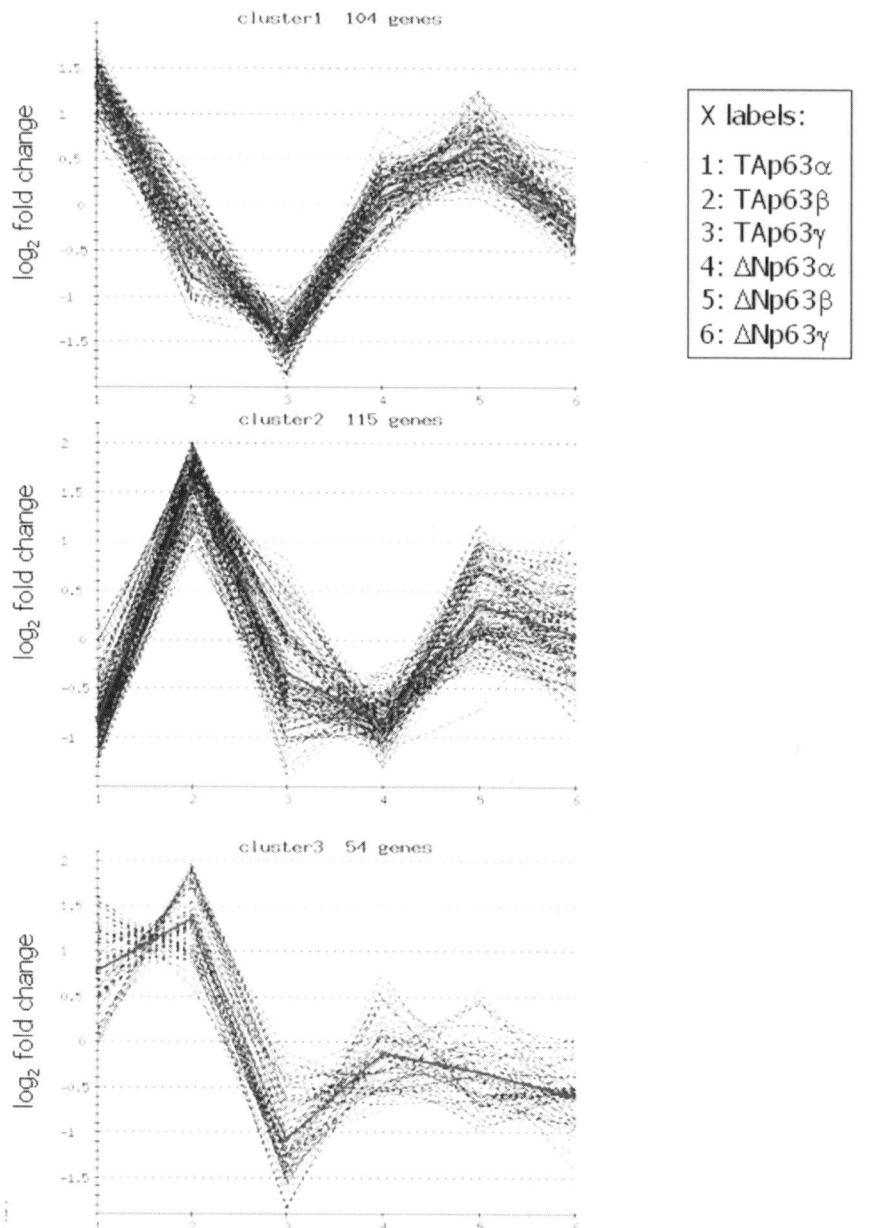

Fig. 7. Clustering results obtained using the tool developed by De Smet et al. *(17)*. The adaptive quality-based clustering was generated using the microarray data derived by transient transfection of six p63 isoforms in a p53 null cell line (Saviozzi, unpublished observations). The clustering procedure produces three clusters, based on a required probability of genes belonging to cluster equal to 0.85. Similar clustering results were obtained using a k-way partition clustering approach (data not shown).

GO can be a valuable tool to link differentially expressed genes to specific functional classes.

GO is divided into three categories:

- Molecular function, which contains information about the tasks performed by individual gene products (e.g., transcription factor, DNA helicase).
- Biological process, which describes broad biological goals, such as mitosis or DNA repair, that are accomplished by ordered assemblies of molecular functions.
- Cellular component, which indicates subcellular structures, locations, and macromolecular complexes (e.g., nucleus, ribosome).

The important point of the GO approach is that each node (ontology term) of the ontology can have more then one parent, and a gene can be associated with more than one node. The availability of more than one GO tag for each gene offers the opportunity to categorize genes on the basis of their GO features. For this reason we have developed a GO clustering tool, written in Java and downloadable at http://www.bioinformatica.unito.it/downloads/clustering/GO-clustering/. The tool relies on the use of GO terms associated with genes annotated in locus link databases (ftp://ftp.ncbi.nih.gov/refseq/LocusLink/ LL.tmpl.gz) and on a k-way clustering tool specifically developed for document clustering (CLUTO, http://www-users.cs.umn.edu/~karypis/cluto/). To cluster genes on the basis of their GO terms, users need to give a list of locus link identifications (LL ids) (*see* **Note 16**) of the genes of interest and set the clustering parameters. The tool will download from the NCBI ftp site the latest version of the LL.tmpl file, and it will associate the GO biological function terms with each of the LL ids supplied by the user. Our tool provides a graphical interface to set the clustering parameters and offers text and graphical visualization of the resulting clusters. A good approach for defining the optimal clustering solution is based on the evaluation of internal cluster similarity (ISim), external similarity (ESim), entropy (Entpy), and purity (Purty) of the clustering solution. ISim displays the average similarity between the objects of each cluster, and ESim displays the average similarity of the objects of each cluster and the rest of the objects. Small entropy values and large purity values indicate good clustering. The optimal clustering solution can be searched for by modifying at least some of the available clustering parameters, (e.g., number of clusters, method to be used for clustering the objects, similarity function to be used for clustering, clustering criterion function, factor by which the program will prune the clustering features before performing the clustering, number of different clustering solutions to be computed by the various partitional algorithms, maximum number of refinement iterations to be performed within each clustering step, and so on). An example of the clustering optimization procedure is as follows:

Step 1. Use default settings and change the number of clusters searching for increasing ISim and decreasing ESim. During this process it is important to keep track of the number of items (genes) present in the clusters.

Step 2. Keep the number of clusters constant and change the clustering methods (*clmethod*) to select the method that improves the clustering quality features (ISim, ESim, Entpy, Purty).

Table 1
Effect of *Colprune* in Clustering Genes by GO Terms

	Colprune		
	1.0	0.9	0.8
% of total items in clusters	0.97	0.59	0.43
Average ISim	0.15	0.51	0.83
Average ESim	0.007	0.026	0.059

Step 3. Repeat **step 1** using the optimal clustering method.

Step 4. Change the similarity function to be used for clustering (*sim*) to define which one improves the clustering quality features. Repeat **step 1**.

Step 5. Change the clustering criterion (*crfun*) and select the one that improves the clustering quality features. Repeat **step 1**.

Step 6. Find the optimal value for the factor (*colprune*) by which the tool will prune the GO features before performing the clustering. This is a number between 0.0 and 1.0 and indicates the fraction of the overall similarity for which the retained GO features must account. This parameter can be useful because some GO features just affect in a negative way the clustering quality features, but pruning GO terms can reduce the number of items that can be clustered. Repeat **step 1**. **Table 1** shows the effect of pruning: reducing the number of GO terms involved in clustering the ISim is improving although the number of clustered genes it is reduced. Pruning in this specific case affects the ESim, in a negative way but the improvement in ISim counterbalances it.

Step 7. Other parameters are available for greater fine tuning of the clustering solution. However, by performing the previously described steps, a reasonable clustering solution is obtained using GO terms.

Figure 8 gives an example of GO clustering done on the full set of annotated genes containing p53-responsive elements in their promoters *(20)*. The results of clustering by GO terms can be a valuable tool to inspect microarray results under a functional perspective and to identify whether specific biological functions are modulated at a transcriptional level. However, GO clustering is strongly influenced by the amount of GO annotation available, which is still a limiting factor because not all gene are annotated.

4.3. Microarray Literature Data Mining

When GO annotations are not available or more information is needed, the published literature might provide a source of information to assist in the interpretation of microarray data. Functional data are rapidly accumulating in the scientific literature, and the biologist needs to retrieve this information, which has been collected by MEDLINE (www.ncbi.nlm.nih.gov), a database that contains over 12,000,000 biomedical journal citations. Finding gene correlations using MEDLINE is time-consuming, even if in recent years some tools have been developed to perform automated information extraction on the MEDLINE database *(21,22)*. For this reason we have

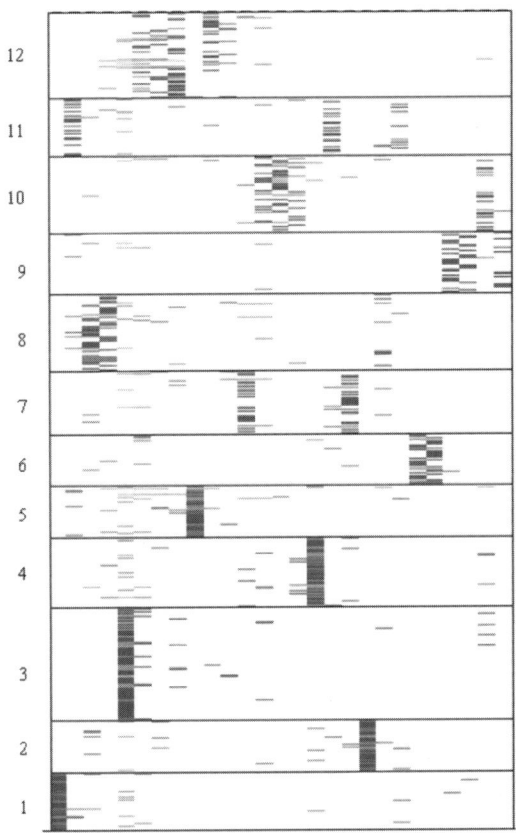

Fig. 8. Genes containing p53-responsive elements were extracted by the Wang et al. *(20)* data set and clustered by their GO annotations. Genes were classified in 12 groups, and a *colprune* of 0.9 was used to optimize the ISim of the clusters. The GO descriptive features of clusters are as follows: *cluster 1*, GO:0006832 (small molecule transport) 98.7%; *cluster 2*, GO:0006366 (transcription from Pol II promoter) 98.0%; *cluster 3*, GO:0007165 (signal transduction) 98.2%; *cluster 4*, GO:0007048 (oncogenesis) 97.4%; *cluster 5*, GO:0007186 (G-protein-coupled receptor protein signaling pathway) 96.7%; *cluster 6*, GO:0006508 (proteolysis and peptidolysis) 52.2%, GO:0006464 (protein modification) 47.1%; *cluster 7*, GO:0007275 (development) 49.7%, GO:0006357 (regulation of transcription from Pol II promoter) 46.6%; *cluster 8*, GO:0007399 (neurogenesis) 47.0%, GO:0007155 (cell adhesion) 46.9%; *cluster 9*, GO:0006629 (lipid metabolism) 38.3%, GO:0006091 (energy pathways) 36.5%, GO:0006899 (nonselective vesicle transport) 24.4%; *cluster 10*, GO:0000074 (regulation of cell cycle) 54.2%, GO:0008283 (cell proliferation) 16.4%, GO:0008285 (negative regulation of cell proliferation) 16.4%, GO:0006468 (protein amino acid phosphorylation) 12.2%; *cluster 11*, GO:0007268 (synaptic transmission) 38.3%, GO:0007345 (embryogenesis and morphogenesis) 35.1%, GO:0007601 (vision) 24.4%; *cluster 12*, GO:0006955 (immune response) 38.3%, GO:0007267 (cell-cell signaling) 21.7%, GO:0007166 (cell surface receptor-linked signal transduction) 19.7%, GO:0006954 (inflammatory response) 13.1%, GO:0006960 (antimicrobial humoral response) 6.1%.

developed a tool (MedMOLE) that makes it simpler to extract functional knowledge by literature abstracts directly/indirectly related to differentially expressed genes identified by microarrays. In our specific application, it was necessary to let documents group on the basis of their content, expressed in terms of nouns, verbs, and adjectives, as the functional meaning of genes should emerge from that context. Indeed, it was necessary to recognize the names of genes inside the texts both for querying purposes and for interpreting the results, i.e., being able to relate each group of documents (or biological function) to the involved genes. The gene name recognition is a particularly difficult task, even for the information extraction tools, as these names do not follow any predefined rule and have many aliases.

We then built a "gene name extractor" based on a gene names dictionary containing official gene names and aliases built on the basis of the LocusLink database. In order to let scientists mine only the subset of MEDLINE abstracts that are of their own interest, we built a web-based application, MedMOLE, that makes available on-line a portion of the MEDLINE database processed by our algorithms and ready for the mining phase (*see* **Note 17**). The dataset was generated by extracting from the NCBI's Pubmed (since 1990) all documents frequently containing gene or protein names by means of the query "(gene OR protein) OR (genes OR proteins)." We obtained about 1.7 million documents containing gene names associated with 9609 LL ids out of the total 14,659 LL ids present in the gene names dictionary at the time this chapter was submitted. Users can mine the gene names-linked MEDLINE dataset using any generic key words or by applying a list of LL ids that interrogates the LocusLink database, implemented in our local SRS server, and builds a MedMOLE query using official and alias gene symbols. MedMOLE outputs (server-based) are made by graphical and textual components (**Fig. 9**). Users can perform the analysis by applying default parameters (30 clusters, clusters based on both frequent and rare words and intracluster similarity set to 0.35) or they might decide to define the number of clusters, select only rare or frequent words as objects of the clustering, and select the minimum similarity between documents inside a cluster. Each analysis has to be repeated various times in order to optimize the clustering solution. After the user has selected the clustering parameters and performed the clustering step, the results are shown as a list of clusters (**Fig. 9A**), each one described by the number of abstracts in the cluster, a link to the abstracts' contents, the cluster number, up to seven key words being the most descriptive of the cluster, and a link to the gene names found in the cluster (report). The report is particularly interesting because it contains a list of the gene names found in the cluster and statistics regarding the gene name features, as well as a list of potential gene associations generated by using the *a priori* algorithm for the induction of

---

Fig. 9. *(opposite page)* (**A**) clusters defined by MedMOLE and listed by decreasing size. (**B**) Graphic representation of the clusters generated by MedMOLE. Segments linking clusters indicate a certain grade of similarity between the linked clusters. The numbers associated to the linking segments indicate the degree of similarity. (Increasing numbers indicate an increase of similarity.) (**C**) The plot describes the frequency of gene names found in the MedMOLE clusters.

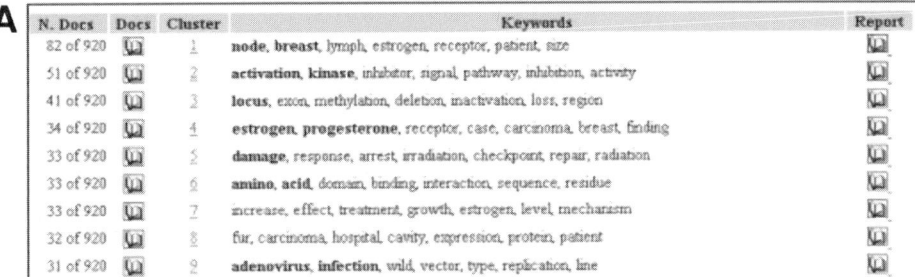
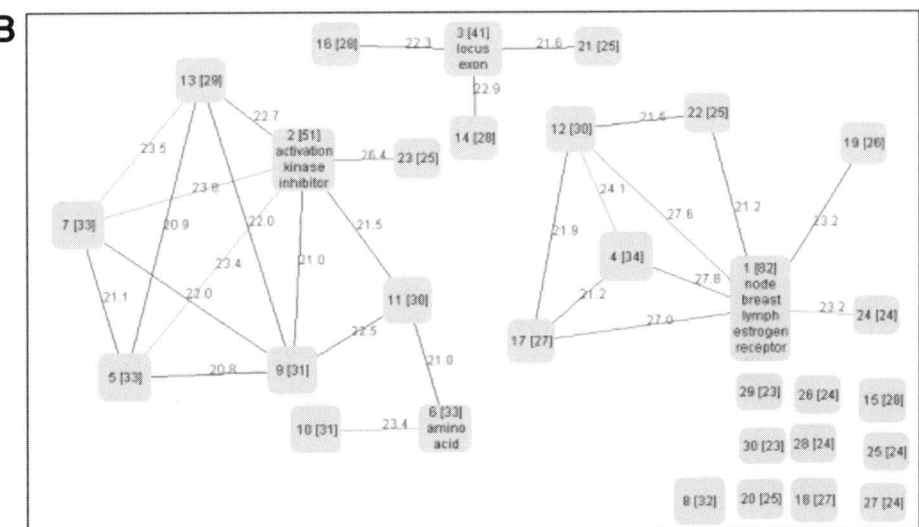
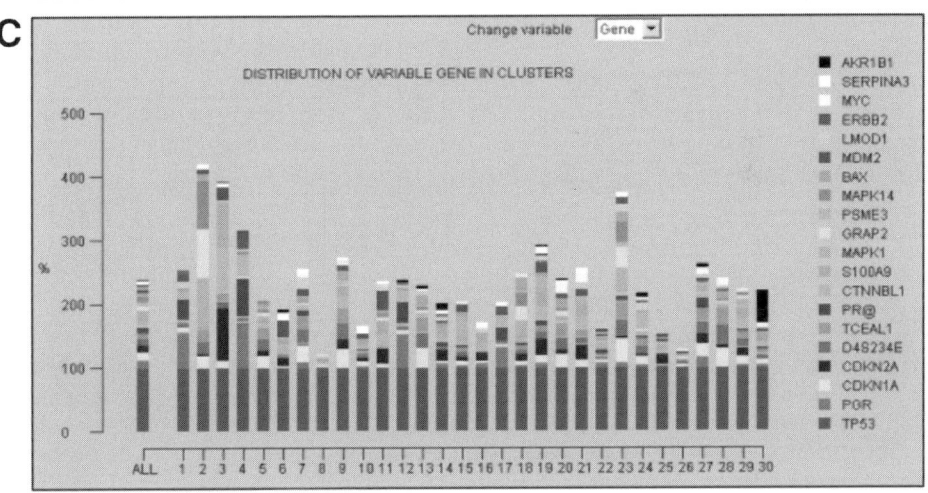

Fig. 9.

association rules *(23,24)*. The Agrawal *A Priori* rules induction algorithm is a powerful method to find regularities in a set of documents/transactions. The tool tries to identify sets of items that are frequently found together, so that from the presence of certain items in a set of documents one can infer that other items are present, e.g., gene A, gene B → gene X (support, confidence). Consider a set (T) of all abstracts containing a group of gene names. The support of a subset of gene names S is the percentage of those abstracts in T containing S. As an example, abstracts contain a given set S of gene names, e.g., S = {AP2, SP1, P53}. If U is the set of all abstracts containing all gene names in S, then support $(S) = (|U|/|T|) * 100\%$, where $|U|$ and $|T|$ are the number of abstracts in U and T, respectively. The confidence of a rule R = "gene A and gene B → gene C" is instead the support of the set of all items that appear in the rule divided by the support of the antecedent of the rule, i.e., confidence $(R) = [\text{support}(\{A, B, C\})/\text{support}(\{A, B\})] * 100\%$.

To give an idea of the potentiality of the tool, we have analyzed with MedMOLE the 38 genes, containing p53-responsive elements, present in the GO cluster 8 (**Fig. 8**; cluster 8 has as main descriptive features the GO terms neurogenesis and cell adhesion). MedMOLE was queried over 1996–2002, using gene LL ids together with the key word p53, in order to identify which of those genes have already been described in association with the p53 gene. From this analysis we obtained 920 documents that were clustered in 30 groups. We observed that only 8 of the 38 genes used in the query were present in the clusters, and 4 of them were associated by the *a priori* rule induction algorithm to p53. So, with few mouse clicks, it was possible to determine that only 10% of the genes related to the GO cluster 8 can be found in abstracts together with p53 and that four of them show a strong statistical association with p53.

In our opinion MedMOLE is a tool that can offer an easy way to navigate the scientific literature related to a group of genes derived by microarray analysis (*see* **Note 18**). In particular, the choice of clustering the abstracts, by their informational content, might help in the identification of a specific biological function bound to differential expression. Furthermore, the availability of the *a priori* rule induction algorithm can simplify the identification of potential functional association between genes.

5. Notes

1. Microarray analysis have been used successfully to define transcriptional signatures to allow for patient-tailored therapy strategy in breast cancer *(25)* or to classify better tumors having no histological counterparts in normal tissues *(26)*, such as synovial sarcomas, which are grouped together as "miscellaneous soft tissue tumors" in the latest edition of the *WHO Soft Tissue Tumor Classification (27)*.
2. Microarray analysis can be a useful tool to identify genes directly activated/repressed by expression of a transcription factor (e.g., STAT1, p53, p63, and so on). Subsequently, primary response genes can be identified by computational searching of factor-specific responsive elements in a DNA region located upstream of genes found to be differentially expressed in microarray experiments.
3. At the time this chapter was submitted, the average knowledge of model organisms (e.g., *Escherichia coli*, *Saccharomyces cerevisiae*) was more than 80%, which means that more than 80% of the genes in *E. coli/S. cerevisiae* have been annotated (i.e., anno-

tated genes are those assigned to a specific biological pathway and/or biochemical function). On the other hand, higher eukaryotes (e.g., *Homo sapiens*, *Mus musculus*) are less characterized, which means that less than 30% of the genome is functionally annotated.

4. Automated partial DNA sequencing on randomly selected tissue-specific complementary DNA (cDNA) clones was used to obtain a collection or short sequences associated with expressed sequence tags (ESTs). ESTs are mainly located at the 3' end of the gene sequence. This fast approach to cDNA characterization facilitated the tagging of most expressed human genes before the genome sequence.

5. Kane et al. *(2)* suggest that the cDNA probe similarity might cause inaccurate expression measurements on cDNA microarray, and spotting more specific cDNA probes of the unique regions from a set of genes and reducing the length of the probe sequences could reduce the potential for cross-hybridization. Probe length may also affect the degree of nonspecific hybridization, as observed in part on cDNA arrays. Therefore, the use of oligonucleotides might overcome the described limitation of cDNA arrays. Furthermore, oligonucleotide synthesis is now less expensive than cDNA preparation. Owing to the larger number of operations required for cDNA preparation with respect to oligonucleotide synthesis, the latter has a limited rate of gene/sequence misleading assignment, and the availability of all genome sequences for various eukaryotic genomes allows the optimization of gene-specific oligonucleotide design.

6. Spotting, hybridizing, and reading instruments are commercially available. However, our current strategy, if low-density arrays are needed, is to buy custom-spotted arrays from specialized companies as the setting up of microarray spotting facilities can be long, frustrating, and expensive.

7. A postlabeling procedure is more suitable because the yield of cDNA is higher than with direct incorporation. Indirect incorporation of the reactive dyes of Cy3 and Cy5 into the probe is greater and more even than CyDye-labeled nucleotide incorporation. The postlabeling method is less prone to incorporating artifacts caused by the size of CyDye nucleotides (e.g., chain termination, proximity quenching, sequence-specific bias).

8. The BioChip Arrayer (Perkin Elmer Life Sciences, Boston, MA) is based on a noncontact dispensing technology. The PiezoTip™ never comes into direct contact with surface, reducing the risk of carryover contamination and enabling high-quality spotting (homogeneous spot size, homogeneous deposition in all spot surface, and so on) at high densities.

9. Absolute expression of a gene is described by two parameters:
    a. Absolute call, which is a qualitative index of gene expression. Absolute call is defined using three letters: P, M, and A, where *P* indicates the presence of gene expression, M indicates borderline expression, and A signifies expression absence.
    b. Absolute intensity, which is a quantitative description of gene expression equal to the fluorescence intensity value measured averaging PM-MM (Microarray Affymetrix Suite V).

10. CyberT is a web-based application; a mirror site is available at www.bioinformatica.unito.it.

11. For more information about the image files manipulation performed by MAS 5.0, check the Affymetrix manual (www.affymetrix.com).

12. Increasing τ can reduce the number of false present calls but may also reduce the number of true present calls.

13. Decreasing the significance level $\alpha 1$ can reduce the number of false detected calls and reduce the number of true detected calls. Increasing the significance level $\alpha 2$ can reduce the number of false undetected calls and reduce the number of true undetected calls.

14. Owing to the limited availability of biological materials or to budget limitations, a large number of replicates (more than four) can rarely be done. However, in our hands a reasonable compromise is to perform experiments as biological triplicates.
15. The *p* value generated by CyberT indicates the probability that a differential expression is caused by chance. In other words the *p* value indicates the probability that the average intensity of the control samples belongs to the same distribution of the experimental samples.
16. LocusLink provides a single-query interface to a curated sequence and descriptive information about genetic loci. It contains information on official nomenclature, aliases, sequence accessions, phenotypes, EC numbers, MIM numbers, UniGene clusters, homology, map locations, and related web sites. Sequence accessions include a subset of GenBank accessions for a locus, as well as the NCBI Reference Sequence (RefSeq). The reference sequences are generated applying a seed sequence as a query in a BLAST analysis. For mRNAs, BLAST results are sorted to identify the longest sequence that retains a gap-free alignment with a minimum of mismatches through the coding region. Sequences that also have a full-length coding region are used to create the predicted and provisional RefSeq mRNA/protein records. These RefSeq records are generated via an automatic process. All provisional RefSeq records will undergo a manual review step. During the review process additional RefSeq records may be generated to represent well-characterized transcript variants.
17. MedMOLE is a web-based tool that can be used upon registration at http://www.cineca.it/HPSystems/Chimica/medmole/index.html. Interfaces for querying MedMOLE by LL ids, gene names, or key words are available at http://www.bioinformatica.unito.it/bioinformatics/medmole/welcome.html.
18. Examples of the use of MedMOLE in the analysis of data derived by microarrays can be found at http://www.bioinformatica.unito.it/bioinformatics/medmole/tutorials.html.

Acknowledgments

This work was partially supported by PRIN2001 2001057147 and FIRB RBAU01JTHS/RBNEO157EH grants of the Italian Ministry of University and S&T Research-MIUR.

References

1. Schena, M., Shalon, D., Davis, R. W., and Brown, P. O. (1995) Quantitative monitoring of gene expression patterns with complementary DNA microarray. *Science* **270,** 467–470.
2. Kane, M. D., Jatkoe, T. A., Stumpf, C. R., Lu, J., Thomas, J. D., and Madore, S. J. (2000) Assessment of the sensitivity and specificity of oligonucleotide (50mer) microarrays. *Nucleic Acids Res.* **28,** 4552–4557.
3. El Atifi, M., Dupre, I., Rostaing, B., Chambaz, E. M., Benabid, A. L., and Berger, F. (2002) Long oligonucleotide arrays on nylon for large-scale gene expression analysis. *Biotechniques* **33,** 612–616.
4. Jin, W., Riley, R. M., Wolfinger, R. D., White, K. P., Passador-Gurgel, G., and Gibson, G. (2001) The contributions of sex, genotype and age to transcriptional variance in *Drosophila melanogaster*. *Nat. Genet.* **29,** 389–395.
5. Dudoit, S., Yang, Y. H., Callow, M. J., and Speed, T. P. Statistical methods for identifying differentially expressed genes in replicated cDNA microarray experiments. Technical report #578 Department of Statistics, UC-Berkeley. August 2000. (<http://www.stat.berkeley.edu/users/terry/zarray/Html/matt.html>http://www.stat.berkeley.edu/users/terry/zarray/Html/matt.html)

6. Li, C. and Wong, W. H. (2001) Model-based analysis of oligonucleotides arrays: expression index computation and outlier detection. *Proc. Natl. Acad. Sci. USA* **98,** 31–36.
7. Hartemink, D. G., Jaakkola, I., and Young, R. (2001) Maximum likelihood estimation of optimal scaling factors for expression array normalization. *Microarrays: Optical Technologies and Informatics (Proceedings of SPIE)*, p. 4266.
8. Rocke, D. M. and Durbin, B. (2001) A model for measurement error for gene expression arrays. *J. Comput. Biol.* **8,** 557–569.
9. Golub, T. R., Slonim, D. K., Tamayo, P., et al. (1999) Molecular classification of cancer: class discovery and class prediction by gene expression monitoring. *Science* **286,** 531–537.
10. Kim, S., Dougherry, E. R., Chen, Y., et al. (2000) Multivariate measurement of gene expression relationships. *Genomics* **67,** 201–209.
11. Celis, J. E., Kruhoffer, M., Gromova, I., et al. (2000) Gene expression profiling: monitoring transcription and translation products using DNA microarrays and proteomics. *FEBS Lett.* **480,** 2–16.
12. Tusher, V. G., Tibshirani, R., and Chu, G. (2001) Significance analysis of microarrays applied to ionizing radiation response. *Proc. Natl. Acad. Sci. USA* **98,** 5116–5121.
13. Baldi, P. and Long, A. D. (2001) A bayesian framework for the analysis of microarray expression data: regularized t-test and statistical inference of gene changes. *Bioinformatics* **17,** 509–519.
14. Li, C. and Wong, W. H. (2001) Model-based analysis of oligonucleotide arrays: model validation, design issues and standard error application. *Genome Biol.* **2,** 32.1–32.11.
15. Tamayo, P., Slonim, D., Merinov, J., et al. (1999) Interpreting patterns of gene expression with self-organizing maps: methods and application to hematopoietic differentiation. *Proc. Natl. Acad. Sci. USA* **96,** 2907–2912.
16. Eisen, M. B., Spellman, P. T., Brown, P. O., and Botstein, D. (1998) Cluster analysis and display of genome-wide expression patterns. *Proc. Natl. Acad. Sci. USA* **95,** 14,863–14,868.
17. De Smet, F., Mathys, J., Marchal, K., Thijs, G., De Moor, B., and Moreau, Y. (2002) Adaptive quality-based clustering of gene expression profiles. *Bioinformatics* **18,** 735–746.
18. Yeung, K. Y., Fraley, C., Murua, A., Raftery, A. E., and Ruzzo, W. L. (2001) Model-based clustering and data transformations for gene expression data. *Bioinformatics* **17,** 977–987.
19. Zhao, Y. and Karypis, G. (2002) Criterion Functions for Document Clustering Experiments and Analysis. *Technical Report* #01-40, 2002. University of Minnesota, Department of Computer Science/Army HPC Research Center, Minneapolis, MN 55455.
20. Wang, L., Wu, Q., Qiu, P., et al. (2001) Analyses of p53 target genes in the human genome by bioinformatic and microarray approaches. *J. Biol. Chem.* **276,** 43,604–43,610.
21. Tanabe, L., Scherf, U., Smith, L. H., Lee, J. K., Hunter, L., and Weinstein, J. N. (1999) MedMiner: an Internet text-mining tool for biomedical information, with application to gene expression profiling. *Biotechniques* **27,** 1210–1217.
22. Hokamp, K. and Wolfe, K. (1999) What's new in the library? What's new in GenBank? Let PubCrawler tell you. *Trends Genet.* **5,** 471–472.
23. Nobata, C., Collier, N., and Tsujii, J. (1999) Automatic term identification and classification in biology texts, in *Proceedings of the Natural Language Pacific Rim Symposium (NLPRS'2000)*, pp. 369–375.

24. Agrawal, R., Imielinski, T., and Swami, A. (1993) Mining association rules between sets of items in large databases, in *Proceedings of the Conference on Management of Data.* ACM Press, pp. 207–216.
25. van't Veer, L. J., Dai, H., van de Vijver, M. J., et al. (2002) Gene expression profiling predicts clinical outcome of breast cancer. *Nature* **415,** 530–536.
26. Nagayama, S., Katagiri, T., Tsunoda, T., et al. (2002) Genome-wide analysis of gene expression in synovial sarcomas using a cDNA microarray. *Cancer Res.* **62,** 5859–5866.
27. Weiss, S. W. and Sobin, L. (1994) Histological typing of soft tissue tumors. In: World Health Organization International Histological Classification of Tumors, 2nd ed.: Springer-Verlag, Berlin.

6

Expression Cloning

Michael J. Lodes, Davin C. Dillon,
Raymond L. Houghton, and Yasir A. W. Skeiky

Abstract

Expression cloning involves the selection of specific polypeptides, generated from a cDNA or genomic DNA library, based on certain characteristics of the expressed proteins, such as antibody or ligand binding, recognition by T-cells, function, or complementation of cell defects. Here we describe the detailed construction of a genomic, random shear lambda expression library, adsorption of anti *Escherichia coli* antibody from antiserum, the screening of an expression library with specific antisera, and the cloning of genes with potential use in the diagnosis of infectious disease. This approach has been used successfully by our laboratory for the discovery of antigenic components of diagnostics and vaccines for several infectious agents including: *Mycobacterium tuberculosis*, *Anaplasma phagocytophila* (formerly *Ehrlichia* spp. or *E. phagocytophila*), *Babesia microti*, *Trypanosoma cruzi*, *Leishmania chagasi*, and *Chlamydia* spp.

Key Words: Expression; cloning; prokaryotic; library; serological; genomic; cDNA; diagnosis.

1. Introduction

Expression cloning covers a wide variety of techniques that utilize both prokaryotic and eukaryotic cells as hosts for the synthesis of proteins from various genomic DNAs or cDNAs *(1)*. These foreign proteins are then detected by several means, including: (1) their ability to bind polyclonal or monoclonal antibodies to known targets or in serum of patients with a particular disease *(2,3)*; (2) their recognition by T-cells *(4–6)*; (3) their function (e.g., receptors, enzymes, ligands, transcription regulators, and so on) *(7–9)*; and (3) their complementation of a specific cell defect *(10)*. We have successfully utilized serologic expression cloning for the detection of antigens that are useful for the diagnosis of infectious organisms including *Mycobacterium tuberculosis (11–13)*, *Anaplasma phagocytophila* (formerly *Ehrlichia* spp. or

From: *Methods in Molecular Medicine, vol. 94: Molecular Diagnosis of Infectious Diseases, 2/e*
Edited by: J. Decker and U. Reischl © Humana Press Inc., Totowa, NJ

E. phagocytophila), the agent of human granulocytic ehrlichiosis *(14)*, *Babesia microti*, the agent of human babesiosis *(15)*, *Trypanosoma cruzi*, the agent of Chagas' disease *(16–20)*, and *Leishmania chagasi (21,22)*. We have also used both serologic and T-cell expression cloning for antigen discovery in the development of vaccines for infectious disease agents including *Leishmania* spp. *(23–25)*, *M. tuberculosis (26–28)*, *Chlamydia* spp., herpes simplex virus type 2 (unpublished data), *Propionibacterium acnes* (acne; unpublished data), and inflammatory bowel disease (unpublished data). We, and many other groups, have also used serologic expression screening to clone genes that are overexpressed in human cancers and that could have utility in diagnosis, vaccine, or therapeutic development *(29)*.

1.1. Expression Libraries

Many types of libraries and expression systems can be used for expression cloning, e.g., lambda phage and phage surface display libraries in prokaryotic cells *(30–36)* and plasmid libraries in eukaryotic cells such as in mammalian expression cloning and yeast two-hybrid screening *(1,7,8,37)*. For the cloning of genes that are relevant for the diagnosis of viral, prokaryotic, fungal, and eukaryotic infectious agents, cDNA and genomic DNA expression libraries in lambda vectors, expressed in *E. coli* host cells, have proved to be effective in our hands. Genomic DNA expression libraries can be used when the organism of interest has a relatively small genome ($<10^7$). For organisms with large genomes ($>10^9$) or genomes containing introns, cDNA libraries might be more useful. However, genomic DNA libraries have an advantage in that they can potentially express all open reading frames; cDNA libraries can express only those genes that are transcribed at a given point in time (a useful characteristic for cloning stage-specific antigens). When screening for a suitable antigen for diagnostic purposes, one would want to cover all possibilities. If one has screened a genomic DNA expression library and suspects that clone insert DNA contains introns or multiple inserts, a cDNA library [or cDNA in combination with polymerase chain reaction (PCR)] might be useful to confirm the sequence that will be used for recombinant protein expression.

Several sources are available for the purchase of kits that can be used to construct cDNA expression libraries (Lambda ZAP II and Lambda ZAP Express from Stratagene; λ SCREEN-1 from Novagen, and others). Genomic DNA expression libraries can be constructed with the same vectors used for cDNA libraries if they are available with the appropriate predigested and dephosphorylated ends. We have routinely used both the Lambda ZAP II and Lambda ZAP Express vectors, predigested with *Eco*RI. Plasmids excised from Lambda ZAP Express contain both prokaryotic and eukaryotic promoters for expression in either bacteria or mammalian cells.

The strategy behind producing an effective genomic DNA library is to create random DNA fragments in a preferred size range (**Fig. 1**). This can be done in several ways including sonication and limited digestion with a frequent-cutting restriction endonuclease. Digestion with endonucleases can be biased by the structure of any given genome in which specific restriction sites might be abundant in most regions but rare in others. Random shearing by sonication appears to be more unbiased, and thus

Expression Cloning

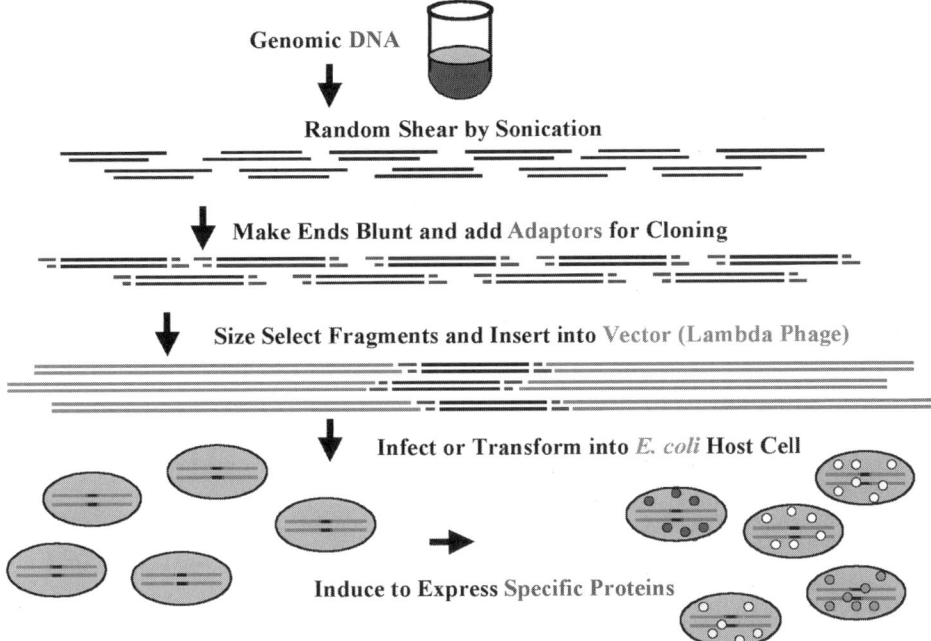

Fig. 1. Schematic representation of the construction of a random shear genomic expression library. Randomly sheared DNA is represented by black double lines, and black double lines with short extensions indicate blunted DNA with adaptors. Lambda phage arms are shown in gray. Ovals with double lines (lower left) indicate bacteria containing lambda phage. Small shaded circles in ovals (lower right) indicate unique proteins that are encoded by the lambda phage and synthesized by the host cells.

one has a better chance of detecting most of the potentially expressible polypeptides. Once the genomic DNA has been sonicated, one can remove very small fragments that are produced by oversonication or from degraded DNA and that can interfere with adaptor and vector ligation, by molecular filtration with an Amicon Microcon filter (model 100) or other suitable filters or columns. The next critical step is blunting the DNA fragments. We have used several enzymes and enzyme combinations including T4 DNA polymerase, DNA pol I, or the Klenow fragment of Pol I in combination with Pfu polymerase for blunting. If a secondary blunting step with Pfu polymerase is incorporated, the reaction must be supplemented with a dNTP mix and the reaction carried out for 30 min at 72°C.

Before the blunted DNA fragments can be incorporated into the lambda vector, adaptors must be ligated to the ends. Adaptors are composed of two complementary oligonucleotides, one phosphorylated to allow for ligation to the blunted fragments, and the other dephosphorylated to prevent the cohesive ends from ligating to each other. Adaptors, as opposed to linkers, do not require a restriction enzyme digestion step that could cut internal sites within the insert. The ligation step is usually carried out for 2–3 d at 4°C with supplemental fresh ligase. Once the adaptors have been added to the fragments, the cohesive ends must be phosphorylated to allow for ligation to the vector. The final step before vector ligation is size selection of the DNA fragments. One can construct a sizing column (see Methods below) or purchase a suitable column such as Invitrogen's cDNA Size Fractionation Columns (cat. no. 18092-015). The fractions are collected, and a portion of each is separated on an agarose gel to visualize the sizes for pooling. All fractions, except those containing free adaptors or very small fragments, can be combined, or two or more libraries can be constructed with the larger and/or smaller fragments. Finally, the ligated library can be packaged with an appropriate packaging extract (i.e., one that lacks restriction activity that would degrade methylated DNA) and introduced into an appropriate host (i.e., one containing the F' episome and that will not degrade methylated DNA).

1.2. Antisera

Selection of antiserum for a screen will depend on the desired characteristics of the final diagnostic assay. If one is using patient serum, one must have sufficient background information on the patient or group of patients. For example, is the disease acute or chronic, latent or active, and how long has the patient been infected? Has the patient been treated with antibiotics or immunosuppressants? Information on the infectious organism is also useful. For example, does the organism pass through multiple developmental stages within the host or does the organism undergo antigenic variation of surface-exposed molecules? This information is necessary because characteristics of the pathogen and the host response will change over time and with treatment.

An appropriate animal model for the infectious disease in question can also be used. Serial bleeds from an infected animal can be used to screen a library to detect antigens that are recognized at various time points in the infection. Antigens with specific characteristics, such as surface exposure or secretion, can be obtained by screening a library with sera from animals immunized with fractionated cells. Antisurface antisera can be obtained by immunizing with outer membrane cell fractions, enzymatically (e.g., trypsin) cleaved surface antigens, or whole organisms (fixed or unfixed). Secreted antigens can be obtained by immunization with culture supernatant containing secreted-excreted antigens or with serum from genetically matched, immunologically compromised animals, such as SCID mice, that have been infected with the organism of interest.

A diagnostic test with a high sensitivity can sometimes be obtained with several rounds of screening or by screening with specific antisera. The initial screening could include a pool of several patients with a high titer to a lysate of the infectious agent. Subsequent rounds of screening can be performed with serum from patients with little or no antibody titer to the antigens obtained from the first round of screening. Alternatively, if the antigens for use in a serodiagnostic test are planned and known antigens and antibody responses are expected, adsorption of sera with the specific, known antigens or use of selected sera with no reactivity to these antigens may reveal additional important antigens for use in a diagnostic. Specific antibody can be adsorbed from sera by using a phage clone, expressing the desired protein, obtained from a previous screen. This phage clone can be used in place of the negative-pick clone (*see* **Subheading 3.2.**) used to adsorb out anti-*E. coli* antibody. The use of sera from different stages or types of disease (e.g., pulmonary TB and extrapulmonary TB) can be used to identify antigens indicative of a specific disease state. A cocktail of recombinant antigens or a recombinant polyprotein (or synthetic peptides) containing the desired epitopes can then be used to develop a highly sensitive and specific assay *(13,19,20)*.

Sera from both patients and lab animals used for immunizations will contain antibody to enteric *E. coli*. This antibody will cause a high background in the screening process and should be removed. Removal can be accomplished by following several protocols including affinity chromatography or pseudoscreening (see below) or by adding an *E. coli* lysate to the serum to block anti-*E. coli* antibody *(38)*.

1.3. Serological Expression Screening

Although many expression systems are available for screening, the advantages of using a prokaryotic expression system and lambda phage vector far outweigh the disadvantages. Advantages include (1) efficiency in packaging lambda particles and infecting the host is high; (2) libraries are easily amplified, stable, and easy to store; and (3) screening plaques is more efficient than screening plasmid colonies *(33)*. Limitations in expressing a eukaryotic library in a prokaryotic host include the following:

1. A requirement for a bacterial ribosome-binding site (RBS). A fraction of recombinant proteins from the library will be fusion proteins and will be expressed using the RBS of the vector fusion gene (e.g., LacZ). Most insert open reading frames (ORFs) will not be in fusion with the vector protein and can initiate internally, using an insert start codon (ATG) with a purine-rich region immediately upstream to serve as the RBS. We find, in certain screenings, that most positive recombinants will have been cloned as a result of internal initiation.
2. The recombinant protein will not fold properly due to its foreign environment. Many B-cell determinants are conformational and thus will not be recognized. Antiserum can be assayed for reactivity to linear determinants by Western blot analysis using reducing sodium dodecyl sulfate-polyacrylamide gel electrophoresis (SDS-PAGE) *(1)*.
3. The bacterial host will not produce the same secondary modifications to the recombinant proteins, such as glycosylation, and thus the recombinant will not be recognized by antiserum specifically reactive to these modifications. Many proteins, however, will contain multiple determinants and will be recognized by polyclonal antiserum.

The screening process involves plating the expression library on a lawn of host cells (**Fig. 2A**). As the host cells multiply, they are lysed by the lambda phage, producing a transparent plaque. This lysate, along with the recombinant protein, is transferred to a nitrocellulose filter, containing the inducing agent isopropyl thiogalactose (IPTG), which is applied to the surface. Plating the library at a density such that (upon overnight incubation) the entire bacterial lawn is lysed can reduce background reactivity. A 150-mm agar plate can be plated with $2.5–3.0 \times 10^4$ plaque-forming units (PFUs) without reducing signal. Plating at too high a density will reduce signal, because plaque

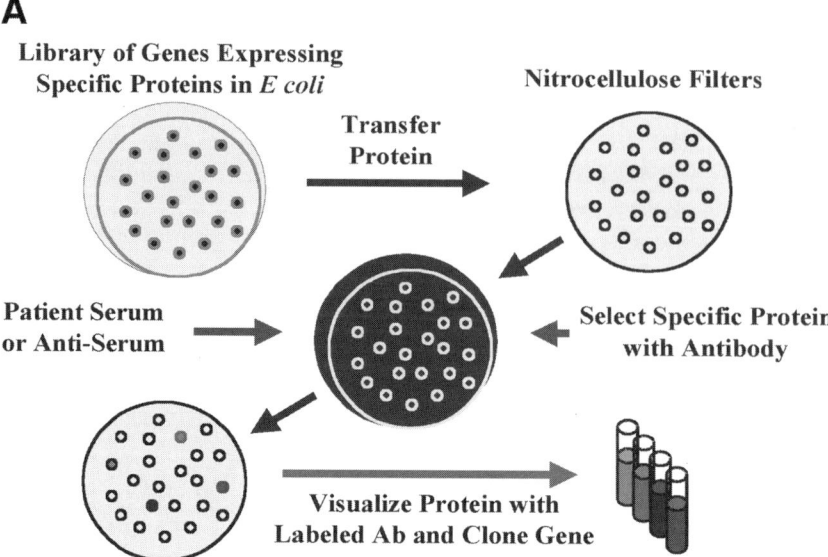

Fig. 2. (**A**) Schematic demonstrating the steps taken to screen an expression library. The upper left figure depicts an LB-agar plate showing bacterial lawn with plaque formation (circles). A filter that has been removed from this plate (upper right) is shown as it progresses through incubation with primary antiserum (middle) and secondary antibody and development (lower left). Tubes shown in the lower right depict the process of cloning and characterization of the genes and proteins that have been identified in the screening process.

growth is limited by available bacterial lawn and will make downstream plaque purification more time-consuming.

When the nitrocellulose filter is removed from the agar plate, it will be treated much like a Western blot filter (**Fig. 2B**). It will be washed in detergent to remove debris, such as bacteria and top agarose, blocked with bovine serum albumin (BSA) or milk protein in detergent and /or 1% Tween 20, washed again, and then incubated with the antiserum. Depending on the expected titer of the antiserum, the blot can be incubated either overnight at 4°C or for 1 h or more at room temperature. The filter is again washed to remove excess antibody and then incubated for 1 h with a secondary antibody that is conjugated with alkaline phosphatase. The secondary antibody can be a combination anti-IgG, IgA, and IgM antibody, or one can use individual secondary antibodies, depending on the desired characteristics of the diagnostic assay. Secondary antibodies from different suppliers will have background characteristics that depend on their purity.

Positive plaques are finally plugged, the phage diluted in buffer, and the entire process repeated until a plaque can be purified and eventually a clone obtained. You should not need to titer each positive pick if your screening conditions are kept constant.

Expression Cloning

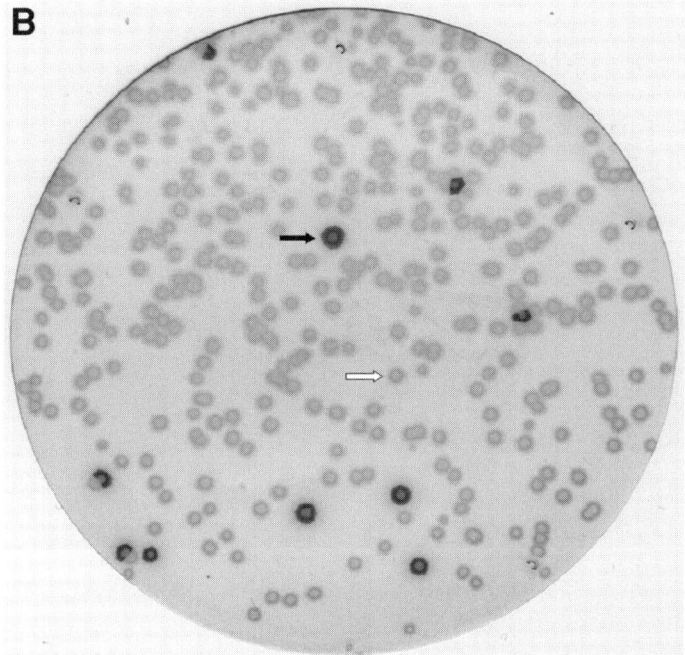

Fig. 2. **(B)** Example of a developed nitrocellulose filter from the secondary purification step of an expression screen. The filter contains both positive (solid arrow) and negative plaques (open arrow) that are evident after development with alkaline phosphatase substrate. Note that some positive plaques, when adjacent to negative plaques, are half-circles. This indicates that plaques can only increase in size when there is a growing bacterial lawn available for infection.

2. Materials
2.1. Construction of a Random Shear, Genomic Expression Library

1. Lambda ZAP Express vector (cat. no. 239211: Stratagene, La Jolla, CA).
2. Gigapack III Gold packaging extract (cat. no. 200201: Stratagene).
3. *Eco*RI adaptors (cat. no. 901110: Stratagene).
4. Sonicator (Sonic Dismembrator 60, Fisher Scientific, Pittsburgh, PA).
5. TE buffer: 10 mM Tris-HCl, 1 mM EDTA, pH 7.5.
6. 10X STE buffer: 1 M NaCl, 200 mM Tris-HCl, pH 7.5, 100 mM EDTA.
7. 10X ligase buffer: 500 mM Tris-HCl (pH 7.5), 70 mM MgCl$_2$, 10 mM dithiothreitol (DTT).
8. 10 mM dNTP mix, 10 mM rATP (Invitrogen, or other supplier).
9. Sephacryl S400 HR (Sigma, Saint Louis, MO).
10. SM buffer: 5.8 g of NaCl, 2.0 g of MgSO$_4$, 50 mL of 1 M Tris-HCl, pH 7.5, 5 mL of 2% gelatin; bring to 1 L with dH$_2$O.
11. Enzymes: T4 DNA polymerase (5 U/μL), T4 DNA ligase (4 U/μL), T4 polynucleotide kinase (10 U/μL), *Eco*RI endonuclease (10 U/μL).
12. Phenol/chloroform/isoamyl alcohol (25:24:1; P/C/I; Invitrogen, cat. no. 15593-031).
13. Glycogen: 20 mg/mL.

14. Host cells/helper phage: XL1-Blue MRF'; XLOLR; ExAssist helper phage (provided with vector kit).
15. Luria-Bertani (LB) broth: 10 g NaCl, 10 g tryptone, 5 g yeast extract; add dH_2O to 1 L, adjust to pH 7.0, and autoclave (for soy medium, substitute 10 g soy hydrolysate for tryptone).
16. LB agar: 1 L of LB broth with 20 g agar; autoclave.
17. LB top agarose: add to 1 L of LB broth 7.0 g agarose and autoclave.
18. 2X YT medium: 5 g sodium chloride, 10 g yeast extract, and 16 g tryptone per liter, pH 7.5 (for soy medium, substitute 16 g soy hydrolysate for tryptone).
19. Reagents for plasmid DNA preparation (Qiaprep spin miniprep kit, cat. no. 27106; Qiagen).
20. Antibiotics: kanamycin at 50 mg/mL, Sigma, cat. no. K-4000; tetracycline hydrochloride at 12.5 mg/mL, Sigma cat. no. T-3383; filter-sterilized.
21. Dimethyl sulfoxide (DMSO); Sigma, cat. no. D-8779.

2.2. Adsorption of Antisera with E. coli Proteins

1. Leupeptin at 10 mg/mL in dH_2O (store at $-20°C$)
2. Gentamycin sulfate: Invitrogen cat. no. 15710-064 (store at $4°C$).
3. 0.1 M PMSF (phenylmethylsulfonyl fluoride) in 100% EtOH (store at $4°C$).
4. 0.5 M EDTA: pH dH_2O to 9.0, dissolve, bring to final vol, pH to 8.0.
5. 5.0% sodium azide in dH_2O (store at room temperature). *Poisonous*, handle with care.
6. See **Subheading 2.3.** for additional reagents.

2.3. Serological Expression Screening

1. Nitrocellulose filters: Protran, cat. no. BA85, 137 mm, 0.45 µm pore (cat. no. 10402548); 82.5 mm (cat. no. 10402579; Schleicher & Schuell, Keene, NH).
2. 10X phosphate-buffered saline (PBS): 11.5 g sodium phosphate, dibasic Na_2HPO_4, 2.3 g sodium phosphate, monobasic NaH_2PO_4, 85 g sodium chloride. Bring volume to 1 L and autoclave (pH of 1X PBS should be 7.4).
3. Tween 20 (polyoxyethylene-sorbitan monolaurate; Sigma, cat. no. P-1379).
4. Secondary antibodies conjugated with alkaline phosphatase (Rockland Immunochemicals, Gilbertsville, PA; Jackson ImmunoResearch, West Grove, PA; Zymed, South San Francisco, CA; or other supplier).
5. Bromochloroindolyl phosphate/nitroblue tetrazolium (BCIP/NBT) combo: Invitrogen, cat. no. 18280-016
6. Alkaline phosphatase buffer: 100 mL of 1 M Tris-HCl pH 10.0, 20 mL of 5 M sodium chloride, and 5 mL of 1 M magnesium chloride in 1 L dH_2O (pH 9.5).
7. Maltose: 20% in dH_2O, filter-sterilized (Sigma, cat. no. M-5885).
8. 1 M magnesium chloride, autoclaved.
9. 1 M IPTG: filtered and stored at $-20°C$ (Sigma, cat. no. I-6758).
10. Oligonucleotide primers for sequencing.
11. Reagents for SDS-PAGE and Western blotting.

3. Methods
3.1. Random Shear Genomic Lambda ZAP Express Library
3.1.1. Sonication

1. Begin the procedure with 20–50 µg genomic DNA in a 1.5-mL Eppendorf tube and bring the volume to 400 µL with TE buffer.

Expression Cloning

2. Sonicate for 4–10 s, depending on desired fragment size, at 30% power (#6 on a Fisher Sonic Dismembrator), and determine fragment sizes by electrophoresing 20 µL on an agarose gel (*see* **Note 1**).
3. Precipitate the sonicated DNA at –20°C by adding 40 µL 3 M sodium acetate and 1100 µL ethyl alcohol (EtOH). Centrifuge for 1 h at 12,800g, 4°C, and wash the pellet with 70% EtOH. Dry the DNA pellet, resuspend in 50 µL dH$_2$O, and quantify the DNA.

3.1.2. Blunting

1. Aliquot 10–15 µg of sonicated DNA (up to 70 µL) into a 1.5-mL Eppendorf tube.
2. Add: 20 µL 5X buffer (included with enzyme), 5 µL 10 mM dNTP mix, 2 µL 0.1 M DTT (included with enzyme if necessary), x µL dH$_2$O (100 µL total), 3 µL T4 DNA polymerase; 5 U/µL).
3. Incubate for 30 min at 37°C in a water bath (*see* **Note 2**).
4. Add 100 µL of TE buffer and extract by adding 200 µL P/C/I. Mix, centrifuge for 2 min, and transfer the top aqueous layer to a new tube. Add 200 µL chloroform, mix, centrifuge for 2 min, and transfer the top aqueous layer to new tube.
5. Precipitate by adding 20 µL of 3 M sodium acetate, 600 µL EtOH, incubate at –20°C, and centrifuge for 1 h at 12,800g, 4°C. Wash the pellet once with 70% EtOH and dry the pellet.

3.1.3. Adaptors

1. Resuspend the dry pellet in 12 µL *Eco*RI adaptors (0.4 µg/µL; Stratagene, cat no. 901110) and incubate at 4°C for 30 min.
2. Add 1.5 µL 10X ligase buffer, 1 µL 10 mM rATP, 1 µL T4 DNA ligase (4 U/µL), and incubate at 4°C overnight.
3. Add an additional 0.5 µL of rATP/ligase mix (1:1) and incubate at 4°C overnight.
4. Heat-inactivate ligase for 30 min at 70°C in a water bath or heat-block.
5. Phosphorylate ligated adaptor ends by adding: 1 µL 10X ligase buffer; 2 µL 10 mM rATP; 6 µL d H$_2$O; 1 µL T4 polynucleotide kinase (PNK; 10 U/µL). Incubate for 30 min at 37°C.
6. Heat-inactivate PNK for 30 min at 70°C in a water bath or heat block.
7. Extract by adding 100 µL 1X STE and 100 µL P/C/I. Mix, centrifuge for 2 min, and transfer upper aqueous layer to new tube. Add 100 µL chloroform, mix, centrifuge for 2 min, and transfer upper aqueous layer to new tube.
8. Precipitate overnight at –20°C by adding 10 µL of 10X STE and 250 µL EtOH. Centrifuge for 1 h at 4°C at 12,800g, wash pellet with 70% EtOH, dry, and resuspend in 100 µL 1X STE.

3.1.4. Size Selection

1. Construct a Sephacryl S400 HR column in a 1-mL syringe by plugging with cotton, packing with Sephacryl, placing in a Falcon 2059 tube and washing three times with 500 µL of 1X STE at 688γ for 2 min. Place a 1.5-mL Eppendorf tube (with lid removed) in the Falcon tube and replace the column (*see* **Note 3**).
2. Add the sample to the column, centrifuge at 688g for 2 min, and collect fraction 1.
3. Add 50 µL 1X STE, centrifuge at 688g for 2 min, and collect fraction 2.
4. Continue until fraction 6 or 7 is collected.
5. Visualize fractionated DNA on an agarose gel (3 µL each) and pool appropriate fractions. Avoid fractions that contain free adaptors and small fragments (<100 bp).
6. Precipitate by adding 2 µL glycogen (20 mg/mL), 2.5 vol EtOH, and incubate at –20°C.

7. Centrifuge for 1 h at 17,500g, 4°C, wash once with 70% EtOH, dry, resuspend in 12 μL dH$_2$O, and quantify DNA (dilute 1 μL in 80 μL dH$_2$O and determine absorbance at 260 nm).

3.1.5. Ligation to Lambda ZAP Express Arms

1. Add 100–300-ng insert in up to 2.5 μL dH$_2$O to a 0.5 μL-Eppendorf tube. Add 0.5 μL 10X ligase buffer, 0.5 μL 10 mM rATP, × μL dH$_2$O (5 μL final vol), 1.0 μL lambda arms (1 μg/μL), 0.5 μL T4 DNA ligase.
2. Mix, incubate overnight at 4°C, and then add 0.5 μL ligase/rATP (1:1).
3. Incubate overnight at 4°C and repeat.

3.1.6. Packaging and Titration

1. Package 2 μL ligation reaction with Gigapack III Gold packaging extract following the manufacturer's recommendations. (Store at 4°C for up to 3 mo.)
2. Titration: (*see* Screening section below for host cell preparation: **Subheadings 3.2.** and **3.3.**).
   Place 200 μL XL1-Blue MRF' cells (OD$_{560}$ = 0.6 in 10 mM MgSO$_4$) in a 1.5-mL tube.
   Add primary library diluted in SM buffer (1, 0.1, 0.01, and 0.001 μL).
   Incubate in a water bath at 37°C for 15 min.
   Add to 4 mL Top Agar at 48°C.
   Spread on 100-mm LB agar plates predried and prewarmed to 42°C.
   Allow agarose to solidify, invert plates, and incubate overnight at 37°C.
   Count plaques and determine PFU per mL SM buffer.

3.1.7. Determination of Library Quality

1. Randomly select 20–50 plaques from titer plates by plugging with a Pasteur pipet (small end) and place into a 1.5-mL tube containing 500 μL SM buffer and 20 μL chloroform. (Store at 4°C overnight or at least 1 h at room temperature; *see* **Note 4**).
2. Excise plasmid following manufacturer's protocol (Stratagene).
3. Culture a single colony in LB broth containing kanamycin at 50 μg/mL (pBK-CMV).
4. Prepare plasmid DNA using a standard kit (Qiagen Qiaprep spin miniprep kit, cat. no. 27106).
5. Digest an aliquot of plasmid DNA with *Eco*RI and visualize inserts on an agarose gel.

3.1.8 Amplification of Primary Library

1. Place 600 μL XL1 Blue MRF' cells at OD$_{560}$ = 0.6 into 1.5-mL tubes (10–20).
2. Add aliquots of primary library containing 5 × 10^4 PFU.
3. Incubate at 37°C for 15 min in a water bath.
4. Add contents of each tube to 8 mL top agarose at 48°C and spread on 150-mm LB agar plates predried and prewarmed to 42°C.
5. Incubate for 6 h at 42°C or until plaques are about 1 mm in size.
6. Add 8 mL of SM buffer to each plate and incubate overnight at 4°C (rocking).
7. Remove and pool SM buffer, add 5.0% chloroform, and incubate 15 min at room temperature.
8. Centrifuge for 15 min at 688g and transfer supernatant to a new tube.
9. Add 0.3% chloroform and 7.0% DMSO; aliquot and store at −80°C.

Expression Cloning

3.2. Adsorption of Antisera with E. coli Proteins

1. Streak a fresh LB agarose-tetracyclin plate (12.5 µg/mL) with XL1-Blue MRF' glycerol stock and incubate at 37°C overnight.
2. Start a 50-mL, single-colony overnight XL1-Blue MRF' culture in LB broth containing 0.2% maltose and 10 mM MgSO$_4$ (500 µL each of filtered 20% maltose and 1 M MgSO$_4$ in 50 mL LB) and incubate at 30°C in a shaking incubator.
3. Centrifuge culture in a 50-mL conical tube at 1876g for 15 min and resuspend bacterial pellet in 30 mL of 10 mM MgSO$_4$ (stock solution stored at 4°C).
4. Dilute bacterial stock solution with 10 mM MgSO$_4$ to an OD$_{560}$ of 0.6 (working solution).
3. Place 600 µL XL1-Blue MRF' cells at OD$_{560}$ = 0.6 (working solution) in 1.5-mL tubes (× 5–10 tubes). Add X µL of negative-pick phage (Lambda ZAP Express vector without insert) to yield approximately 25–30,000 PFU/plate. Incubate at 37°C for 15 min and add to 8.0 mL melted top agarose at 48°C. Pour on 5–10 prewarmed 150-mm LB agar plates and incubate at 42°C for 4–6 h or until plaques begin to form. Place nitrocellulose filters on plates and incubate at 37°C overnight (*see* **Note 5**).
4. Wash filters three times for 10 min each with PBS/0.1% Tween 20 (PBST), block filters with PBS/1.0% Tween 20 for 1 h, and then wash filters three times for 10 min with PBST. Mix the following solutions and pour into labeled 150-mm Petri dishes. If less serum is used, adjust the volume of PBST.

Serum	Follow-up wash	Reagent
100 µL	100 µL	0.5 M EDTA
200 µL	200 µL	0.1 M PMSF
100 µL	100 µL	10 mg/mL leupeptin
100 µL	100 µL	10 mg/mL gentamycin
7.5 mL	9.5 mL	PBST
2.0 mL	—	Serum
10.0 mL	10.0 mL	

5. Place a single filter into the Petri dish containing the serum dilution, lysate side up, and rock for 40 min at room temperature. Remove first filter, transfer to follow-up wash for 40 min, add a new filter to serum, and rock for 40 min. Repeat until all filters have been added to both serum and follow-up wash. The combined serum and follow-up wash will have a final dilution of 1:10 and can be aliquoted and stored at –80°C. After dilution of the serum with 0.1% PBST for screening, add sodium azide to a final volume of 0.05% and store at 4°C [*see* **Note 6**: alternative methods *(38)*].

3.3. Serological Expression Screening

3.3.1. Primary Screen

1. Add 600 µL XL1-Blue MRF' cells in 10 mM MgSO$_4$ at OD$_{560}$ = 0.6 (from an overnight LB culture containing 0.2% maltose, 10 mM MgSO$_4$; see working solution **Subheading 3.2.4.**) to 5–10 1.5-mL Eppendorf tubes and add X µL of primary or amplified library to each tube to produce 20–30,000 PFU/plate (*see* **Note 7**).
2. Incubate tubes for 15 min at 37°C and then add to 8 mL top agarose at 48°C.
3. Pour onto 150-mm LB agarose plates prewarmed to 42°C and incubate at 42°C for approximately 5–6 h or until plaques begin to form.
4. Label X nitrocellulose filters with a permanent marker or pencil and then immerse filters in 10 mM IPTG in dH$_2$O and let dry on filter paper.

5. Add labeled filters to plates and mark in three or five locations at the edge with a hot needle (18-gauge needle heated to red with a Bunsen burner).
6. Invert plates and incubate overnight in an incubator at 37°C.
7. Cool plates at 4°C for 15 min and remove filters with fine forceps, being careful not to remove top agarose with the filter (*see* **Note 8**).
8. Wash filters three times in PBST and then block filters for 1 h in 1% BSA in PBST. A second 1-h block with 1% Tween 20 in PBS can help reduce a high background.
9. Wash filters three times in PBST and then incubate with primary antibody dilution overnight at 4°C with rocking. Alternatively, the filters can be incubated at room temperature for 1 h or more if the serum titer is high (*see* **Note 9**). Incubate one or two filters in approx 15–20 mL diluted serum in 150 × 20-mm Petri dishes.
10. Again wash filters three times for 30 min in PBST and then incubate with secondary antibody conjugated with alkaline phosphatase for 1 h (1:1000–1:10,000 dilution in PBST, following the manufacturer's instructions).
11. Wash filters three times in PBST and then wash filters twice for 10 min in alkaline phosphatase buffer, pH 9.5.
12. Develop filters with NBT/BCIP (Invitrogen) following the manufacturer's instructions and then dry on blotting paper. (Positive spots will become purple upon development.)
13. Plug any positive plaques with the large end of a Pasteur pipet and place the plug in 1 mL SM buffer containing 30 µL of chloroform. Align filters and plates by first circling positive spots on damp filters with a ballpoint pen. Turn filters over, trace circle, and then allow filters to dry completely. Place the plate on the respective filter, align holes in filter and plate, and then plug the agar over the circled positive plaque. Use a sterile toothpick to remove the plug and place it in SM buffer. Store plugs and plates at 4°C and do not freeze, as this will destroy the phage.

3.3.2. Secondary Screen (Plaque Purification)

1. Place 200 µL working dilution XL1-Blue MRF' cells at $OD_{560} = 0.6$ into a 1.5-mL Eppendorf tube.
2. Add 2–5 µL of phage diluted 1:500 in SM buffer (from the large plug in 1 mL SM buffer). Phage can also be titered so that approximately 100–200 PFU can be plated on each plate.
3. Incubate tubes for 15 min at 37°C and add to 4 mL top agar at 48°C.
4. Pour onto 100-mm LB agar plates prewarmed to 42°C and spread.
5. Invert plates and incubate at 42°C for approx 5–6 h or until plaques begin to form.
6. Prewet labeled nitrocellulose filters in 10 m*M* IPTG and dry on filter paper.
7. Add labeled filters to plates and mark with a hot needle.
8. Incubate plates overnight at 37°C.
9. Process filters as in **Subheading 3.3.1.7.** above, except plug secondary picks with the small end of a Pasteur pipet and place plugs in 500 µL SM buffer containing 20 µL chloroform (*see* **Note 4**).

3.3.3. Tertiary Screen (Plaque Purification)

1. If a tertiary plaque purification is necessary, add 200 µL of working dilution XL1-Blue MRF' cells at $OD_{560} = 0.6$ to a 1.5-mL Eppendorf tube. (Tertiary purification is necessary only if a clean plaque cannot be selected from the secondary purification plate).
2. Add 5 µL of phage, diluted 1:250 in SM buffer (small plug in 0.5 mL SM buffer from the secondary purification).
3. Incubate for 15 min at 37°C and add to 4 mL top agar at 48°C.

Expression Cloning

4. Pour onto 100-mm LB agar plates prewarmed to 42°C and spread.
5. Incubate at 42°C for approx 5–6 h.
6. Continue as in **Subheading 3.3.2.6.**

3.3.4. Excision of Plasmid

1. Follow the Stratagene Lambda ZAP Express vector protocol for excision of phagemid using the host strains XL1-Blue MRF' and XLOLR and the ExAssist helper phage.
2. Prepare plasmid DNA following **Subheading 3.1.7.3.**
3. Sequence DNA using the vector primers BK Reverse or the T3 20-mer primer and M13-20 or the T7 22-mer primer and specific internal primers.

3.3.5. Induction of Protein from Plasmid Clones

1. Transform cloned plasmid DNA into competent XL1-Blue host cells and plate on LB agar-kanamycin plates.
2. Pick multiple colonies and grow 10 mL overnight cultures in LB-kanamycin medium.
3. Start fresh 10 mL 2X YT-kanamycin cultures with 100–200 µL of overnight culture.
4. When cultures have grown to an OD_{560} of 0.4–0.6, remove 1 mL of culture (T0 sample).
5. Centrifuge T0 sample and resuspend pellet in X µL of 1X SDS-PAGE buffer (X = OD_{560} × 150 µL).
6. Induce cultures with X µL IPTG (X = mL 2X YT × 2 µL 1 M IPTG).
7. Remove 1 mL T1–T6 (hour) samples, centrifuge, and resuspend as in **step 5** above.
8. Resolve proteins by SDS-PAGE, stain with Coomassie Blue, and transfer to nitrocellulose if necessary for Western blots.

4. Notes

1. Because sonicators may vary, begin with a test of the genomic DNA of interest or a genomic DNA that is similar in nucleotide content (e.g., AT- or GC-rich). Sonicate at 30–50% power for 5 s; remove a 20-µL sample, and continue to sonicate for an additional 5 s. Visualize these samples plus a DNA ladder on an agarose gel. Choose the suitable sonicator settings and time for the desired fragment sizes. An alternative mechanical shearing process involves passaging the genomic DNA sample through a fine-gauge needle. Size assessment is performed after several hundred passages.
2. We have successfully used T4 DNA polymerase and DNA pol I or the Klenow fragment in combination with pfu polymerase for the blunting reaction. If two sequential blunting steps are used, a buffer that is compatible with both enzymes, such as Stratagene's Universal Buffer (0.5X), should be used. The pfu polymerase reaction should be supplemented with dNTP mix and incubated at 72°C for no more than 30 min.
3. Size selection of insert DNA can be accomplished in several ways including construction of a column with Sephacryl or purchasing an appropriate column such as Invitrogen's cDNA Size Fractionation Columns (cat. no. 18092-015). Two important considerations are (1) small fragments and unligated adaptors must be removed before ligation to vector; and (2) the precipitation step following fraction collection and pooling must be accomplished without additional salt (1X STE) as sodium ions will interfere with the ligation step. The precipitation of DNA can be enhanced by the addition of glycogen.
4. Positive plaques can be located on the plate by circling all positive spots on the moist filter with a ballpoint pen. Invert the filter and recircle all positives on the backside. The agar plates can now be placed on the dry filters and the positioning marks aligned. For primary picks, plug the agar with the large end of a Pasteur pipet and remove the plug

with a sterile toothpick. For secondary and tertiary picks, use the small end of a Pasteur pipet attached to a pipet pump (VWR cat. no. 53502-233). Slowly draw in the plug and then, if the plug cannot be expelled with the pump, draw in enough SM buffer to cover the plug and then expel into the Eppendorf tube. Depending on alignment accuracy, several plugs may be taken in the same area, creating a 3–4-mm plug. This larger plug increases the chance of recovery of the positive plaque, but dilutes the percentage of positive plaques, increasing the possibility of having no positives on a secondary screen, especially if too few PFUs are plated.

5. The Negative Pick lambda phage, used to lyse the *E. coli* lawn for serum adsorption, is a lambda phage vector with either no insert or an irrelevant or nonexpressing insert. This phage can usually be obtained when checking the library for quality (*see* **Subheading 3.1.7.**). If an additional Negative Pick phage is needed, it can be amplified following the protocol in **Subheading 3.1.8.** If one needs to adsorb a specific antibody from a serum sample, the Negative Pick phage can be substituted with a specific phage clone that expresses the protein of interest. In this case, the nitrocellulose filters must be prewetted with 10 mM IPTG to induce the production of recombinant protein.

6. Anti-*E. coli* antibody in your antisera will produce a high background staining during screening and should be removed or blocked. Alternate methods include affinity chromatography or the addition of *E. coli* lysate to the serum to block anti-*E. coli* antibody. Affinity chromatography can be accomplished by binding an *E. coli* lysate to cyanogen bromide-activated Sepharose 4B *(38)*. An *E. coli* lysate can be produced by freeze–thawing a bacterial pellet followed by sonication *(38)*.

7. When plating your library for the primary screen, an optimal number of PFU (20–30,000 PFU/150-mm Petri dish) will prevent downstream problems. Too few PFU can produce background staining upon filter development due primarily to the unlysed bacterial lawn. Too many PFU will make subsequent plaque purification more complicated and can cause the loss of some positive signals, as plaques can increase in size only when a bacterial lawn is present (*see* **Fig. 2B**, plaques at lower left). Also, proteins bind more efficiently to the filter while the lawn is lysing. Once a plaque has formed, protein will not bind effectively to the filter, hence a light center appears in some positive spots on the developed filter (*see* **Fig. 2B**). Plate enough PFU to lyse the lawn completely upon overnight incubation.

8. Under optimal conditions, the bacterial lawn should be opaque and well developed, and the plaques should be clear and well defined. If the lawn is light and/or the plaques are not distinct, or are cloudy or speckled, the problem could be caused by a contaminated XL1-Blue MRF' overnight culture. Restreak the glycerol stock on fresh tetracycline plates and replace sterile water and all media, including LB medium, maltose, and $MgSO_4$.

9. If sera availability is limited, used sera can be stored and reused multiple times. Storage is best at 4°C for periods of up to 2–3 mo when 0.05% sodium azide is added as a preservative. For longer periods, diluted sera can be frozen, but this can result in increased background. Reuse of sera also has the advantage of continuing to reduce background due to the removal of *E. coli*-specific antibody that had not previously been removed by the earlier adsorption step.

References

1. Seed, B. (1995) Developments in expression cloning. *Curr. Opin. Biotechnol.* **6,** 567–573.
2. Young, R. A. and Davis, R. W. (1983) Efficient isolation of genes by using antibody probes. *Proc. Natl. Acad. Sci. USA* **80,** 1194–1198.

3. Mierendorf, R. C., Percy, C., and Young, R. A. (1987) Gene isolation by screening lambda gt11 libraries with antibodies. *Methods Enzymol.* **152,** 458–469.
4. Shastri, N. (1996) Needles in haystacks: identifying specific peptide antigens for T cells. *Curr. Opin. Immunol.* **8,** 271–277.
5. Alderson, M. R., Bement, T., Day, C. H., et al. (2000) Expression cloning of an immunodominant family of *Mycobacterium tuberculosis* antigens using human CD4+ T cells. *J. Exp. Med.* **191,** 551–559.
6. Skeiky, Y. A. W., Ovendale, P. J., Jen, S., et al. (2000) T cell expression cloning of a *Mycobacterium tuberculosis* gene encoding a protective antigen associated with the early control of infection. *J. Immunol.* **165,** 7140–7149.
7. Chien, C.-T., Bartel, P. S., Sternglanz, R., and Fields, S. (1991) The two-hybrid system: a method to identify and clone genes for proteins that interact with a protein of interest. *Proc. Natl. Acad. Sci. USA* **88,** 9578–9582.
8. Simonsen, H. and Lodish, H. F. (1994) Cloning by function: expression cloning in mammalian cells. *Trends Pharmacol. Sci.* **15,** 437–441.
9. Stark, G. R. and Gudkov, A. V. (1999) Forward genetics in mammalian cells: functional approaches to gene discovery. *Hum. Mol. Genet.* **8,** 1925–1938.
10. Steimle, V. and Mach, B. (1995) Complementation cloning of mammalian transcriptional regulators: the example of MHC class II gene regulators. *Curr. Opin. Genet. Dev.* **5,** 646–651.
11. Dillon, D. C., Alderson, M. R., Day, C. H., et al. (2000) Molecular and immunological characterization of *Mycobacterium tuberculosis* CFP-10, an immunodiagnostic antigen missing in *Mycobacterium bovis* BCG. *J. Clin. Microbiol.* **38,** 3285–3290.
12. Lodes, M. J., Dillon, D. C., Mohamath, R., et al. (2001) Serological expression cloning and immunological evaluation of MTB48, a novel *Mycobacterium tuberculosis* antigen. *J. Clin. Microbiol.* **39,** 2485–2493.
13. Houghton, R. L., Lodes, M. J., Dillon, D. C., et al. (2002) Use of multiepitope polyproteins in serodiagnosis of active tuberculosis. *Clin. Diagn. Lab. Immunol.* **9,** 883–891.
14. Lodes, M. J., Mohamath, R., Reynolds, L. D., et al. (2001) Serodiagnosis of human granulocytic ehrlichiosis by using novel combinations of immunoreactive recombinant proteins. *J. Clin. Microbiol.* **39,** 2466–2476.
15. Lodes, M. J., Houghton, R. L., Bruinsma, E. S., et al. (2000) Serological expression cloning of novel immunoreactive antigens of *Babesia microti. Infect. Immun.* **68,** 2783–2790.
16. Burns, J. M. Jr., Shreffler, W. G., Rosman, D. E., Sleath, P. R., March, C. J., and Reed, S. G. (1992) Identification and synthesis of a major conserved antigenic epitope of *Trypanosoma cruzi. Proc. Natl. Acad. Sci. USA* **89,** 1239–1243.
17. Skeiky, Y. A. W., Benson, D. R., Guderian, J., et al. (1993) *Trypanosoma cruzi* acidic ribosomal P protein gene family: novel P proteins encoding unusual cross-reactive epitopes. *J. Immunol.* **151,** 5504–5515.
18. Skeiky, Y. A. W., Benson, D. R., Elwasila, M., Badaro, R., Burns, J. M., and Reed, S. G. (1994) Shared antigens between *Leishmania* and *Trypanosoma cruzi*: characterization of the *Leishmania chagasi* acidic ribosomal protein P0. *Infect. Immun.* **62,** 1643–1651.
19. Houghton, R. L., Benson, D. R., Reynolds, L. D., et al. (1999) A multi-epitope synthetic peptide and recombinant protein for the detection of antibodies to *Trypanosoma cruzi* in radioimmunoprecipitation-confirmed and consensus-positive sera. *J. Infect. Dis.* **179,** 1226–1234.
20. Houghton, R. L., Benson, D. R., Reynolds, L., et al. (2000) Multiepitope synthetic peptide and recombinant protein for the detection of antibodies to *Trypanosoma cruzi* in patients with treated or untreated Chagas' disease. *J. Infect. Dis.* **181,** 325–330.

21. Skeiky, Y. A., Guderian, J. A., Benson, D. R., et al. (1995) A recombinant *Leishmania* antigen that stimulates human peripheral blood mononuclear cells to express a Th1-type cytokine profile and to produce interleukin 12. *J. Exp. Med.* **181**, 1527–1537.
22. Skeiky, Y. A. W., Coler, R. N., Brannon, M., et al. (2002) Protective efficacy of a tandemly linked, multi-subunit recombinant leishmanial vaccine (Leish-111f) formulated in MPL adjuvant. *Vaccine* **20**, 3292–3303; erratum **20**, 3783.
23. Burns, J. M. Jr., Shreffler, W. G., Benson, D. R., Ghalib, H. W., Badaro, R., and Reed, S. G. (1993) Molecular characterization of a kinesin-related antigen of *Leishmania chagasi* that detects specific antibody in African and American visceral leishmaniasis. *Proc. Natl. Acad. Sci. USA* **90**, 775–779.
24. Skeiky, Y. A. W., Benson, D. R., Guderian, J., Bacelar, O., Carvalho, E. M., and Reed, S. G. (1995) Leishmaniasis patient immune responses to *Leishmania* and human heat shock proteins. *Infect. Immun.* **63**, 4105–4114.
25. Bhatia, A., Daifalla, N. S., Jen, S., Badaro, R., Reed, S. G., and Skeiky, Y. A. (1999) Cloning, characterization and serological evaluation of K9 and K26: two related hydrophilic antigens of *Leishmania chagasi*. *Mol. Biochem. Parasitol.* **102**, 249–261.
26. Webb, J. R., Vedvick, T. S., Alderson, M. R., et al. (1998) Molecular cloning, expression, and immunogenicity of MTB12, a novel low-molecular-weight antigen secreted by *Mycobacterium tuberculosis*. *Infect. Immun.* **66**, 4208–4214.
27. Dillon, D. C., Alderson, M. R., Day, C. H., et al. (1999) Molecular characterization and human T-cell responses to a member of a novel *Mycobacterium tuberculosis mtb39* gene family. *Infect. Immun.* **67**, 2941–2950.
28. Skeiky, Y. A. W., Lodes, M. J., Guderian, J. A., et al. (1999) Cloning, expression, and immunological evaluation of two putative secreted serine protease antigens of *Mycobacterium tuberculosis*. *Infect Immun.* **67**, 3998–4007.
29. Sahin, U., Tureci, O., Schmitt, H., et al. (1995) Human neoplasms elicit multiple specific immune responses in the autologous host. *Proc. Natl. Acad. Sci. USA* **92**, 11,810–11,813.
30. Frischauf, A. M., Lehrach, H., Poustka, A., and Murray, N. (1983) Lambda replacement vectors carrying polylinker sequences. *J. Mol. Biol.* **170**, 827–842.
31. Murray, N. E. (1991) Special uses of lambda phage for molecular cloning. *Methods Enzymol.* **204**, 280–301.
32. Chauthaiwale, V. M., Therwath, A., and Deshpande, V. V. (1992) Bacteriophage lambda as a cloning vector. *Microbiol. Rev.* **56**, 577–591.
33. Christensen, A. C. (2001) Bacteriophage lambda-based expression vectors. *Mol. Biotechnol.* **17**, 219–224.
34. Crameri, R. and Kodzius, R. (2001) The powerful combination of phage surface display of cDNA libraries and high throughput screening. *Comb. Chem. High Throughput Screen.* **4**, 145–155.
35. Walter, G., Konthur, Z., and Lehrach, H. (2001) High-throughput screening of surface displayed gene products. *Comb. Chem. High Throughput Screen.* **4**, 193–205.
36. Rhyner, C., Kodzius, R., and Crameri, R. (2002) Direct selection of cDNAs from filamentous phage surface display libraries: potential and limitations. *Curr. Pharmaceut. Biotechnol.* **3**, 13–21.
37. Song, O. and Fields, S. (1989) A novel genetic system to detect protein-protein interactions. *Nature* **340**, 245–246.
38. Sambrook, J., Fritsch, E. F., and Maniatis, T. (1989) *Molecular Cloning, A Laboratory Manual*, 2nd ed. Cold Spring Harbor Laboratory Press, Cold Spring Harbor, NY.

III

IDENTIFICATION OF EPITOPES AND IMMUNOMODULATORY COMPONENTS

7

Determination of Epitopes by Mass Spectrometry

Christine Hager-Braun and Kenneth B. Tomer

Abstract

As a response to an infection, the immune system produces antibodies. The determination of the antigenic structure recognized by the antibody through epitope mapping provides information about the interaction between antigen and antibody for the diagnosis of a disease on a molecular level, for characterizing the pathogenesis of the infectious material, and for the development of interfering drugs or preventative vaccines. Here we present the determination of the fine structure of the linear epitope located on the gp41 protein of the human immunodeficiency virus recognized by the monoclonal antibody 2F5. In this approach we coupled the antigen SOSgp140 to the antibody 2F5, which was covalently linked to an Fc-specific antibody immobilized on cyanogen bromide (CNBr)-activated Sepharose beads. Digestion of the antigen with endoproteinase LysC resulted in an affinity-bound peptide whose fine structure was characterized by digestion with carboxypeptidase Y and aminopeptidase M. All steps of this method were monitored by matrix-assisted laser desorption/ionization mass spectrometry (MALDI/MS). The epitope recognized by 2F5 was identified to be the 16-mer peptide with the sequence NEQELLELDKWASLWN.

Key Words: Epitope mapping; fine structure; linear epitope; secondary Fc-specific antibody; primary antigen-recognizing antibody; human immunodeficiency virus.

1. Introduction

According to the World Health Organization's AIDS Epidemic Update Report for December 2002 *(1)*, approx 42 million people worldwide are infected with the human immunodeficiency virus (HIV). An epidemic like this makes it necessary to investigate the infection and its subsequent steps at the biochemical level in order to understand the process and to find means to prevent or interfere with an infection. As a response to an infection, the immune system develops antibodies directed against the antigenic structures or epitopes. Determination of the epitopes provides invaluable knowledge for understanding the pathogenesis and developing of vaccines against

From: *Methods in Molecular Medicine, vol. 94: Molecular Diagnosis of Infectious Diseases, 2/e*
Edited by: J. Decker and U. Reischl © Humana Press Inc., Totowa, NJ

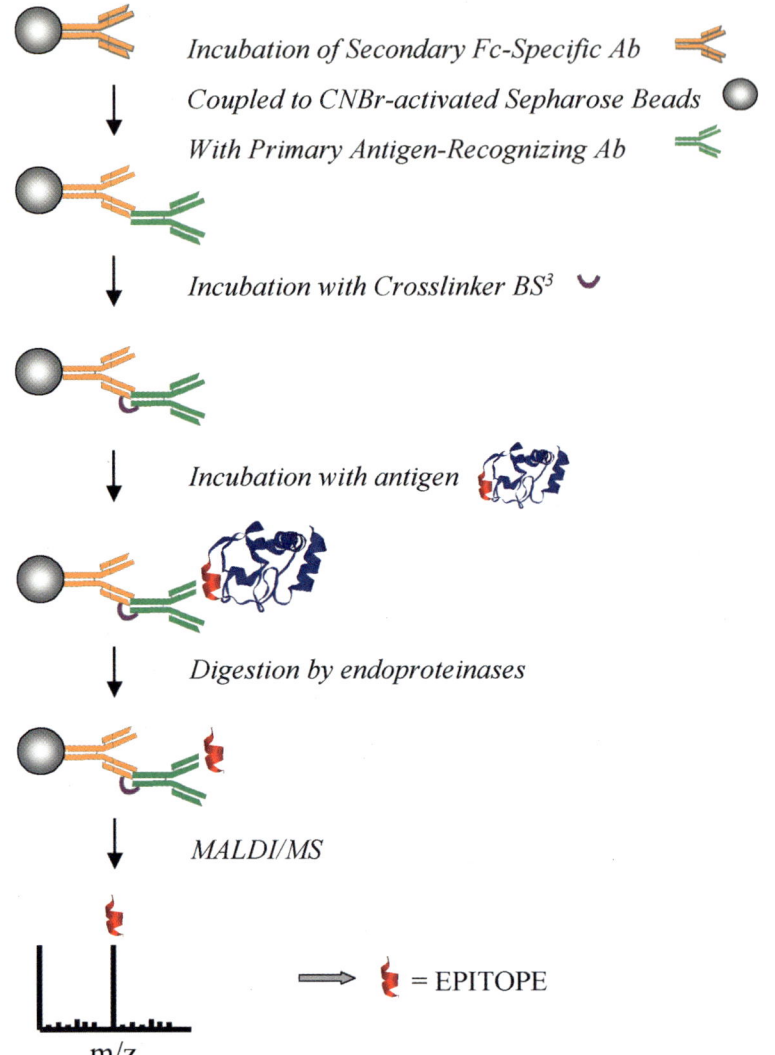

Fig. 1. Schematic outline of the epitope excision method followed by mass spectrometry. Ab, antibody; BS^3, *bis*(sulfosuccinimidly)suberate; CNBr, cyanogen bromide; MALDI/MS, matrix-assisted laser desorption/ionization mass spectrometry.

HIV. Mapping the epitope on the gp41 glycoprotein of the human immunodeficiency virus recognized by the neutralizing monoclonal antibody 2F5 *(2)* illustrates an approach that is also applicable to other antigen/antibody complexes in characterizing immune responses or autoimmune diseases. Briefly, the antigen-recognizing primary antibody is covalently coupled to an immobilized Fc-specific secondary antibody. Subsequently, the antigen is bound to the primary antibody and proteolytically digested. The individual steps are monitored by matrix-assisted laser ionization mass spectrometry (MALDI/MS). The peptide that remains affinity-bound throughout the digestion steps reflects the epitope *(2,3)*. A schematic outline of this approach is depicted in **Fig. 1**.

Although a variety of alternative techniques, such as peptide reactivity assays, mutation analyses, antigen proteolysis, crystallography, or nuclear magnetic resonance (NMR) spectroscopy, are available to determine epitopes *(4–12)*, limitations to these approaches reduce the rate of success. Crystallography requires high concentrations of antigen and antibody and the formation of diffracting crystals. For NMR spectroscopy, the complex of antibody and antigen often exceeds the mass that can be analyzed. Peptide reactivity assays, either based on enzyme-linked immunosorbent assay (ELISA) or surface plasmon resonance imaging, are only applicable to linear epitopes but not to discontinuous epitopes or antigens containing post-translational modifications. Mutation analyses focus only on specific amino acids by point mutations and are, therefore, tedious. Moreover, an altered binding affinity of the antigen to the antibody can be false positive, when a distant mutation results in a conformational change of the antigenic structure.

The method of enzymatic proteolysis of the antibody-bound antigen followed by MALDI/MS analysis as described here allows mapping of linear epitopes with three-dimensional conformation or post-translational modification at a low sample consumption *(2,13–15)*.

2. Materials

1. Cyanogen bromide (CNBr)-activated Sepharose 4B beads (Pharmacia Biotech, Piscataway, NJ).
2. Compact Reaction Columns (CRCs) with 35-μm column filters (USB, Cleveland, OH).
3. 1 mM HCl.
4. 0.1 M NaHCO$_3$, pH 8.2.
5. Coupling buffer: 0.1 M NaHCO$_3$, 150 mM NaCl, pH 8.2.
6. Anti-human Fc-specific IgG antibody (Sigma, St. Louis, MO) as secondary antibody.
7. Quenching buffer: 0.1 M Tris-HCl, pH 8.0.
8. 0.1 M sodium acetate, 0.5 M NaCl, pH 4.0.
9. Monoclonal antibody 2F5, IgG1 isotype, κ-chain (AIDS Research and Reference Reagent Program, Division of AIDS, National Institute of Allergy and Infectious Diseases, National Institutes of Health) as the primary antibody.
10. Phosphate-buffered saline (PBS), PBS, pH 7.2: 100 mM Na-phosphate buffer, 150 mM NaCl, pH 7.2.
11. 10 mM *bis*(sulfosuccinimidyl)suberate (BS^3; Pierce, Rockford, IL) in PBS, pH 7.2.
12. SOSgp140 (JR-FL) glycoprotein, expressed as a gp120 glycoprotein covalently linked to the ectodomain of gp41 (gift from N. Schülke, Progenics Pharmaceuticals, Tarrytown, NY; *16*).
13. Endoproteinase Lys-C (Wako, Dallas, TX).
14. 50 mM Tris-HCl, pH 8.0.
15. Carboxypeptidase Y (Roche Diagnostics, Indianapolis, IN).
16. PBS, pH 6.1: 100 mM Na-phosphate buffer, 150 mM NaCl, pH 6.1.
17. Aminopeptidase M (Roche Diagnostics).
18. Saturated solution of recrystallized α-cyano-4-hydroxycinnamic acid in ethanol/water/concentrated formic acid 45/45/10 (v/v/v).

3. Methods
3.1. Preparation of Immobilized Antibody Columns
1. Fill approx 0.2 g of dry CNBr-activated Sepharose 4B beads into a Falcon tube.
2. Mix the beads gently with 10 mL 1 mM HCl (*see* **Note 1**). Let the slurry equilibrate at room temperature for about 15 min, and then decant most of the supernatant.
3. Transfer 20 μL of beads slurry into each of two compact reaction columns (CRC).
4. Wash both CNBr-activated Sepharose columns with 6X 0.8 mL 1 mM HCl and 6X 0.4 mL 0.1 M $NaHCO_3$, pH 8.2.
5. To the drained beads in each CRC add 80 μL of 0.1 M $NaHCO_3$/150 mM NaCl, pH 8.2, and 20 μL of anti-human Fc-specific secondary antibody (48 μg) (*see* **Note 2**). Incubate at room temperature with slow rotation for 1.5 h.
6. Drain the beads and block any unreacted sites with 0.1 M Tris-HCl, pH 8.0, by rinsing the beads once with 0.4 mL and then incubating them in 0.4 mL for 1 h at room temperature with slow rotation.
7. Remove any antibody not covalently linked to the beads by a sequence of washing steps thereby alternating between 0.1 M sodium acetate/0.5 M NaCl, pH 4.0, and 0.1 M Tris-HCl, pH 8.0 with a volume of 0.4 mL each for three times. Finally, wash three times with 0.4 mL PBS.
8. Obtain a MALDI spectrum of the beads. Essentially, no ions should be detected at this point as the secondary antibody is covalently linked to the Sepharose beads (**Fig. 2A**).
9. Add 50 μL equivalent to 50 μg of the primary antibody 2F5 to one of the anti-human Fc-specific secondary antibodies containing CRC (= vial "Fc-specific Ab + 2F5"). To the other CRC, which will serve as the control (= vial "Fc-specific Ab"), add 50 μL PBS. Incubate both CRCs for 1 h at room temperature with slow rotation.
10. Drain the beads and save the drain in case the yield of captured primary antibody is low and a longer incubation time would be necessary.
11. Wash the beads three times with 0.4 mL PBS.
12. Check a 0.5 μL-aliquot of the beads by MALDI/MS to determine whether or not the primary antibody was successfully affinity-bound to the secondary antibody (**Fig. 2B**; *see* **Note 3**).
13. To stabilize the complex between the anti-human Fc-specific secondary antibody and the 2F5 primary antibody, the proteins are crosslinked. For this, add 10 μL of freshly prepared 10 mM BS^3 in PBS to vial "Fc-specific Ab + 2F5" and treat the control vial "Fc-specific Ab" the same way. Incubate both CRCs for 45 min at room temperature in the dark with slow rotation (*see* **Note 4**).
14. To quench the crosslinking reaction, drain the beads and wash twice with 100 μL 0.1 M Tris-HCl, pH 8.0.
15. Wash the beads three times with 0.4 mL PBS and obtain a spectrum of a 0.5 μL-aliquot of the beads by MALDI/MS. As the secondary and the primary antibody in the vial "Fc-specific Ab + 2F5" are now covalently linked, no ions from the antibodies should be observed in the spectrum (**Fig. 2C**).
16. From vial "Fc-specific Ab + 2F5," now crosslinked, transfer about one-fourth of the beads into a new CRC and add 50 μL of PBS. This vial will serve as the control for the proteolysis experiments.
17. Incubate the remaining beads of the vial "Fc-specific Ab + 2F5" with the antigen SOSgp140 using 50 μg of antigen solution (200 μL). Rotate slowly at room temperature for 2 h. In parallel, incubate the control beads without antigen.

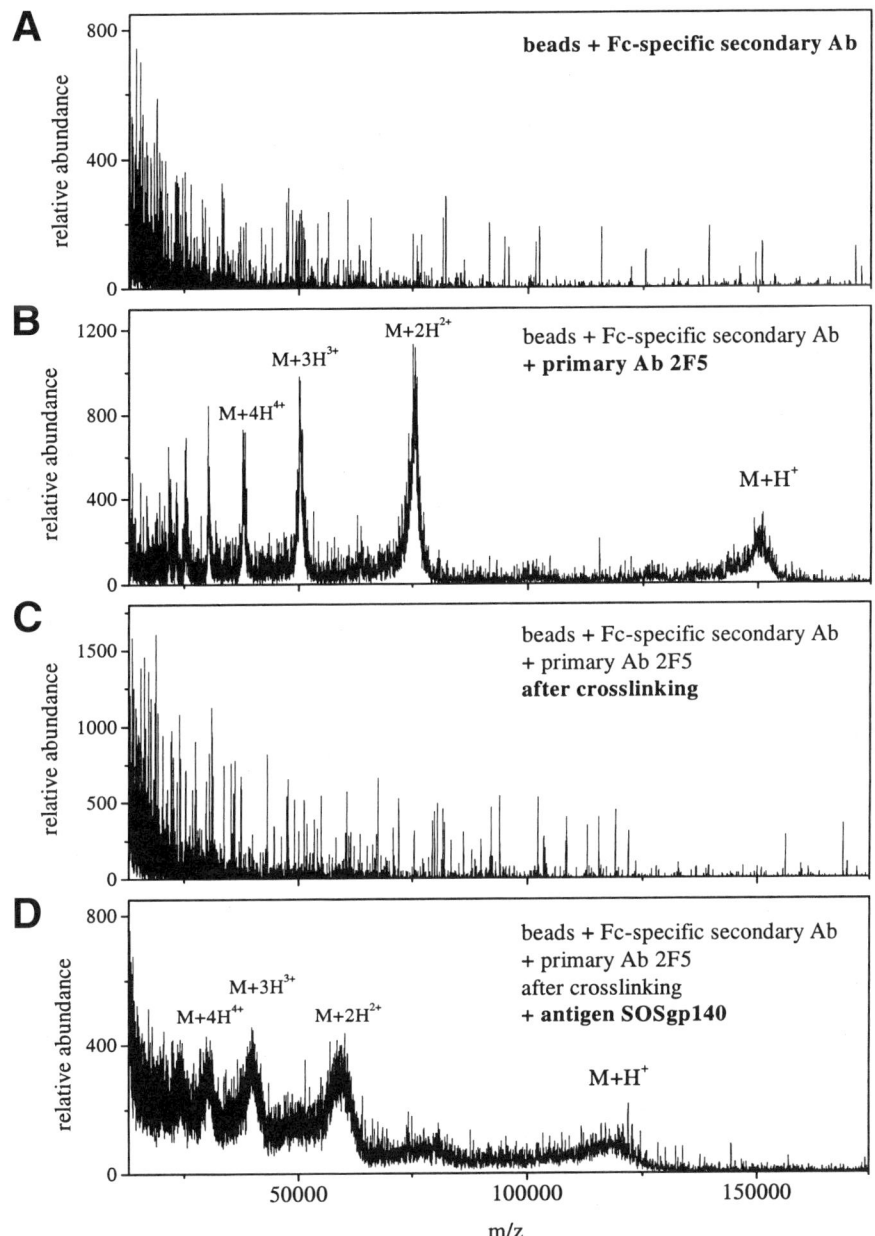

Fig. 2. MALDI/MS spectra of the Fc-specific secondary antibody (Ab) coupled to the CNBr-activated Sepharose beads (**A**) followed by the addition of the primary antibody 2F5 before (**B**) and after (**C**) crosslinking with BS^3. (**D**) MALDI/MS spectrum of the affinity-bound antigen SOSgp140. The spectra were recorded in linear mode. (Adapted with permission from ref. *2*. Copyright 2001 American Society for Microbiology.)

18. Drain the beads and rinse three times with 0.4 mL PBS.
19. Obtain a MALDI/MS spectrum of a 0.5 µL-aliquot of the antigen-containing beads and of the control beads (**Fig. 2D**).

3.2. Proteolysis of the Affinity-Bound Antigen SOSgp140

3.2.1. Endoproteinase LysC

1. Prepare a fresh solution of endoproteinase LysC (0.1 µg/µL) in 50 mM Tris-HCl, pH 8.0, and store it at 0°C (*see* **Note 5**).
2. Resuspend the beads containing the antigen as well as the control beads in 50 µL 50 mM Tris-HCl, pH 8.0 (*see* **Note 6**).
3. Add 50 µL of LysC solution (5 µg) to both CRCs.
4. Incubate both CRCs for 2.5 h at 37°C with slow rotation.
5. Wash the beads three times with 0.4 mL PBS.
6. Analyze a 0.5 µL-aliquot of both the antigen-containing beads and the control beads by MALDI/MS.

3.2.2. Carboxypeptidase Y

1. Prepare a stock solution of carboxypeptidase Y with a concentration of 0.5 µg/µL in deionized water just prior to use and store it on ice.
2. Resuspend the beads containing the antigen as well as the control beads in 50 µL PBS, pH 6.1.
3. Add 50 µL of carboxypeptidase Y solution to both CRCs.
4. Incubate both vials at 37°C with slow rotation.
5. After a 1-min incubation, drain the beads, wash them three times with PBS, pH 6.1, resuspend them in 50 µL PBS, pH 6.1, and remove a 0.5-µL aliquot of the beads for immediate MALDI/MS analysis.
6. Add fresh buffer (50 µL PBS, pH 6.1) and carboxypeptidase Y (50 µL) to the beads, continue incubation, and remove aliquots of washed beads at various time points like 1-min intervals (here: 2, 3, 4, 5, and 6 min) for MALDI/MS analyses. Although during the first 5 min a progress in digestion could be monitored (**Fig. 3**), at 6 min of incubation with carboxypeptidase Y, further digestion products were not detected (spectrum not shown). **Table 1** lists the ions observed in the MALDI/MS spectrum after 5 min of carboxypeptidase Y incubation and the corresponding amino acid sequences.

3.2.3. Aminopeptidase M

1. Wash beads containing the epitope and the control beads three times with 0.4 mL PBS, pH 7.2.
2. Resuspend the beads of both CRCs in 49 µL PBS, pH 7.2.
3. Add 1 µL of 5 µg/µL aminopeptidase M stock solution (as supplied by the manufacturer in ammonium sulfate solution) to both vials.
4. Incubate at 37°C with slow rotation.
5. At various time points like 1-h intervals (here: 1, 4, and 7 h), wash the beads with PBS and remove an aliquot for MALDI/MS analysis (**Fig. 4**). After the analysis, add new buffer and aminopeptidase to the beads and continue incubation. **Table 2** gives an overview of the ions found by MALDI/MS after 7 h of digestion and the assignments to the amino acid sequences.

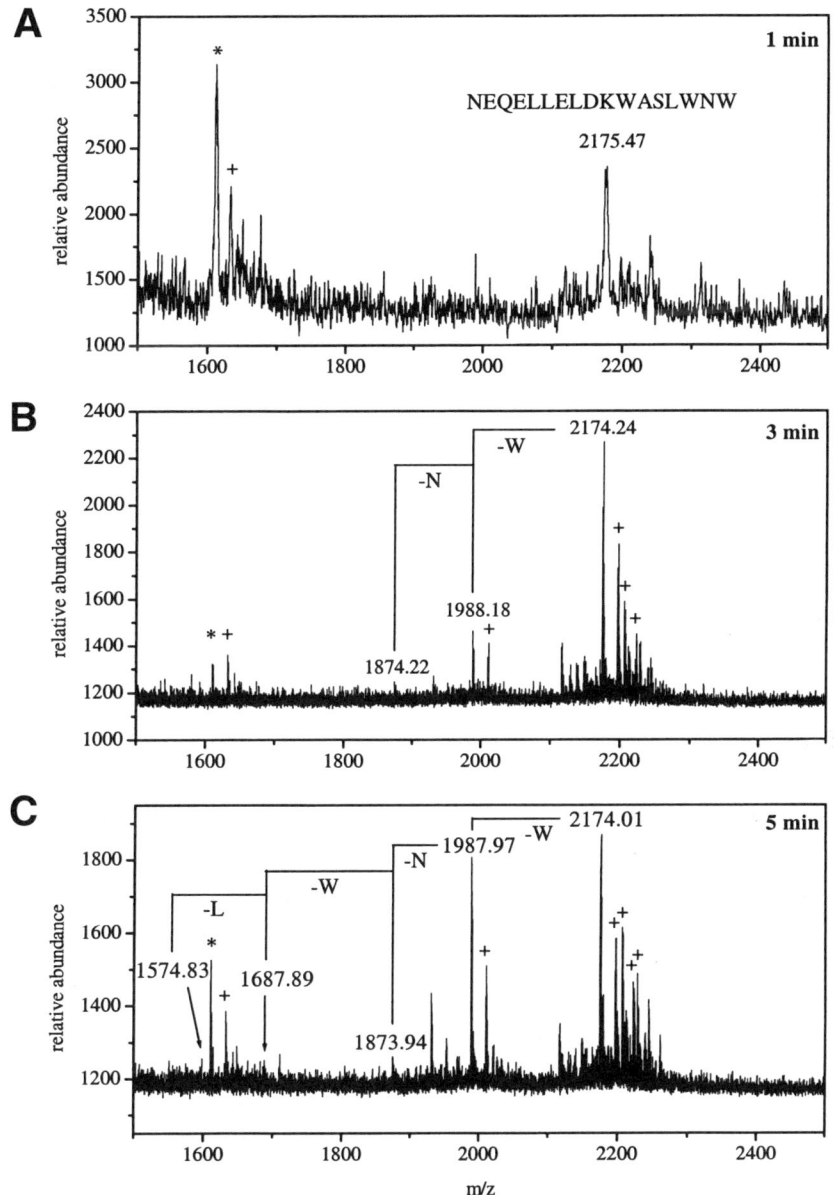

Fig. 3. MALDI/MS spectra obtained from the rinsed beads after digestion of affinity-bound SOSgp140 with endoproteinase LysC for 2.5 h followed by digestion with carboxypeptidase Y for 1 min (**A**), 3 min (**B**), and 5 min (**C**) to determine the fine structure of the epitope on the C-terminal end. Spectrum A was recorded in linear mode and spectra B and C in reflector mode of the MALDI instrument. The ion labeled with an asterisk represents a background ion, and ions labeled with a + represent Na^+-adducts and/or oxidized ions. (Adapted with permission from ref. *2*. Copyright 2001 American Society for Microbiology.)

Table 1
Products Found after Digestion of the Affinity-Bound SOSgp140 with LysC for 2.5 h Followed by Carboxypeptidase Y for 5 Min as Detected by MALDI/MS[a]

Amino acid sequence	Calculated M+H⁺	Measured M+H⁺
NEQELLELDKWASLWNW	2174.06	2174.01
NEQELLELDKWASLWN	1987.98	1987.97
NEQELLELDKWASLW	1873.94	1873.94
NEQELLELDKWASL	1687.86	1687.89
NEQELLELDKWAS	1574.77	1574.83

[a]*See* **Fig. 3C**.

Table 2
Products Found after Digestion of the Affinity-Bound SOSgp140 with LysC for 2.5 h Followed by Carboxypeptidase Y for 6 Min and Aminopeptidase M for 7 h as Detected by MALDI/MS[a]

Amino acid sequence	Calculated M+H⁺	Measured M+H⁺
NEQELLELDKWASLW*	1873.94	1874.17
NEQELLELDKWASLWN*	1987.98	1988.23
NEQELLELDKWASLWNW*	2174.06	2174.30
EQELLELDKWASLWN	1873.94	1874.17
EQELLELDKWASLW	1759.89	1760.16

[a]*See* **Fig. 4B**. The amino acid sequences marked with an asterisk were already detected after the carboxypeptidase Y digest.

3.3. Mass Spectrometric Analysis

A 0.5-μL aliquot of the bead slurry or the drained liquid was spotted on a stainless-steel target and mixed with an equal volume of the saturated α-cyano-4-hydroxycinnamic acid solution. The target was left at room temperature to dry on air. MALDI mass spectra were recorded on a Voyager DE-STR (Applied Biosystems, Framingham, MA). Similar results should be obtained with similar equipment. External calibration of the mass range of interest was used.

3.4. Data Interpretation

1. Knowing the amino acid sequence of the antigen under investigation allows the researcher to determine which endoproteinase is best for digestion. A computer-based *in silico* digest gives a list of expected peptides with a molecular weight assigned. Choose the most suitable proteinase based on these *in silico* digest masses to allow MALDI/MS measurements with high mass accuracy on your specific instrument.
2. The amino acid sequence that contains the epitope should be protected by the antibody. Cleavage sites within the antigenic structure should not be accessible to the endoproteinases, and, therefore, only adjacent cleavage sites will be used. To determine the fine structure of the epitope, digestion with carboxypeptidase Y will give information

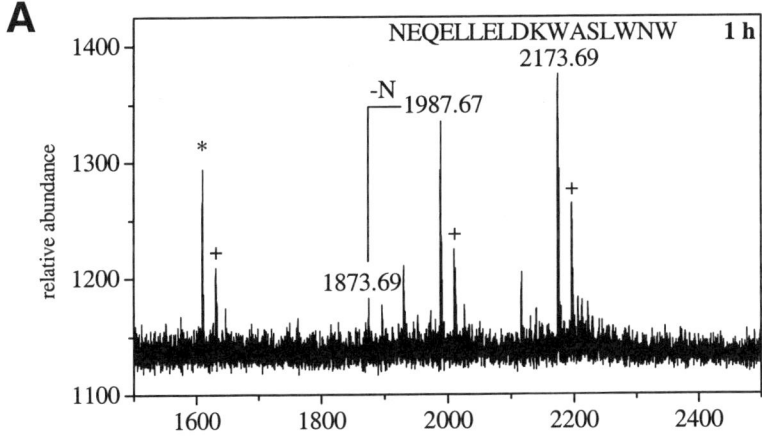
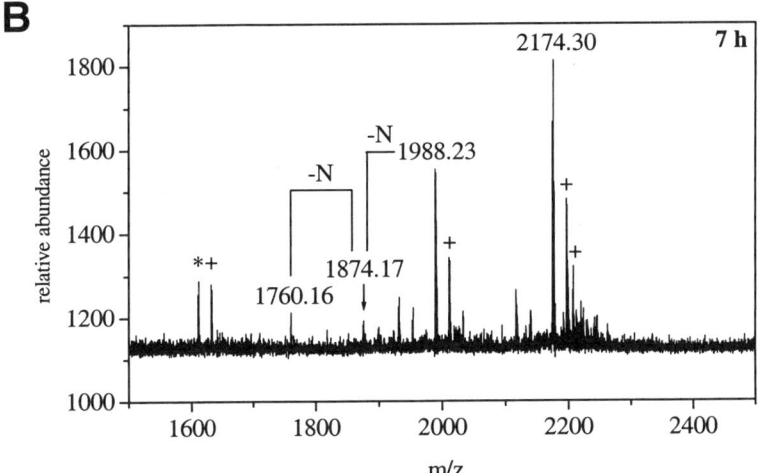

Fig. 4. MALDI/MS spectra obtained from the rinsed beads after digestion of affinity-bound SOSgp140 with endoproteinase LysC for 2.5 h followed by digestion with carboxypeptidase Y for 6 min and subsequent digestion with aminopeptidase M for 1 h (**A**) and 7 h (**B**) to determine the fine structure of the epitope on the N-terminal end. The spectra were recorded in reflector mode. The ion labeled with an asterisk represents a background ion, and ions labeled with a + represent Na⁺-adducts and/or oxidized ions. (Adapted with permission from ref. *2*. Copyright 2001 American Society for Microbiology.)

about the C-terminal end of the epitope, whereas digestion with aminopeptidase M will characterize the N-terminal side. The exoproteinases should only cleave "overhanging" ends of the affinity-bound peptide.

3. The ion detected in the MALDI spectrum after 2.5 h of LysC digestion and 1 min of carboxypeptidase Y digestion could be assigned to the amino acid sequence NEQELL

ELDKWASLWNW (**Fig. 3A**). Further incubation with carboxypeptidase showed the loss of W, N, W, and L from the C-terminus (**Fig. 3B** and **C**). The low abundances of the ions after the loss of N, W, and L, respectively, compared with the high abundance of the ion at m/z 1987.97 after the initial loss of W indicates that amino acids N, W, and L are partially protected by the antibody against proteolysis.

4. Using a heterogeneous sample after the carboxypeptidase Y digest for subsequent characterization of the N-terminus of the epitope results in a relatively complicated interpretation of the MALDI spectra. The ion at 1874.17 in **Fig. 4B** could be the result of the carboxypeptidase Y digest (NEQELLELDKWASLW) or could reflect the loss of the N-terminal asparagine residue from the ion at 1988.23 (NEQELLELDKWASLWN → EQELLELDKWASLWN) by the activity of the aminopeptidase M. The ion at 1760.16 indicates the removal of asparagine from the N-terminus of the peptide NEQELL ELDKWASLW with m/z = 1874.17 originating from the carboxypeptidase Y digest.

5. Based on the LysC digest, the carboxypeptidase Y digest, and the aminopeptidase M digest, the epitope recognized by the monoclonal antibody 2F5 is NEQELL ELDKWASLWN and is located near the C-terminus of gp41.

4. Notes

1. Handle the beads gently. Avoid any shaking, vortexing, or fast centrifugation, which might cause the beads to crush. All centrifugation steps to drain the beads are carried out in an Eppendorf (Westburg, NY) microcentrifuge at low force only (2–3 min at approx 80g). For MALDI/MS analysis, only 10–20 beads are necessary to give sufficient results. Check under a microscope if enough beads were spotted for analysis.

2. The CNBr-activated Sepharose beads will react with primary amines such as the C-terminus of the heavy and light chains of the antibody as well as the ε-amines in lysine residues. Reactions are also possible with amines from buffers such as Tris, so in some cases it may be necessary to exchange the buffer of the antibody solution, for example, by gel filtration.

3. In the MALDI spectrum, IgG molecules will give a singly charged ion at about 150 kDa. However, doubly and triply charged ions will be more abundant. Do not use dithiothreitol to reduce disulfide bridges as this will cause the antibody to dissociate into its heavy and light chains. In case no ions can be detected in the MALDI spectrum of the beads, check the drain for presence of the antibody. If most of the antibody is detected in the drain, add the drain to the beads and continue to incubate at room temperature with slow rotation.

4. BS^3 [*bis*(sulfosuccinimidyl)suberate] is a homobifunctional N-hydroxysuccinimide ester that reacts with the α-amine at the N-terminus of a protein and the ε-amine of lysine residues. This crosslinker is water-soluble, and the storage of a stock solution is not recommended.

5. If using a different endoproteinase, a normally suggested protein/substrate ratio of 1:20 or higher is recommended. This ratio is higher than for in-solution digests but takes into consideration the possibility that the cleavage sites of the affinity-bound antigen in its native conformation are less accessible than in solution under denaturing conditions. Nevertheless, test the activity of the endoproteinase with the antigen in solution (or a readily available protein with similar molecular weight) to determine the rate of digestion.

6. In general you should wash the beads with the buffer that is recommended for the proteinase that you use. However, compatible components should be substituted for buffers and

additives that are incompatible with subsequent MALDI/MS analysis. Avoid high concentrations of salt, as this can result in poor MALDI/MS signals, avoid glycerol, which will interfere with the formation of crystals for MALDI/MS analysis, and avoid the use of detergents because their presence will result in MALDI/MS ions with high abundance over a wide mass range, possibly suppressing the ions of the peptides of interest.

References

1. World Health Organization's AIDS Epidemic Update Report for December 2002 as found on the internet: http://www.unaids.org/barcelona/presskit/barcelona%20report/contents_html.html.
2. Parker, C. E., Deterding, L. J., Hager-Braun, C., et al. (2001) Fine definition of the epitope on the gp41 glycoprotein of human immunodeficiency virus type 1 for the neutralizing monoclonal antibody 2F5. *J. Virol.* **75,** 10,906–10,911.
3. Peter, J. F. and Tomer, K. B. (2001) A general strategy for epitope mapping by direct MALDI-TOF mass spectrometry using secondary antibodies and crosslinking. *Anal. Chem.* **73,** 4012–4019.
4. Zwick, M., Labrijn, A. F., Wang, M., et al. (2001) Broadly neutralizing antibodies targeted to the membrane-proximal external region of human immunodeficiency virus type 1 glycoprotein gp41. *J. Virol.* **75,** 10,892–10,905.
5. Reineke, U., Ivascu, C., Schlief, M., et al. (2002) Identification of distinct antibody epitopes and mimotopes from a peptide array of 5520 randomly generated sequences. *J. Immunol. Methods* **267,** 37–51.
6. Wegner, G. J., Lee, H. J., and Corn, R. M. (2002) Characterization and optimization of peptide arrays for the study of epitope-antibody interaction using surface plasmon resonance imaging. *Anal. Chem.* **74,** 5161–5168.
7. Thali, M., Moore, J. P., Furman, C., et al. (1993) Characterization of conserved human immunodeficiency virus type 1 gp120 neutralization epitopes exposed upon gp120-CD4 binding. *J. Virol.* **67,** 3978–3988.
8. Trkola, A., Purtschner, M., Muster, T., et al. (1996) Human monoclonal antibody 2G12 defines a distinctive neutralizing epitope on the gp120 glycoprotein of human immunodeficiency virus type 1. *J. Virol.* **70,** 1100–1108.
9. Jemmerson, R. and Paterson, Y. (1986) Mapping epitopes on a protein antigen by the proteolysis of antigen-antibody complexes. *Science* **232,** 1001–1004.
10. Jensen, T. H., Jensen, A., Szilvay, A. M., and Kjems, J. (1997) Probing the structure of HIV-1 Rev by protein footprinting of multiple monoclonal antibody-binding sites. *FEBS Lett.* **414,** 50–54.
11. Kwong, P. D., Wyatt, R., Robinson, J., Sweet, R. W., Sodroski, J., and Hendrickson, W. A. (1998) Structure of an HIV gp120 envelope glycoprotein in complex with the CD4 receptor and a neutralizing human antibody. *Nature* **393,** 648–659.
12. Sharon, M., Gorlach, M., Levy, R., Hayek, Y., and Anglister, J. (2002) Expression, purification, and isotope labeling of a gp120 V3 peptide and production of a Fab from a HIV-1 neutralizing antibody for NMR studies. *Protein Expres. Purif.* **24,** 374–383.
13. Papac, D. I., Hoyes, J., and Tomer, K. B. (1994) Epitope mapping of the gastrin releasing peptide/anti-bombesin monoclonal antibody complex by proteolysis followed by matrix-assisted laser desorption mass spectrometry. *Protein Sci.* **3,** 1488–1492.
14. Parker, C. E., Papac, D. I., Trojak, S. K., and Tomer, K. B. (1996) Epitope mapping by mass spectrometry: determination of an epitope on HIV-1$_{IIIB}$ p26 recognized by a monoclonal antibody. *J. Immunol.* **15,** 198–206.

15. Jeyarajah, S., Parker, C. E., Sumner, M. T., and Tomer, K. B. (1998) MALDI/MS mapping of HIV-gp120 epitopes recognized by a limited polyclonal antibody. *J. Am. Soc. Mass Spectrom.* **9,** 157–165.
16. Schülke, N., Vesanen, M. S., Sanders, R. W., et al. (2002) Oligomeric and conformational properties of a proteolytically mature, disulfide-stabilized human immunodeficiency virus type 1 gp140 envelope glycoprotein. *J. Virol.* **76,** 7760–7776.

8

Identification of T-Cell Epitopes Using ELISpot and Peptide Pool Arrays

Timothy W. Tobery and Michael J. Caulfield

Abstract

Here we describe a method for T-cell epitope identification using a modified ELISpot assay that is both simple and efficient. By using a carefully constructed array of pools of overlapping peptides spanning the entire antigen sequence to stimulate T-cell responses, we are able to detect antigen-specific cytokine responses by both CD8+ and CD4+ T cells and identify the specific peptides to which the cells are responding. Additionally, by performing magnetic bead depletion of either CD8+ or CD4+ cells prior to the assay, we are able to determine the phenotype of the responding cells to each of the peptide epitopes identified. Use of this method will allow the identification of both CD4+ and CD8+ T-cell epitopes without the need for MHC allele-matched reagents and without the need for highly specialized instrumentation. By using an array of peptide pools, this method also dramatically reduces the number of immune cells required to test the entire antigen sequence, often a limiting factor in vaccine testing and other studies.

Key Words: T cell; epitope; peptide; ELISpot.

1. Introduction

Cell-mediated immune responses play a vital role in providing immunity to viral infections. CD8+ T-cells are able to clear viruses by the direct lysis of virally infected cells and/or the secretion of antiviral cytokines. CD4+ T-cells also act to control viral infections through the secretion of cytokines that promote and maintain a strong antiviral CD8+ T-cell response. The activation of these potent antiviral responses is mediated through recognition by the T-cell receptor of virally encoded peptide epitopes presented on MHC class I (CD8) or class II (CD4) molecules on the surface of antigen-presenting cells (1,2). Thus, the recognition of viral epitopes is of great importance, and much effort has gone into the identification of the specific peptide epitopes that induce potent antiviral T-cell responses.

From: *Methods in Molecular Medicine, vol. 94: Molecular Diagnosis of Infectious Diseases, 2/e*
Edited by: J. Decker and U. Reischl © Humana Press Inc., Totowa, NJ

Many techniques exist for the monitoring of T-cell responses and the identification of the epitopes recognized by those T-cells. Classical approaches such as the measurement of direct lysis in bulk killing *(3–6)* and limiting dilution analysis assays *(4,7)* for CD8+ cytotoxic T-lymphocyte (CTL) activity require the use of radioactively labeled MHC-matched target cells and are cumbersome and quite labor-intensive. Recent improvements to these assays have incorporated fluorescent labels rather than radioactivity as a readout of target cell lysis *(8,9)*, but this does not reduce the effort required to perform them, nor does it obviate the need for MHC-matched target cells. Additionally, whereas bulk killing and limiting dilution analysis assays are extremely specific, their sensitivity is quite low *(10)*.

Recently developed MHC I tetramer staining assays, on the other hand, are highly sensitive, but they too require the prior knowledge of the MHC haplotype of the responding T-cells and, more important, measure responses directed against only the peptide loaded into the MHC during formation of the tetrameric complex *(11–13)*. This severely limits the utility of the tetramer staining assay as a screening method for the identification of novel peptide epitopes. In addition, tetramer staining measures the ability of the T-cell receptor to bind a peptide/MHC I complex, but does not measure T-cell function. To do so, tetramer staining must be combined with additional techniques such as intracellular cytokine staining to measure the function of the T-cells that are able to recognize the peptide–MHC complex *(14)*.

Intracellular cytokine staining (ICS) assays detect the expression of effector cytokines following stimulation of T-cells with peptides or pools of peptides *(15–19)* or intact antigens *(20)*. They are quite sensitive and have the additional advantage of being able to detect the phenotype of the responding T-cells by staining for the presence of cell surface markers such as CD4 or CD8 *(21,22)*. However, whereas multiple phenotypes of responding cells can be detected by ICS in a single reaction tube by using antibodies tagged with different fluorophores, each stimulation condition (i.e., a potential peptide epitope or pool of peptides) must be tested in separate tubes, making these assays somewhat labor-intensive and time-consuming. Efforts are currently under way to address these concerns by developing methods that will allow the use of the ICS assay in a high-throughput setting (i.e., 96-well plate formats with the use of multichannel pipetors), but these advances are still forthcoming.

ELISpot assays *(23–27)* also have the same advantages of the ICS assay, namely, no requirement for prior knowledge of epitope sequences or MHC alleles, no need for MHC-matched target cells, and relative ease compared with classical techniques. However, ELISpot assays are much more efficient in that multiple conditions can be tested in the same assay, reducing both time and effort required for the generation of results *(28)*. They can also be performed without highly specialized and expensive instrumentation such as a flow cytometer, making them more readily accessible than the ICS assay. The identification of peptide epitopes by ELISpot can be further streamlined by using an array of peptide pools consisting of overlapping peptides *(29)*, dramatically reducing the number of conditions that need to be tested in order to screen all the potential epitopes in a given viral antigen. Finally, by separating or depleting T-cell subsets prior to adding immune cells to the assay, the phenotype of the response may be determined *(30,31)*.

The identification of human papillomavirus (HPV)16E1 epitopes in BALB/c mice immunized with a DNA plasmid expressing HPV16E1 will be used to demonstrate the results that can be obtained using interferon-γ (IFN-γ) ELISpot assays with peptide pool arrays to identify both CD8+ and CD4+ T cell epitopes.

2. Materials

1. 96-well multiscreen membrane plate (cat. no. SEM004M99, Millipore).
2. Coating antibody: purified anti-mouse IFN-γ, rat IgG1, clone R4-6A2 (cat. no. 18181D, PharMingen).
3. Second antibody: biotin anti-mouse IFN-γ, rat IgG1, clone XMG1.2 (cat. no. 18112D, PharMingen)
4. Streptavidin–alkaline phosphatase (AP) conjugate (cat. no. 13043E, PharMingen).
5. 1-Step nitroblue tetrazolium/bromochloroindolyl phosphate (NBT/BCIP) (cat. no. 34042, Pierce).
6. Concanavalin A 1 mg/mL in saline (cat. no. C0412, Sigma).
7. 20mer peptide stocks (lyophilized).
8. Dimethyl sulfoxide (DMSO) (cat. no. D128-1, Fisher).
9. Phosphate-buffered saline (PBS without Ca and Mg).
10. Polyoxyethylenesorbitan monolaurate (Tween-20; cat. no. P-2287, Sigma).
11. Dynabeads Mouse CD4 (cat. no. 114.06, Dynal).
12. Dynabeads Mouse CD8 (cat. no. 114.08, Dynal).
13. ACK lysing buffer (formula no. 79-0422DG, Gibco BRL).
14. Trypan blue: stock 0.4%, dilute to 0.2% in PBS (cat. no. 15250-012, Gibco BRL).
15. Second antibody buffer: PBS + 0.005% Tween-20, 5% heat-inactivated fetal bovine serum (ΔFBS).
16. RPMI-10 medium: RPMI-1640 + 10% ΔFBS, 2 mM L-glutamine, 10 mM HEPES, 1X Pen/Strep, 50 μM β-mercaptoethanol, filter sterilized:
    a. FBS (cat. no. SH30070-03, Hyclone).
       Heat-inactivate FBS: 56°C (±2°C) for 30–45 min.
       Aliquot ΔFBS into 50-mL aliquots.
       Store frozen at –20°C (±5°C).
    b. RPMI-1640 medium (cat. no. 11875-093, Gibco-BRL).
    c. 1 M HEPES (cat. no. 15630-080, Gibco-BRL).
    d. 200 mM L-glutamine (cat. no. 25030-081, Gibco-BRL).
    e. Penicillin/streptomycin (cat. no. 15140-122, Gibco-BRL).
    f. β-mercaptoethanol (cat. no. M7522, Sigma).
    g. 0.2 μm filter units (cat. no. 25952-1L, Corning)

3. Methods

3.1. Preparation of Peptide Stocks and Peptide Pool Array

1. Obtain a complete panel of 20mer peptides overlapping by 10 amino acids that covers the entire sequence of the antigen of interest. Peptides should be at least minimally purified (>70%) to remove excess reagents (salts and so on) and lyophilized. Highly high-performance liquid chromatography (HPLC)-purified peptides (>90%) will significantly improve results but are not necessary. Peptide libraries prepared at small scale using "pin" technologies (e.g., www.mimetopes.com), rather than traditional peptide synthesis, can be used as an economical source of peptides. Store lyophilized peptides at –20°C.

	C1	C2	C3	C4	C5	C6	C7	C8
R1	1	3	5	7	9	11	13	15
R2	31	17	19	21	23	25	27	29
R3	45	47	33	35	37	39	41	43
R4	59	61	63	49	51	53	55	57
R5	10	12	14	16	2	4	6	8
R6	24	26	28	30	32	18	20	22
R7	38	40	42	44	46	48	34	36
R8	52	54	56	58	60	62	64	50

Fig. 1. HPV16E1 peptide pool array. An array of eight column pools and eight row pools, each containing eight non-overlapping peptides, was generated by combining equal amounts of each peptide stock and adjusting the volume in each pool to give a 1:10 final dilution of each peptide, resulting in pools containing each peptide at 2 mg/mL.

2. Solubilize each peptide at 20 mg/mL in DMSO. Number these peptide stocks sequentially (peptide 1 is amino acids 1–20, peptide 2 is amino acids 11–30, and so on). Store peptide stocks at –70°C (*see* **Note 1**).
3. Map a two-dimensional array of the individual peptides such that pools of the peptides along an axis of the array do not contain any overlapping peptides (**Fig. 1**). For HPV16E1, there are 64 peptides in the panel, so an 8 × 8-array was constructed. Generate the peptide pools by adding equal quantities of each peptide stock (equal volumes if all stocks are at 20 mg/mL) and a volume of DMSO sufficient to result in a 1:10 dilution of each peptide in the pool (final pool concentration is 2 mg/ml for each constituent peptide). Store peptide pools at –70°C.
4. Generate a total peptide pool by combining equal quantities of each peptide and adjusting the total volume to yield a 1:100 (200 μg/mL) dilution in DMSO. Store pool at –70°C.

3.2. Preparation of Immune Cells

Note: The first step of Subheading 3.3. must be performed the day before the assay is to be set up.

3.2.1. Isolation of Murine Splenocytes (see **Note 2**)

1. Sacrifice immune mice and harvest each spleen into 3 mL RPMI in a 15-mL conical tube.
2. Cut spleen into small pieces using aseptic technique and grind pieces through sterile stainless-steel mesh.
3. Allow large pieces of connective tissue to settle at 1g for 2 min at ambient temperature.
4. Transfer supernatant containing splenocytes to fresh 15-mL conical tube.
5. Pellet cells at 250g for 8–10 min at ambient temperature.
6. Discard supernatant and resuspend pellet in 3 mL ACK lysing buffer. Incubate for 3 min at ambient temperature.
7. Add 10 mL RPMI-10 to tube and invert to mix.
8. Pellet cells as in **step 5** and discard supernatant.
9. Resuspend cells in 10 mL RPMI-10 and count cells using hemacytometer.

T-Cell Epitope Identification 125

10. Pellet cells as in **step 5** and resuspend cells at 10^7 cells/mL in RPMI-10. If performing depletion of cell subsets, proceed to **Subheading 3.2.2.**; if not, skip to **Subheading 3.3**.

3.2.2. Magnetic Bead Depletion of Cell Subsets (see Note 3)

1. Calculate the number of depleted cells that are required for the assay (*see* **Subheading 3.3.7.**) and transfer a volume of cell suspension containing three times the number of required cells to a 12 × 75-mm snap cap polypropylene tube.
2. Pellet cells as in **Subheading 3.2.1., step 5**.
3. During pelleting step, transfer the required volume of anti-CD4- or anti-CD8-coated magnetic beads to a 15-mL conical tube. Use 5 beads/cell/sample. Place tube containing beads on magnetic particle concentrator (MPC) for 2 min at ambient temperature. Aspirate and discard supernatant. Wash beads with 10 mL RPMI-10 by resuspending and then placing on MPC for 2 min. Resuspend beads in 1 mL/sample RPMI-10.
4. Aspirate supernatant away from cell pellet and resuspend in 1 mL RPMI-10.
5. Add 1 mL of bead suspension to cells and cap tube tightly. Incubate at 4°C with gentle rocking for 30 min.
6. Place tube containing cells and beads on MPC for 2 min at ambient temperature. Collect supernatant containing depleted cells into fresh 12 × 75-mm snap cap tube.
7. Wash beads with 2 mL RPMI-10 and add wash supernatant to cells. Discard washed beads.
8. Count depleted cells with hemacytometer. Pellet cells as in **Subheading 3.2.1., step 5** and discard supernatant.
9. Resuspend depleted cells at 10^7 cells/mL in RPMI-10 and proceed to **Subheading 3.3**.

3.3. IFN-γ ELISpot Assay (see Note 4)

Caution: Perform steps 1–7 in biological safety cabinet.

1. **To be performed the day prior to the assay:** dilute anti-IFN-γ monoclonal antibody to a final concentration of 10 µg/mL in sterile PBS. Add 100 µL/well of diluted antibody to assay plates and incubate for 16–24 h at 4°C. (Plates may be coated up to 3 d before the assay setup and stored at 4°C.)
2. Wash coated plates three times by hand with 200 µL/well of RPMI-10.
3. Block plate with 200 µL/well of RPMI-10 for 2 h at 37°C.
4. Prepare peptide pools and controls as follows: dilute peptide pool stocks from **Subheading 3.1.** in RPMI-10 to a final concentration of 4 µg/mL. Make a dilution of DMSO in RPMI-10 that will match the final DMSO concentration in the peptide pools. (For example, if peptide stocks are at 200 µg/mL, a 1:50 dilution is made to generate a 4 µg/mL working solution. Therefore a 2% DMSO solution is used as a control). Make a 5 µg/mL solution of concanavalin A (ConA) in RPMI-10 for murine splenocytes for use as positive controls.
5. Wash blocked plates once by hand with 200 µL/well of RPMI-10.
6. Add 50 µL/well of peptide pools or controls to plates, using at least 2 replicate wells/condition.
7. Add 50 µL/well of cell suspensions at 10^7 cells/mL to wells containing the stimuli for a total volume of 100 µL/well. Incubate plates for 18–22 h at 37°C.
8. Wash plates six times with 300 µL/well of PBS + 0.005% Tween using a plate washer (Titertek MAP or equivalent) or by hand.
9. Dilute biotinylated anti-IFN-γ antibody to a working concentration of 20 µg/mL in second antibody buffer (*see* **Subheading 2.**). Add 50 µL/well to plate (1 µg/well final). Incubate for 16–24 h at 4°C.

pool	SFC/10⁶ cells	pool	SFC/10⁶ cells
DMSO	3	**pooled E1**	**406**
C1	16	R1	6
C2	6	R2	12
C3	**200**	**R3**	**232**
C4	30	R4	8
C5	**286**	R5	4
C6	6	**R6**	**362**
C7	18	**R7**	**70**
C8	**152**	R8	28

Fig. 2. HPV16E1-specific IFN-γ ELISpot responses to HPV16E1 peptide pool array in immune BALB/c mice. Values represent the average of two wells/pool corrected for number of IFN-γ spot-forming cells (SFC)/million cells. Positive responses are indicated in bold type. DMSO, dimethyl sulfoxide.

10. Repeat **step 8**.
11. Dilute streptavidin-AP 1:2000 in second antibody buffer. Add 100 µL/well to plates and incubate for 1.5–2 h at ambient temperature.
12. Wash plates three times with 300 µL/well of PBS + 0.005% Tween followed by three washes with 300 µL/well of PBS using a plate washer (Titertek MAP or equivalent) or by hand.
13. Add 100 µL/well of one-step NBT/BCIP reagent. Incubate for 5–15 min in the dark at ambient temperature. Stop reaction by rinsing plate under gently running tap water.
14. Allow plates to dry completely before enumerating spots.

3.4. Epitope Identification

1. Count spots in each well using a dissecting microscope (*see* **Note 5**). Take the average of all replicate wells to obtain the average number of spots/well (5×10^5 cells/well) and multiply by 2 to obtain the number of IFN-γ spot-forming cells (SFCs)/million cells for each stimulation condition (*see* **Note 6**).
2. Identify wells that have a positive IFN-γ ELISpot response (>50 SFCs/million cells above the DMSO control wells) and the corresponding peptide pools used to stimulate the cells in that well (**Fig. 2**) (*see* **Note 7**).
3. Using the peptide pool array diagram, identify constituent peptides that are present in the positive pools (**Fig. 3**). Each peptide is present in one row and one column pool, so a response specific for a single peptide will score as positive for one row and one column pool.
4. In a subsequent assay, test each of the potential 20mer peptides individually, including T-cell subset depletions if phenotyping of the response is desired. If T-cell subset depletions were performed, identify the peptides that stimulate positive IFN-γ ELISpot responses for both the undepleted and depleted cells. If a peptide is positive for the undepleted cells, but negative following the T-cell subset depletion, the responding T-cells must belong to that subset. Conversely, if the response is not affected by the depletion, then the responding T-cells must not belong to that subset (**Fig. 4**).

	C1	C2	C3	C4	C5	C6	C7	C8
R1	1	3	5	7	9	11	13	15
R2	31	17	19	21	23	25	27	29
R3	45	47	33	35	37	39	41	43
R4	59	61	63	49	51	53	55	57
R5	10	12	14	16	2	4	6	8
R6	24	26	28	30	32	18	20	22
R7	38	40	42	44	46	48	34	36
R8	52	54	56	58	60	62	64	50

Fig. 3. Identification of potential HPV16E1 20mer peptides containing T-cell epitopes. The individual 20mer peptides contained in each positive peptide array pool that stimulate a IFN-γ ELISpot response are predicted by the intersection of rows and columns in the array. Positive row and column pools and the 20mer peptides contained at their intersections are shaded. In the case of HPV16E1, there are a minimum of three and a maximum of nine 20mer peptides that are recognized in immunized BALB/c mice.

	SFC/10^6 cells	
20mer peptide	undepleted cells	CD4+ depleted cells
peptide 22	162	160
peptide 28	6	10
peptide 32	172	186
peptide 33	172	180
peptide 36	4	3
peptide 37	0	0
peptide 42	2	0
peptide 43	0	6
peptide 46	86	48

Fig. 4. IFN-γ ELISpot responses to individual HPV16E1 20mer peptides. The nine potential 20mer peptides identified from the peptide pool array were tested individually in an ELISpot assay using undepleted and CD4+ T-cell-depleted cells from DNA-HPV16E1 immunized Balb/c mice. Four of the nine peptides stimulated positive IFN-γ responses. Three of the peptides, peptides 22, 32, and 33, were unaffected by depletion of CD4+ T-cells, indicating a CD8+ T-cell response against these peptides. Two of these, peptides, 32 and 33, are overlapping peptides and may share a single epitope contained in the overlap region. The fourth peptide, number 46, showed a partial reduction in signal, indicating a mixed CD4+/CD8+ response to this peptide.

position		1	2	3	4	5	6	7	8	9											
H2-K^d binding motif			Y F	N I L	P	M	K	T F N		I L											
HPV16E1 peptide 22 (a.a. 211-230)	f	k	e	l	Y	g	v	s	F	s	e	L	v	r	p	f	k	s	n	k

	SFC/10^6 cells	
peptide	undepleted cells	CD8+ depleted cells
DMSO	2	1
HPV16E1 211-230 (20mer peptide 22)	133	3
HPV16E1 214-222	125	2

Fig. 5. (A) Prediction of minimal epitopes when MHC I haplotype is known. Using the known binding motif for H2-K^d, it is possible to predict the most likely minimal epitope in HPV16E1 peptide 22 that is recognized in BALB/c mice. For this 20mer, there is only one 9mer peptide (amino acids 214–222) that contains the appropriate anchor residues. Dominant anchor residues are in boldface, and auxiliary anchor residues are capitalized; shaded residues indicate matches between the HPV16E1 peptide 22 sequence and the H2-K^d binding motif. (B) Confirmation of minimal epitope by ELISpot assay. The predicted 9mer epitope (amino acids 214–222) contained in HPV16E1 peptide 22 was tested individually against the 20mer peptide in an IFN-γ ELISpot assay. Both peptides stimulate equivalent responses in undepleted and CD8+ T-cell-depleted cells from E1-immune BALB/c mice. For both peptides, the response is completely lost after depletion of CD8+ cells, confirming the phenotype of the responding cells. DMSO, dimethyl sulfoxide; SFC, spot-forming cells.

5. For most applications, identification of the 20mer peptide containing an epitope without identifying the minimal epitope is sufficient. At the peptide concentrations (2 mg/mL) used in this method, IFN-γ ELISpot responses to the 20mer peptide and the minimal 9mer epitope are equivalent (**Fig. 5B**). If it is necessary to define the minimal epitopes recognized, an additional ELISpot assay must be performed. In this additional assay, a series of shorter length peptides (9mers) derived from the reactive 20mer peptide sequence and overlapping by a single amino is tested individually in the ELISpot assay. Alternatively, if the MHC haplotype is known, the minimal epitope may be predicted by a number of algorithms based on the presence of anchor residues and other commonly occurring motifs in the peptides. This is relatively simple for a handful of 20mer peptides but is quite difficult for prediction over the entire amino acid sequence of the protein.

4. Notes

1. Rarely, peptides may be insoluble in DMSO at 20 mg/mL and should then be used at 10 mg/mL and volumes adjusted accordingly. Solubilization of peptides may be enhanced by agitation using a vortex mixer and/or prolonged incubation at ambient temperature. Peptide stock solutions should not be heated above ambient temperature to facilitate solubilization.
2. Isolation of primate peripheral blood mononuclear cells (PBMCs). This method may also be used for primate immunogenicity studies. In this case, the source of immune cells is peripheral blood. An alternate protocol for the isolation of these cells is given here:
   a. Draw peripheral blood from test subjects into Vacutainer tubes (Becton-Dickinson) containing an anticoagulant such as heparin or EDTA. Yields of PBMCs are on average 1–3 million cells/mL blood.
   b. Transfer blood to 50-mL conical tubes and dilute with an equal volume of Hanks' balanced salt solution (HBSS). Gently underlay the blood with a 2:1 ratio of Ficoll-Paque using a 10-mL serological pipet.
   c. Spin Ficoll tubes in a swinging bucket rotor at $500g$ for 30–45 min at ambient temperature with the brake off.
   d. Carefully remove buffy coat using a 10-mL serological pipet and transfer to a 50-mL conical tube. Add sterile PBS to a total volume of 40 mL and mix gently.
   e. Pellet cells at $500g$ for 15 min at ambient temperature. Discard supernatant.
   f. Resuspend cells in 10 mL RPMI-10 and pellet as in **Subheading 3.2.1., step 5**. Discard supernatant.
   g. If no red blood cells are present in the pellet by visual inspection, skip to the next step. If significant red blood cells are present, resuspend pellet in 3 mL ACK lysing buffer and incubate for 2–3 min at ambient temperature. Add 10 mL RPMI-10, invert to mix, pellet as in **step 5**, and discard supernatant.
   h. Resuspend PBMC pellet in 10 mL RPMI-10 and count cells using hemocytometer.
   i. Pellet cells as in **Subheading 3.2.1., step 5** and resuspend cells at 10^7 cells/mL in RPMI-10. If performing depletion of cell subsets, proceed to **Subheading 3.2.2.**; if not, skip to **Subheading 3.3**.
3. If you are using primate immune cells, Dynabeads M-450 CD4 (cat. no. 111.06, Dynal) and Dynabeads M-450 CD8 (cat. no. 111.08, Dynal) specific for human CD4+ and CD8+ T-cells will crossreact with rhesus CD4+ and CD8+ T-cells and can be used for the depletions. The remainder of the protocol is the same for primate or murine cells.
4. If you are using primate immune cells, the following anti-IFN-γ antibodies may be used for the ELISpot assay: purified anti-human rIFN-γ (cat. no. 1598-00, R&D Systems) for coating and biotinylated anti-human rIFN-γ (cat. no. BAF285, R&D Systems) for detection. Also, instead of using Con A, make a 10 μg/mL solution of PHA-M (cat. no. L-8902, Sigma) in RPMI-10 to use as a positive control for primate PBMCs. The remainder of the protocol is the same for primate or murine cells.
5. Enumeration of spots can be difficult and subjective. Here are some guidelines to improve the accuracy and precision of manual spot counts:
   a. The well membranes should be free from blue background color, although slight discoloration of the membrane will not impede spot enumeration. If significant background color is present, reduce the concentration of the streptavidin–AP conjugate and/or the incubation time with the NBT/BCIP substrate. If these changes do not improve the results, the biotinylated antibody may need to be retitrated.

b. If all spots are faint, increase the concentration of streptavidin–AP and/or incubation time with substrate. If these changes do not improve the results, the biotinylated antibody may need to be retitrated.

c. True spots will be dark with slightly fuzzy edges. Blue-colored regions with very sharp edges are caused by debris and are not true spots. Pinpoint spots regardless of darkness or fuzziness are not true spots, but their presence should be noted. Dark, fuzzy spots with white centers are true spots in which the substrate has detached from the membrane due to overloading in response to a very high IFN-γ concentration.

d. Inclusion of a rectilinear or polar grid in the microscope optics may improve the accuracy of counts if the number of spots/well is large.

e. Resolution of individual spots by manual counts may not be possible if the number of spots is high (>300 spots/well). In this case a designation of too numerous to count is used to score the well. Reducing the number of cells/well will decrease the number of spots, but this is not advised since the ability of responder T-cells to contact antigen-presenting cells in the wells, and thus to secrete cytokines, is impaired if the cell density is too low. If it is necessary to reduce the cell density, a careful titration of cells to ensure linearity of the response over a range of cell densities is suggested.

6. The positive control wells should contain >300 spots/well. Significantly fewer than this number of spots indicates a problem with the preparation of the immune cells or an assay failure.

7. Occasionally a sample will have very high numbers of spots in the negative control wells. This high background response may indicate a problem with the preparation of the immune cells or may reflect a high degree of nonspecific activation of T-cells. Samples with high negative control spot counts should be noted and should not be scored as positive even if the criteria of >50 spots above background is met.

Acknowledgments

The authors thank Dr. William McClements for construction of the peptide pool array, Michael P. Neeper and Xin-Min Wang for generation of HPV plasmids for immunization of mice, and Su Wang for technical assistance with the ELISpot assays.

References

1. Braciale, T. J., Morrison, L. A., Sweetser, M. T., Sambrook, J., Gething, M. J., and Braciale, V. L. (1987) Antigen presentation pathways to class I and class II MHC-restricted T lymphocytes. *Immunol. Rev.* **98,** 95–114.
2. Abbas, A., Lichtman, A., and Pober, J. (2000) *Cellular and Molecular Immunology,* 4th ed. Philadelphia: WB Saunders.
3. Engers, H. D., Thomas, K., Cerottini, J. C., and Brunner, K. T. (1975) Generation of cytotoxic T lymphocytes in vitro. V. Response of normal and immune spleen cells to subcellular alloantigens. *J. Immunol.* **115,** 356–360.
4. Coligan, J., Kruisbeek, A., Margulies, D., Shevach, E., and Strober, W., eds. *Current Protocols in Immunology.* New York: Wiley Interscience.
5. Koenig, S., Fuerst, T. R., Wood, L. V., et al. (1990) Mapping the fine specificity of a cytolytic T cell response to HIV-1 nef protein. *J. Immunol.* **145,** 127–135.
6. Kurokohchi, K., Arima, K., and Nishioka, M. (2001) A novel cytotoxic T-cell epitope presented by HLA-A24 molecule in hepatitis C virus infection. *J. Hepatol.* **34,** 930–935.

7. Yamada, A., Ziese, M. R., Young, J. F., Yamada, Y. K., and Ennis, F. A. (1985) Influenza virus hemagglutinin-specific cytotoxic T cell response induced by polypeptide produced in *Escherichia coli*. *J. Exp. Med.* **162,** 663–674.
8. Liu, L., Chahroudi, A., Silvestri, G., et al. (2002) Visualization and quantification of T cell-mediated cytotoxicity using cell-permeable fluorogenic caspase substrates. *Nat. Med.* **8,** 185–189.
9. Sheehy, M. E., McDermott, A. B., Furlan, S. N., Klenerman, P., and Nixon, D. F. (2001) A novel technique for the fluorometric assessment of T lymphocyte antigen specific lysis. *J. Immunol. Methods* **249,** 99–110.
10. Murali-Krishna, K., Altman, J. D., Suresh, M., et al. (1998) Counting antigen-specific CD8 T cells: a reevaluation of bystander activation during viral infection. *Immunity* **8,** 177–187.
11. Altman, J. D., Moss, P. A. H., Goulder, P. J. R., et al. (1996) Phenotypic analysis of antigen-specific T lymphocytes [published erratum appears in *Science* (1998) **280,** 1821]. *Science* **274,** 94–96.
12. Chen, F. E., Aubert, G., Travers, P., Dodi, I. A., and Madrigal, J. A. (2002) HLA tetramers and anti-CMV immune responses: from epitope to immunotherapy. *Cytotherapy* **4,** 41–48.
13. Kwok, W. W., Ptacek, N. A., Liu, A. W., and Buckner, J. H. (2002) Use of class II tetramers for identification of CD4+ T cells. *J. Immunol. Methods* **268,** 71–81.
14. Appay, V. and Rowland-Jones, S. L. (2002) The assessment of antigen-specific CD8+ T cells through the combination of MHC class I tetramer and intracellular staining. *J. Immunol. Methods* **268,** 9–19.
15. Betts, M. R., Ambrozak, D. R., Douek, D. C., et al. (2001) Analysis of total human immunodeficiency virus (HIV)-specific CD4(+) and CD8(+) T-cell responses: relationship to viral load in untreated HIV infection. *J. Virol.* **75,** 11,983–11,991.
16. Kern, F., Surel, I. P., Brock, C., et al. (1998) T-cell epitope mapping by flow cytometry. *Nat. Med.* **4,** 975–978.
17. Kern, F., Faulhaber, N., Frommel, C., et al. (2000) Analysis of CD8 T cell reactivity to cytomegalovirus using protein-spanning pools of overlapping pentadecapeptides. *Eur. J. Immunol.* **30,** 1676–1682.
18. Maecker, H. T., Maino, V. C., and Picker, L. J. (2000) Immunofluorescence analysis of T-cell responses in health and disease. *J. Clin. Immunol.* **20,** 391–399.
19. Maecker, H. T., Dunn, H. S., Suni, M. A., et al. (2001) Use of overlapping peptide mixtures as antigens for cytokine flow cytometry. *J. Immunol. Methods* **255,** 27–40.
20. Asanuma, H., Sharp, M., Maecker, H. T., Maino, V. C., and Arvin, A. M. (2000) Frequencies of memory T cells specific for varicella-zoster virus, herpes simplex virus, and cytomegalovirus by intracellular detection of cytokine expression. *J. Infect. Dis.* **181,** 859–866.
21. Maino, V. C. and Picker, L. J. (1998) Identification of functional subsets by flow cytometry: intracellular detection of cytokine expression. *Cytometry* **34,** 207–215.
22. Maecker, H. T., Auffermann-Gretzinger, S., Nomura, L. E., Liso, A., Czerwinski, D. K., and Levy, R. (2001) Detection of CD4 T-cell responses to a tumor vaccine by cytokine flow cytometry. *Clin. Cancer Res.* **7,** 902s–908s.
23. Czerkinsky, C., Andersson, G., Ekre, H. P., Nilsson, L. A., Klareskog, L., and Ouchterlony, O. (1988) Reverse ELISpot assay for clonal analysis of cytokine production. I. Enumeration of gamma-interferon-secreting cells. *J. Immunol. Methods* **110,** 29–36.
24. Lalvani, A., Brookes, R., Hambleton, S., Britton, W. J., Hill, A. V., and McMichael, A. J. (1997) Rapid effector function in CD8+ memory T cells. *J. Exp. Med.* **186,** 859–865.

25. Larsson, M., Jin, X., Ramratnam, B., et al. (1999) A recombinant vaccinia virus based ELISPOT assay detects high frequencies of Pol-specific CD8 T cells in HIV-1-positive individuals. *AIDS* **13,** 767–777.
26. Currier, J. R., deSouza, M., Chanbancherd, P., Bernstein, W., Birx, D. L., and Cox, J. H. (2002) Comprehensive screening for human immunodeficiency virus type 1 subtype- specific CD8 cytotoxic T lymphocytes and definition of degenerate epitopes restricted by HLA-A0207 and -C(W)0304 alleles. *J. Virol.* **76,** 4971–4986.
27. Kumar, A., Weiss, W., Tine, J. A., Hoffman, S. L., and Rogers, W. O. (2001) ELISPOT assay for detection of peptide specific interferon-gamma secreting cells in rhesus macaques. *J. Immunol. Methods* **247,** 49–60.
28. Anthony, D. D., Valdez, H., Post, A. B., Carlson, N. L., Heeger, P. S., and Lehmann, P. V. (2002) Comprehensive determinant mapping of the hepatitis C-specific CD8 cell repertoire reveals unpredicted immune hierarchy. *Clin. Immunol.* **103,** 264–276.
29. Tobery, T. W., Wang, S., Wang, X. M., et al. (2001) A simple and efficient method for the monitoring of antigen-specific T cell responses using peptide pool arrays in a modified ELISpot assay. *J. Immunol. Methods* **254,** 59–66.
30. Steele, J. C., Roberts, S., Rookes, S. M., and Gallimore, P. H. (2002) Detection of CD4(+)- and CD8(+)-T-cell responses to human papillomavirus type 1 antigens expressed at various stages of the virus life cycle by using an enzyme-linked immunospot assay of gamma interferon release. *J. Virol.* **76,** 6027–6036.
31. Geginat, G., Schenk, S., Skoberne, M., Goebel, W., and Hof, H. (2001) A novel approach of direct ex vivo epitope mapping identifies dominant and subdominant CD4 and CD8 T cell epitopes from *Listeria monocytogenes*. *J. Immunol.* **166,** 1877–1884.

9

Virus-like Particles

A Novel Tool for the Induction and Monitoring of Both T-Helper and Cytotoxic T-Lymphocyte Activity

Ludwig Deml, Jens Wild, and Ralf Wagner

Abstract

T-cells play a crucial role in the control of various viral infections such as HIV and herpes viruses. Thus, the development of advanced techniques for the stimulation and measurement of both antigen-specific T-helper and CTL responses is one of most meaningful objectives in vaccinology. Herein, we present HIV-1 Pr55^{gag} lipoprotein particles (VLPs) to be a potent antigen for introducing epitopes into the MHC-class-I and -II processing and presentation pathway. These VLPs can easily be produced in insect cells by using the baculovirus expression system. Immunization studies in mice revealed the strong capacity of these VLPs to stimulate Gag-specific T-helper-1 cell-biased humoral and cellular immune responses. In addition, these VLPs can be used as a stimulator antigen for the detection of Gag-specific T-helper and CTL responses, as determined by conventional ELISA, ELISpot, FACS, and ^{51}Cr-release assays. These results strongly underline the value of VLPs as a stimulator of MHC-class-I and -II mediated epitope presentation for preventive, therapeutic, and diagnostic purposes.

Key Words: Virus-like particles; vaccination; protein expression and purification; measurement of T cell responses.

1. Introduction

For several prevalent viral pathogens such as most human herpes viruses and the human immunodeficiency virus (HIV), cellular immunity appears to play a pivotal role in host defense mechanisms *(1–6)*. For HIV, abundant clinical and experimental evidence emphasizes CD8+ cytotoxic T-lymphocytes (CTLs) as the key protective immune parameter in the delay of disease progression and perhaps the resistance to viral infection *(7–9)*. Gag-, Pol-, Env-, and Nef-specific CTLs have been detected early in the course of HIV-1 infection, prior to the appearance of humoral response, and are presumed to control the initial viremia *(8,10,11)*. In addition, in persons with chronic infection, a decline of Gag-specific CTL precursors was shown to coincide with a CD4 drop (increasing virus load) and disease progression *(12)*. The usefulness of Gag immunogens for vaccine development and immunotherapeutic interventions is further supported by the fact that the protein is relatively conserved among diverse HIV-1 subtypes. Broad cross-clade CTL recognition directed against Gag-specific targets has been well documented *(13–15)* and has been suggested to contribute to long-term control of virus replication *(16)*. Thus, within the last decades, enormous efforts have been made to develop advanced strategies to stimulate and monitor virus-specific T-cell reactivities for preventive, therapeutic, and diagnostic applications. Several approaches have been evolved to deliver exogenous antigens to the MHC class I processing and presentation pathway *(17)* as a prerequisite for CTL activation. For example, polypeptides have been combined with appropriate immunostimulatory adjuvants such as monophosphory lipid A (MLP) *(18)*, saponin *(19)*, emulsions such as complete *(20)* and incomplete Freund's adjuvant *(21)*, lipid particles such as immunostimulating complexes (ISCOMs) *(22)*, and liposomes *(23)*, as well as microparticulate antigens such as poly(lactide-co-glycolide) (PLG) *(24)*. However, none of these substances is currently licensed for use in human vaccines.

Alternatively, DNA vaccines as well as various recombinant viral and bacterial gene transfer vectors have been approved to induce strong CTL responses, but the safety and efficacy of these methods in humans remains to be determined more precisely. Thus, the development of a safe strategy to induce CTL responses with nonreplicating antigens is still an important prerequisite for the design of novel vaccines.

In recent years, several groups have recognized that particulate antigens generally induce more effective cellular and humoral immune responses than soluble proteins. Virus-like particles (VLPs) have been shown to possess a number of advantages over conventional protein immunogens for vaccine purposes. Most of these VLPs can easily be produced and purified in large quantities in heterologous expression systems *(25)*. In addition, these VLPs do not replicate and are noninfectious.

Capsid as well as envelope proteins from various infectious agents possess the ability to self-assemble into highly organized particulate structures. For example, hepatitis B virus (HBV) core and surface antigen *(26–29)*, papilloma virus L1 or L1/L2 *(30,31)*, parvovirus VP2 *(32,33)*, and yeast retrotransposon Ty-p1 protein-based virus-like particles *(34,35)* have been constructed for vaccine purposes. In addition, retroviral Gag precursor proteins have been previously shown by others and us to include all the information required for particle assembly *(36,37)*. For HIV, expression of Pr55^{gag} in mammalian and insect cells results in the formation of intact VLPs that morphologically and antigenically resemble native immature virions *(38–41)*. Recombinant HIV-1 Pr55^{gag} VLPs can easily be produced in insect cells by using the baculovirus expression system, as described previously *(42)*. Briefly, recombinant Gag proteins were expressed by infecting HighFive insect cells with the Gag-recombinant baculovirus AcPr55 *(43)*. Three days post infection, the cell culture supernatants were precleared by low-speed centrifugation, and particles were concentrated by centrifu-

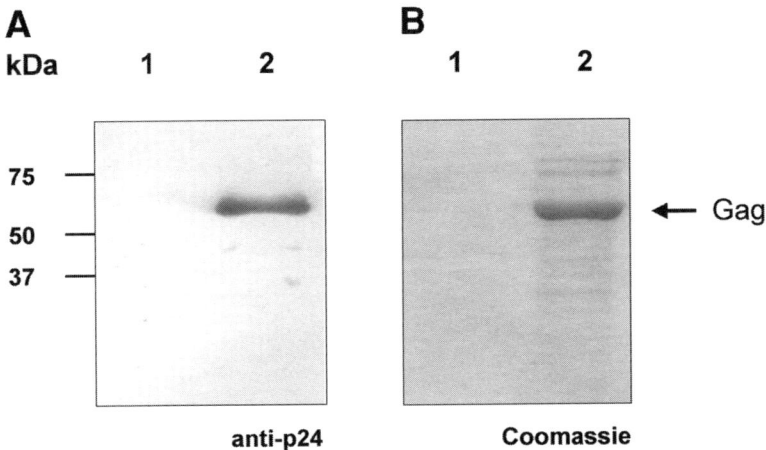

Fig. 1. Biochemical analysis of Pr55^{gag} VLPs. Precleared cell culture supernatants of AcPr55 and noninfected HighFive cells (negative control) were concentrated by sedimentation through a 30% sucrose cushion, and pelleted material was separated on a 10–50% sucrose gradient. The p24-positive fractions (lanes 2) as well as the corresponding fractions of the control (lanes 1) were combined, and aliquots of well-concentrated samples were separated by 12.5% SDS-PAGE and analyzed either by (**A**) immunoblotting with a p24-specific monoclonal antibody or (**B**) staining with Coomassie brilliant blue. Arrows at the right indicate the positions of the HIV Gag protein. The positions of the markers are given on the left (in kilodaltons).

gation through a 30% sucrose cushion. Then, sedimented VLPs were further purified by banding on a 10–60% sucrose step gradient. Pr55^{gag}-positive fractions were combined and dialyzed against phosphate-buffered saline. The identity of the proteins obtained was verified by immunoblotting using the p24-specific monoclonal antibody (16/4/2) *(44)* (**Fig. 1A**), as described previously *(45)*. The resulting VLP preparations were over 90% pure (**Fig. 1B**), as calculated by Coomassie staining.

Immunological testing of these particulate antigens in mice revealed their potency to induce strong Th1-biased humoral and cellular immune responses, characterized by the induction of both IgG1 and IgG2a isotypes and the activation of interferon-γ (IFN-γ) producing Th-1 cells and Gag-specific CTL (**Fig. 2A–C**) *(42,46,47)*. These data encouraged us to test the potential use of these VLPs as stimulators for the detection of Gag-specific T-cell responses for HIV-infected individuals.

Over the past decade, various novel experimental techniques have been applied that permit quantitative evaluation of polypeptide-specific CD4+ or CD8+ T-cells. These methods include ELISpot analysis as well as flow cytometry–based intracellular cytokine and tetramer staining. All these procedures rely on the measurement of IFN-γ production by polypeptide-activated CTLs. ELISpot analysis allows the direct visualization of single reactive cells by labeling of captured cytokine on a membrane surface. However, the ELISpot assay provides no information about the phenotype of the reactive cell population. Flow cytometry is the basis for use of peptide–MHC tet-

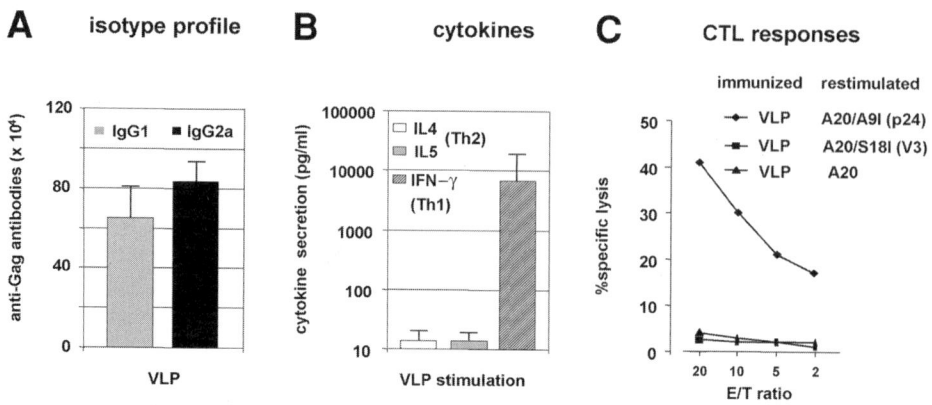

Fig. 2. Strength of Gag-specific humoral and cellular immune responses in BALB/c mice induced following subcutaneous (sc) injection of Pr55^{gag} virus-like particles (VLPs). BALB/c mice were sc immunized with 10 µg of Pr55^{gag} VLPs and boostered at wk 3 and 6 with the same dose of VLP antigens. Serum samples as well as splenic cells were recovered from sacrified mice at wk 2 after the second booster injection and analyzed for Gag-specific immune responses. **(A)** Gag-specific IgG1 and IgG2a antibodies were measured in serum by ELISA and are expressed as the reciprocal of the highest plasma dilution that resulted in an adsorbance value (OD = 495) three times greater than that of the same dilution of the corresponding preimmune serum with a cutoff value of 0.05. Each bar represents the group mean (n = 5) for anti-Gag titers, and vertical lines represent the SD. The numbers above each bar pair represent the calculated ratio of IgG1 to IgG2a anti-Gag isotypes. **(B)** Cytokine profile of in vitro Gag-stimulated splenocytes from mice immunized sc with purified VLP preparations. The values represent the means + SD of splenocytes of five mice per experiment. IL, interleukin; IFN, interferon. **(C)** Cytotoxic T-lymphocyte (CTL) activity in splenocytes from mice immunized im by needle injection or id by particle gun with the indicated Gag expression vectors. Lymphoid cells obtained from mice 5 d after the booster injection were cocultured with Gag peptide-pulsed syngeneic P815 mastocytoma cells (irradiated with 20,000 rads). Control assays included splenocytes of nonimmunized mice stimulated in vitro with peptide-pulsed P815 cells. Cytotoxic effector populations were harvested after 5 d of in vitro culture. The cytotoxic response was read against 9mer Gag-peptide-pulsed A20 cells and untreated A20 negative target cells in a standard ^{51}Cr release assay. The data shown are mean values of triplicate cultures. The SEMs of triplicate data were always less than 15% of the mean. E/T, effector/target.

ramers to detect epitope-specific T-cells via labeling of the cognitive T-cell receptor (TCR) *(48)*. This assay, however, is restricted to the knowledge of defined peptide epitopes and the corresponding human leukocyte antigen (HLA) haplotype. Although this method is elegant for analysis and tracking of defined cellular immune responses, it is rather impractical for cohort studies designed to assess the total cellular immune response to a complex antigen or immunogen during infection or vaccination *(49)*. An alternative assay format for determining polypeptide-specific T-cell responses is the multiparameter flow cytometric determination of intracellularily accumulated IFN-γ

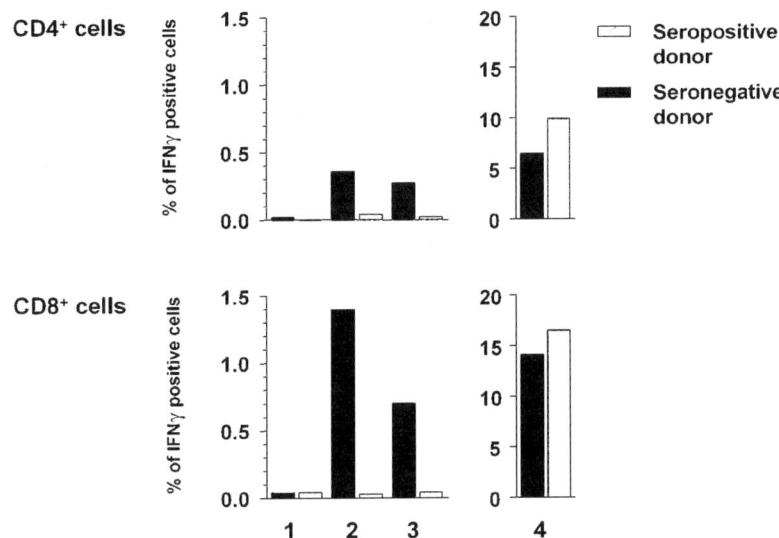

Fig. 3. Quantitation of HIV-specific CD8+ and CD4+ T-cells in an HIV-seropositive subject. Whole blood cells of a seropositive and a seronegative donor were stimulated by using (2) 5 μg/mL Pr55^{gag} virus-like particles (VLP) or (3) SDS-treated soluble Pr55^{gag} protein, respectively. (1) Negative control stimulations were carried out using equivalent quantities of wild-type baculovirus particles. (4) Positive controls comprised *Staphylococcus aureus* enterotoxin B (SEB). The frequency of antigen-specific T-cells is given as the relative number of CD4+ or CD8+ cells positive for both CD69 and IFN-γ.

after in vitro stimulation of blood cells in the presence of intracellular transport blockers such as Brefeldin A.

However, to date no experimental assay has been described that allows the simultaneous analysis of antigen-specific CD4+ and CD8+ T-cell frequency and function. This would be of particular interest because the outcome/prognosis of a virus infection depends on effective T-cell cooperativity *(50)*. Furthermore, the recent progress and need for immunotherapeutic strategies in the treatment of virus infections such as HIV require a rapid and simple assay that allows follow-up of cohorts and can be applied in a clinical setting. On the basis of recent observations on the detection of antigen-induced intracellular cytokine induction *(51)*, such an assay was developed. Pr55^{gag} VLPs were used to stimulate and subsequently quantify HIV-specific CD4+ and CD8+ T-cells from the whole blood of HIV-infected individuals *(52,53)*. Therefore, whole blood of HIV-infected individuals was specifically stimulated by HIV-1 Pr55^{gag}, and activation-induced intracellular cytokine expression in CD4+ and CD8+ T-cells was analyzed by flow cytometry. HIV-1-specific CD8+ and CD4+ T-cells can be quantified simultaneously (**Fig. 3**). As specific antigen, HIV-1 Pr55^{gag} virus-like particles were superior to soluble protein, especially for the activation of CD8+ T-cells. In untreated individuals, a high frequency of HIV-specific T-cells was observed.

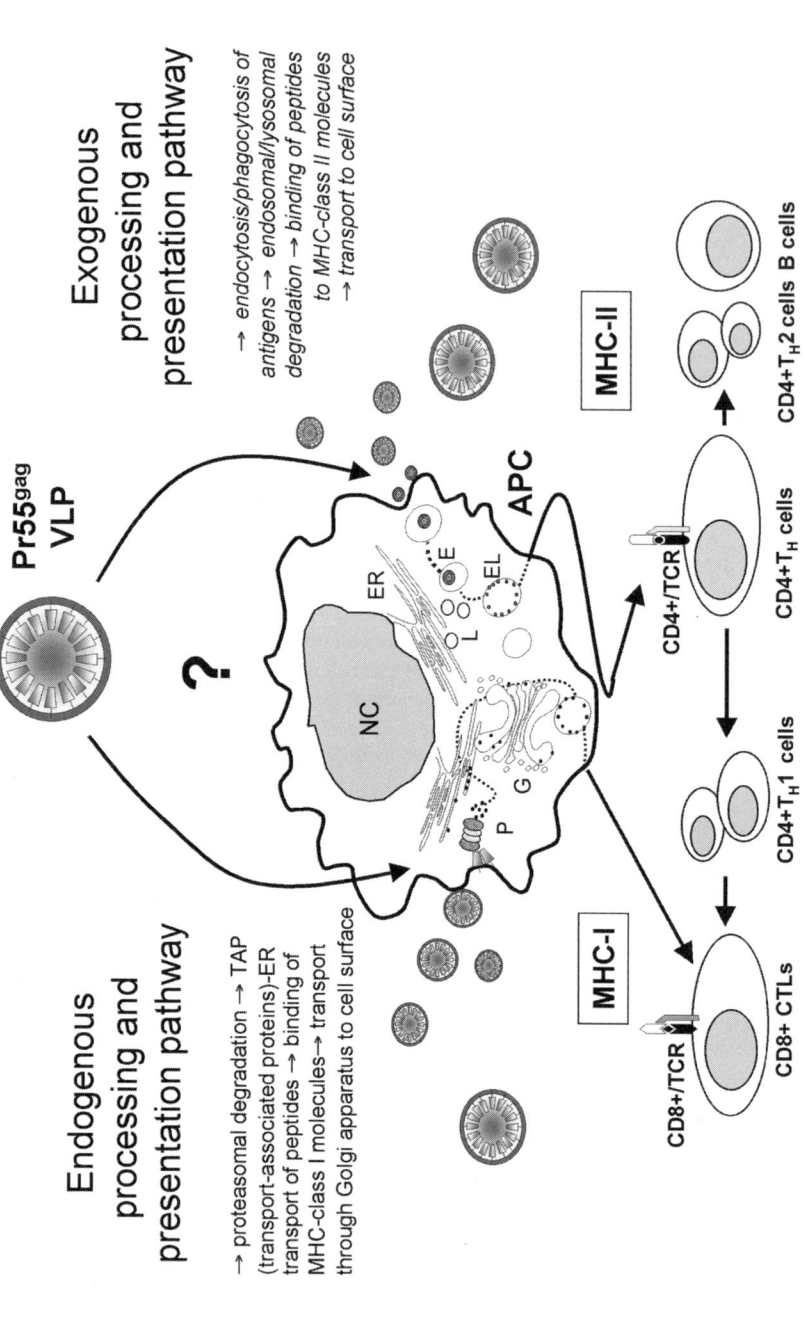

Fig. 4. Schematic drawing of supposed pathways for uptake, processing, and MHC class I and II presentation of virus-like particle-derived polypeptides resulting in the induction of both CD4+ and CD8+ T-cell responses. APC, antigen-presenting cell; CTL, CD8+ cytotoxic T-cells; E, endosome; EL, endolysosome; ER, endoplasmatic reticulum; G, Golgi apparatus; L, lysosome; MHC, major histocompatibility complex; NC, nucleus; P, proteosome; TCR, T-cell receptor; VLP, virus-like particles.

The frequency of CD8+ T-cells was consistently higher than the respective CD4+ T-cell response, thus demonstrating a dominance in CD8+ T-cell expansion in persistent HIV infection. Thus, the whole blood assay provides a rapid estimate of the total antiviral T-cell resources and is highly suited for a clinical setting. It may therefore have widespread applications for the evaluation of vaccination strategies and immunotherapy.

In summary, these investigations clearly demonstrate the potency of VLPs to introduce polypeptides into the MHC class I and MHC class II procession and presentation pathways, thus mediating the induction of strong humoral and cellular immune responses (**Fig. 4**). However, the mechanisms underlying these observations are yet not well understood and need to be investigated in more detail.

2. Materials
2.1. Expression of Pr55^{gag} VLPs Using the Baculovirus Expression System
2.1.1. Propagation of Insect Cells
1. *Spodoptera frugiperda* (SF9) insect cells.
2. *Trichoplusia ni* (HighFive) insect cells.
3. Insect-XPRESS medium (BioWhittaker, Walkersville, MA).
4. TC100 1X medium (Gibco/BRL, Eggenstein-Leopoldshafen, Germany).
5. Fetal calf serum (FCS) (Gibco/BRL).
6. Antibiotics. Stock solutions: Pen/Strep (10,000 IU/mL penicillin/10 mg/mL streptomycin) (PAN Biotech, Aidenbach, Germany).
7. Complete Insect-XPRESS medium: Insect-XPRESS medium supplemented with 1% (v/v) Pen/Strep.
8. Complete TC100 medium: TC100 medium supplemented with 10% FCS and 1% (v/v) Pen/Strep.
9. Tissue culture flasks, dishes, and plates (Falcon, Becton Dickinson Labwares, Bedford, MA).
10. 100- and 500-mL spinner flasks (Bellco, Vineland, NJ).
11. Magnetic stirrer (H + P Labortechnik, München, Germany).
12. Temperature-controlled incubator (Heraeus Instruments, Hanau, Germany).
13. Phosphate-buffered saline (PBS): 150 mM NaCl, 3 mM KCl, 4 mM Na$_2$HPO$_4$, 1.5 mM KH$_2$PO$_4$.
14. Trypane blue. Prepare a 0.4% (w/v) solution in PBS. Alternatively, the solution can be purchased ready-made (Gibco/BRL).
15. "Neubauer improved" hemocytometer.
16. Inverted tissue culture microscope (Helmut Hund, Wetzlar, Germany).

2.1.2. Construction of Plasmids
1. The baculovirus transfer vector *pVL1392* (Invitrogen, San Diego, CA). This plasmid is suitable for the expression of foreign genes using the baculovirus expression system. Heterologous genes are inserted into the polylinker downstream of the polyhedrin promoter and must provide their own ATG start codon.
2. Nucleic acids and oligonucleotides: genomic DNA of an HIV-positive donor or an appropriate vector containing the coding region of the HIV Pr55^{gag} polyprotein, a dNTP solution (20 mM) containing all four dNTPs, appropriate oligonucleotide primers (10 µM) in Tris-EDTA (TE) buffer (10 mM Tris-HCl, 1 mM EDTA, pH 8.0).
3. Enzymes: T4-DNA polymerase, *Eco*RI, *Bcl*I, *Bam*HI, T4-DNA ligase, calf intestinal phosphatase (all Roche Applied Science, Mannheim, Germany) and the appropriate 10X buffers; all used according to the manufacturer's instructions.

4. DNA sample loading dye: 0.25% (v/w) bromphenol blue, 0.25% (v/w) xylene cyanol, 50 mM Tris-HCl, pH 7.6, 60% (v/v) glycerol.
5. Agarose (FMC BioProducts, Rockland, ME).
6. TBE buffer: 40 mM Tris, 100 mM boric acid, 2 mM Na$_2$EDTA, pH 8.5.
7. Ethidium bromide solution: 0.5 mg/L ethidium bromide dissolved in H$_2$O; store in the dark.
8. Lambda *Bst*EII molecular weight markers (New England Biolabs, Schwalbach, Germany).
9. QIAquick PCR Purification Kit (Qiagen, Hilden, Germany).
10. QIAquick Gel Extraction Kit (Qiagen).
11. Chemically competent *Escherichia coli* K-12 derivatives such as DH5α, XL1-blue, JM83, and JM110 (Stratagene, La Jolla, CA).
12. LB (Luria-Bertani) broth: 10 mg/L tryptone, 5 mg/L yeast extract, 10 mg/L NaCl.
13. LB$_{amp}$: LB broth containing 100 µg/mL ampicillin.
14. LB agar plates. Media and agar plates were prepared according to standard protocols as described in detail *(54)*.
15. Alkaline lysis buffers for minipreparations: Solution I: 50 mM glucose, 25 mM Tris-HCl, pH 7.5, 10 mM EDTA, 100 µg/mL RNAse. Solution II: 0.2 N NaOH, 1% sodium dodecyl sulfate (SDS). Solution III: 3 M potassium acetate, pH 4.8.
16. Ethanol [100% and 70% (v/v) in H$_2$O].
17. Qiagen EndoFree Midi Kit (Qiagen).
18. Isopropanol.

2.1.3. Generation of Recombinant Baculoviruses

1. *S. frugiperda* (SF9) insect cells.
2. Complete TC100 medium (Gibco/BRL).
3. 6-well tissue culture plates (Falcon).
4. Baculoviral DNA (Invitrogen).
5. Sterile microcentrifuge tubes (Eppendorf).
6. 2.5 M CaCl$_2$; sterilize by filtration; aliquot and store at –20°C.
7. 2X HEPES-buffered saline (HeBs): 350 mM NaCl, 6.25 mM HEPES, 1.8 mM Na$_2$HPO$_4$ · 2 H$_2$O; adjust exactly to pH 7.05 with NaOH. Sterilize by filtration through a 0.22-µm nitrocellulose filter (Pall, Ann Arbor, MA) and store at –20°C until use.
8. Temperature-controlled incubator (Heraeus Instruments).

2.1.4. Determination of Viral Titers by Plaque Assay

1. 6-well tissue culture plates (Falcon).
2. 15-mL Falcon tubes.
3. Sterile 100-mL glass bottles.
4. Water bath.
5. Exponential culture of SF9 insect cells.
6. Agarose (BioWhittaker, Rockland, ME).
7. Complete TC100 medium.
8. 1.3X TC100 medium (Gibco BRL).
9. FCS.
10. Pen/Strep.
11. Complete 1.3X TC100 medium: 1.3X TC100 medium supplemented with 10% FCS and 1% (v/v) Pen/Strep.

12. Soft agar: 0.6% *low melting point* agarose (Gibco/BRL) in TC100 medium, 10% FCS.
13. Neutral red stock solution: dissolve 2 g neutral red in 250 mL 75% ethanol; after 2 days fill up to 1 L with H_2O.
14. Neutral red soft agar: add 4 mL neutral red stock solution to 100 mL soft agar.

2.1.5. Infection of Insect Cells with Recombinant Viruses

1. Exponential culture of HighFive insect cells.
2. Complete Insect-XPESS medium.
3. 75-cm^2 cell culture flasks.
4. Rocking platform (model 3016; GFL, Burgwedel, Germany).
5. Temperature-controlled incubator.

2.1.6. Purification and Biochemical Characterization of VLPs

1. 50-mL plastic centrifuge tubes (Greiner, Frickenhausen, Germany).
2. Cell centrifuge (Rotixa 120R; Hettich, Tuttlingen, Germany).
3. Sucrose solutions: prepare various sucrose solutions, containing 10, 20, 30, 40, and 50% (w/v) sucrose in PBS. Sterilize the solutions by filtration through 0.22-μM nitrocellulose filters (Pall).
4. Beckman ultracentrifuge.
5. Beckman SW28 rotor and centrifuge tubes (cat. no. 344058).
6. PBS.
7. Bio-Rad D_c protein assay (Bio-Rad, Hercules, CA).
8. An appropriate enzyme-linked immunosorbent assay (ELISA) kit for the quantitation of HIV p24 capsid protein (p24CA) (e.g., Coulter HIV p24 Ag assay, Beckman Coulter, Krefeld, Germany).
9. Spectrometer (e.g., Uvikon Spectrophotometer 930; Kontron Instruments, München, Germany).
10. Disposable polystyrene cuvets with 1-cm path length (Bio-Rad).
11. Appropriate buffers and solutions for SDS-polyacrylamide gel electrophoresis (PAGE) of proteins (*see* **Note 1**).
12. Coomassie Brilliant Blue R-250.
13. Methanol/acetic acid solution: mix 500 mL methanol and 400 mL H_2O bidest and fill up to 1 L with glacial acetic acid.
14. Appropriate buffers, solutions, and equipment for immunoblotting of proteins.
15. Monoclonal mouse anti-Gag antibody (16/4/2) *(44)*.

2.2. Evaluation of Immunogenicity of Pr55^{gag} in a Mouse Model

2.2.1. Immunization of Mice

1. Naive female BALB/c or C57BL/6 mice at an age of 44–60 d (Charles River, Sulzfeld, Germany or Harlan, Borchen, Germany) (*see* **Note 2**).
2. Hairclipper.
3. 1-mL syringe with 27-G needle (Becton Dickinson).
4. 70% ethanol.

2.2.2. Preparation of Single-Cell Cultures of Murine Splenocytes

1. 70% ethanol.
2. Equipment for euthanasia.
3. Dissecting instruments: sterile scissors and tweezers.
4. 15- and 50-mL plastic centrifuge tubes (Greiner).

5. Wash buffer: PBS, 1% FCS (Pan Biotech, Aidenbach, Germany).
6. Pipets: 5, 10, and 25 mL serological plastic pipets (Sarstedt, Nümbrecht, Germany).
7. 100-μm nylon cell strainer (Becton Dickinson Labwares).
8. 5-mL syringe (Becton Dickinson Labwares).
9. ACK hemolysis buffer: 150 mM NH$_4$Cl, 1 mM KHCO$_3$, 0.1 mM EDTA (all from Sigma-Aldrich), pH 7.2–7.4.
10. Splenocyte medium: RPMI-1640 with L-glutamine and 2 g/L NaHCO$_3$, supplemented with 5% FCS (PAN Biotech, Aidenbach, Germany), 20 mM HEPES (Sigma-Aldrich), and 0.1% 500 mM 2-mercaptoethanol, (Merck, Darmstadt, Germany) with or without 1% Pen/Strep solution (PAN Biotech).
11. "Neubauer improved" hemocytometer.

2.2.3. Determination of T-Cell Responses

2.2.3.1. ELISA

1. 96-well plates (F96 Maxisorb plates, Nunc, Wiesbaden, Germany).
2. Various OptEIA™ ELISA kits for the detection of interleukin (IL)-2, IL-4, IL-5, IL-6, IL-10, IL-12, tumor necrosis factor-α (TNF-α), IFN-γ, each containing a capture antibody, a detection antibody, an avidin-horseradish peroxidase conjugate, and a recombinant standard protein (BD, Heidelberg, Deutschland). The measurement of cytokines was performed according to the manufacturer's protocol.
3. Coating buffer: 0.1 M carbonate, pH 9.5, or 0.2 M sodium phosphate, pH 6.5.
4. Stimulator immunogen: purified Pr55^{gag} VLPs at a concentration of 2 μg/μL.
5. PBS.
6. Assay diluent: PBS, 10% FCS (Pan Biotech).
7. Wash buffer: PBS, 0.05% Tween 20.
8. TMB substrate solution (BD).
9. Stop solution: 2 N H$_2$SO$_4$.
10. ELISA reader (SLT Spectra, SLT Labinstruments, Germany).
11. ELISA software (WSOFTmax).

2.2.3.2. ELISPOT

1. 96-well plates: MAHA Multiscreen-HA plate (0.45 μm surfactant-free mixed cellulose ester membrane, Millipore, Bedford, MA).
2. Anti-IFN-γ antibody; (clone R4-6A2; BD).
3. PBS.
4. Wash buffer: PBS, 10% FCS.
5. Splenocyte medium: RPMI-1640, 5% FCS, 1% glutamine, 1% Pen/Strep (v/v), 20 mM HEPES, 50 μM 2-mercaptoethanol.
6. Stimulator polypeptides: purified Pr55^{gag} VLPs at a concentration of 2 μg/μL in PBS as well as the 9mer peptide A9I (AMQMLKETI), representing a D^d-restricted murine CTL epitope within the p24 capsid domain of the HIV-1$_{LAI}$ Gag polyprotein *(55)*, and the 18mer V3 peptide S18I (SIRIQRGPGRAFVTIGKI), which represents a defined D^d-restricted murine CTL epitope of HIV-1$_{LAI}$ gp160 *(56)* in BALB/c mice (Toplab, Martinsried, Germany). Dissolve the peptides to 10 μg/μL in dimethyl sulfoxide (DMSO) and store appropriate aliquots of the peptide solutions at –20°C until use.
7. Ionomycin stock solution: 1 mg/mL ionomycin dissolved in DMSO. Store aliquots at 4°C.

Virus-Like Particles

8. Lysis buffer: PBS, 0.05% Tween 20.
9. Biotinylated anti-IFN-γ antibody (clone XMG1.2; BD).
10. Dilution buffer: PBS, 10% FCS, 0.05% Tween 20.
11. Streptavidin-AP (Roche).
12. Nitroblue tetrazolium/bromochloroindolyl phosphate (NBT/BCIP) stock solution (Roche).
13. AP buffer: 100 mM Tris-HCl, pH 9.5, 100 mM NaCl, 50 mM MgCl$_2$.
14. Dissecting microscope (Helmut Hund).
15. ELISpot reader (Bioreader 2000, BIO-SYS, Karben, Germany).

2.2.3.3. ^{51}CHROMIUM RELEASE ASSAY

1. Splenocyte medium: RPMI-1640, 5% FCS, 1% glutamine, 1% Pen/Strep, 20 mM HEPES, 50 μM 2-mercaptoethanol.
2. CytoTox medium: α-MEM-Glutamax (Gibco/BRL), 10% FCS, 1% glutamine, 1% Pen/Strep.
3. Stimulator medium: α-MEM medium (Gibco-BRL), 10% FCS, 50 μM 2-mercaptoethanol. Add 1–10% of a selected batch of concanavalin A-stimulated rat spleen cell supernatants as a source of growth factors.
4. The p24-peptide A9I (AMQMLKETI) and the 18mer V3 peptide S18I (SIRIQRGPG RAFVTIGKI) at a concentration of 10 μg/mL in DMSO.
5. 25-cm^2 cell culture flasks (Nunc, Wiesbaden, Germany).
6. 15-mL Falcon tubes (Becton Dickinson Labwares).
7. Stimulator cells: single-cell suspension of murine H-2^d mastocytoma cell line P815 (TIB 64; ATCC, Rockville, MD).
8. Wash buffer: PBS, 1% FCS.
9. Source of radiation.
10. 96-well round-bottomed plates (BD).
11. Target cells: P815 cells.
12. 5-mL round-bottomed polystyrene tubes (Falcon).
13. 1 mCi/mL Na$_2$^{51}CrO$_4$ in isotonic medium (Amersham, Uppsala, Sweden).
14. PBS, 5% Triton X-100.
15. 96-well LumaPlates (Packard LumaPlate-96; Lumac-LSC 4096) (Canberra Packard, Schwadorf, Austria).
16. Packard Topcount (Canberra Packard).

2.2.3.4. INTRACELLULAR STAINING OF CYTOKINES BY FACS ANALYSIS

1. 96-well plates, round-bottomed and flat-bottomed (Nunc).
2. Splenocyte medium.
3. Brefeldin A (BFA) stock solution: 5 mg/mL BFA (Sigma-Aldrich) dissolved in ethanol. Freeze aliquots at –20°C.
4. Phorbol 12-myristate 13-acetate (PMA) stock solution: 0.1 mg/mL PMA (Sigma-Aldrich, Taufkirchen, Germany) dissolved in DMSO. Freeze aliquots at –20°C.
5. Ionomycin stock solution: 1 mg/mL ionomycin dissolved in DMSO. Store aliquots at 4°C.
6. PBS.
7. Fluorescence-activated cell sorting (FACS) buffer 1: PBS, 1% FCS, 1 mg/mL NaN$_3$.
8. FACS antibodies: anti-mouse CD8α-APC (clone 53-6.7), anti-mouse CD4-fluorescein isothiocyanate (FITC) (clone RM4-5), anti-mouse CD16/32 (clone 2.4G2), anti mouse IFN-γ-PE (cat. no. 554412) (all from BD).
9. FACS buffer 2 (PBS, 1 mg/mL NaN$_3$).

10. 4% paraformaldehyde (PFA) solution: dissolve 4 g PFA in 100 mL prewarmed PBS; store at 4°C.
11. Cytofix/Cytoperm (BD): 4% PFA, 1% saponin (Sigma).
12. Perm/wash buffer: PBS, 1% saponin.
13. 5-mL polystyrene round-bottomed tubes (Falcon cat. no. 35 2054; BD).
14. FACS Calibur, BD.

2.3. Measurement of Human Gag-Specific T-Cell Responses by FACS Analysis

1. Polypropylene tubes.
2. Antibodies: anti-CD28 (clone L293); anti-CD49d (clone L25.3); anti-CD4-PerCP (clone SK3); anti-CD8-PerCP (clone SK1); anti-CD69-PE (clone L78); anti-IFN-γ-FITC (clone 4S.B3) (all from BD).
3. *Staphylococcus aureus* enterotoxin (SEB) (Sigma).
4. Brefeldin A (Sigma).
5. Lysing solution (BD).
6. FACS buffer: PBS, 5% FCS, 0.5% BSA, 0.07% NaN_3.
7. 1% paraformaldehyde solution.

3. Methods

Methods that are not specifically described in detail can be performed as previously described by Sambrook and co-workers *(54)* or according to the manufacturer's protocol.

3.1. Expression of Pr55^{gag} VLPs Using the Baculovirus Expression System

The following section details the methods for producing Pr55^{gag} VLPs using the baculovirus expression system.

3.1.1. Propagation of Insect Cells: General Handling Techniques

Both *T. ni*–derived HighFive and *S. frugiperda*–derived SF9 insect cells are routinely cultured at 26°C in a temperature-controlled incubator and can be grown at temperatures from 24 to 27°C. The cells do not require CO_2 supplementation and can be easily transferred between monolayer and suspension cultures without loss in viability or growth. HighFive cells are routinely propagated in plastic flasks using serum-free Insect-XPRESS medium supplemented with an antibiotic (e.g., Pen/Strep) and should be subcultured two or three times a week. HighFive cells can be split 1:3–1:10 and still retain their ability to grow well.

SF9 cells are routinely grown in suspension cultures using TC100 medium supplemented with 10% FCS and an appropriate antibiotic (e.g., Pen/Strep). Cells are seeded at a density of 2–3×10^6 cells/mL and reach a maximum density of approximately 2–3×10^7 cells/mL within 3 days of culture. For propagation of recombinant baculoviruses, cells are seeded in plastic flasks with a density of 9×10^6 cells/75-cm^2 flask.

3.1.1.1. SUBCULTURING IN FLASKS

1. Warm up the medium to room temperature before use.
2. Resuspend HighFive cells from a nearly confluent culture by gently punching against the plastic flask and repeated pipeting of the obtained cell suspension against the bottom of the flask.

3. Transfer 2–3 mL of the cell culture to a new 75-cm² flask containing 10 mL of fresh complete Insect-XPRESS medium.
4. Incubate at 26°C in a temperature-controlled incubator.
5. Subculture the cells every third to fourth d as described in **steps 2–4**.

3.1.1.2. SUSPENSION CULTURE

1. Determine the SF-9 (>95% viable; *see* **Note 3**) cell density using a hemocytometer.
2. Add an appropriate volume of prewarmed complete TC100 medium to the spinner flask. Inoculate cells to a starting density of about $2–3 \times 10^6$ cells/mL.
3. Incubate the spinner at 26°C with constant stirring at 50–60 rpm. Leave the lid of the vessel open one-fourth of a turn to ensure good aeration.
4. For routine maintenance, subculture the cells at a density of about $1–2 \times 10^7$ cells/mL (about twice a week).
5. To subculture, remove 80% or more of the suspension culture and replace with an equal volume of fresh medium (*see* **Note 4**).

3.1.2. Construction of Plasmids

1. Generate an *Eco*RI/*Bcl*I fragment encoding *gag* by specific polymerase chain reaction (PCR) amplification (*see* **Note 5**).
2. Ligate the *Eco*RI/*Bcl*I digested *pVL1392* expression vector with the PCR-derived Gag-coding region. The resulting vector is designated *pVL1392-gag*.
3. Transform *E. coli* DH5α cells and plate on LB agar plates containing 50 µg/mL ampicillin. Incubate overnight at 37°C.
4. Test individual colonies for the presence of the desired foreign gene by digestion of "miniprep" DNA with appropriate restriction enzymes.
5. Produce appropriate amounts of endotoxin-free lots of the expression plasmid *pVL1392-gag* (*see* **Note 6**).

3.1.3. Generation of Recombinant Baculoviruses

1. The evening before transfection, seed 2×10^5 SF9 cells/well in a 6-well plate, and incubate cells over night at 26°C in an incubator.
2. Replace cell culture supernatant by 3 mL fresh medium 30–60 min before transfection.
3. Add 30 µL 2.5 M CaCl$_2$ to a mixture of 2 µg wild-type baculovirus DNA and 13 µg of plasmid *pVL1392-gag* in 270 µL sterile H$_2$O.
4. To form the calcium phosphate–DNA co-precipitate, gently vortex at half-maximal speed 300 µL 2X HeBs buffer and add 300 µL of the DNA/CaP mixture of in a dropwise fashion. Incubate the tube for 15–20 min at room temperature.
5. Pipet the transfection solution slowly and dropwise to the cells.
6. Carefully shake the plates and incubate the cells for 10–16 h at 26°C.
7. Replace the medium from each dish of transfected cells with 5 mL fresh complete TC100 medium.
8. Incubate for 5–7 d at 26°C.
9. Harvest the supernatant and remove cellular debris by low-speed centrifugation at 300*g* for 5 min. Store the virus stocks at 4°C until use. Within standard serum-supplemented medium, the viruses are quite stable and will maintain infectiousness for several months.

3.1.4. Determination of Viral Titers by Plaque Assay

Determination of titers of recombinant baculoviruses may be accomplished by plaque formation in an immobilized monolayer culture.

1. Plate each 1×10^6 SF9 cells/well of three 6-well plates in 2 mL complete TC100 medium. Use cells that are in midlog phase of growth (*see* **Note 7**).
2. Allow cells to attach at 26°C for at least 1 h.
3. Place a 100-mL flask containing 40 mL of a 4% low-melting-point agarose solution in TC100 in a 70°C water bath. Place an empty 100-mL bottle and a bottle of 1.3X complete TC100 medium in a 40°C water bath.
4. Following a 1-h incubation of the plates, observe monolayers to cell attachment and confluency (*see* **Note 8**).
5. Produce eight-log serial dilutions of the virus stock by sequentially diluting 0.5 mL of the previous dilutions in 4.5 mL of complete TC100 medium in 12-mL disposable tubes. Conclude with eight tubes containing each 5 mL each of a 10^{-8}–10^{-1} dilution of the original virus stock.
6. Label the plates equivalent to the compounded dilutions. Analyze each dilution in duplicate experimental setups.
7. Sequentially replace the supernatant of each duplicate well with 1 mL of the respective virus dilution. For negative control, add virus-free medium to seeded SF9 cells. Incubate for 1 h at 26°C.
8. Prepare TC100 plaquing overlay: move bottles from water bath (from **step 3**) to a sterile laminar flow when agarose has liquefied (20–30 min). Quickly dispense 10 mL of the 4% agarose solution and 30 mL of complete 1.3X TC100 medium to the empty bottle and mix gently. Return the bottle with the plaquing overlay to the 40°C water bath until use.
9. Sequentially (from high to low dilution) remove the virus inoculum from each well, and place with 2 mL of the diluted agarose. Work quickly to avoid desiccation of the monolayer.
10. Allow the gel to solidify for 10–20 min.
11. Incubate for 3 d at 26°C.
12. Place 100 mL neutral red soft agar in a 60°C water bath till agarose has liquified (20–30 min).
13. Add 2 mL of neutral red soft agar to each well and incubate for additional 3–10 d.
14. Monitor plaques daily until the number of plaques counted does not change for two consecutive days. Plaques can be identified using an inverted light microscope (*see* **Note 9**). The titer (plaque-forming units [PFU]/mL) may be calculated by the following formula: PFU/mL = 1/dilution factor × number of plaques × 1/(mL of inoculum/plate).

3.1.5. Infection of Insect Cells with Recombinant Viruses

1. Seed approximately 9×10^6 HighFive cells/75-cm² flask in 12 mL serum-free Insect-XPRESS medium (*see* **Note 8**). Use cells that are in midlog phase of growth.
2. Allow cells to attach for at least 2 h at 26°C.
3. In between, prepare a virus dilution containing 0.5 mL of a stock of recombinant baculovirus AcPr55 (containing approx 5×10^7 infectious viruses/mL) and 3 mL complete Insect-XPRESS medium per 75-cm² flask (*see* **Note 10**).
4. Replace the medium from each 75-cm² flasks with 3.5 mL of infection solution. Carefully rotate the flasks to spread the inoculum over the complete monolayer of cells.
5. Incubate the flasks on a slowly rocking platform at room temperature for 30 min (*see* **Note 11**).
6. Add 9 mL Insect-XPRESS medium/75-cm² flask (*see* **Note 12**) and incubate the flasks for 3–4 d at 26°C in a temperature-controlled incubator.

3.1.6. Purification of VLPs

1. Transfer the cell culture supernatants of AcPr55-infected HighFive insect cells in 50-mL Falcon tubes and remove contamination cells and cell debris by low-speed centrifugation (300g for 15 min at room temperature).
2. Combine the precleared cell culture supernatants in a plastic flask and store them on ice until use.
3. Prepare a 30% sucrose cushion in a clear ultracentrifuge tube. Therefore, pipet 20 mL of the cell culture supernatant in a 38-mL centrifuge tube. Place in a 10-mL plastic pipet 5 mL of a 30% sucrose solution below the supernatant and fill up the tube with an additional 10 mL of cell culture supernatant.
4. Centrifuge the tubes at 83,000g (25,000 rpm in a Beckman SW28 rotor) for 2.5 h at 16°C.
5. Pour out the complete liquid, add 100–200 µL sterile PBS to the remaining pellet, and seal the tube with a Parafilm.
6. Allow the pellet to swell in PBS overnight on ice.
7. Resuspend the pellet with a pipet and combine the suspensions in a 2-mL Eppendorf tube.
8. Prepare a 12 mL 10–50% sucrose gradient in a 13-mL centrifuge tube and gently layer 100 µL of the suspension on the top of the gradient.
9. Centrifuge the tubes at approximately 100,000g (28,000 rpm in a Beckman TFT 41.14 rotor) for 2.5 h at 16°C.
10. Use a 1000 µL pipet to collect 600-µL fractions from the top of the gradient.
11. Analyze the content of Pr55^{gag} antigen using a commercial p24 capture assay.
12. Combine p24-positive samples and dialyze them against PBS in a molecular porous membrane tubing with a molecular weight cutoff of 20 kDa.
13. Determine protein concentration of collected fractions using a Bio-Rad D_c protein assay according to the manufacturer's instructions.
14. Analyze the identity and purity of the purified VLP preparation by immunoblotting and Coomassie staining of 12.5% SDS-PAGE, as previously described by Sambrook and coworkers *(54)* or according to the manufacturer's protocol.

3.2. Evaluation of Immunogenicity of Pr55^{gag} in a Mouse Model

3.2.1. Immunization of Mice

1. Dissolve the purified VLPs at a concentration of 10 µg/mL in sterile PBS and homogenize the injectable by vortexing.
2. Use a clipper to remove hair from an area at the base of the tail and wipe with 70% ethanol on a gauze swap.
3. Fill syringe with injectable and remove air bubbles by gentle tapping with a finger against the syringe.
4. Inject 100 µL of the injectable (in absence of additional adjuvant) under the skin at the base of the tail. Inject the immunogen with moderate pressure and speed.
5. Apply booster injections at wk 3 and 6 after the primary immunization.

3.2.2. Preparation of Single-Cell Cultures of Murine Splenocytes

1. Euthanize the BALB/c or C57BL/6 mice by cervical dislocation.
2. Remove the spleens aseptically by necrosection.
3. Place the spleens in 5 mL wash buffer per spleen in a 50-mL Falcon tube. Keep the spleens at room temperature.

Perform the following steps under a laminar flow bench; use sterile plastic serological pipets only in order to prevent potential cell damage by residual traces of detergent within cleaned glass pipets.

4. Place a cell strainer on a 50-mL Falcon tube.
5. Remove almost all medium from the 50-mL Falcon tube, thereby retaining the spleen.
6. Transfer the spleen into the cell strainer placed on a 50-mL Falcon tube.
7. Generate a single-cell suspension by grinding the spleen against the cell strainer with the plunger of a 5-mL syringe until mostly fibrous tissue remains.
8. Aspirate the remaining cells from the cell strainer by repeatedly adding 2 mL wash buffer and short-time lifting the cell strainer from the Falcon tube.
9. Discard cell strainer and the fibrous tissue.
10. Centrifuge the obtained cell suspension for 5 min at 300g at room temperature and discard the supernatant.
11. Resuspend the pellet in 5 mL ACK hemolysis buffer by gently but thoroughly pipeting with a 10-mL plastic pipet.
12. Directly centrifuge the cell suspension at 300g for 5 min at room temperature.
13. Discard the complete supernatant.
14. Resuspend the pellet by repeated rubbing the 50-mL Falcon tube against the grid of the laminar flow bench.
15. Add 5 mL wash buffer to the resuspended splenocytes and mix gently.
16. Incubate at room temperature for 1 min and separate the cell suspension from sedimented fibrous tissue.
17. Centrifuge the remaining cell suspension at approx 300g for 5 min at room temperature.
18. Discard most of the supernatant, but leave 0.1 mL medium on the pellet.
19. Repeat washing of cells (*see* **steps 15–19**) twice.
20. Resuspend the pellet in 5 mL of splenocyte medium.
21. Adjust the cell number to a final concentration of 2×10^7 cells/mL in splenocyte medium.
22. Preserve the cells at 37°C in a humid atmosphere until needed.

3.2.3. Determination of T-Cell Responses
3.2.3.1. ELISA

1. Fill up 5–10 µg of purified Pr55^{gag} VLPs to a total volume of 100 µL with ice-cold PBS in an 1.5 mL Eppendorf tube.
2. Add 100 µL of the splenocyte suspension (2×10^6 cells) to 100 µL of the suspension with stimulator and fill up to 1 mL with splenocyte medium.
3. Mix once by gently pipeting the suspension up and down.
4. Transfer the mixture into one of the inner eight wells of a 24-well tissue culture plate (*see* **Note 13**).
5. Fill the 16 outer wells of the 24-well tissue culture plate with sterile PBS and tape the plate to the lid on the longer sides with Scotch tape to prevent evaporation of the inner wells.
6. Incubate the plate at 37°C for 4–36 h (*see* **Note 14**).
7. Transfer the cell culture supernatant from tissue culture plate into a 1.5-mL Eppendorf tube.
8. Centrifuge for 10 min at 1300g at room temperature.
9. Transfer the supernatant into a new 1.5-mL Eppendorf tube and store at –80°C until needed.

3.2.3.2. ELISpot

The protocol listed below is a modified version of a manual provided by Millipore (Bedford, MA). Perform consecutive **steps 1–8** under a sterile laminar flowhood.

1. Coating of ELISpot plates: dilute the coating antibody (anti-IFN-γ antibody, clone R4-6A2; BD Pharmingen) to 5–10 μg/mL in sterile PBS.
2. Add 100 μL to each well of a MAHA-S45 plate and incubate overnight in a wet chamber at 4°C.
3. Stimulation and ELISpot analysis: obtain single-cell culture of splenic cells as described in **Subheading 3.2.2., steps 1–22**.
4. Prepare stimulators as follows:
   Unstimulated control: 100 μL splenocyte medium.
   Specific stimulation: to 100 μL splenocyte medium, add appropriate concentrations (1–10 μg/mL) polypeptide or appropriate volumes of a stimulator suspension.
   Positive control: to 100 μL splenocyte medium add 1 μL/well of each PMA and ionomycin.
5. Aspirate the wells and wash 5–10 times with 300-μL/well wash buffer. After the last wash, invert the plate and beat on adsorbent paper to remove residual buffer.
6. Block plates with 300 μL/well PBS, 5% FCS. Incubate at room temperature for 1–2 h.
7. Remove the blocking buffer and add 100 μL of the cell suspension to each well.
8. Add 100 μL of one of the stimulator solutions.
9. Incubate for 24 h at 37°C and 5% CO_2 in a humidified atmosphere.
10. Remove cells by washing six times with 200 μL PBS, 0.05% Tween 20.
11. Dilute biotinylated anti-IFN-γ monoclonal antibody in PBS, 10% FCS, 0.05% Tween 20 to 1 μg/mL.
12. Add 100 μL of the antibody solution to each well and incubate for 2 h at room temperature.
13. Wash the plates 6 times with 200 μL PBS, 0.05% Tween 20.
14. Add 100 μL streptavidin–alkaline phosphatase (diluted 1:1000 in PBS, 0.5% FCS) and incubate for 1 h at room temperature.
15. Wash the plates a minimum of 10 times with 200 μL PBS.
16. Add 100 μL of a substrate solution (e.g., BCIP/NBT; Roche Diagnostics) and incubate until dark spots emerge.
17. Stop color development by washing in running water.
18. Leave the plates to dry, inspect, and count spots in a dissection microscope (×40) or in an ELISpot reader.

3.2.3.3. ^{51}Chromium Release Assay

1. For the preparation of stimulator cells, grow adequate amounts of syngeneic P815 cells in splenocyte medium.
2. Harvest cells by centrifugation at 300*g* for 5 min at room temperature.
3. Resuspend the pellet in CytoTox medium.
4. Count cells and adjust to 10^7 cells/mL in CytoTox medium.
5. Transfer the needed volume of cell suspension (use 10^6 for each stimulation) into a 25-cm^2 cell culture flask, add 1 μL/mL cell suspension of an appropriate peptide solution, and mix the cells by gently flipping the flask.
6. Incubate the flask upright for 1 h at 37°C and 5% CO_2 in a humid incubator.
7. In parallel, prepare nonstimulated P815 cells as negative control.

8. Radiate the cell suspensions with 200 gray (20,000 rad) according to the manufacturer's manual.
9. Sediment cells at $300g$ for 5 min at room temperature.
10. Count cells and adjust to 10^6 cells/mL in CytoTox medium, and store intermediately at 37°C.
11. For the in vitro restimulation of splenic cells, generate cultures of splenic cells as described in **Subheading 3.2.2**. Adjust the cells to a final concentration of 2×10^7/mL in splenocyte medium.
12. Mix 1.5 mL (3×10^7) of spleen cell suspension with 1 mL (10^6) stimulator cells (approach each with either pulsed or unpulsed P815 cells) and fill up to 10 mL with stimulator medium.
13. Incubate the culture flasks upright for 5 d at 37°C and 5% CO_2 in a humidified incubator.
14. Transfer effector cells to a 15-mL conical tube, thereby making sure that you do not loose adherent cells.
15. Sediment cells by centrifugation for 5 min at $300g$ at room temperature, and resuspend the cell pellet in CytoTox medium.
16. Adjust to 2×10^7 cells/mL in a tightly capped tube and incubate cells at 37°C and 5% CO_2 in a humidified incubator until determination of CTL activity by a ^{51}Cr release assay.
17. For the determination of CTL activity by a chromium release assay, grow adequate numbers of P815 cells in splenocyte medium (for each 96-well plate you need 3×10^6 P815 cells in minimum).
18. Harvest cells by centrifugation at $300g$ for 5 min at room temperature.
19. Resuspend cell pellet in CytoTox medium.
20. Count cells and adjust to 5×10^6 cells/mL in CytoTox medium.
21. Transfer 200 µL of the cell suspension to a 5-mL Falcon tube.
22. Label cells by adding 250 µCi of $^{51}NaOCr_2$. Close the cap only loosely to allow air exchange.
23. Incubate cells for 1.5 h at 37°C and 5% CO_2 in a humidified incubator.
24. Wash cells twice with 3 mL of CytoTox medium by centrifuging them at $300g$ at room temperature for 5 min (approximately 10% of the radioactivity will be taken up by the cells at this time point.)
25. Count cells again and adjust to 10^6 cells/mL in CytoTox medium.
26. Transfer 250 µL of cell suspension in a new Falcon tube.
27. Add 1 µL of peptide solution, mix by gently by flipping the flask, and incubate for 1 h at 37°C. Unpulsed P815 cells used as negative controls to calculate the background CTL activity also have to be prepared according to **steps 13–22**.
28. During the incubation steps, prepare 96-well plates for making serial dilutions of the effector cells as follows. Pipet 100 µL of CytoTox medium in wells 4–12 of line A–G as well as in wells 1–6 of line H (Wells 1–6 of line H are used to determine the spontaneous release, and wells 7–12 are used to measure the total amount of radioactivity.)
29. Pipet 100 µL of effector cell suspension in wells 1–6 of line A–G. Prepare serial 1:2 dilutions in triplicate using a multichanel pipet as follows. Mix the cell suspension in lane 4 and transfer 100 µL to lane 7; repeat mixing and transfer 100 µL to lane 10; repeat mixing and discard 100 µL.
30. Repeat these steps for lanes 5/8/11 and 6/9/12.
31. Incubate the plate at 37°C in a humidified, temperature-controlled incubator with an atmosphere of 5% CO_2 until the labeled target cells are ready to use.

32. For each 96-well plate, pipet 2×10^5 target cells and 2×10^6 nonpulsed nonlabeled P815 cells in a Petri dish and fill up to 12.5 mL with CytoTox medium (see **Note 15**).
33. Pipet 100 µL to each well including line H of the prepared 96-well plates.
34. Incubate the plate for 3.5 h in a humidified 37°C, 5% CO_2 incubator.
35. Add 100 µL of PBS with 5% Triton X-100 to wells 7–12 in line H and roughly mix the suspension by pipeting up and down several times.
36. Carefully pipet 100 µL of each well to a 96-well LumaPlate by using a multichannel pipet without touching the cell pellets at the bottom of the wells. Alternatively, plates can be centrifuged at 300g for 1 min to attach the cells at the bottom of the wells.
37. Dry the plates overnight at room temperature.
38. Seal the plate with an adhesive plastic foil.
39. Count ^{51}Cr in a gamma scintillation counter according to the counter manual.
40. Calculate corrected percent lysis for each concentration of effector cells, using the mean cpm from each replicate of wells.
41. Corrected % lysis = [(measured ^{51}Cr release – spontaneous ^{51}Cr release)/(total ^{51}Cr amount – spontaneous ^{51}Cr release)] × 100 (see **Note 16**).
42. Represent CTL data in lytic units, graphs, or lysis values

3.2.3.4. INTRACELLULAR STAINING OF CYTOKINES BY FACS ANALYSIS

Obtain single cell culture of splenic cells described in **Subheading 3.2.2., steps 1–22**.

1. Aliquot 100 µL (= 2×10^6 cells) of cell suspension/well into a 96-well flat-bottomed tissue culture plate.
2. Prepare stimulators as follows:
   Unstimulated control: to 100 µL splenocyte medium add 2 µL/mL BFA stock solution.
   Specific stimulation: to 100 µL splenocyte medium add 10 µg/mL of an appropriate polypeptide and 2 µL/ml BFA stock solution.
   Positive control: to 100 µL splenocyte medium add 1 µL/ml PMA, 1 µL/mL ionomycin, and 2 µL/mL BFA (each of the described stock solution).
3. Add 100 µL of the appropriate activator to the center of each well.
4. Incubate for 6 h at 37°C.
5. Centrifuge at 300g for 5 min. Decant by turning the plate upside down; then resuspend the cells by smooth shaking of the plate.
6. Wash the cells twice each with 200 µL FACS buffer 1 per well. Transfer the cells after the first wash step to 96-well round-bottomed plates. Spin the plates as described in **step 5** and decant the wash buffer.
7. For blocking of Fc receptors, add 100 µL FACS buffer 1 with 1 µL anti-CD16/32 antibody.
8. Incubate plates for 10 min at 4°C.
9. Add 10 µL of a 1:20 dilution (in FACS buffer 1) of staining antibody (anti-CD4-FITC or anti-CD8-APC) to the appropriate wells.
10. Incubate for 20–25 min on ice in the dark.
11. Wash the cells twice with FACS buffer 2 as described in **step 6**.
12. Add 250 µL Cytofix/Cytoperm to each well and resuspend cells gently by flipping against the tube.
13. Incubate 20 min at room temperature in the dark.
14. Wash twice with Perm/wash; centrifuge with 500g for 4–5 min (see **Note 17**).

15. Prepare intracellular staining antibody (anti-IFN-γ-PE) by diluting 1:100 in Perm/wash buffer. Resuspend cells by flipping against the tube.
16. Incubate for 25 min on ice in the dark.
17. Wash three times with 200 μL Perm/wash as described in **step 14**.
18. Add 200 μL of FACS buffer 1 to each well.
19. Store light protected for up to 48 h.
20. When ready for acquisition, resuspend cells with a pipet and transfer to FACS tubes.
21. Measure the number and identity of IFN-γ producing cells with a FACSscan analytical instrument [e.g., FACS Calibur (BD) or Coulter EPICS (Beckman Coulter)] and appropriate software for data acquisition and analysis.

3.3. Measurement of Human Gag-Specific T-Cell Responses by FACS Analysis

1. Collect 5–10 mL whole blood from HIV-seropositive and, for control, -seronegative donors in sterile tubes containing sodium heparine (at a concentration of approx 20 U/mL; *see* **Note 18**).
2. Stimulate at least 150 μL heparinizied blood per sample with either 5 μg/mL VLP or 10 μg/mL *Staphylococcus aureus* enterotoxin B (SEB; positive control) or an appropriate volume of sample buffer without stimulator (negative control), all in the presence of 1 μg/mL of the costimulatory antibodies anti-CD28 and anti-CD49d.
3. Incubate the samples for up to 2 h at 37°C in a humidified incubator with a 5% CO_2 atmosphere.
4. Add 10 μg/mL Brefeldin A to each sample.
5. Incubate for up to 4 h at 37°C and 5% CO_2 in a humidified incubator.
6. Add 16 μL of a 20 mM EDTA solution to each tube (for a final concentration of 2 mM EDTA), vortex vigorously, and incubate for 15 min at room temperature.
7. Fix and lyse the cells for 10 min at room temperature using 1X lysing solution (BD) according to the manufacturer's instructions.
8. Centrifuge at 300*g* for 5 min.
9. Discard the supernatant and resuspend the cell pellet in FACS buffer.
10. Centrifuge again at 300*g* for 5 min and discard the supernatant.
11. Resuspend the pellet in 2 mL FACS buffer.
12. Store samples at 4°C until staining procedure.
13. Centrifuge the cell suspension at 300*g* for 5 min and discard the supernatant.
14. Add 2 mL FACS buffer containing 0.1% saponin.
15. Incubate for 10 min at room temperature.
16. Add 10 μL each of anti-CD69-PE and anti-IFN-γ-FITC and either anti-CD4-PerCP or anti-CD8-PerCP to obtain an antibody cocktail.
17. Incubate for 30 min at room temperature in the dark.
18. Wash cells twice with 3 mL FACS buffer as described in **step 13**.
19. Resuspend the pellet in 1 mL of 1% paraformaldehyde solution.
20. Store light protected for up to 24 h.
21. When ready for acquisition, resuspend cells with a pipet and transfer to FACS tubes. Measure the number and identity of IFN-γ producing cells with a FACSscan analytical instrument [e.g., FACS Calibur (BD) or Coulter EPICS (Beckman Coulter)] and an appropriate software for data acquisition and analysis. Analyze at least 30,000 CD4+ or CD8+ T-lymphocytes.

4. Notes

1. All buffers and solutions are prepared according to standard protocols *(54)*.
2. Mice were purchased from commercial breeders (e.g., Charles River, Harland) at an age of 44 d and housed under specific pathogen-free conditions. All immunological investigations were performed with 6–12 wk-old mice.
3. Cell viability can be determined by adding 180 µL of trypan blue solution to 20 µL of cell suspension and direct examination under an inverted tissue culture microscope at low magnification. Cells that soak up trypan blue are considered nonviable. Cell viability should be at least 95% for healthy log-phase cultures.
4. Aeration may be required in large flasks for optimal growth of the cells. The optimal aeration conditions have to be determined individually.
5. All PCR techniques are performed according to standard protocols *(57)*.
6. Perform all DNA purification steps as recommended by the Qiagen Plasmid Purification handbook and enclosed in the EndoFree Kits (Qiagen, Hilden, Germany).
7. Cells used for transfection, virus amplification, or protein expression by infection with recombinant viruses should be at midlog or exponential growth phase to achieve optimal virus yield and protein expression.
8. Infect the cells with recombinant baculoviruses when the monolayer has reached a confluency of 50–60% to achieve optimal protein production.
9. Plaques from wild-type baculoviruses tend to be nearly white in appearance, and cells within the plaques will contain large angular inclusion bodies. In contrast, recombinant viruses produce bright plaques that do not include inclusion bodies. To determine the virus titer of the inoculum, an optimal range of count is 5–20 plaques/well of a 6-well plate.
10. The quality and yields of the polypeptides produced vary between cell lines, and infection kinetics of recombinant viruses may be dependent on multiple parameters such as the multiplicity of infection (MOI) of the recombinant virus as well as the incubation time post infection. Thus, for evaluation of the optimal parameters of polypeptide production, we recommend the use of varying MOIs (e.g., 0.1, 0.5, 1, 5, 10) and time intervals (e.g., 24, 36, 48, 72, and 96 h).
11. During the incubation, the rocking plate temperature should not surpass 30°C. Therefore, we recommend incubating the plates on a wood-fiber board, taking care that the growth surface is always wet.
12. Pour in the medium with a plastic pipet at the upside of the flask to avoid washing up the adherent cells.
13. Use only the inner eight wells of the 24-well plate to avoid inconsistent results, owing to enhanced evaporation of liquid from the outer wells.
14. We recommend a short incubation for the detection of IL-12 and TNF-α and a long incubation for analyzing other cytokines such as IL-2, IL-4, IL-5, IL-6, IL-10, and IFN-γ.
15. Nonpulsed, nonlabeled cells are utilized to reduce the background activity of lysis as a result of unwanted recognition P815 cells epitopes by the effector cells after 5 d of in vitro restimulation.
16. Release of ^{51}Cr from target cells incubated with Triton X-100 is referred to as maximum release; ^{51}Cr release from target cells incubated with medium indicates spontaneous release.
17. At this point cells are permeabilized and require stronger centrifugation to sediment.
18. **Caution:** When working with the blood of human HIV-infected and noninfected donors, biosafety practices must be followed.

Acknowledgments

The excellent technical assistance of Simon Bredl, Denijal Kosovac, and Juha Lindner is appreciated.

References

1. Rook, A. H. (1988) Interactions of cytomegalovirus with the human immune system. *Rev. Infect. Dis.* **10,** 460–467.
2. Arvin, A. M. (1992) Cell-mediated immunity to varicella-zoster virus. *J. Infect. Dis.* **166,** 35–41.
3. Simmons, A. and Tscharke, D. C. (1992) Anti-CD8 impairs clearance of herpes simplex virus from the nervous system: implications for the fate of virally infected neurons. *J. Exp. Med.* **175,** 1337–1344.
4. Gotch, F., Gallimore, A., and McMichael, A. (1996) Cytotoxic T cells—protection from disease progression—protection from infection. *Immunol. Lett.* **51,** 125–128.
5. Khanna, R., Moss, D. J., and Burrows, S. R. (1999) Vaccine strategies against Epstein-Barr virus-associated diseases: lessons from studies on cytotoxic T-cell-mediated immune regulation. *Immunol. Rev.* **170,** 49–64.
6. Khanna, R. and Burrows, S. R. (2000) Role of cytotoxic T lymphocytes in Epstein-Barr virus-associated diseases. *Annu. Rev. Microbiol.* **54,** 19–48.
7. Pantaleo, G., Demarest, J. F., Soudeyns, H., et al. (1994) Major expansion of CD8+ T cells with a predominant V beta usage during the primary immune response to HIV. *Nature* **370,** 463–467.
8. Koup, R. A., Safrit, J. T., Cao, Y., et al. (1994) Temporal association of cellular immune responses with the initial control of viremia in primary human immunodeficiency virus type 1 syndrome. *J. Virol.* **68,** 4650–4655.
9. Rinaldo, C., Huang, X. L., Fan, Z. F., et al. (1995) High levels of anti-human immunodeficiency virus type 1 (HIV-1) memory cytotoxic T-lymphocyte activity and low viral load are associated with lack of disease in HIV-1-infected long-term nonprogressors. *J. Virol.* **69,** 5838–5842.
10. Koup, R. A. and Ho, D. D. (1994) Shutting down HIV. *Nature* **370,** 416.
11. Safrit, J. T., Andrews, C. A., Zhu, T., Ho, D. D., and Koup, R. A. (1994) Characterization of human immunodeficiency virus type 1-specific cytotoxic T lymphocyte clones isolated during acute seroconversion: recognition of autologous virus sequences within a conserved immunodominant epitope. *J. Exp. Med.* **179,** 463–472.
12. Klein, M., Van Baalen, C., Holwerda, A., et al. (1995) Kinetics of Gag-specific cytotoxic T lymphocyte responses during the clinical course of HIV-1 infection: a longitudinal analysis of rapid progressors and long-term asymptomatics. *J. Exp. Med.* **181,** 1365–1372.
13. Betts, M. R., Krowka, J., Santamaria, C., et al. (1997) Cross-clade human immunodeficiency virus (HIV)-specific cytotoxic T-lymphocyte responses in HIV-infected Zambians. *J. Virol.* **71,** 8908–8911.
14. Durali, D., Morvan, J., Letourneur, F., et al. (1998) Cross-reactions between the cytotoxic T-lymphocyte responses of human immunodeficiency virus-infected African and European patients. *J. Virol.* **72,** 3547–3553.
15. Lynch, J. A., deSouza, M., Robb, M. D., et al. (1998) Cross-clade cytotoxic T cell response to human immunodeficiency virus type 1 proteins among HLA disparate North Americans and Thais. *J. Infect. Dis.* **178,** 1040–1046.
16. Wagner, R., Leschonsky, B., Harrer, E., et al. (1999) Molecular and functional analysis of a conserved CTL epitope in HIV-1 p24 recognized from a long-term nonprogressor: con-

straints on immune escape associated with targeting a sequence essential for viral replication. *J. Immunol.* **162,** 3727–3734.
17. Cox, J. C. and Coulter, A. R. (1997) Adjuvants—a classification and review of their modes of action. *Vaccine* **15,** 248–256.
18. Gustafson, G. L. and Rhodes, M. J. (1992) Bacterial cell wall products as adjuvants: early interferon gamma as a marker for adjuvants that enhance protective immunity. *Res. Immunol.* **143,** 483–488.
19. Wu, J. Y., Gardner, B. H., Murphy, C. I., et al. (1992) Saponin adjuvant enhancement of antigen-specific immune responses to an experimental HIV-1 vaccine. *J. Immunol.* **148,** 1519–1525.
20. Ke, Y., Li, Y., and Kapp, J. A. (1995) Ovalbumin injected with complete Freund's adjuvant stimulates cytolytic responses. *Eur. J. Immunol.* **25,** 549–553.
21. Schulz, M., Zinkernagel, R. M., and Hengartner, H. (1991) Peptide-induced antiviral protection by cytotoxic T cells. *Proc. Natl. Acad. Sci. USA* **88,** 991–993.
22. Takahashi, H., Takeshita, T., Morein, B., Putney, S., Germain, R. N., and Berzofsky, J. A. (1990) Induction of CD8+ cytotoxic T cells by immunization with purified HIV-1 envelope protein in ISCOMs. *Nature* **344,** 873–875.
23. Zhou, F., Rouse, B. T., and Huang, L. (1992) Prolonged survival of thymoma-bearing mice after vaccination with a soluble protein antigen entrapped in liposomes: a model study. *Cancer Res.* **52,** 6287–6291.
24. Maloy, K. J., Donachie, A. M., O'Hagan, D. T., and Mowat, A. M. (1994) Induction of mucosal and systemic immune responses by immunization with ovalbumin entrapped in poly(lactide-co-glycolide) microparticles. *Immunology* **81,** 661–667.
25. VanCott, T. C., Kaminski, R. W., Mascola, J. R., et al. (1998) HIV-1 neutralizing antibodies in the genital and respiratory tracts of mice intranasally immunized with oligomeric gp160. *J. Immunol.* **160,** 2000–2012.
26. von Brunn, A., Fruh, K., Muller, H. M., Zentgraf, H. W., and Bujard, H. (1991) Epitopes of the human malaria parasite *P. falciparum* carried on the surface of HBsAg particles elicit an immune response against the parasite. *Vaccine* **9,** 477–484.
27. Schlienger, K., Mancini, M., Riviere, Y., Dormont, D., Tiollais, P., and Michel, M. L. (1992) Human immunodeficiency virus type 1 major neutralizing determinant exposed on hepatitis B surface antigen particles is highly immunogenic in primates. *J. Virol.* **66,** 2570–2576.
28. Michel, M. L., Mancini, M., Schlienger, K., and Tiollais, P. (1993) Recombinant hepatitis B surface antigen as a carrier of human immunodeficiency virus epitopes. *Res. Virol.* **144,** 263–267.
29. Schirmbeck, R., Melber, K., Kuhrober, A., Janowicz, Z. A., and Reimann, J. (1994) Immunization with soluble hepatitis B virus surface protein elicits murine H-2 class I-restricted CD8+ cytotoxic T lymphocyte responses in vivo. *J. Immunol.* **152,** 1110–1113.
30. Breitburd, F., Kirnbauer, R., Hubbert, N. L., et al. (1995) Immunization with viruslike particles from cottontail rabbit papillomavirus (CRPV) can protect against experimental CRPV infection. *J. Virol.* **69,** 3959–3963.
31. Kirnbauer, R. (1996) Papillomavirus-like particles for serology and vaccine development. *Intervirology* **39,** 54–61.
32. Sedlik, C., Saron, M., Sarraseca, J., Casal, I., and Leclerc, C. (1997) Recombinant parvovirus-like particles as an antigen carrier: a novel nonreplicative exogenous antigen to elicit protective antiviral cytotoxic T cells. *Proc. Natl. Acad. Sci. USA* **94,** 7503–7508.
33. Lo-Man, R., Rueda, P., Sedlik, C., Deriaud, E., Casal, I., and Leclerc, C. (1998) A recombinant virus-like particle system derived from parvovirus as an efficient antigen carrier to elicit a polarized Th1 immune response without adjuvant. *Eur. J. Immunol.* **28,** 1401–1407.

34. Griffiths, J. C., Berrie, E. L., Holdsworth, L. N., et al. (1991) Induction of high-titer neutralizing antibodies, using hybrid human immunodeficiency virus V3-Ty viruslike particles in a clinically relevant adjuvant. *J. Virol.* **65,** 450–456.
35. Harris, S. J., Gearing, A. J., Layton, G. T., Adams, S. E., and Kingsman, A. J. (1992) Enhanced proliferative cellular responses to HIV-1 V3 peptide and gp120 following immunization with V3:Ty virus-like particles. *Immunology* **77,** 315–321.
36. Delchambre, M., Gheysen, D., Thines, D., et al. (1989) The GAG precursor of simian immunodeficiency virus assembles into virus-like particles. *EMBO J.* **8,** 2653–2660.
37. Wills, J. W. and Craven, R. C. (1991) Form, function, and use of retroviral gag proteins. *AIDS* **5,** 639–654.
38. Gheysen, D., Jacobs, E., de Foresta, F., et al. (1989) Assembly and release of HIV-1 precursor Pr55gag virus-like particles from recombinant baculovirus-infected insect cells. *Cell* **59,** 103–112.
39. Vernon, S. K., Murthy, S., Wilhelm, J., et al. (1991) Ultrastructural characterization of human immunodeficiency virus type 1 Gag-containing particles assembled in a recombinant adenovirus vector system. *J. Gen. Virol.* **72,** 1243–1251.
40. Royer, M., Cerutti, M., Gay, B., Hong, S. S., Devauchelle, G., and Boulanger, P. (1991) Functional domains of HIV-1 gag-polyprotein expressed in baculovirus-infected cells. *Virology* **184,** 417–422.
41. Wagner, R., Fliessbach, H., Wanner, G., et al. (1992) Studies on processing, particle formation, and immunogenicity of the HIV-1 gag gene product: a possible component of a HIV vaccine. *Arch. Virol.* **127,** 117–137.
42. Wagner, R., Deml, L., Schirmbeck, R., Niedrig, M., Reimann, J., and Wolf, H. (1996) Construction, expression and immunogenicity of chimeric HIV-1 virus-like particles. *Virology* **220,** 128–140.
43. Deml, L., Schirmbeck, R., Reimann, J., Wolf, H., and Wagner, R. (1997) Recombinant human immunodeficiency Pr55gag virus-like particles presenting chimeric envelope glycoproteins induce cytotoxic T-cells and neutralizing antibodies. *Virology* **235,** 26–39.
44. Wolf, H., Modrow, S., Soutschek, E., Motz, M., Grunow, R., and Döbl, H. (1990) Production, mapping and biological characterisation of monoclonal antibodies to the core protein (p24) of the human immunodeficiency virus type 1. *AIFO* **1,** 24–29.
45. Deml, L., Kratochwil, G., Osterrieder, N., Knuchel, R., Wolf, H., and Wagner, R. (1997) Increased incorporation of chimeric human immunodeficiency virus type 1 gp120 proteins into Pr55gag virus-like particles by an Epstein-Barr virus gp220/350-derived transmembrane domain. *Virology* **235,** 10–25.
46. Wagner, R., Teeuwsen, V. J., Deml, L., et al. (1998) Cytotoxic T cells and neutralizing antibodies induced in rhesus monkeys by virus-like particle HIV vaccines in the absence of protection from SHIV infection. *Virology* **245,** 65–74.
47. Paliard, X., Liu, Y., Wagner, R., Wolf, H., Baenziger, J., and Walker, C. M. (2000) Priming of strong, broad, and long-lived HIV type 1 p55gag-specific CD8+ cytotoxic T cells after administration of a virus-like particle vaccine in rhesus macaques. *AIDS Res. Hum Retroviruses* **16,** 273–282.
48. Ogg, G. S., Jin, X., Bonhoeffer, S., et al. (1998) Quantitation of HIV-1-specific cytotoxic T lymphocytes and plasma load of viral RNA. *Science* **279,** 2103–2106.
49. Betts, M. R., Casazza, J. P., Patterson, B. A., et al. (2000) Putative immunodominant human immunodeficiency virus-specific CD8(+) T-cell responses cannot be predicted by major histocompatibility complex class I haplotype. *J. Virol.* **74,** 9144–9151.

50. Zajac, A. J., Murali-Krishna, K., Blattman, J. N., and Ahmed, R. (1998) Therapeutic vaccination against chronic viral infection: the importance of cooperation between CD4+ and CD8+ T cells. *Curr. Opin. Immunol.* **10,** 444–449.
51. Waldrop, S. L., Pitcher, C. J., Peterson, D. M., Maino, V. C., and Picker, L. J. (1997) Determination of antigen-specific memory/effector CD4+ T cell frequencies by flow cytometry: evidence for a novel, antigen-specific homeostatic mechanism in HIV-associated immunodeficiency. *J. Clin. Invest.* **99,** 1739–1750.
52. Sester, M., Sester, U., Kohler, H., et al. (2000) Rapid whole blood analysis of virus-specific CD4 and CD8 T cell responses in persistent HIV infection. *AIDS* **14,** 2653–2668.
53. Heintel, T., Sester, M., Rodriguez, M. M., et al. (2002) The fraction of perforin-expressing HIV-specific CD8 T cells is a marker for disease progression in HIV infection. *AIDS* **16,** 1497–1501.
54. Sambrook, J., Fritsch, E. F., and Maniatis, T. (1989) *Molecular Cloning: A Laboratory Manual.* Cold Spring Harbor Laboratory Press, Cold Spring Harbor, NY.
55. Qiu, J. T., Song, R., Dettenhofer, M., et al. (1999) Evaluation of novel human immunodeficiency virus type 1 Gag DNA vaccines for protein expression in mammalian cells and induction of immune responses. *J. Virol.* **73,** 9145–9152.
56. Takahashi, H., Cohen, J., Hosmalin, A., et al. (1988) An immunodominant epitope of the human immunodeficiency virus envelope glycoprotein gp160 recognized by class I major histocompatibility complex molecule-restricted murine cytotoxic T lymphocytes. *Proc. Natl. Acad. Sci. USA* **85,** 3105–3109.
57. White, B. A. (1993) *Methods in Molecular Biology: PCR Protocols, Current Methods and Applications.* Humana, Totowa, NJ.

10

Application of Single-Cell Cultures of Mouse Splenocytes as an Assay System to Analyze the Immunomodulatory Properties of Bacterial Components

Ludwig Deml, Michael Aigner, Alexander Eckhardt, Jochen Decker, Norbert Lehn, and Wulf Schneider-Brachert

Abstract

Recently, various bacterial components have been suggested as initiating and modulating immune activation, thereby substantially affecting the complex and dynamic host/pathogen interactions. Herein, we present a valuable and simple methodology for determining the capacity of bacteria as well as defined bacterial structures to stimulate cellular effectors of the innate and cognate immune system. This assay format is based on the exposure of freshly prepared single-cell cultures of splenic cells derived from naive mice with the immunogen of interest. Herein, the determination of exclusive panels of cytokines by the ELISA, ELISpot, and FACS technology will serve as an indicator for the activation of defined arms of the immune system. An increased knowledge about microbial components with immunomodulatory properties will substantially contribute to a more detailed understanding of the dynamic interplay between the host and potential pathogens and, based on this knowledge, to the development of novel substances for the prevention and therapy of microbial infections.

Key Words: Splenic cell culture; immunomodulation; bacterial components; innate immunity; determination of immune activation; cytokines.

1. Introduction

One of the most fascinating objectives in immunology is the understanding of how the vertebrate immune system almost exclusively recognizes non-self structures and patterns associated with pathogens to initiate immune defense responses to combat infectious agents. In general, immune activation by pathogens is initiated by the interaction of defined microbial components with two distinct sets of host recognition molecules, namely, the microbial pattern recognition receptors (PRR) and the antigen receptors of T- and B-lymphocytes *(1,2)*. These receptors play a crucial role in the activation of innate and cognate immune responses, respectively. The strategy of initial activation of innate immune responses is based on the detection of constitutive and conserved components or products, which are common in defined groups of microbes but not present in the host. The recognition repertoire of vertebrates described here is limited to a variety of pathogen-associated molecular patterns (PAMPs) such as lipopolysaccharides (LPS), peptidoglycans, lipoteichonic acid, glucans, mannans, and nucleic acids such as unmethylated CpG-positive bacterial DNA and double-stranded (ds) viral RNA *(3)*, as well as microbial polypeptides such as the baculovirus envelope protein gp67 *(4)*, bacterial flagellin *(5,6)*, and protozoan LeIF protein *(7)*. Recently, it has been shown that cellular immune activation in response to PAMPs is mediated by a family of vertebrate transmembrane pattern recognition receptors termed Toll-like receptors (TLRs) *(8,9)*. Through the recognition of pathogens or their components, TLRs can promote the survival, maturation, and functional activation of crucial components of the immune system, e.g., antigen-presenting cells (APCs) *(10)*. In addition, PAMP recognition induces the upregulation of costimulatory molecules, including B7-1 (CD80) and B7-2 (CD86) and the production of instructive cytokines such as interleukin-12 (IL-12) and IL-18, thereby promoting the development of strong and broad T-helper 1 (Th1)-biased cellular and humoral immune responses, which are compulsory to control a variety of microbial infections. This results in the activation of APCs to produce a panel of different cytokines involved in inflammatory reactions and an enhanced presentation of antigenic epitopes to the highly specialized cognitive components of the immune system.

In summary, these data underline the extraordinary importance of microbial PRR-mediated pathogen recognition for the stimulation of both innate and cognate immune responses. Thus, an increased knowledge of microbial components, which are able to act as "danger signals" for immune activation, will substantially contribute to a more detailed understanding of pathogen/host interactions and will also support the development of novel substances that could have wide preventive and therapeutic value as immunomodulators and adjuvants for vaccines against infectious diseases, cancer, and allergy.

Here we present a valuable and simple assay format for analyzing the capacity of live bacteria and bacterial components to activate the innate immune system. This system is based on the incubation of single-cell cultures of splenocytes derived from naive mice with a bacterium or bacterial component of interest. The secretion of defined cytokines upon stimulation serves as a measure of the immunomodulatory properties of these bacterial immunogens.

Helicobacter pylori was analyzed in these studies because it has been identified as the causative agent of inflammatory gastroduodenal pathology. The presence of *H. pylori* invariably causes a continuous gastric inflammation (although in most cases clinically silent). This immune response initially comprises a predominant recruitment of neutrophils, followed by the infiltration of T- and B-lymphocytes, plasma cells, and macrophages. In addition, increased numbers of APCs such as monocytes and gastric mucosal dendritic cells are present. Apart from conventional APCs, epithelial cells from gastric biopsies have also been shown to express both B7-1 and B7-2 constitutively, indicating that they can provide T-cell costimulation in vivo. Attachment of *H. pylori* to the gastric epithelium induces enhanced levels of cytokines and chemokines [IL-1β, IL-2, IL-6, IL-8, tumor necrosis factor (TNF), GROα, ENA-78, RANTES] by activation of nuclear factor–κB (NF-κB). In addition, lymphoid follicles consisting of B-cells surrounded mostly by CD4+ T-cells were present in the lamina propria in *H. pylori*–induced gastritis. These data indicate a strong lymphocytic immune response leading to the presence of high titers of *H. pylori*-specific mucosal antibodies. However, despite this strong long-term local and systemic immune response against *H. pylori* immunogens, in most cases the infection is not eradicated and the bacterium persists virtually lifelong (reviewed in refs. *11–16*).

Therefore, it is tempting to speculate whether *H. pylori* mediates specific escape mechanisms to outwit the host's immune system (as they are successfully used by certain viruses to evade immune surveillance). The T-lymphocytes present in the inflammatory infiltrate are predominantly of a CD3+CD4+ phenotype. Immature CD4+ cells can be functionally polarized in Th1 cells, mainly secreting IL-2 and interferon-γ (IFN-γ), and Th2 cells, which are characterized by IL-4, IL-5, IL-10, and IL-13 production. Although Th2 cells promote humoral immune responses upon contact with extracellular pathogens, Th1 cells are induced mostly in response to intracellular pathogens *(16)*. The paradox of *H. pylori* is that although it is a noninvasive extracellular bacterium and induces a strong humoral response, it predominantly induces a Th1 cell-biased cellular immune response. Given that Th1 and Th2 cells reciprocally inhibit each other by the secretion of their signature cytokines IFN-γ and IL-4, respectively, it was speculated that *H. pylori* directs a Th1 cell-biased CD4+ cell response by inducing an overproduction of IFN-γ from natural killer (NK) cells and/or T-cells *(15)*. Crude preparations of *H. pylori*–induced proliferation and IFN-γ secretion by human peripheral blood mononuclear cells (PBMCs) or mouse splenocytes. Interestingly, responsiveness occurred in both cell populations irrespective of whether or not the donors were infected with *H. pylori*. These findings suggest the possibility that the observed response of naive donors toward *H. pylori* antigens may reflect prior encounters with microorganisms, which share common T-cell determinants *(15)*. Alternatively, components of *H. pylori* may act as PAMPs to stimulate the activation of the innate immune system.

To understand the mechanisms for *H. pylori*–induced Th1 cell-biased immune response, we analyzed the role of the innate immune system in *H. pylori* infection in more detail. For this purpose we chose splenocytes from naive mice as a model system to investigate the immune activation in response to coincubation with *H. pylori* or an *H. pylori*–specific protein that has been recombinantly expressed in *E. coli* or stable transfected *Drosophila* Schneider-2 (DS-2) insect cells (**Fig. 1**). The *H. pylori* cysteine-rich protein (HcpA) attracted our particular attention as a possible virulence factor because of its particular biochemical and immunological properties. HcpA is a secretory protein that is resistant to low pH values and stimulates the production of high titers of specific antibodies in *H. pylori*–infected individuals with gastritis or peptic ulcer disease (unpublished observations).

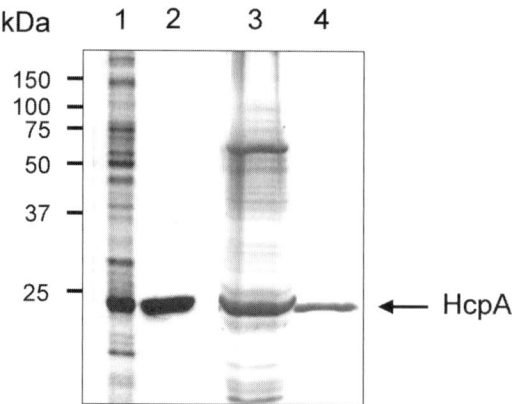

Fig. 1. Analysis of purified *H. pylori* cysteine-rich protein (HcpA). HcpA was expressed by either *E. coli* strain JM83 (lanes 1–2) or stable transfected DS-2 cells (lanes 3–4). Purification of HcpA was performed by heat precipitation and cation exchange chromatography and monitored by Coomassie blue-stained SDS-PAGE. Lanes contain the periplasm of induced JM83 harboring *pRBI-PDI-hcpA* (lane 1) as well as purified *E. coli*-derived HcpA (lane 2) and supernatant from stable transfected DS-2 cells at day 7 post induction (lane 3) as well as purified DS-2 derived HcpA (lane 4).

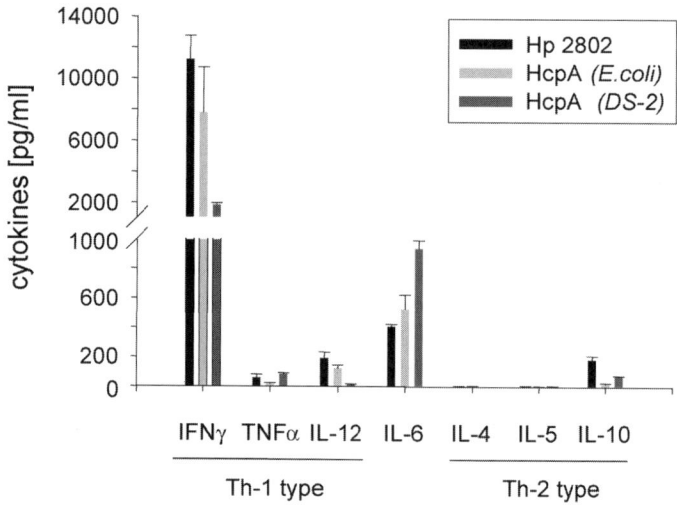

Fig. 2. Pattern of released cytokines by murine splenocytes upon stimulation with purified DS-2 and *E. coli*-derived *H. pylori* cysteine-rich protein (HcpA) as well as live *H. pylori* strain (Hp 2802) detected by ELISA. Levels of interleukin-12 (IL-12) and tumor necrosis factor-α (TNF-α) were determined 4 h post stimulation, whereas concentrations of IL-4, IL-5, IL-6, IL-10, and interleukin-γ (IFN-γ) were measured at h 24. Each value represents the mean cytokine value ± the SD of three independent experiments.

As depicted in **Fig. 2**, a typical Th1-type cytokine profile was initiated by cocultivation of splenocytes from uninfected mice with live *H. pylori*, as clearly shown by the production of IL-12 and IFN-γ upon stimulation. Interestingly, treatment with recombinant HcpA, independently of the source (*E. coli*- or DS-2-derived material) induced a similar cytokine pattern as live *H. pylori* (**Fig. 2**). In contrast, only trace quantities of IL-10 were observed, in accordance with the fact that IL-12 reciprocally inhibits IL-10-driven stimulation of Th2 cells. This suggests that the host mounts an inherent response upon the first exposure to *H. pylori* antigen, apparently dictating the type of long-term Th1-type immune response that develops. Analysis of the initial interplay of *H. pylori* with the cells of the innate immune system may help to disclose the early steps of *H. pylori*–mediated immunopathology. Treatment of splenocytes with recombinant *H. pylori* proteins offers the opportunity to study specific bacterial virulence factors individually independent of the influence of the whole bacterium or bacterial components present during infection with live *H. pylori*.

2. Materials

2.1. Construction of Expression Vectors

1. The bacterial *pRBI-PDI* plasmid *(17)* carries a bicistronic expression unit containing an α-amylase/trypsin inhibitor gene, 5′ fused to a bacterial periplasmic secretion signal sequence and a disulfide isomerase (*dsbC*) gene under the transcriptional control of the lac promotor (*see* **Note 1**).
2. The *Drosophila* expression vector *pMt/BiP/V5-HisA* (Invitrogen, San Diego, CA) includes an inducible metallothionein promoter in conjunction with a BiP signal sequence and an SV40 late polyadenylation signal as well as plasmid *pA5c-DHFR* that contains the *Drosophila* dihydrofolate reductase (*dhfr*) selectable marker gene under the transcriptional control of an insect cell's constitutively active actin 5c distal promoter *(18)* (*see* **Note 2**).
3. Nucleic acids and oligonucleotides: genomic *H. pylori* DNA or an appropriate vector containing the coding region for the HcpA, a dNTP solution (20 mM) containing all four dNTPs, appropriate oligonucleotide primers (10 μM) in Tris-EDTA (TE) buffer (10 mM Tris-HCl, 1 mM EDTA, pH 8.0).
4. Enzymes: T4-DNA polymerase, *Stu*I, *Hind*III, *Bgl*II, *Age*I, T4-DNA ligase, calf intestinal phosphatase (all Roche Applied Science, Mannheim, Germany) and the appropriate 10X buffers; all used according to the manufacturer's instructions.
5. DNA sample loading dye: 0.25% (v/w) bromphenol blue, 0.25% (v/w) xylene cyanol, 50 mM Tris-HCl, pH 7.6, 60% (v/v) glycerol.
6. Agarose (FMC BioProducts, Rockland, ME).
7. TBE buffer: 40 mM Tris, 100 mM boric acid, 2 mM Na$_2$EDTA dissolved in H$_2$O.
8. Ethidium bromide solution: 0.5 mg/L ethidium bromide dissolved in H$_2$O; store at 4°C in the dark.
9. Lambda *Bst*EII molecular weight markers (New England Biolabs, Schwalbach, Germany).
10. QIAquick PCR Purification Kit (Qiagen, Hilden, Germany).
11. QIAquick Gel Extraction Kit (Qiagen).
12. Chemically competent *E. coli* K-12 derivatives such as DH5α, XL1-blue, JM83, and JM110 (Stratagene, La Jolla, CA).
13. LB (Luria-Bertani) broth (10 mg/L tryptone, 5 mg/L yeast extract, 10 mg/L NaCl).

14. LB$_{amp}$: LB broth containing 100 µg/mL ampicillin.
15. LB agar plates. Media and agar plates were prepared according to standard protocols as described in detail *(19)*.
16. Alkaline lysis buffers for minipreparations: Solution I: 50 m*M* glucose, 25 m*M* Tris-HCl, pH 7.5, 10 m*M* EDTA, 100 µg/mL RNAse. Solution II: 0.2 *N* NaOH, 1% sodium dodecyl sulfate (SDS). Solution III: 3 *M* potassium acetate, pH 4.8.
17. Water-saturated phenol/chloroform/isoamyl alcohol (25/24/1).
18. Ethanol [100% and 70% (v/v) in H$_2$O].
19. EndoFree Midi Kit (Qiagen).
20. Isopropanol (100%).

2.2. Expression of Recombinant HcpA Protein

2.2.1. Expression in E. coli

1. 1 *M* IPTG (isopropyl-β-D-thiogalactopyranoside) stock solution (Sigma-Aldrich) in ddH$_2$O; store at –20°C.
2. 1 *M* NAcCys (*N*-acetyl-L-cysteine) stock solution (Sigma-Aldrich) in ddH$_2$O; prepare fresh.
3. LB broth.
4. Spectrometer (e.g., Uvikon Spectrophotometer 930; Kontron Instruments, München, Germany).
5. Polymyxin B buffer: 10 m*M* MOPS (4-morpholinepropanesulfonic acid), pH 6.8, 150 m*M* NaCl, 5 m*M* Na$_2$EDTA, 1 g/L polymyxin B (Riedel-de Haën, Seelze, Germany) dissolved in ddH$_2$O.

2.2.2. Expression in Drosophila Schneider-2 (DS-2) Cells

2.2.2.1. GENERATION OF STABLE TRANSFECTED DS-2 CELLS

1. DS-2 cells (Invitrogen, Karlsruhe, Germany).
2. DS-2 medium (Invitrogen).
3. Fetal calf serum (FCS; Invitrogen).
4. Kanamycin, penicillin/streptomycin, or gentamycin (PAN Biotech, Aidenbach, Germany).
5. Various tissue culture flasks and plastic tubes (Becton Dickinson Labwares, Bedford, MA).
6. DOTAP liposomes (Roche Diagnostics, Mannheim, Germany).
7. 20 m*M* HEPES, 150 m*M* NaCl, pH 7.4.
8. 10 m*M* NaOH.
9. Methotrexate stock solution: 2–4 mg/mL methotrexate dissolved in 10 m*M* NaOH.
10. Stereo dissecting microscope (Helmut Hund, Wetzlar, Germany).
11. "Neubauer improved" hemocytometer.

2.2.2.2. INDUCTION OF ANTIGEN PRODUCTION

1. Insect Express HighFive medium (PAA Laboratories, Linz, Austria).
2. Kanamycin (Pan Biotech).
3. 20 m*M* CuSO$_4$ (100X stock solution; Merck, Darmstadt, Germany).

2.3. Purification of HcpA Protein

1. Cell centrifuge (Rotixa 120R; Hettich, Tuttlingen, Germany).
2. Spectrapor molecular porous membrane tubing with a 3500-Daltons cutoff (Spectrum Laboratories, Rancho Dominguez, CA).
3. 50 m*M* MES (4-morpholine-ethanesulfonic acid) (Sigma-Aldrich) in dissolved ddH$_2$O, pH 6.4; prepare fresh (*see* **Note 3**).

4. BioCAD 700E FPLC (PerSeptive Biosystems, Framingham, MA).
   5. Poros CM FPLC column (PerSeptive Biosystems).
   6. Elution buffer: salt gradient 0–1 M NaCl.
   7. Tris buffer: 20 mM Tris-HCl, pH 7.5, 50 mM NaCl.
   8. Bio-Rad D_c protein assay (Bio-Rad, Hercules, CA).
   9. Polyclonal anti-HcpA rabbit serum (generated by David's Biotechnologie, Regensburg, Germany).

2.4. Propagation of H. pylori

   1. WC Wilkins-Chalgrens Agar (Oxoid, Basingstroke, UK), supplemented with 10% defibrinated horse blood (Elocin-Lab, Mühlheim, Ruhr, Germany) and 25 mg/L DENT (Oxoid).
   2. Phosphate-buffered saline (PBS) with bivalent ions: Solution A: 150 mM NaCl, 3 mM KCl, 4 mM Na$_2$HPO$_4$, 1.5 mM KH$_2$PO$_4$. Solution B: 1.5 mM CaCl$_2$, 625 µM MgCl$_2$. Add 800 mL H$_2$O to 100 mL solution A, mix well and fill up to 1000 mL with solution B.
   3. Cotton swabs.
   4. Uvikon Spectrophotometer 930 (Kontron Instruments).

2.5. Preparation of Single-Cell Cultures of Murine Splenocytes

   1. Naive female BALB/c or C57BL/6 mice at an age of 44–60 d (Charles River, Sulzfeld, Germany or Harlan, Borchen, Germany) (*see* **Note 4**).
   2. 70% ethanol.
   3. Equipment for euthanasia.
   4. Dissecting instruments: sterile scissors and tweezers.
   5. 15- and 50-mL plastic centrifuge tubes (Greiner, Frickenhausen, Germany).
   6. PBS: 150 mM NaCl, 3 mM KCl, 4 mM Na$_2$HPO$_4$, 1.5 mM KH$_2$PO$_4$.
   7. Wash buffer: PBS, 1% FCS.
   8. Pipets: 5-, 10-, and 25-mL serological plastic pipets (Sarstedt, Nümbrecht, Germany).
   9. 100-µm nylon cell strainer (Becton Dickinson Labwares).
  10. 5-mL syringe (Becton Dickinson Labwares).
  11. ACK hemolysis buffer: 150 mM NH$_4$Cl, 1 mM KHCO$_3$, 0.1 mM EDTA (all from Sigma-Aldrich), pH 7.2–7.4.
  12. Splenocyte medium: RPMI-1640 with L-glutamine and 2 g/L NaHCO$_3$, supplemented with 5% FCS (PAN Biotech), 20 mM HEPES (Sigma-Aldrich), and 0.5 mM 2-mercaptoethanol (Merck) with or without 1% Pen/Strep solution (PAN Biotech) (*see* **Note 5**).
  13. "Neubauer improved" hemocytometer.

2.6. Stimulation Experiments

   1. 24-well tissue culture plate; flat bottom with lid (Costar, Corning, NY).

2.7. Determination of Cytokine Production

2.7.1. ELISA-Assay

   1. 96-well plates (F96 Maxisorb plates, Nunc, Wiesbaden, Germany).
   2. Various OptEIA™ ELISA kits for the detection of IL-2, IL-4, IL-5, IL-6, IL-10, IL-12, TNF-α, and IFN-γ, each containing a capture antibody, a detection antibody, an avidin–horseradish peroxidase conjugate, and a recombinant standard protein (BD, San Diego, CA). The measurement of cytokines was performed according to the manufacturer's protocol.

3. Coating buffer: 0.1 M carbonate, pH 9.5, or 0.2 M sodium phosphate, pH 6.5.
4. PBS.
5. Assay diluent: PBS, 10% FCS.
6. Wash buffer: PBS, 0.05% Tween 20.
7. TMB substrate solution (BD).
8. Stop solution: 2 N H_2SO_4.
9. ELISA reader (SLT Spectra, SLT Labinstruments, Germany).
10. ELISA software (WSOFTmax).

2.7.2. FACS Assay

1. 96-well plates, round bottom and flat bottom (Nunc).
2. Splenocyte medium.
3. Brefeldin A (BFA) stock solution: 5 mg/mL BFA (Sigma-Aldrich) dissolved in ethanol. Freeze aliquots at –20°C.
4. Phorbol 12-myristate 13-acetate (PMA) stock solution: 0.1 mg/mL PMA (Sigma-Aldrich, Taufkirchen, Germany) dissolved in dimethyl sulfoxide (DMSO). Freeze aliquots at –20°C.
5. Ionomycin stock solution: 1 mg/mL ionomycin dissolved in DMSO. Store aliquots at 4°C.
6. PBS.
7. Fluorescein-activated cell sorting (FACS) buffer 1: PBS, 1% FCS, 1 mg/mL NaN_3.
8. FACS antibodies: anti-mouse CD8a-APC (clone 53-6.7), anti-mouse CD4-FITC (clone RM4-5), anti-mouse CD16/32 (clone 2.4G2), and anti-mouse IFN-γ-PE (cat. no. 554412; all from BD).
9. FACS buffer 2: PBS, 1 mg/mL NaN_3.
10. 4% paraformaldehyde (PFA) solution: dissolve 4 g PFA in 100 mL prewarmed PBS; store at 4°C.
11. Cytofix/Cytoperm (BD): 4% PFA, 1% saponin (Sigma).
12. Perm/wash buffer: PBS, 1% saponin.
13. 5-mL polystyrene round-bottom tubes (Falcon 35 2054; BD).
14. FACS Calibur (BD).

2.7.3. ELISpot

1. 96-well plates: MAHA Multiscreen-HA plate (0.45-μm surfactant-free mixed cellulose ester membrane, Millipore, Bedford, MA).
2. Anti-IFN-γ antibody (clone R4-6A2; BD).
3. PBS.
4. Wash buffer: PBS, 10% FCS.
5. Splenocyte medium.
6. IL-2 (Roche, Mannheim, Germany).
7. PMA-ionomycin.
8. Lysis buffer: PBS, 0.05% Tween 20.
9. Biotinylated anti-IFN-γ antibody (clone XMG1.2; BD).
10. Dilution buffer: PBS, 10% FCS, 0.05% Tween 20.
11. Streptavidin-AP (Roche).
12. Nitroblue tetrazolium/bromomethylindolyl phosphate (NBT/BCIP) stock solution (Roche).
13. AP buffer: 100 mM Tris-HCl, pH 9.5, 100 mM NaCl, and 50 mM $MgCl_2$.

3. Methods

Methods that are not specifically described in detail can be performed as previously described by Sambrook and coworkers *(19)* or according to the manufacturer's protocol.

3.1. Construction of Expression Vectors

3.1.1. E. coli Expression Plasmid

1. Generate a polymerase chain reaction (PCR) fragment encoding *hcpA* by specific PCR amplification (*see* **Note 6**). The forward primer should contain two additional residues at the 5' end, and the reverse primer should contain at least one stop codon (*see* **Note 7**).
2. Digest the *pRBI-PDI* expression vector with *Stu*I/*Hind*III, and incubate the resulting linearized plasmid with T4 DNA polymerase to generate blunt ends (*see* **Note 8**).
3. Ligate the linearized *pRBI-PDI* expression vector with the PCR-derived Hcpa-coding region. The resulting vector is designated *pRBI-PDI-hcpA*.
4. Transform *E. coli* JM83 cells and plate on LB_{amp} agar plates. Incubate overnight at 37°C (*see* **Note 9**).
5. Test individual colonies for presence of the *hcpA* gene by digestion of "miniprep" DNA with appropriate restriction enzymes.
6. Verify the orientation and identity of *hcpA* by sequencing.

3.1.2. DS-2 Expression Vectors

1. Generate a *Bgl*II/*Age*I fragment encoding *hcpA* by specific PCR amplification (*see* **Note 6**).
2. Ligate the *Bgl*II/*Age*I-digested *pMt/BiP/V5-HisA* expression vector with the PCR-derived HcpA coding region. The resulting vector is designated *pMt-hcpa*.
3. Transform *E. coli* XL1-blue cells and plate on LB agar plates containing 50 µg/mL ampicillin. Incubate overnight at 37°C.
4. Test individual colonies for the presence of the desired foreign gene by digestion of "miniprep" DNA with appropriate restriction enzymes.
5. Produce appropriate amounts of endotoxin-free lots of the expression plasmid *pMt-hcpA* and the selection vector *pA5cDHFR* (*see* **Note 10**).

3.2. Expression of Recombinant HcpA Protein

3.2.1. Expression in E. coli

1. Inoculate a small volume of LB_{amp} medium (5 mL) with 1 colony of *E. coli* JM83 containing the *pRBI-PDI-hcpA* expression vector. Incubate the culture overnight at 37°C in a shaking incubator (200 rpm).
2. Inoculate LB_{amp} medium with a 1:100 dilution of the overnight culture. Incubate the cultures at 37°C and 200 rpm until cells reach midlog growth (OD_{580} of 0.7).
3. Then, incubate the cultures at room temperature until cells reach an OD_{580} of 1.1.
4. Induce the cell culture by adding 1 m*M* IPTG and 5 m*M* *N*-acetyl-L-cysteine (final concentrations) and continue incubation overnight at room temperature on a shaker (200 rpm).
5. Measure the OD_{580} and centrifuge the culture broth for 15 min at 5000*g* and 4°C.
6. Remove the supernatant by aspiration and resuspend the pellet in 4 mL of ice-cold polymyxin B buffer per 1 OD_{580}/1L culture broth, and incubate for 1 h on ice to lyse the periplasm selectively (*see* **Note 11**).

7. Sediment the bacterial cells from the released periplasm by centrifugation for 30 min at 15,000g and 4°C.
8. Transfer the supernatant in a new tube and discard the pellet.

3.2.2. Expression in Drosophila Schneider-2 Cells

General cultivation and transfection procedures for *Drosophila* Schneider-2 cells as well as the selection of cultures of stable transfected DS-2 cells were performed as described previously in detail *(20)*.

3.2.2.1. GENERATION OF STABLE TRANSFECTED *DROSOPHILA* SCHNEIDER-2 CELLS FOR THE PRODUCTION OF HCPA PROTEINS

1. Carry out all transfections as described in the manufacturer's protocols (Roche, Mannheim, Germany). Briefly, for each transfection, seed 2×10^6 DS-2 cells in each well of a 6-well plate. Allow the cells to attach fully to the bottom of the well.
2. For each transfection, set up the following transfection mixture: 4 µg of highly pure transfer plasmids [expression unit (*pMt-hcpA*):selection unit (*pA5cDHFR*) in a molar ratio of 20:1] (*see* **Note 12**), 1 mL of DS-2 medium, and 20 µL of vigorously vortexed DOTAP liposomes.
3. Vortex vigorously for 10 s and incubate the mixture for 15 min at room temperature.
4. Remove all media containing FCS from the wells and replace it with 2 mL fresh and prewarmed DS-2 medium without any supplements. Wash the cells by gently rocking the plate.
5. Remove all DS-2 media from the monolayer and add the transfection mixture (*see* **step 2**) dropwise. Take care that the bottom surface of the well is completely covered with liquid.
6. Incubate the plates on a slowly rocking platform at room temperature for 4 h (*see* **Note 13**).
7. Following the 4-h incubation period, replace transfection mixture in each well by 5 mL of complete DS-2 medium including 10% FCS and an antibiotic (preferentially kanamycin).
8. Select stable transfected cell lines over a course of 3–6 wk by growing the cultures in medium, which contains increasing concentrations of methotrexate (1–8 µg/mL) as described previously in detail *(20)*.

3.2.2.2. INDUCTION OF ANTIGEN PRODUCTION

1. Induce one flask of stable transfected cells by adding fresh DS-2 medium supplemented with 10% FCS, 50 µg/mL kanamycin, and 200 µM CuSO$_4$ (*see* **Note 14**).
2. After 7–10 d of induction, collect HcpA-positive cell-culture supernatants by low-speed centrifugation (300g for 10 min at room temperature).

3.3. Purification of HcpA

1. Incubate the obtained HcpA-positive supernatants (*see* **Subheading 3.2.1.** for HcpA expression in *E. coli* and **Subheading 3.2.2.** for HcpA production in stable transfected DS-2 cells) in a water bath at 40°C for 30 min.
2. Sediment the heat-precipitated material at 15,000g for 30 min at 4°C.
3. Retain the supernatant and repeat heat precipitation as described in **step 2** but at 50°C and 60°C.
4. Dialyze the supernatant overnight at 4°C against three changes of 50 mM MES buffer in a molecular porous membrane tubing with a molecular weight cutoff of 3.5 kDa.

5. Sediment the heat-precipitated material at 15,000g for 20 min at 4°C.
6. Purify the HcpA by cationic exchange chromatography; elute the bound protein with 50 mM MES buffer and a linear gradient (0–1 M) of NaCl (*see* **Note 15**).
7. Determine protein concentration of collected fractions using a Bio-Rad D_c protein assay according to the manufacturer's instructions.
8. Load aliquots of each fraction on a 12.5% SDS-PAGE and run the gel until the bromophenol blue reaches the bottom of the gel.
9. Stain the gel with silver or carry out an immunoblot to visualize the purified protein (*see* **Note 16**).
10. Dialyze the HcpA-positive fractions against 20 mM Tris-HCl, pH 7.5, 50 mM NaCl in a molecular porous membrane tubing with a molecular weight cutoff of 3.5 kDa.

3.4. Propagation of H. pylori for Stimulation Assays

1. Cultivate *H. pylori* on WC-DENT agar plates at 36°C under microaerophilic conditions (11% O_2, 9% CO_2, 80% N_2).
2. Harvest one plate of logarithmically growing *H. pylori* after 48 h of incubation by scraping off the bacteria with a sterile cotton swap.
3. Suspend the collected *H. pylori* by stirring the cotton swab in 1 mL ice-cold PBS.
4. Repeat **steps 2–3** using the same cotton swab, and keep the resulting bacterial suspension on ice.
5. Centrifuge the *H. pylori* suspension at 1700g for 6 min at 4°C.
6. Discard the supernatant and resuspend the pellet in 1 mL ice-cold PBS.
7. Incubate the suspension for 30 s on ice to sediment insoluble impurities such as agar clots.
8. Transfer the resulting supernatant to a fresh 1.5-mL Eppendorf tube.
9. Repeat **steps 5–8**.
10. Transfer the supernatant to a fresh 1.5-mL Eppendorf tube, and keep it on ice.
11. Dilute the *H. pylori* suspension 1:10 in ice-cold PBS.
12. Determine the OD_{580} of the suspension against PBS.
13. Prepare serial dilutions of the *H. pylori* suspension ranging from 1×10^3 to 1×10^8 bacteria/mL in ice-cold PBS.
14. Plate 100 µL of each dilution on two WC-DENT agar plates and incubate for 72 h at microaerophilc conditions.
15. Count the colonies on the appropriate plates (usually 1×10^5 to 1×10^7 colonies/plate).
16. Calculate the number of *H. pylori* cells/1 OD_{580} (*see* **Note 17**).

3.5. Preparation of Single Cell Cultures of Murine Splenocytes

1. Euthanize the BALB/c or C57BL/6 mice by cervical dislocation.
2. Remove the spleens aseptically by necrosectomy.
3. Place the spleens in 5 mL splenocyte medium per spleen in a 50-mL Falcon tube. Keep the spleens at room temperature.

Perform the following steps under a laminar flow bench, and use sterile plastic serological pipets only in order to prevent potential cell damage by residual traces of detergent within cleaned glass pipets.

4. Place a cell strainer on a 50-mL Falcon tube.
5. Remove almost all medium from the 50-mL Falcon tube, thereby retaining the spleen.
6. Transfer the spleen into the cell strainer placed on a 50-mL Falcon tube.

7. Generate a single-cell suspension by grinding the spleen against the cell strainer with the plunger of a 5-mL syringe until mostly fibrous tissue remains.
8. Aspirate the remaining cells from the cell strainer by repeatedly adding 2 mL wash buffer and short-time lifting the cell strainer from the Falcon tube.
9. Discard cell strainer and the fibrous tissue.
10. Centrifuge the obtained cell suspension for 5 min at 300g at room temperature and discard the supernatant.
11. Resuspend the pellet in 5 mL ACK hemolysis buffer by gently but thoroughly pipeting with a 10- or 25-mL plastic pipet.
12. Directly centrifuge the cell suspension at 300g for 5 min at room temperature.
13. Discard the complete supernatant.
14. Resuspend the pellet by repeated rubbing of the 50-mL Falcon tube against the grid of the laminar flow bench.
15. Add 5 mL wash buffer to the resuspended splenocytes and mix gently.
16. Incubate at room temperature for 1 min and separate the cell suspension from sedimented fibrous tissue.
17. Centrifuge the remaining cell suspension at 300g for 5 min at room temperature.
18. Discard most of the supernatant, but leave 0.1 mL medium on the pellet.
19. Repeat washing of cells (*see* **steps 15–19**) twice.
20. Resuspend the pellet in 5 mL of splenocyte medium.
21. Adjust the cell number to a final concentration of 2×10^7 cells/mL in splenocyte medium.
22. Preserve the cells at 37°C in a humid atmosphere until needed.

3.6. Stimulation Experiments
3.6.1. Infection with H. pylori

1. Transfer an appropriate volume of the *H. pylori* into a 1.5-mL Eppendorf tube.
2. Fill up the tubes to a total volume of 100 μL with ice-cold PBS.
3. Add 100 μL of the splenocyte suspension (= 2×10^6 splenocytes) to 100 μL of the bacterial suspension. Fill up to 1 mL with splenocyte medium.
4. Mix once by gently pipeting the suspension up and down.
5. Transfer the mixture into one of the inner eight wells of a 24-well tissue culture plate (*see* **Note 18**).
6. Fill the 16 outer wells of the 24-well tissue culture plate with sterile PBS and seal the plate at the longer sides with Scotch tape to prevent evaporation.
7. Incubate the plate at 37°C for 4-36 h (*see* **Note 19**).
8. Transfer the cell culture supernatant from tissue culture plate into a 1.5-mL Eppendorf tube.
9. Centrifuge for 10 min and 1300g at room temperature.
10. Transfer supernatant into a new 1.5-mL Eppendorf tube, and store at –80°C until needed.

3.6.2. Stimulation with Recombinant Antigens

1. Pour 5–10 μg of purified HcpA into 1.5-mL Eppendorf tubes.
2. Proceed as described in **Subheading 3.6.1.**, **steps 2–10**.

3.7. Determination of Cytokine Production
3.7.1. ELISA

1. The two-site enzyme-linked immunosorbent assay (ELISA; BD) was employed to assay the levels of IFN-γ, TNF-α, IL-4, IL-5, IL-6, IL-10, and IL-12. All kits were used according to a slightly modified version of the manufacturer's protocol.

2. Briefly, coat each required well of the microtiter plate with 100 µL of capture antibody diluted in the recommended coating buffer (*see* **Note 20**). Seal plate and incubate overnight at 4°C.
3. Aspirate the wells and wash three to five times with 300 µL/well wash buffer. After the last wash, invert the plate and beat on adsorbent paper to remove residual buffer.
4. Block plates with 300 µL/well assay diluent. Incubate at room temperature for 1–2 h.
5. Aspirate and wash plates as described in **step 3**.
6. Prepare standard and sample dilutions in assay diluent as described in the manufacturer's protocol.
7. Pipet 100 µL of each standard, sample, and control into appropriate wells. Seal plates and incubate for 2 h at room temperature.
8. Aspirate and wash plates five times for all test kits in **step 3**.
9. Add 100 µL of working detector (detection antibody and avidin–HRP reagent) to each well. Seal plate and incubate for 1 h at room temperature.
10. Aspirate and wash plates as described in **step 3**, but with 10 or more washes per well.
11. Add 100 µL of substrate solution to each well and incubate for 20–30 min at room temperature in the dark.
12. Add 50 µL stop solution to each well.
13. Read adsorbance at 450 nm within 30 min of stopping reaction; if available, subtract adsorbance at 570 nm from adsorbance of 450 nm.
14. Calculate the cytokine concentrations using an appropriate calculation software.

3.7.2. FACS Analysis

Obtain single-cell culture of splenic cells as described in **Subheading 3.5., steps 1–22**.

1. Aliquot 100 µL (= 2×10^6 cells) of cell suspension/well into a 96-well flat bottom tissue culture plate.
2. Prepare stimulators as follows:
   Unstimulated control: to 100 µL RPMI add 2 µL/mL BFA stock solution.
   Specific stimulation: to 100 µL RPMI add 10 µg/mL protein or appropriate volumes of a *H. pylori* suspension and add 2 µL/mL BFA stock solution.
   Positive control: to 100 µL RPMI add 1 µL/ml PMA, 1 µL/mL ionomycin, and 2 µL/mL BFA (each of the described stock solution).
3. Add 100 µL of the appropriate activator to the center of each well.
4. Incubate for 6 h at 37°C.
5. Centrifuge at 300*g* for 5 min. Decant by turning the plate upside down; then resuspend the cells by smooth shaking of the plate.
6. Wash the cells twice with each 200 µL FACS buffer 1 per well. Transfer the cells after the first wash step to 96-well round-bottom plates. Spin the plates as described in **step 5** and decant the wash buffer.
7. For blocking of Fc-receptors, add 100 µL FACS buffer 1 with 1 µL anti-CD16/32 antibody.
8. Incubate for 10 min at 4°C.
9. Add 10 µL of a 1:20 dilution (in FACS buffer 1) of staining antibody (anti-CD4-FITC or anti-CD8-APC) to the appropriate wells.
10. Incubate for 20–25 min on ice in the dark.
11. Wash the cells twice with FACS buffer 2 as described in **step 6**.
12. Add 250 µL Cytofix/Cytoperm to each well and resuspend cells gently by flipping against the tube.

13. Incubate for 20 min at room temperature in the dark.
14. Wash twice with Perm/wash; centrifuge at 500g for 4–5 min (*see* **Note 21**).
15. Prepare intracellular staining antibody (anti-IFN-γ-PE) by diluting 1:100 in Perm/wash buffer. Resuspend cells by flipping against the tube.
16. Incubate for 25 min on ice in the dark.
17. Wash three times with 200 µL Perm/wash as described in **step 14**.
18. Add 200 µL of FACS buffer 1 to each well.
19. Store light protected for up to 24 h.
20. When ready for acquisition, resuspend cells with a pipet and transfer to FACS tubes.
21. Measure the number and identity of IFN-γ producing cells with a FACSscan analytical instrument [e.g., FACS Calibur (BD) or Coulter EPICS (Beckman Coulter)] and an appropriate software for data acquisition and analysis.

3.7.3. ELISpot

The protocol listed below is a modified version of a manual provided by Millipore (Bedford, MA). Perform consecutive **steps 1–8** under a sterile laminar flowhood.

1. Coating of ELISpot plates: dilute the coating antibody (anti-IFN-γ antibody, clone R4-6A2; BD Pharmingen) to 5–10 µg/mL in sterile PBS.
2. Add 100 µL to each well of a MAHA-S45 plate and incubate overnight in a wet chamber at 4°C.
3. Stimulation and ELISpot analysis: obtain single-cell culture of splenic cells as described in **Subheading 3.5., steps 1–22**.
4. Prepare stimulators as follows:
5. Unstimulated control: 100 µL RPMI.
6. Specific stimulation: to 100 µL RPMI add appropriate concentrations (1–10 µg/mL) polypeptide or appropriate volumes of an *H. pylori* suspension.
7. Positive control: to 100 µL RPMI add PMA-ionomycin.
8. Aspirate the wells and wash 5–10 times with 300 µL/well wash buffer. After the last wash, invert the plate and beat on adsorbent paper to remove residual buffer.
9. Block plates with 300 µL/well PBS, 5% FCS. Incubate at room temperature for 1–2 h.
10. Remove the blocking buffer and add 100 µL of the cell suspension to each well.
11. Add 100 µL of one of the stimulator solutions.
12. Incubate for 24 h at 37°C and 5% CO_2 in a humidified atmosphere.
13. Remove cells by washing six times with 200 µL PBS, 0.05% Tween 20.
14. Dilute biotinylated anti-IFN-γ monoclonal antibody in PBS, 10% FCS, 0.05% Tween 20 to 1 µg/mL.
15. Add 100 µL of the antibody solution to each well and incubate for 2 h at room temperature.
16. Wash the plates six times with 200 µL PBS, 0.05% Tween 20.
17. Add 100 µL streptavidin-alkaline phosphatase (diluted 1:1000 in PBS, 0.5% FCS) and incubate for 1 h at room temperature.
18. Wash the plates for a minimum of 10 times with 200 µL PBS.
19. Add 100 µL of a substrate solution (e.g., BCIP/NBT; Roche Diagnostics) and incubate until dark spots emerge.
20. Stop color development by washing in pouring water.
21. Leave the plates to dry, inspect, and count spots in a dissection microscope (×40) or in an ELISpot reader.

4. Notes

1. The periplasmic protein disulfide isomerase (DsbC) mediates shuffling of cysteine bridge formation. The pRBI-PDI expression vector was kindly provided by Stefan Strobl, MPI für Biochemie (München, Germany).
2. The basic pHG vector for the construction of the pA5c-DHFR plasmid was kindly provided by Dr. Martin Rosenberg (Smith Klein & French Laboratories, King of Prussia, PA).
3. Prepare a 500 mM MES stock solution. Store at 4°C for up to 6 mo.
4. Mice were purchased from commercial breeders (e.g., Charles River, Harland) at an age of 44 d and housed under specific pathogen-free conditions.
5. Use splenocyte medium with antibiotics (Pen/Strep or kanamycin) for assaying immunostimulatory properties of bacterial components, but use splenocyte medium without antibiotics for assaying the stimulatory properties of live *H. pylori* strains.
6. All PCR techniques are performed according to standard protocols *(21)*.
7. The two additional residues at the 5' end of the PCR product are necessary to clone the coding region of hcpA in frame with the pRBI-PDI plasmid-encoded bacterial periplasmic secretion signal sequence. The stop codon at the 3' end has to be added, since the stop codon of the generic RBI gene of the pRBI-PDI plasmid is lost in consequence of the restriction digest.
8. All DNA manipulation and transformation experiments are performed according to standard cloning protocols *(19)*.
9. Standard transformation protocols *(22)* are used.
10. Perform all DNA purification steps as recommended by the Qiagen Plasmid handbook and enclosed in the EndoFree Midi Kit (Qiagen, Hilden, Germany).
11. Under the described mild conditions the bacteria will be lysed only partially by polymyxin B treatment, without releasing the cytoplasm from the bacterial cell. In addition, polymyxin B treatment leads to complexation and inactivation of unwanted *Escherichia coli* LPS.
12. We tested different transfection protocols to obtain stable transfected cells that exhibited a high yield of production of the reporter construct. Independent cotransfections were performed with 4 µg of DNA using pMt-hcpA to pA5c-DHFR ratios of 50:1, 20:1, 1:1, and 1:20. In each case metothrexate-resistant cell lines were generated. The highest yields of HcpA were detected from the supernatant of cells cotransfected with pMt-hcpA to pA5c-DHFR at a ratio of 20:1.
13. During the incubation, the rocking plate temperature should not surpass 30°C. Therefore, we recommend incubating the plates on a wood-fiber board, taking care that the growth surface is always wet.
14. Cadmium ($CdCl_2$) is also an efficient inducer of the Mtn promoter even at concentrations as low as 10 µM. In comparison, $ZnSO_4$ was found to be only a weak inducer of the Mtn promoter.
15. HcpA elutes as a single peak at 0.5 M NaCl.
16. Molecular biological analysis of purified HcpA by silver staining of polyacrylamide gels or by immunoblotting according to standard protocols *(19)*.
17. This step should be carried out for every single *H. pylori* strain separately, since the value can differ significantly between different strains. For the clinical *H. pylori* strain HP 2802, 1 OD_{600} corresponds to 2.28×10^8 cells/mL.
18. Use only the inner 8 wells of the 24-well plate to avoid inconsistent results.

19. We recommend short-time incubation for the detection of IL-12 and TNF-α and elongated incubation for analyzing other cytokines such as IL-2, IL-4, IL-5, IL-6, IL-10, and IFN-γ.
20. The manufacturer (BD) recommends different buffers (0.1 M carbonate, pH 9.5 or 0.2 M sodium phosphate, pH 6.5) for the coating of different capture antibodies.
21. At this point cells are fixed and require stronger centrifugation to sediment.

References

1. Janeway, C. A. Jr. (1989) The role of CD4 in T-cell activation: accessory molecule or co-receptor? *Immunol. Today* **10,** 234–238.
2. Janeway, C. A. Jr. (1992) The immune system evolved to discriminate infectious nonself from noninfectious self. *Immunol. Today* **13,** 11–16.
3. Medzhitov, R. and Janeway, C. A. Jr. (2000) Innate immune recognition: mechanisms and pathways. *Immunol. Rev.* **173,** 89–97.
4. Gronowski, A. M., Hilbert, D. M., Sheehan, K. C., Garotta, G., and Schreiber, R. D. (1999) Baculovirus stimulates antiviral effects in mammalian cells *J. Virol.* **73,** 9944–9951.
5. Hayashi, F., Smith, K. D., Ozinsky, A., et al. (2001) The innate immune response to bacterial flagellin is mediated by Toll-like receptor 5. *Nature* **410,** 1099–1103.
6. Smith, K. D. and Ozinsky, A. (2002) Toll-like receptor-5 and the innate immune response to bacterial flagellin. *Curr. Top. Microbiol. Immunol.* **270,** 93–108.
7. Borges, M. M., Campos-Neto, A., Sleath, P., et al. (2001) Potent stimulation of the innate immune system by a *Leishmania brasiliensis* recombinant protein. *Infect. Immun.* **69,** 5270–5277.
8. Rock, F. L, Hardiman, G., Timans, J. C., Kastelein, R. A, and Bazan, J. F. (1998) A family of human receptors structurally related to *Drosophila* Toll. *Proc. Natl. Acad. Sci. USA* **95,** 588–593.
9. Aderem, A. and Ulevitch, R. J. (2000) Toll-like receptors in the induction of the innate immune response. *Nature* **406,** 782–787.
10. Hartmann, G., Weiner, G. J., and Krieg, A. M. (1999) CpG DNA: a potent signal for growth, activation, and maturation of human dendritic cells. *Proc. Natl. Acad. Sci. USA* **96,** 9305–9310.
11. Dundon, W. G., de Bernard, M., and Montecucco, C. (2001) Virulence factors of *Helicobacter pylori*. *Int. J. Med. Microbiol.* **290,** 647–658.
12. Atherton, J. C. (1998) *H. pylori* virulence factors. *Br. Med. Bull.* **54,** 105–120.
13. Bodger, K. and Crabtree, J. E. (1998) *Helicobacter pylori* and gastric inflammation. *Br. Med. Bull.* **54,** 139–150.
14. Shimoyama, T. and Crabtree, J. E. (1998) Bacterial factors and immune pathogenesis in *Helicobacter pylori* infection. *Gut* **43(suppl 1),** S2–5.
15. Zevering, Y., Jacob, L., and Meyer, T. F. (1999) Naturally acquired human immune responses against *Helicobacter pylori* and implications for vaccine development. *Gut* **45,** 465.
16. Suerbaum, S. and Michetti, P. (2002) *Helicobacter pylori* infection. *N. Engl. J. Med.* **347,** 1175–1186.
17. Wunderlich, M. and Glockshuber, R. (1993) In vivo control of redox potential during protein folding catalyzed by bacterial protein disulfide-isomerase (DsbA). *J. Biol. Chem.* **268,** 24547–24550.
18. Deml, L., Wolf, H., and Wagner, R. (1999) High level expression of hepatitis B virus surface antigen in stably transfected *Drosophila* Schneider-2 cells *J. Virol. Methods* **79,** 191–203.

19. Sambrook, J., Fritsch, E. F., and Maniatis, T. (1989) *Molecular Cloning: A Laboratory Manual*. New York, Cold Spring Harbor Laboratory.
20. Deml, L. and Wagner, R. (1998) Stable transfected *Drosophila* Schneider-2 cells as a novel tool to produce recombinant antigens for diagnostic, therapeutic and preventive purposes, in U Reischl (ed), *Methods in Molecular Medicine. Molecular Diagnosis of Infectious Diseases*. Humana, Totowa, NJ, pp. 185–200.
21. White, B. A. (1993) *Methods in Molecular Biology: PCR Protocols, Current Methods and Applications*. Humana, Totowa, NJ.
22. Inoue, H. and Okayama, H. (1990) High efficiency transformation of *Escherichia coli* with plasmids. *Gene* **96,** 23–28.

IV

EXPRESSION OF RECOMBINANT PROTEINS

11

High-Throughput Expression and Purification of 6xHis-Tagged Proteins in a 96-Well Format

Jutta Drees, Jason Smith, Frank Schäfer, and Kerstin Steinert

Abstract

By using automation and affinity-tag technologies, analysis of the large number of ORFs generated by genome-sequencing projects is greatly accelerated. Protocols describing culture of *E. coli* in automation-compatible formats and subsequent micro- to large-scale automated purification of 6xHis-tagged proteins are presented.

Key Words: His-Tag; automation; affinity-tag; Ni-NTA; QIAexpress.

1. Introduction

The completion of several genome-sequencing projects has led to the identification of many thousands of open reading frames (ORFs). A major research effort is now under way to clone, express, and analyze the proteins encoded by these ORFs. The sheer scale of the genomes demands high-throughput methods to characterize the large numbers of proteins. Molecular biology allows the expression of ORFs fused to an affinity tag that can be used for a single-step purification on an affinity matrix. Using QIA*express*® vectors to place a 6xHis tag at the N- or C-terminus of a protein overexpressed in *E. coli* allows efficient one-step purification using Ni-NTA technology under native or denaturing conditions.

QIAGEN (*see* **Note 1**) offers products for automated protein purification for each step of a proteomics or functional genomics study *(1)*. For initial expression-clone screening, proteins can be purified in microscale amounts (1–15 μg) using Ni-NTA Magnetic Agarose Beads. Initial functional studies can be carried out using the up to 2 mg protein obtained in the medium-scale Ni-NTA Superflow 96 BioRobot® Kit procedure. Large-scale purification using Ni-NTA Superflow Columns provides enough protein (up to 15 mg) for structural analyses using X-ray crystallography or nuclear magnetic resonance (NMR).

Protocols are presented here for expression of 6xHis-tagged proteins in *E. coli* in 96- and 24-well blocks and shake flasks. Subsequent automated purification using

Ni-NTA Magnetic Agarose Beads, Ni-NTA Superflow BioRobot Kits, and Ni-NTA Superflow Columns is described.

2. Materials

1. pQE expression vector encoding 6xHis-tagged protein of interest (QIAGEN, Valencia, CA). Vectors are available that place the 6xHis tag at the N- or C-terminus, or contain special features (e.g., protease recognition sites, additional epitopes for immunodetection).
2. *Escherichia coli* strains.
3. Luria-Bertani (LB) medium: 10 g/L tryptone; 5 g/L yeast extract; 10 g/L NaCl.
4. Kanamycin (stock solution 25 mg/mL, working concentration 25 µg/mL) and ampicillin (stock solution 100 mg/mL, working concentration 100 µg/mL); both dissolved in H_2O; sterile filter and store in aliquots at –20°C.
5. Agar plates with selective antibiotic.
6. IPTG (isopropyl-β-D-thiogalactopyranoside): stock solution 1 M in H_2O; sterile filter; store in aliquots at –20°C.
7. 96-well blocks, 24-well blocks, or shake flasks for bacterial cultivation.
8. Microporous tape sheets (e.g., AirPore™ Tape Sheets, QIAGEN).
9. Lysis buffer for purification under native or denaturing conditions. Buffer compositions are provided in the handbook supplied with all QIAGEN kits.
10. Wash buffer for purification under native or denaturing conditions. Buffer compositions are provided in the handbook supplied with all QIAGEN kits.
11. Elution buffer for purification under native or denaturing conditions. Buffer compositions are provided in the handbook supplied with all QIAGEN kits.
12. Lysozyme solution (10 mg/mL in distilled water).
13. Benzonase®.
14. Ni-NTA Magnetic Agarose Beads—for microscale purification, 1–15 µg (QIAGEN).
15. Ni-NTA Superflow 96 BioRobot Kit—for medium-scale purification, up to 2 mg (QIAGEN).
16. Ni-NTA Superflow Columns 1.5 mL—for large-scale purification, up to 15 mg (QIAGEN).

3. Methods

Protocols are provided for cultivation of *E. coli* in 96- and 24-well blocks and shake flasks, and protocols for microscale (1–15 µg), medium-scale (up to 2 mg), and large-scale (up to 15 mg) automated purification of 6xHis-tagged proteins are described.

3.1. Cultivating E. coli *Harboring pQE Expression Constructs*

3.1.1. pQE Expression Vectors

pQE expression vectors belong to the pDS family of plasmids *(2)* (**Fig. 1**) and contain a powerful phage T5 promoter that is repressed by two *lac* operator sequences and the β-lactamase gene *bla* conferring resistance to ampicillin. The extremely high transcription rate initiated at the T5 promoter can only be efficiently regulated and repressed by the presence of high levels of the *lac* repressor protein, which can be expressed in *cis* or *trans*. The low-copy plasmid pREP4, which confers kanamycin resistance and constitutively expresses the *lac*I gene *(3)*, can be used to express *lac* repressor protein in *trans*. Multiple copies of pREP4 in host strain cells ensure the

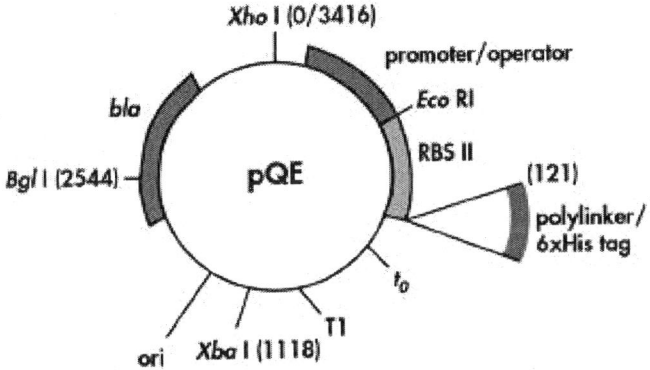

Fig. 1. Basic features of pQE vectors used for expressing 6xHis-tagged proteins.

production of high levels of the *lac* repressor protein that tightly regulates recombinant protein expression. The pREP4 plasmid is compatible with all plasmids carrying the ColE1 origin of replication and is maintained in *E. coli* in the presence of kanamycin at a concentration of 25 μg/mL. The *cis*-repressed vectors pQE-80L, -81L, and -82L from QIAGEN contain the *lacI*^{q} gene and do not require the presence of pREP4.

Expression of recombinant proteins encoded by pQE vectors is rapidly induced by the addition of isopropyl-β-D-thiogalactoside (IPTG), which binds to the *lac* repressor protein and inactivates it. Once the *lac* repressor is inactivated, the host cell's RNA polymerase can transcribe the sequences downstream from the promoter. The transcripts produced are then translated into the recombinant protein. The special "double operator" system in the pQE expression vectors, in combination with the high levels of the *lac* repressor protein generated by pREP4 or the *lacI*^{q} gene on pQE-80L, pQE-81L, or pQE-82L, ensure tight control at the transcriptional level.

3.1.2. E. coli Host Strains

Any *E. coli* host strain containing both the expression (pQE) and the repressor (pREP4) plasmids can be used for the production of recombinant 6xHis-tagged proteins. pREP4 must be maintained by selection for kanamycin resistance. *E. coli* strains that harbor the *lacI*^{q} mutation, such as XL1 Blue, JM109, and TG1, produce enough *lac* repressor to block transcription efficiently and are ideal for storing and propagating pQE plasmids. These strains can also be used as expression hosts for expressing nontoxic proteins, but they may be less efficient, and expression may be regulated less tightly than in strains harboring the pREP4 plasmid. If the expressed protein is toxic to the cell, "leaky" expression before induction may result in poor culture growth or in the selection of deletion mutants that grow faster than bacteria containing the correct plasmid. *E. coli* M15[pREP4] is the recommended strain for the expression of 6xHis-tagged proteins using pQE vectors. The *cis*-repressed pQE-80L series of vectors can easily be used with any *E. coli* host strain, and kanamycin selection is not necessary.

3.1.3. Transformation of E. coli

1. Clone protein expression constructs into pQE vectors using standard recombinant DNA methods *(4)* and transform them into M15 *E. coli* carrying pREP4.
2. Plate them onto selective medium containing relevant antibiotic(s) and incubate them overnight at 37°C.
3. Inoculate precultures by picking single colonies using a sterile toothpick, inoculating a single well or vessel containing LB medium and the relevant antibiotic(s), allowing them to stand for 5 min, and shaking briefly. Remove the toothpicks and cover the vessel.
4. Shake the precultures at 150 rpm and incubate them overnight at 37°C (multiwell blocks) or 20–25°C (shake flasks).

3.1.4. Growing Expression Cultures in 96-Well Square-Well Blocks for Microscale Purification Using Ni-NTA Magnetic Agarose Beads

1. Prepare a 96-well square-well block containing 1 mL LB medium per well with appropriate antibiotics under sterile conditions.
2. Inoculate expression cultures by adding 25 µL of overnight preculture from the 96-well square-well block to obtain an OD_{600} of approx 0.05.
3. Cover 96-well block with a microporous tape sheet and incubate cultures at 37°C, with shaking at 150 rpm, for 2–3 h or until an OD_{600} of 0.6–0.8 is reached.
4. Induce expression of 6xHis-tagged proteins by adding 25 µL of 0.1 M IPTG (final concentration: 0.5 mM IPTG) and incubate the cultures overnight or for 16 h at 37°C and 150 rpm, or until an OD_{600} of 6 is reached.
5. Pellet cells by centrifuging a 96-well block at 4000g for 10 min in a centrifuge suitable for 96-well plates, remove supernatant, and freeze cell pellets at −20°C.
6. Proceed with one of the microscale purification protocols in the *Ni-NTA Magnetic Agarose Beads Handbook*.

3.1.5. Growing Expression Cultures in 24-Well Blocks for Medium-Scale Purification Using the Ni-NTA Superflow 96 BioRobot Kit

1. Prepare four 24-well blocks containing 5 mL LB medium with appropriate antibiotics under sterile conditions.
2. Inoculate expression cultures in 24-well blocks by adding 25–100 µL of overnight preculture from the 96-well square-well block to obtain an OD_{600} of approx 0.05.
3. Cover 24-well blocks with a porous tape sheet and incubate cultures at 37°C, with shaking at 150 rpm, for 2–3 h or until an OD_{600} of 0.6–0.8 is reached.
4. Induce expression of 6xHis-tagged proteins by adding 25 µL of 0.1 M IPTG (final concentration: 0.5 mM IPTG) and incubate the cultures overnight or for 16 h at 37°C and 150 rpm, or until an OD_{600} of 6 is reached. To avoid exceeding a final OD_{600} of 6, we recommend monitoring growth in several wells of the 24-well blocks in pilot experiments under the chosen growth conditions. Shorten the induction phase of the expression cultures if overnight growth leads to OD_{600} values greater than 6. If the cultures are grown to higher optical densities, the TurboFilter® Plate can become blocked during clearing of the crude lysate.
5. Pellet cells by centrifuging 24-well blocks at 1000g for 10 min in a centrifuge suitable for 96-well plates, remove supernatant, and freeze cell pellets at −20°C.

6. Proceed with one of the medium-scale purification protocols in the *Ni-NTA Superflow 96 BioRobot Handbook*.

3.1.6. Growing 15–25 mL Expression Cultures in Shake Flasks for Medium-Scale Purification Using the Ni-NTA Superflow 96 BioRobot Kit High-Yield Protocol

1. Prepare the required number of 150-mL shake flasks containing 15–25 mL LB medium with appropriate antibiotics under sterile conditions.
2. Inoculate expression cultures in shake flasks by adding a volume of preculture to obtain an OD_{600} of approx 0.05.
3. Incubate the cultures at 37°C, with shaking at 150 rpm, for 2–3 h or until an OD_{600} of 0.6–0.8 is reached.
4. Induce expression of 6xHis-tagged proteins by addition of IPTG to a final concentration of 1 mM and incubate cultures for 4 h at 37°C and 150 rpm.
5. Pellet cells by repeated centrifugation into the wells of 24-well blocks, and freeze cell pellets at –20°C (e.g., to process 15-mL culture volume, centrifuge 7.5 mL, discard supernatant, add and centrifuge the remaining 7.5 mL, and discard supernatant).
6. Proceed with one of the high-yield medium-scale purification protocols in the *Ni-NTA Superflow 96 BioRobot Handbook*.

3.1.7. Growing 1-L Expression Cultures in Shake Flasks for Large-Scale Purification Using Ni-NTA Superflow Columns

1. Inoculate 1-L expression cultures in shake flasks by adding a volume of preculture to obtain an OD_{600} of approx 0.05.
2. Incubate the cultures at 37°C with vigorous shaking, for 2–3 h or until an OD_{600} of 0.6 is reached.
3. Induce expression of 6xHis-tagged proteins by addition of IPTG to a final concentration of 1 mM and incubate cultures for 4–5 h at 37°C.
4. Harvest the cells by centrifugation at 4000g for 20 min.
5. Freeze cells in dry ice/ethanol or liquid nitrogen, or store cell pellet overnight at –20°C.
6. Proceed with one of the large-scale purification protocols in the *Ni-NTA Superflow 96 BioRobot Handbook*.

3.2. Microscale Purification of 6xHis-Tagged Proteins Using Ni-NTA Magnetic Agarose Beads

1. Add Ni-NTA Magnetic Agarose Beads to crude or cleared *E. coli* cell lysates in a 96-well microplate, where they bind 6xHis-tagged proteins.
2. Then place the plate onto a 96-well magnet where the beads are held in place while wash steps are carried out.
3. After removal from the magnet, add elution buffer and return the microplate to the magnet where eluted proteins are collected (**Fig. 2**).

The fully automated purification procedure was carried out on a BioRobot workstation according to the protocol supplied that can also be found in the *Ni-NTA Magnetic Agarose Beads Handbook*. **Figure 3** shows a Coomassie-stained sodium dodecyl sulfate polyacrylamide gel electrophoresis (SDS-PAGE) gel of eluates obtained using the microscale purification procedure under denaturing conditions.

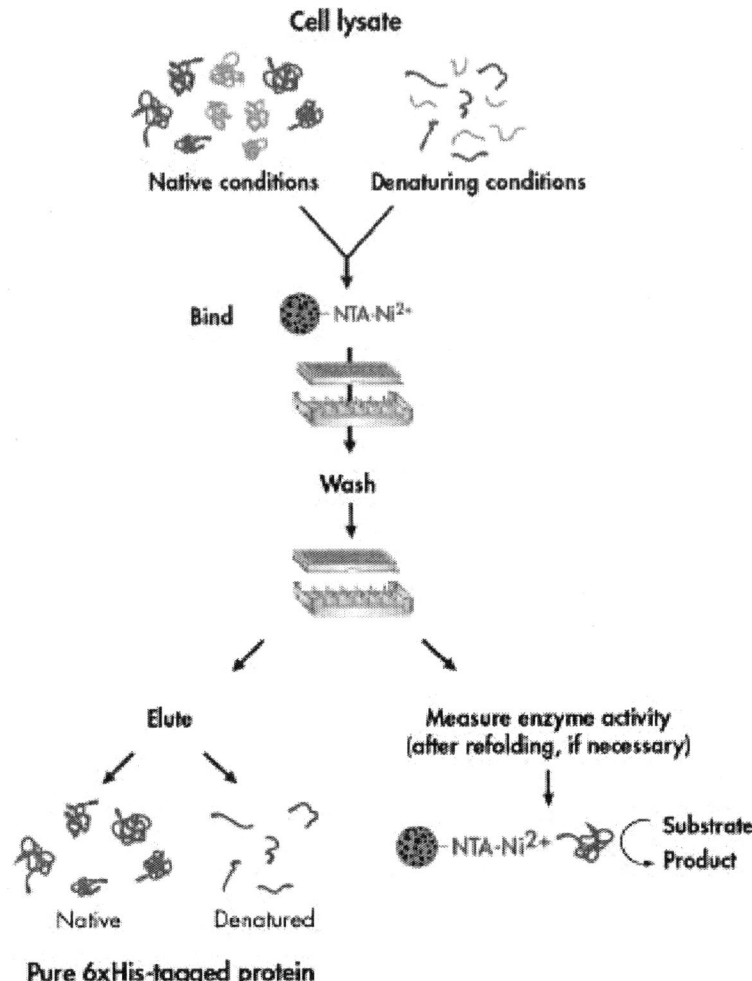

Fig. 2. Microscale 6xHis-tagged protein purification using Ni-NTA Magnetic Agarose Beads.

3.3. Medium-Scale Purification of 6xHis-Tagged Proteins Using the Ni-NTA Superflow 96 BioRobot Kit (BioRobot 9600 and 3000)

In the standard protocol, 5-mL overnight cultures are harvested by centrifugation, frozen, and resuspended in lysis buffer. In the high-yield protocol, *E. coli* cultures of up to 25 mL (native conditions) or 15 mL (denaturing conditions) can be processed.

1. Pipet lysates onto a TurboFilter® 96 Plate that is placed directly above a QIAfilter™ 96 Plate into which Ni-NTA Superflow has been pipeted, forming 96 mini-chromatography columns. Lysates are drawn by vacuum through the TurboFilter Plate—where cell debris

Expression and Purification of 6xHis-Tagged Proteins

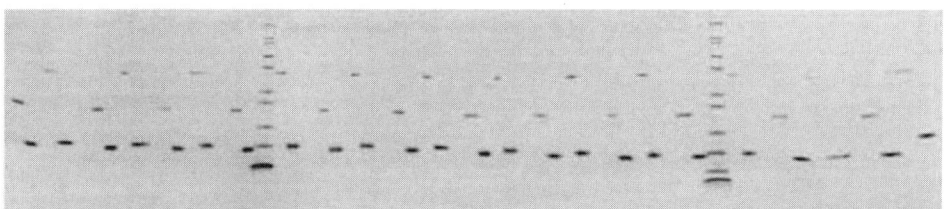

Fig. 3. Simultaneous purification of 6xHis-tagged protein under denaturing conditions from *E. coli* cells expressing either thioredoxin (T), dihydrofolate reductase (D), chaperonin (C), or protein M50 (M), or from cells lacking any expression plasmid (–) using Ni-NTA Magnetic Agarose Beads.

Table 1
Yields of 6xHis-Tagged Proteins
Using the High-Yield Ni-NTA Superflow 96 BioRobot Protocol

6xHis-tagged protein	Total yield per well (μg)[a]	Protein concentration (mg/mL)[b]
Green fluorescent protein	4000	6.0
T7 RNA polymerase	1000	1.4
GroES	300	0.4
Chloramphenicol acetyltransferase	2400	4.4
GroEL	740	1.0
Tumor necrosis factor-α	1600	2.5
GroES/GroEL	1200	1.5
Cpn-10	170	0.3

[a]Yield obtained in two 550-μL elution fractions (average of six independent purifications); 80% of the protein elutes in the first 550 μL.
[b]Protein concentration in the first 550-μL elution fraction.

is removed—and flow through the mini Ni-NTA Superflow columns where 6xHis-tagged proteins bind.

2. After washing, elute 6xHis-tagged proteins into a 96-well elution vessel (**Fig. 4**).

The fully automated purification procedure was carried out on a BioRobot workstation according to the supplied protocol that can also be found in the *Ni-NTA Superflow BioRobot Handbook*. **Figure 5** shows a Coomassie-stained SDS-PAGE gel of eluates obtained using the Ni-NTA Superflow 96 BioRobot Kit high-yield protocol under native conditions. The total amount of protein obtained per well is given in **Table 1**.

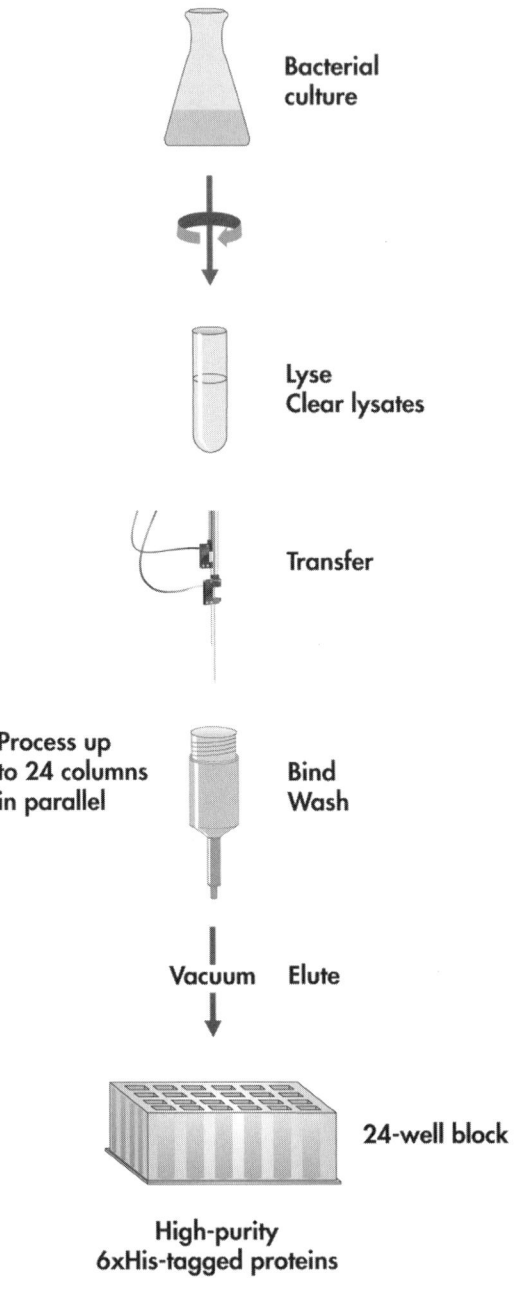

Fig. 4. Medium-scale 6xHis-tagged protein purification using the Ni-NTA Superflow 96 BioRobot Kit.

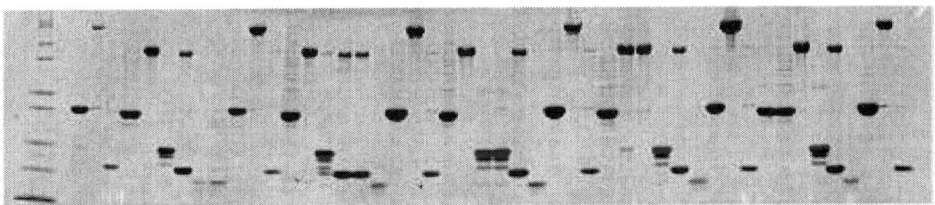

Fig. 5. Vector constructs for the expression of 6xHis-tagged proteins were transformed into *E. coli* and plated on selective medium, and colonies were picked for inoculating 25-mL cultures. Expression of 6xHis-tagged proteins was induced with IPTG for 2–4 h. Cells were pelleted in 24-well blocks and processed on the BioRobot 3000 using 200 μL Ni-NTA Superflow resin per well. Five microliters (0.9%) of the first elution fraction was loaded for SDS-PAGE, and proteins were visualized by Coomassie staining. G, green fluorescent protein (29 kDa); T, T7 RNA polymerase (100 kDa); S, *E. coli* GroES (12 kDa). Some endogenous GroEL is copurified; C, *E. coli* chloramphenicol acetyltransferase (28 kDa); L, *E. coli* GroEL (60 kDa); α, human tumor necrosis factor-α (18 kDa); E, *E. coli* GroES purified as a complex with co-overexpressed nontagged GroEL (12 and 60 kDa); 10, *Saccharomyces cerevisiae* Cpn-10 (10 kDa); M, markers.

3.4. Large-Scale Purification of 6xHis-Tagged Proteins Using Ni-NTA Superflow Columns

1. Lyse cells from up to 1 L of bacterial culture using a buffer containing nucleases, and prepare cleared lysates by centrifugation.
2. Transfer cleared lysates to 14-mL tubes and place them along with the relevant number of Ni-NTA Superflow Columns (up to 12 per manifold) on the BioRobot worktable for processing. The automated procedure then begins.
3. After column equilibration, the BioRobot 3000 workstation pipets cleared lysates onto Ni-NTA Superflow Columns. 6xHis-tagged proteins bind strongly and selectively to Ni-NTA, while contaminants are removed in the flow-through fractions.
4. After two wash steps, elute highly pure 6xHis-tagged proteins into 24-well blocks using 3 mL elution buffer (**Fig. 6**).

Figure 7 shows a Coomassie-stained SDS-PAGE gel of eluates obtained using the Ni-NTA *Superflow Columns* large-scale purification procedure under native conditions. The total amount of protein obtained per column is given in **Table 2**.

4. Notes

1. The literature that QIAGEN supplies with its protein products gives comprehensive background information and numerous hints and tips on how to optimize procedures. All literature is also available online at www.qiagen.com/literature/. The *QIAexpressionist*™ is a handbook for high-level expression and purification of 6xHis-tagged proteins. The *QIAexpress Detection and Assay Handbook* contains protocols for detection of 6xHis-tagged proteins and ELISA procedures using antibodies directed against 6xHis-

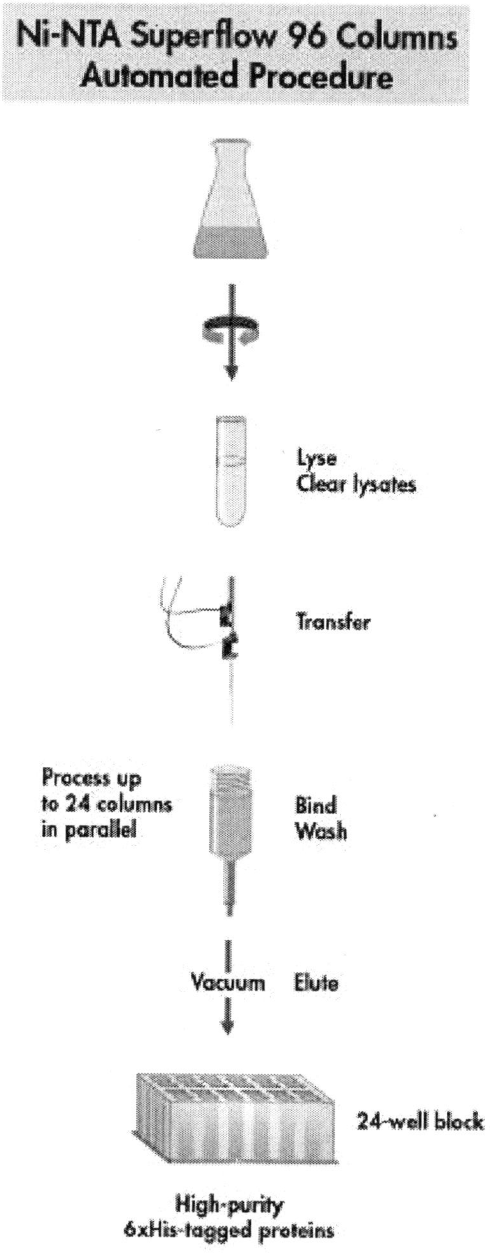

Fig. 6. Large-scale purification of 6xHis-tagged protein using Ni-NTA Superflow Columns.

Expression and Purification of 6xHis-Tagged Proteins

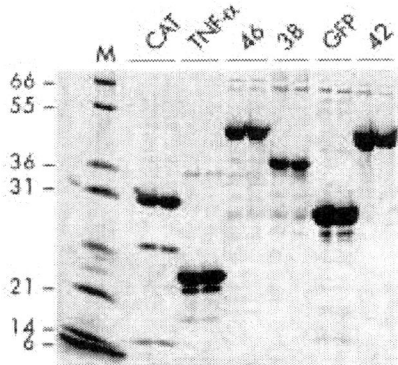

Fig. 7. SDS-PAGE analysis of 12 6xHis-tagged protein preparations performed in parallel using Ni-NTA Superflow Columns. For *E. coli* culture volumes and yields, *see* **Table 1**. CAT, 6xHis-tagged chloramphenicol acetyl-transferase; TNF-α, 6xHis-tagged tumor necrosis factor-α; 46, 6xHis-tagged 46-kDa protein; 38, 6xHis-tagged 38-kDa protein; GFP, 6xHis-tagged green fluorescent protein; 42, 6xHis-tagged 42-kDa protein. Aliquots (5 µL) of the final eluates (3 mL) were loaded onto the gel. Proteins were visualized by Coomassie staining. M, markers.

Table 2
Yields of 6xHis-Tagged Proteins Using Ni-NTA Superflow Columns

6xHis-tagged protein	Average yield per column (mg)[a]	*E. coli* culture processed per column (mL)
Chloramphenicol acetyltransferase	12.7	100
Tumor necrosis factor-α	24.9	100
46-kDa protein	12.5	200
38-kDa protein	11.1	200
Green fluorescent protein	23.3	100
42-kDa protein	15.1	200

[a]Preparations were performed in duplicate.

tagged proteins. The most important points for automated purification are summarized below.
  a. If the 6xHis-tagged protein does not bind to the Ni-NTA matrix, the following possible causes should be investigated.
    i. **The 6xHis-tag is not present.** Sequence your expression construct to ensure that the reading frame is correct, and check the sequence for possible internal translation starts (N-terminal tags) or premature termination sites (C-terminal tags).
    ii. **6xHis tag is inaccessible or has been degraded.** If, after folding, the 6xHis tag lies within the body of the protein, binding to Ni-NTA may be compromised. Move the 6xHis tag to the opposite end of the protein, or purify under denaturing

conditions. Check that the 6xHis tag is not associated with a portion of the protein that undergoes post-translational processing (e.g., signal peptide cleavage).
  iii. **Binding conditions are incorrect.** Check pH values and composition of all buffers prior to use. This is especially important when purifying under denaturing conditions, as dissociation of urea often causes a shift in pH. In addition, the presence of chelating or reducing agents or excess imidazole can preclude efficient binding. Make sure that the Ni-NTA matrix is fully resuspended before it is pipeted.
 b. If no protein is found in the eluate, the following possible causes should be investigated:
  i. **6xHis-tagged protein elutes in the wash buffer.** The stringency of the wash may be too high. Reduce the concentration of imidazole or increase the pH slightly. Check pH values and composition of wash buffers prior to use.
  ii. **Protein precipitates during purification.** Try performing the protein purification at room temperature or adding solubilization reagents to buffers. Add up to 0.1% (v/v) Triton X-100 or Tween 20, up to 20 mM β-mercaptoethanol, or up to 2 M NaCl to all buffers to keep proteins in solution. Try adding stabilizing cofactors, such as 1–10 mM Mg^{2+}.
  iii. **6xHis-tagged protein does not elute.** Increase imidazole concentration or reduce pH of the elution buffer. EDTA can be used for elution, but bear in mind that eluted proteins will be complexed to nickel ions.

References

1. Schäfer, F., Römer, U., Emmerlich, M., Blümer, J., Lubenow, H., and Steinert, K. (2002) Automated high-throughput purification of 6xHis-tagged proteins. *J. Biomolec. Tech.* **12,** 131.
2. Bujard, H., Gentz, R., Lanzer, M., et al. (1987) A T5 promoter based transcription-translation system for the analysis of proteins in vivo and in vitro. *Methods Enzymol.* **155,** 416.
3. Farabaugh, P. J. (1978) Sequence of the Iac I gene. *Nature* **274,** 765.
4. Ausubel, F. M., Brent, R., Kingston, R. E., et al, eds. (1995) *Current Protocols in Molecular Biology.* John Wiley & Sons, New York.

12

Production of Antigens in *Chlamydomonas reinhardtii*

Green Microalgae as a Novel Source of Recombinant Proteins

Markus Fuhrmann

Abstract

Recombinant small-scale proteins are produced in a number of systems, from bacteria like *Escherichia coli*, through lower eukaryotes like baker's yeast, up to mammalian cell cultures. However, the need for safe and cheap sources of large amounts of recombinant proteins for different purposes, including material sciences, diagnostics, and, of course, medical therapy, has forced the development of alternative production systems. Green microalgae are cheap and easily grown and offer a high protein content, which would seem to make them ideal hosts for the large-scale sustainable production of recombinant proteins in the future. In selected species, recombinant DNA can be introduced into the genomes of the nucleus, the chloroplast, and even the mitochondria, and thus the system offers both prokaryotic (chloroplast, mitochondria) and eukaryotic translation systems for a tailored expression of virtually any protein.

Key Words: Green algae; edible vaccines; molecular farming; recombinant; surface display; synthetic genes; secretion pathway; membrane proteins.

1. Introduction

Eukaryotic microalgae, like *Chlamydomonas reinhardtii* and *Chlorella vulgaris*, are a group of organisms that combine several features that make them an attractive host for high-level protein production in general. In contrast to bacterial or yeast expression systems, green algae are fast and easily grown in huge amounts on pure mineral medium, with daylight as the only source of energy and CO_2 as the carbon source. Even though they are easily cultured and manipulated, green algae have a complex machinery for post-translational modifications like complex glycosylations, similar to higher plants. The vision of recombinant protein production for therapeutic purposes in higher plants is already widely known as *molecular farming* and has now been developed for more than a decade. For algae, especially in the Asian hemisphere, there is an established infrastructure and techniques for cultivation, because several algae are traditionally consumed as food in this part of the world. An intriguing possibility that has raised interest for some time is the development of edible vaccines from recombinant plants. This kind of vaccination should be especially desirable, if one keeps in mind the need for large amounts of vaccines against infectious diseases in the third world. Because in such countries, not only does the vaccine have to be affordable, but also it should be stable at room temperature, to avoid the necessity of a continuous cooling chain. Additionally, it also should be administered orally because sterile injection tools cannot always be guaranteed in those countries. A novel approach in this field uses green algae, which carry recombinant proteins on their surface, to immunize humans and animals against certain infectious diseases by simply eating or inhaling whole microalgae. This chapter briefly describes the methodology of transformation and enhanced foreign gene expression by the use of synthetic genes and endogenous regulatory sequences.

2. Materials

1. Strains of *Chlamydomonas reinhardtii*, preferentially cell wall–deficient mutants, like cw15 or cw2, because transformation with exogenous DNA is much more efficient with wall-less cells: *C. reinhardtii* CC-425 arg2 cw15 sr-u-2-60 mt+ (available through the core culture collection at the Chlamydomonas Genetics Center, Durham, NC; http://www.biology.duke.edu/chlamy).
2. Growth medium: Tris-acetate phosphate medium (TAP) + arginine (50 mg/L); described in ref. *1*:
   a. TAP salts: NH_4Cl (15 g), $MgSO_4 \cdot 7\, H_2O$ (4.0 g), $CaCl_2 \cdot 2\, H_2O$ (2.0 g). Add water to 1 L.
   b. Phosphate solution (1 *M* potassium phosphate, pH 7.0): K_2HPO_4 (28.8 g), KH_2PO_4 (14.4 g). Add water to 100 mL.
   c. Hutner trace elements (a modified version derived from ref. *2* and described in ref. *1*. For 1 L of trace elements mix, first completely dissolve each compound separately in the volume of water indicated:

Na_2–EDTA	50 g	250 mL H_2O (use boiling water)
$ZnSO_4 \cdot 7\, H_2O$	22 g	100 mL H_2O
H_3BO_3	11.4 g	200 mL H_2O
$MnCl_2 \cdot 4\, H_2O$	5.06 g	50 mL H_2O
$CoCl_2 \cdot 6\, H_2O$	1.61 g	50 mL H_2O
$CuSO_4 \cdot 5\, H_2O$	1.57 g	50 mL H_2O
$(NH_4)_6Mo_7O_{24} \cdot 4\, H_2O$	1.10 g	50 mL H_2O
$FeSO_4 \cdot 7\, H_2O$	4.99 g	50 mL H_2O

   Mix all solutions except EDTA, and heat this mixture. Then add the EDTA solution to the boiling solution. The color should turn to dark green. After everything is completely dissolved, cool down the mixture to 70°C, and adjust the pH to 6.7 with 80–90 mL of hot 20% KOH (!) with an appropriately standardized pH meter. Finally, bring the solution to a volume of 1 L. Loosely close the bottle to allow aeration of the solution, which should turn purple within about 2 wk. If a brownish precipitate appears, it can be removed by filtration. The stock solution can be stored at 4°C or frozen for more than a year.
   d. For each liter of medium, use 2.42 g Tris-HCl, 25 mL TAP salts, 0.375 mL phosphate solution, 1.0 mL Hutner trace elements, and 1.0 mL glacial acetic acid. The complete

mixture can be sterilized by autoclaving. L-arginine (50 mg per liter of medium) has to be added after autoclaving, for cultivation of the *CC-425* strain. TAP agar plates should be prepared with 1.5% agar. For selection purposes, plates should be prepared without arginine, and, when necessary the selective antibiotic. If bacterial contamination is a problem, 100 μg/mL ampicillin can be added without affecting the algal growth. The plates and liquid medium can be stored for several months at 4°C.
3. Exogenous DNA: (Linearized or supercoiled) plasmid DNA containing selectable marker, e.g., pArg 7.8 *(3)*. Linear expression cassette as restriction fragment or polymerase chain reaction (PCR) product, e.g., pMF124GFP *(4)*.
4. Glass Beads: portions of 300 mg acid-washed glass beads with a diameter of approx 0.5 mm (Sigma, G8772) in conical 1.5-mL plastic tubes. Sterilize by autoclaving before use, and store at ambient temperature.
5. Polyethylene glycol (PEG) 6000: 20%, sterilized by autoclaving; store in aliquots of 1 mL at −20°C.

3. Methods

The method described essentially resembles the transformation with glass beads given in ref. *5*, which has been used extensively for transformation of cell wall–deficient strains of *Chlamydomonas*. The method offers the advantage of good transformation rates, and no specialized equipment is needed. Other methods, like biolistic transformation *(6)* or electroporation *(7)* offer advantages in special applications but require specialized equipment that is not needed for general production of recombinant proteins in this green alga.

1. Grow cells in 50 mL TAP + R, in 100-mL Erlenmeyer flasks, on a rotary shaker with 120 rpm at 25°C and under continuous illumination using moderate light intensities of cool fluorescent white (45 μE m^{-2} s^{-1}).
2. Grow the culture until the cell density reaches 5×10^5 per mL. Transformation efficiency is mostly dependent on the culture conditions (*see* **Note 1**).
3. Spin down the culture in conical tubes at moderate speed for 5 min at a relative centrifugal force of 800*g*. Discard the supernatant.
4. Carefully resuspend the green cell pellet in 500 μL of TAP (without arginine).
5. Add 100 μL of 20% PEG 6000 solution to the tube with the prepared glass beads.
6. Take 300 μL of the cell suspension and add it to the PEG/glass bead tube.
7. Add the exogenous DNA (1–2 μg each) and close the tube.
8. Vigorously mix the tubes on a Fisher Vortex Genie 15 s. Efficiency of transformation may vary on the speed settings (*see* **Note 2**).
9. Let the glass beads settle down and transfer the supernatant to 2 mL of fresh TAP medium.
10. Wash the glass beads twice with 1 mL of TAP medium to recover most of the cells from the tube and combine the wash fractions with the supernatant from **step 9**.
11. Spread 1 mL out of the 4-mL cell suspension carefully on selective TAP plates (without arginine), leaving the surface clearly wet. Other plating methods include embedding in soft agar or starch suspensions (*see* **Note 3**).
12. Dry the plates under a sterile hood until the surface has just dried, usually about 30 min to 1 h.
13. Close the plates, wrap them with Parafilm, and place them upside down in the light at 25°C.
14. Visible colonies should appear within 6 to 8 d.
15. Screen individual clones for the presence of the expression cassette (genomic PCR), mRNA (RT-PCR) and the recombinant protein (activity assays or immunological tests; *see* also **Note 4**).

4. Notes

1. The most critical parameter for achievement of high transformation rates is the culture conditions, especially the time the culture has been growing in liquid culture.
2. Another factor that influences transformation rates is the vortex used. Usually the best speed setting for each machine has to be determined empirically in a series of transformations.
3. Particularly if one uses antibiotic-resistance marker genes available for *Chlamydomonas* transformation (like *ble* for zeocin or *aphVIII* for paromomycin), cell density on the final selection plate tends to influence the number of recovered clones. Usually one transformation should be divided and plated on at least four agar plates with a diameter of 80 mm. Several different methods for plating have been described, taking into account that the cell wall–deficient strains used are more labile to drought and mechanical stress than the wild type. Among the most common and protective plating methods are embedding the cells in soft agar, starch suspensions, or air drying the loosely spread cell suspensions under a sterile hood. Although the starch suspension has been reported to be the most efficient plating methodology in a comparative analysis, it is sometimes difficult to avoid contaminations using ethanol-suspended starch. Soft agar can decrease the number of survivors, if the cells are plated with too large volumes of agar above 42°C. Finally, drying the plates under a sterile hood requires the least material preparation; however, one has to leave the plates uncovered until the liquid has completely been absorbed by the plate. Prolonged drying of the plate may also lead to a significant loss of survivors.
4. Probably the most important point is the design of the expression cassette. A number of trials were undertaken using viral promoters to drive foreign gene expression in *Chlamydomonas*, but the only really successful expressions were achieved using endogenous promoters. In particular, a modified truncated version of the endogenous *rbcS2* promoter, encoding the small subunit of the abundant enzyme Rubisco *(8)*, and the promoter of the *hsp70A* gene, encoding a cytoplasmic heat shock protein of the Hsp70 type *(9)*, have proved to be valuable tools for gene expression in *Chlamydomonas*. Recently, the set of available promoters has been extended by the promoter of the *psaD* gene, encoding an abundant subunit of the photosystem I complex *(10)*. Nearly all *Chlamydomonas* genes, with *psaD* as a prominent exception, contain at least one intron that often fulfils important functions in regulation of gene expression and recruitment of transcriptional activators. The intron 1 of the *rbcS2* gene has been analyzed in detail for its positive influence on foreign gene expression in the case of *ble* and *aphVIII (11,12)*. In addition, the adaptation of foreign genes to the preferred codon usage of highly expressed nuclear (or chloroplast) genes from *Chlamydomonas* has improved recombinant protein expression levels significantly, as demonstrated for green fluorescent protein and human single-chain antibodies *(4,13,14)*. General information about molecular tools for *Chlamydomonas* nuclear transformation and chloroplast engineering has recently been reviewed *(15,16)*.

References

1. Harris, E. H. (1989) *The* Chlamydomonas *Sourcebook: A comprehensive guide to biology and laboratory use.* Academic Press, San Diego.
2. Hutner, S. H., Provasoli, L., Schatz, A., and Haskins, C. P. (1950) Some approaches to the study of the role of metals in the metabolism of microorganisms. *Proc. Am. Philos. Soc.* **94,** 152–170.
3. Debuchy, R., Purton, S., and Rochaix, J. D. (1989) The argininosuccinate lyase gene of *Chlamydomonas reinhardtii*: an important tool for nuclear transformation and for correlating the genetic and molecular maps of the ARG7 locus. *EMBO J.* **8,** 2803–2809.

4. Fuhrmann, M., Oertel, W., and Hegemann, P. (1999) A synthetic gene coding for the green fluorescent protein (GFP) is a versatile reporter in *Chlamydomonas reinhardtii*. *Plant J.* **19,** 353–361.
5. Kindle, K. L. (1990) High-frequency nuclear transformation of *Chlamydomonas reinhardtii*. *Proc. Natl. Acad. Sci. USA* **87,** 1228–1232.
6. Kindle, K. L., Schnell, R. A., Fernandez, E., and Lefebvre, P. A. (1989) Stable nuclear transformation of *Chlamydomonas* using the *Chlamydomonas* gene for nitrate reductase. *J. Cell Biol.* **109,** 2589–2601.
7. Shimogawara, K., Fujiwara, S., Grossman, A., and Usuda, H. (1998) High-efficiency transformation of *Chlamydomonas reinhardtii* by electroporation. *Genetics* **148,** 1821–1828.
8. Stevens, D. R., Rochaix, J. D., and Purton, S. (1996) The bacterial phleomycin resistance gene *ble* as a dominant selectable marker in *Chlamydomonas*. *Mol. Gen. Genet.* **251,** 23–30.
9. Schroda, M., Blocker, D., and Beck, C. F. (2000) The HSP70A promoter as a tool for the improved expression of transgenes in *Chlamydomonas*. *Plant J.* **21,** 121-131.
10. Fischer, N. and Rochaix, J. D. (2001) The flanking regions of PsaD drive efficient gene expression in the nucleus of the green alga *Chlamydomonas reinhardtii*. *Mol. Genet. Genomics* **265,** 888–894.
11. Lumbreras, V., Stevens, D. R., and Purton, S. (1998) Efficient foreign gene expression in *Chlamydomonas reinhardtii* mediated by an endogenous intron. *Plant J.* **14,** 441–447.
12. Sizova, I., Fuhrmann, M., and Hegemann, P. (2001) A *Streptomyces rimosus aphVIII* gene coding for a new type phosphotransferase provides stable antibiotic resistance to *Chlamydomonas reinhardtii*. *Gene* **277,** 221–229.
13. Franklin, S., Ngo, B., Efuet, E., and Mayfield, S. P. (2002) Development of a GFP reporter gene for *Chlamydomonas reinhardtii* chloroplast. *Plant J.* **30,** 733–744.
14. Mayfield, S. P., Franklin, S. E., and Lerner, R. A. (2003) Expression and assembly of a fully active antibody in algae. *Proc. Natl. Acad. Sci. USA* **100,** 438–442.
15. Fuhrmann, M. (2002) Expanding the molecular toolkit for *Chlamydomonas reinhardtii*—from history to new frontiers. *Protist* **153,** 357–364.
16. Rochaix, J. D., Goldschmidt-Clermont, M., and Merchant, S., eds. (1998) *The Molecular Biology of Chloroplasts and Mitochondria in* Chlamydomonas. Kluwer Academic Publishers, Dordrecht.

13

Codon-Optimized Genes that Enable Increased Heterologous Expression in Mammalian Cells and Elicit Efficient Immune Responses in Mice after Vaccination of Naked DNA

Marcus Graf, Ludwig Deml, and Ralf Wagner

Abstract

Many of the problems related with mammalian gene expression, such as low translation efficiency and mRNA halflife, can be solved by means of a rational gene design, based on modern bioinformatics, followed by the *de novo* generation of a synthetic gene. Moreover, high expression rates and prolonged mRNA stability are not only crucial for heterologous mammalian expression, but, in particular, are important for the generation of effective DNA vaccines. In this chapter we show that an optimized synthetic gene encoding the HIV-1 Pr55^{gag} outperforms wild-type gene driven expression by several orders of magnitude. RNA analysis revealed that this positive effect was mostly due to increased mRNA stability of the optimized transcripts. Moreover, mice vaccinated with the optimized gag gene elicited a much stronger immune response against Pr55^{gag} than the control groups immunized with the respective wild-type gene.

Key Words: Synthetic gene; gene optimization; RNA optimization; mammalian expression; DNA vaccine; antibody.

1. Introduction

Both the design and generation of recombinant bioproducts for research and development critically depend on the availability of naturally occurring templates. Conventional strategies supporting a rational modification of these sequences such as site-directed mutagenesis or established cloning strategies are time-consuming and make strong demands on laboratory personnel. Novel approaches that aim at avoiding biological limitations such as low RNA stability, inefficient nuclear export, antitermination, or insufficient translational efficiency critically depend on technological innovation.

Many of the above limitations my be solved by means of a rational gene design, based on modern bioinformatics, followed by the *de novo* generation of a "synthetic" gene. The construction of synthetic genes relies solely on the sequence information that is usually available via publicly accessible or commercial databases without depending on a physical template. The "gene synthesis" approach allows a precise adaptation of the desired product to the requirements of the final application without any need to consider technical limitations resulting from complex cloning strategies. In the past, gene synthesis has successfully been used to increase translational activity by adapting the codon usage of the gene to be expressed to the t-RNA frequencies of the heterologous production system.

Using a rational gene design and applying state-of-the-art bioinformatic software tools and algorithms, we would like to show that gene synthesis allows optimization of a gene for any application. In particular, we want to show in this chapter that optimized synthetic genes are superior to wild-type genes, when used for vaccination of naked plasmid DNA. Direct injection of DNA is a novel and promising method of delivering specific antigens for immunization *(1–4)* without the need for purified antigens or life vectors with uncertain safety profiles. Plasmid DNA immunization has potential advantages compared with traditional protein vaccination owing to the strong T-helper 1 (Th1) and cytotoxic T-lymphocyte (CTL) responses induced, the prolonged antigen expression, and the long-lived effector activity *(2,5,6)* and thus can be used for vaccination *(2,7,8)*. In animal models, DNA vaccination has been shown to induce protective immunity against a variety of viral *(4,9–13)*, bacterial *(14,15)*, and parasitic pathogens *(16–18)*.

One major prerequisite of an effective DNA vaccine is to ensure a prolonged and strong heterologous antigen expression within the vaccinated host. However, many immunologically relevant genes of pathogens show a codon usage very different from that of mammals and may contain cryptic splice sites or premature polyadenylation, significantly limiting long-term, high-level expression in mammalian cells. This is especially true for procaryotic genes, such as *Helicobacter pylori* or mycobacteria, malaria genes, or many viral genes such as those from hepatitis C virus or retroviruses such as HIV. Regarding the latter, analysis of the HIV-1 group-specific antigen Gag, encoding a polyprotein comprising the viral capsid, the nucleic capsid, and the matrix protein, predicted major difficulties for heterologous expression in conventional mammalian expression and/or DNA vaccination vectors. To improve heterologous expression and thus immunogenicity, a fully synthetic gene was created by adapting codon bias and GC content to humans, while avoiding negative sequence elements such as splice sites and premature poly(A) sites or mRNA secondary structures. The codon-optimized gene was readily produced at high levels in different mammalian cells and elicited a strong Th1-like immune response, whereas its respective wild-type gene (*wtgag*), encoding the very same Pr55^{gag} viral antigen, did not.

2. Materials

1. 96-well MaxiSorp enzyme-linked immunosorbent assay (ELISA) plates (Nunc, Wiesbaden, Germany).
2. Agarose and polyacrylamide gel equipment.
3. Syngag (Geneart, Germany).
4. Chemically competent *E. coli* strain DH5α, Luria-Bertani (LB) media, ampicillin.

5. Electroblotting equipment.
6. ELISA reader.
7. Female BALB/c mice (Charles River, Sulzfeld, Germany).
8. Hand-held Accel® gene gun device, syringes.
9. Horseradish peroxidase (HRP)-conjugated goat anti-mouse IgG1, IgG2a, and IgE isotypes.
10. Oligonucleotides (Invitrogen, UK).
11. pcDNA 3.1 mammalian expression system (Invitrogen, USA).
12. Radiolabeled ^{35}S-methionine, α-^{32}P-CTP.
13. Taq Plus precision system (Stratagene, Heidelberg, Germany), T4-ligase, restriction enzymes (New England Biolabs, Frankfurt am Main, Germany).
14. Whatman filter paper (Maidstone, England).
15. Cloned β-actin cDNA (pGEM-β-actin, kindly provided by Thomas Dobner, RIMMH, Germany).
16. Cloned ERR tag (pCMV-ERR, kindly provided by Kurt Bieler, RIMMH).
17. Bio-Rad Protein-Assay (Bio-Rad, Munich, Germany).
18. RNeasy-Kit (Qiagen, Hilden, Germany).
19. Buffer for Northern blots: 0.04 M 4-morpholinepropanesulfonic acid (MOPS), 0.01 M NaAc, 0.001 M EDTA, 6.5% formaldehyde, pH 7.0.
20. Molecular Analyst Software (Bio-Rad).
21. Ladderman Labeling Kit (Takara Biochemical, Berkeley, CA).
22. Lysis buffer: 0.5% Triton X-100, 100 mM Tris-HCl (pH 7.4).
23. p24 capture-ELISA antibodies (kindly provided by Dr. Matthias Niedrig, RKI, Berlin, Germany) *(19)*.
24. Carbonate buffer: 0.1 M (pH 9.5).
25. Wash buffer: 0.1% Tween 20, 300 mM NaCl, 10 mM Na$_2$HPO$_4$, 1.5 mM NaH$_2$PO$_4$.
26. OPD solution (Abbott, Wiesbaden, Germany).
27. TNT Coupled Reticulocyte Lysate System (Promega, Madison, WI).
28. HeLa Cell Extract Transcription System (Promega).
29. Negatively charged nylon membrane (Boeringer-Mannheim, Mannheim, Germany).
30. DNA-coated gold particles: 0.95-μm particles, 2 μg DNA/mg gold.
31. HRP-conjugated goat anti-mouse IgG1, IgG2a, and IgE isotypes: 1:2000 in PBS, 2% Tween 20, 3% fetal calf serum (FCS; PharMingen, Hamburg, Germany).

3. Methods

This chapter reports on the gene analysis and design of synthetic HIV-1 gag (*see* **Subheading 3.1.**) and the verification of the desired properties (*see* **Subheadings 3.2.–3.10.**). Results are summarized in **Subheading 3.11**.

3.1. Gene Analysis and Design

The *in silico* gene expression analysis and multiparameter gene optimization of HIV-1 gag was performed using GeneOptimizer software (Geneart, Germany; *see* **Note 1**). The gene is optimized with respect to the following factors: (*see* **Notes 2 and 3**):

1. Increased GC content.
2. Avoidance of negatively *cis*-acting sequences.
3. Codon adaption to frequently used human codons.

3.2. Cloning

Polymerase chain reaction (PCR) amplification of HX10 proviral DNA using the primers gag-1 (5'-gcg GGT ACC GAA TTC agg aga gag atg ggt gcg aga gcg tca gta tta agc-3') and gag-2 gag (5'-gcc GAG CTC CTC GAG GGA TCC tta ttg tga cga ggg gtc gtt gcc aaa gag-3') revealed a 1537-nt fragment that was cloned into pcDNA 3.1 (Invitrogen) under the transcriptional control of the cytomegalovirus (CMV) immediate-early promotor using *Kpn*I and *Xho*I cloning sites. A synthetic sequence coding for the HIV-1$_{IIIB}$ Pr55^{gag} polyprotein was constructed, cloned, and sequence-verified. Finally, this gene was named *syngag* and was subcloned into pcDNA 3.1 under the transcriptional control of the CMV immediate-early promotor using *Kpn*I and *Xho*I cloning sites. Both the wild-type and the synthetic *gag* gene encode the same amino acid sequence derived from the HIV-1$_{IIIB}$ proviral clone HX10, but they show a diverse primary DNA sequence. To be able to detect both mRNAs at the same time, a short DNA tag (ERR) was cloned downstream of the coding regions within an untranslated region upstream of the pcDNA3.1 poly(A) site using *Bam*HI and *Xho*I cloning sites.

3.3. Cell Culture and Transfections

Cos-7 and human H1299 cells *(20)* were transfected by the calcium coprecipitation technique *(21)*:

1. The day prior to transfection, plate 2×10^6 cells on 100-mm-diameter culture dishes and incubate for 24 h.
2. Transfect with 45 μg of Pr55^{gag}-expressing plasmids.
3. Harvest cells 48 h post transfection.
4. Wash twice in phosphate-buffered saline (PBS) and store for further analysis.

3.4. Western Blot Analysis

1. Lyse harvested cells in cell lysis buffer.
2. Subject to repeated freeze/thaw cycles.
3. Clear by centrifugation (20,800*g* for 5 min).
4. Measure the total amount of protein using the Bio-Rad Protein Assay following the manufacturer's instructions.
5. Separate 50 μg total protein by electrophoresis on denaturing sodium dodecyl sulfate (SDS) 12.5% polyacrylamide gels and transfer onto nitrocellulose membrane by electroblotting.
6. Detect expression of Pr55^{gag} by an HIV-1 p24-specific monoclonal antibody 13-5 *(22)* and visualize by chromogenic staining.

3.5. Northern Blot Analysis

Basically, Northern blot analysis was performed as described elsewhere (*RNA Applications Guide*, Promega).

1. Synthesize an ERR and β-actin-specific radiolabeled DNA probe by random priming and elongation of the β-actin cDNA and ERR tag using the Ladderman Labeling Kit following the manufacturer's instructions. The β-actin cDNA was released by *Hin*dIII and *Eco*RI digestion of pGEM-β-actin and gel-purified.
2. Prepare total RNA from 10^7 transfected cells using the RNeasy-Kit following the manufacturer's instructions.

3. Separate RNA preparations on a 1% agarose gel and capillary-blot onto a negatively charged nylon membrane.
4. Hybridize the membrane overnight at 50°C to a radiolabeled probe.
5. Visualize hybridization by exposure to a Phosphor-Imager Screen and analyze using Molecular Analyst Software.

3.6. Capture ELISA

1. Coat 96-well MaxiSorp ELISA plates overnight with 100 µL of the p24-specific monoclonal antibody 11-G-7 diluted 1:170 with carbonate buffer.
2. Wash six times with wash buffer.
3. Add different amounts of cell lysates diluted in 0.5% bovine serum albumin (BSA) in wash buffer to the wells.
4. Incubate overnight at 4°C with a second HRP-conjugated monoclonal antibody (diluted 1:600 in 0.5% BSA in wash buffer) recognizing a different epitope within p24.
5. Wash the plates six times with wash buffer.
6. Stain antibody conjugates with OPD solution and measure absorption OD (495 nm) with an ELISA reader.
7. Determine the concentration of $Pr55^{gag}$ by a calibration curve using different concentrations of purified $Pr55^{gag}$, produced in insect cells and using the baculovirus expression system *(23)*.

3.7. In Vitro Translation

Use the TNT Coupled Reticulocyte Lysate System following the manufacturer's instructions:

1. Incubate a mixture of reticulocyte lysate, T7-polymerase, 1 µg of plasmid DNA, amino acids, and ^{35}S-methionine (40 µCi) for 60 min at 30°C.
2. Subject the probes to a SDS 12.5% polyacrylamide gel electrophoresis.
3. Dry on Whatman filter paper and detect translational activity by autoradiography.

3.8. In Vitro Transcription

Use the HeLa Cell Extract Transcription System following the manufacturer's instructions:

1. Perform transcription with a nuclear extract of HeLa cells, in the presence of nucleotides, 3 m*M* $MgCl_2$, 500 ng *Xho*I linearized plasmid DNA and α-^{32}P-CTP (10 µCi).
2. Separate the transcribed RNA on a 1% agarose gel and capillary-blot onto a negatively charged nylon membrane.
3. Detect transcripts by autoradiography, visualize by exposure to a Phosphor-Imager Screen, and analyze with Molecular Analyst Software.

3.9. DNA Vaccination of Mice

Female BALB/c mice were housed under specific pathogen-free conditions and injected at the age of 6–12 wk:

1. Immunize groups of five mice each in both tibialis anterior muscles with 80 µg endotoxin-free plasmid DNA. Add all component solutions at the same time, mix by vortexing, and leave at room temperature prior to immunization.
2. Alternatively, gene-gun-inoculate on shaved abdominal skin using plasmid DNA-coated gold particles (0.5 µg gold shot, 50–71% coating efficiency) and the hand-held Accel gene gun device employing compressed helium (400 psi) as the particle motive force.
3. Inoculate/boost mice three times at intervals of 3 wk.

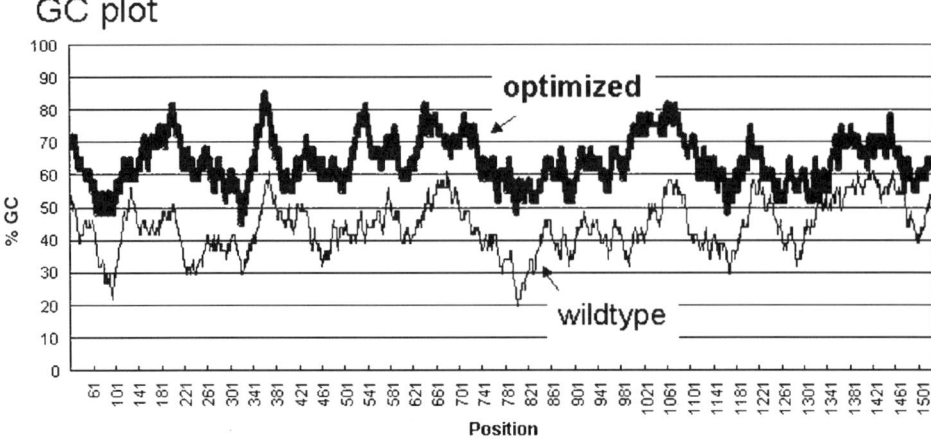

Fig. 1. GC analysis of wild-type and optimized genes. Plot represents the average GC content of 40 nucleotides at the indicated sequence positions.

3.10. Evaluation of Antibody Responses

Antibodies specific for Pr55^{gag} polyproteins were quantified by an end-point dilution ELISA assay (in duplicate) on samples from individual animals:

1. Recover serum from mice by tail bleed at the indicated time points after the boost injection.
2. Capture Gag-reactive antibodies from the serum to a solid phase of Prep cell-purified Pr55^{gag} polyproteins (100 μL of 1 μg/mL per well coated to 96-well MaxiSorp ELISA plates overnight at 4°C in a refrigerator) for 2 h at 37°C.
3. Wash the plates six times with wash buffer.
4. Detect captured antibodies with 100 μL/well HRP-conjugated goat anti-mouse IgG1, IgG2a, and IgE antibodies.
5. Wash the plates six times with wash buffer.
6. Stain antibody conjugates with OPD solution (100 μL/well, 20 min at room temperature in the dark) and measure absorption OD (495 nm) with an ELISA reader.

End-point titers were defined at the highest serum dilution that resulted in an adsorbance value (OD 492 nm) three times greater than that of the same dilution of a nonimmune serum. The serum of each mouse was assayed, and these values were used to calculate the mean and standard deviation for each group of five mice.

3.11. Results

3.11.1. Bioinformatic Analysis of the Wild-Type and Optimized Coding Region

With respect to increasing the overall GC content, avoidance of negatively *cis*-acting sequence motifs, and codon adaptation to frequently used human codons, a synthetic sequence encoding gag was created. Both the wild-type gag (wtgag) and the synthetic gag (syngag) were analyzed for their GC content (**Fig. 1**), codon bias (**Fig. 2**),

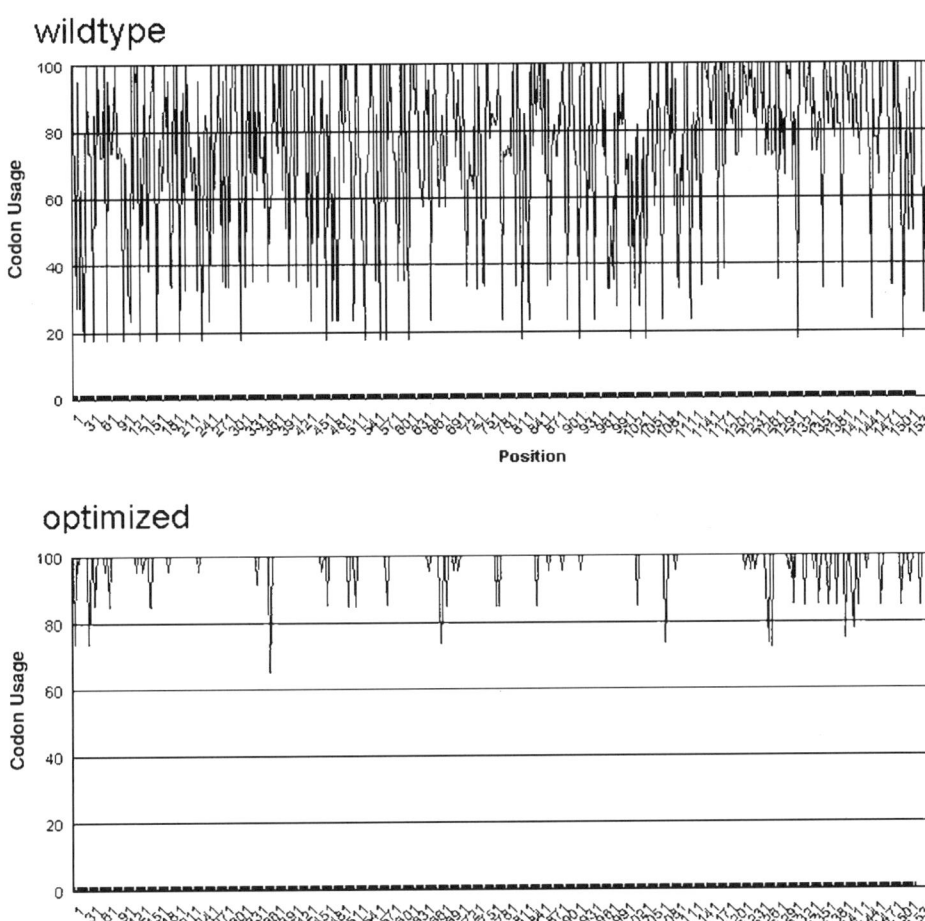

Fig. 2. Codon usage analysis of wild-type and optimized genes for expression in human cells. The most frequently used codon in humans was set to 100 and the remaining were scaled accordingly. This procedure allows us to compare the adaptiveness of different codons relative to each other (*relative adaptiveness [24]*). Plots represent the relative adaptiveness of a given codon at the indicated codon position.

and presence of splice sites (consensus sequences and derivatives thereof), poly(A) sites (consensus sequences), and instability elements, such as the granulocyte/macrophage colony-stimulating factor (GM-CSF) adenine-rich element AUUUA. Within the synthetic reading frame, all *cis*-acting sites that may have a negative influence on expression rates were removed (**Table 1**).

Moreover, the overall GC content was increased from 44.1 to 62.7%, which increases the overall mRNA stability of the gag-encoding synthetic gene, and there is no sequence stretch within the gene showing an average GC content below 50% (**Fig. 1**).

Table 1
***cis*-Active Sequence Motifs**

	Wild type	Optimized
Procaryotic inhibitory motifs	1	0
Repeat sequences and secondary stretches	1	0
Splice donor and acceptor sites, branch points	0	0
Adenine-rich instability element	2	0
UA dinucleotide	96	20
Poly(A) sites	2	0

In contrast to the wild-type gene, the codon-optimized syngag genes show a codon bias in humans and do not use any rarely used codon (**Fig. 2**). This is also reflected by their codon adaption index (CAI), which is a measurement of the relative adaptiveness of the codon usage of a gene compared with the codon usage of highly expressed genes *(24)*. An ideally biased gene would shows a CAI of 1.0, even though no natural human gene reaches this theoretical value. However, the codon-optimized syngag gene shows a CAI of 0.99, whereas the wild-type gene reaches a CAI of only 0.69.

3.11.2. Expression Analysis of Optimized and Wild-Type Genes in Mammalian Cells

To ensure the sequence integrity of both reading frames, wild-type and synthetic plasmid constructs were tested for in vitro transcription and translation under the transcriptional control of the T7 promotor. The highly artificial in vitro assay using T7-polymerase and reticulocyte lysates resulted in a full length of both Gag encoding constructs (**Fig. 3A**). Although it was unlikely, we further excluded the possibility that the altered codon usage directly affected the transcriptional activity of the CMV promotor enhancer unit. For that purpose HeLa-derived nuclear extracts were used to compare transcription of wild-type and synthetic gag gene-derived RNAs in vitro. As shown in **Fig. 3B**, in vitro transcription yielded Gag-encoding RNAs of similar length and comparable amounts irrespective of the codon bias of the expression construct (*see* **Note 4**).

In the next series of experiments, we tested the expression constructs in relevant mammalian cell lines using transient transfection. As expected from the *in silico* analysis, synthesis of Pr55^{gag} from the wild-type gag gene (*wtgag*) was extremely low. In contrast, high-level expression of Pr55^{gag} was achieved after transient transfection of the syngag-encoding plasmid and could be readily detected with Pr55^{gag}-specific antibodies (**Fig. 4A**). The differences in expression levels were reflected by the amount of mRNA detected by Northern blot analysis. The levels of syngag RNAs exceeded those of wtgag RNA by several orders of magnitude, compared with the message of the housekeeping gene β-actin (**Fig. 4B**). These results were confirmed using various cell lines (Cos7, HeLa), ruling out that cell-type specific factors critically contributed to the effects observed (data not shown; *see* **Note 5**).

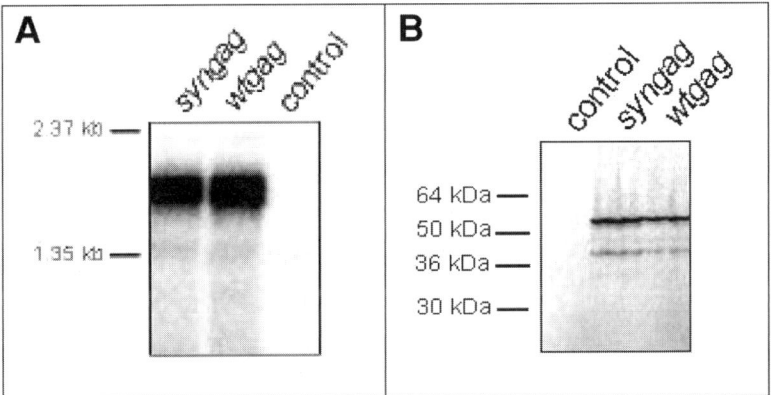

Fig. 3. (**A**) Translational assays based on T7-promoted transcription. Coupled in vitro transcription/translation was used to compare translational activity of wild-type and synthetic Pr55^{gag}-encoding plasmids. Translational products were separated by polyacrylamide gel electrophoresis and detected by autoradiography. (**B**) In vitro transcription using HeLa nuclear extracts. Linearized plasmid DNA (1 μg) was incubated in the presence of a HeLa nuclear extract and radioactively labeled nucleotides. Transcripts were separated on a agarose gel, blotted to a nylon membrane, and detected by autoradiography.

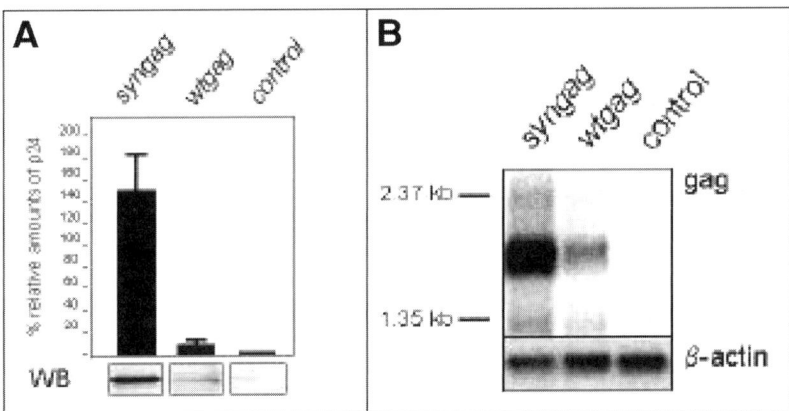

Fig. 4. H1299 cells were transiently transfected with the indicated plasmids. (**A**) Cells were harvested 48 h post transfection and lysed, and 50 μg of total protein was subjected to Western blot analysis (WB). Yields of Pr55^{gag} were measured by testing different dilutions of cell-lysate in a capture-ELISA using purified Pr55^{gag} for standardization (upper panel). Bars represent relative Pr55^{gag} expression levels and are given as the mean of triplicate determinations. (**B**) H1299 cells were transfected with the indicated constructs and harvested 48 h post transfection. Cells were lysed, and total RNA was isolated and subjected to Northern blot analysis. Pr55^{gag}-encoding transcripts were detected by a radiolabeled probe and standardized by the amount of β-actin RNA detected (lower panel). The Northern blot analyses were repeated several times with comparable results.

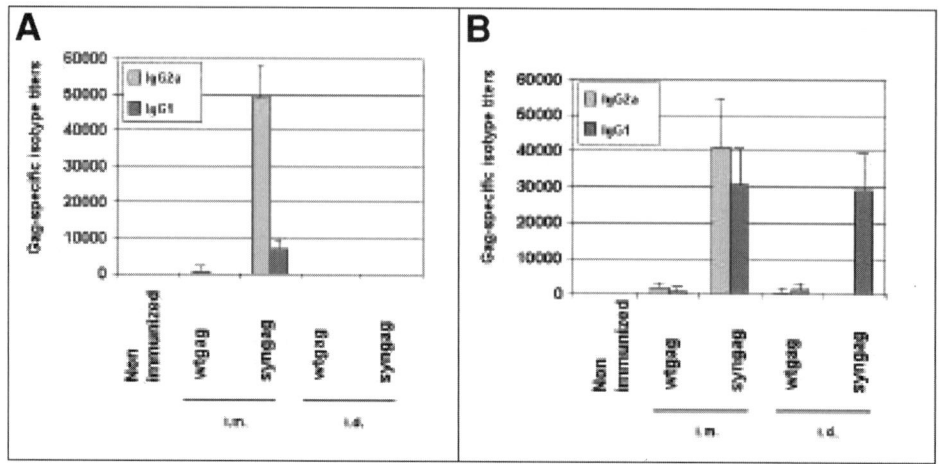

Fig. 5. Influence of immunization route on the antibody isotype responses to gag induced by gag expression vectors. Mice were immunized either by i.m. injection or gene gun immunization and boostered twice at 3-wk intervals. Serum IgG1 and IgG2a antibody titers were taken 1 wk after the first (**A**) and second (**B**) booster immunization of mice that were immunized with gag expression plasmids, as indicated. Gag-specific serum IgG1 and IgG2a antibodies were measured by ELISA and expressed as the reciprocal of the highest plasma dilution that resulted in an adsorbance value (OD 492 nm) 3 times greater than that of a preimmune serum, with a cutoff value of 0.05. Each bar represents the group mean ($n = 5$) for anti-gag titers, and vertical lines represent the SEM.

3.11.3. Isotype Responses of Mice Immunized with Wild-Type and Optimized DNA Vaccines

The method of DNA delivery has a major impact on the effectiveness and type of immune response elicited after immunization of naked DNA. We therefore compared wild-type versus optimized DNA vaccination vectors using intramuscular DNA immunization and bombardment of abdominal skin with gold particles coated with DNA (gene gun; *see* **Note 6**).

Antibody isotype responses to Gag were assessed 2 wk after the first and 1 wk after the second booster immunization. Mice that were im immunized with the syngag plasmid produced a significant Th1-mediated anti-Gag response, characterized by high titers of IgG2a response with substantial reciprocal mean titers of 1:49,000, but only minute quantities of Gag-reactive IgG2a antibodies (1:7,000; IgG2a/IgG1 ratio of 7; **Fig. 5A**). In contrast, the antibody response in mice vaccinated with wtgag expression constructs showed substantially decreased IgG2a and IgG1 isotypes. At that time point, no anti-Gag response was detectable from mice immunized with syngag or wtgag by particle gun.

One week after the second booster immunization, the syngag group of mice developed increased levels of IgG1 isotypes (1:13,650), but decreased IgG1 isotypes (10,240), with an IgG2a/IgG1 ratio of 1.3 (**Fig. 5B**). In contrast, immunization of gag expression plasmids with the particle gun resulted in a Th2-mediated antibody response. Thus, high titers of anti-gag IgG1 isotypes (9600) but no specific IgG2a antibodies (<40) were detectable from the serum of mice immunized with syngag by particle gun (**Fig. 5B**). Furthermore, significant titers of IgG1 antibodies (1:1360), but low levels of IgG2a antibodies (1:360) were also detectable from the serum of mice immunized by particle gun. At no time points was a Gag-specific antibody measurable from the serum of nonimmunized mice (*see* **Note 7**).

4. Notes

1. The safety profile and yields of proteinaceous bioproducts to be produced in various cell lines can be significantly improved by altering or modifying the coding sequence. Advantages regarding safety of rationally designed and fully synthetically constructed genes are manifold. Moreover, synthetic genes can nowadays be ordered like any other laboratory commodity and are easily available. Because synthetic genes are constructed from chemically synthesized oligonucleotides, there is no hazard in transferring infectious agents like virons or prions from native starter material to "down the road" applications like recombinant producer cell lines, gene therapy, or DNA vaccination vectors; such genes will certainly contribute toward developing novel safety standards for approval of bioproducts for clinical trials. The opportunity to design and produce optimized genes rationally without any need of a "wet" template may also contribute toward reducing the time span between the identification of genes by sequencing projects or functional genomic programs (differential display, proteomics) and the biochemical analysis and production of the corresponding products for preventive or therapeutic purposes. In addition, in some cases the original template itself is not even available or has to be isolated under very stringent safety precautions, for instance, when working with L3 or L4 organisms.
2. The coding region of HIV-1 gag was optimized for expression in human cells by performing a multiparameter optimization, which utilizes a "sliding combination window" comprising all possible combinations of synonymous codons that are generated and validated by a complex quality function. This approach ensures that the calculated sequence is definitely optimal with respect to these local properties and their user-defined weighting.
3. It is generally accepted that increased expression levels are due to a combined effect of (a) increased mRNA half-life, (b) faster tRNA availability due to codon optimization, and (c) absence of negative sequence elements interfering with nuclear export and/or cellular maintenance of the Gag-encoding mRNAs. These results are consistent with previous investigations demonstrating that expression of HIV-1 structural genes *(25–29)* can be drastically increased by codon optimization.
4. The differences in expression rates between the wild-type and the synthetic gene could not be related to premature translational termination, since coupled transcription/translation control experiments proved that both wild-type and synthetic genes can be readily translated to their full length, at least in vitro. Moreover in vitro transcription using HeLa cell lysates revealed that both genes are transcribed at similar rates.
5. The RNA data obtained by Northern blot analysis, showing significantly increased amounts of Gag-encoding RNAs in cells transfected with the optimized gene, indicate that the optimized GC-rich mRNA has a prolonged lifetime compared with AT-rich wild-type transcripts. Consequently, expression of the synthetic Gag exceeds those obtained using the wild-type gene by several orders of magnitude. However, this dramatic increase

in recombinant protein production cannot be solely explained by higher levels of Gag-encoding transcripts in the cells transfected with the optimized gene. Underlying a linear correlation between mRNA levels and protein biosynthesis, one would expect an increase of 5–10-fold.
6. It is known that the mode of application of naked DNA has a marked influence on the quality of the immune response *(30–33)*. Currently, the two methods of administration that are most widely used are direct injection into the muscle or gene gun administration into the dermis *(4,34)*.
7. Taken together, these data strongly support the use of synthetic genes for heterologous gene expression in mammalian cells and in particular their use in designing novel generations of candidate DNA vaccines. Regarding the latter, the opportunity of designing (a) the genes of interest including (b) the plasmid backbone on a rational basis will allow us to increase and to modulate immune responses as desired toward Th1- or Th2-type effector functions. On the other hand, sequences encoded by any type of viral vector used for gene therapy purposes may be rendered immunosilent by removing known immunostimulatory CpG sequence motifs and introducing inhibitory sequence elements.

References

1. Davis, H. L., Michel, M. L., and Whalen, R. G. (1993) DNA-based immunization induces continuous secretion of hepatitis B surface antigen and high levels of circulating antibody. *Hum. Mol. Genet.* **2,** 1847.
2. Tang, D. C., DeVit, M., and Johnston, S. A. (1992) Genetic immunization is a simple method for eliciting an immune response. *Nature* **356,** 152.
3. Wang, B., Boyer, J., Srikantan, V., et al. (1993) DNA inoculation induces neutralizing immune responses against human immunodeficiency virus type 1 in mice and nonhuman primates. *DNA Cell Biol.* **12,** 799.
4. Xiang, Z. Q., Spitalnik, S., Tran, M., Wunner, W. H., Cheng, J., and Ertl, H. C. (1994) Vaccination with a plasmid vector carrying the rabies virus glycoprotein gene induces protective immunity against rabies virus. *Virology* **199,** 132.
5. Michel, M. L., Davis, H. L., Schleef, M., Mancini, M., Tiollais, P., and Whalen, R. G. (1995) DNA-mediated immunization to the hepatitis B surface antigen in mice: aspects of the humoral response mimic hepatitis B viral infection in humans. *Proc. Natl. Acad. Sci. USA* **92,** 5307.
6. Yankauckas, M. A., Morrow, J. E., Parker, S. E., et al. (1993) Long-term anti-nucleoprotein cellular and humoral immunity is induced by intramuscular injection of plasmid DNA containing NP gene. *DNA Cell Biol.* **12,** 771.
7. Robinson, H. L., Hunt, L. A., and Webster, R. G. (1993) Protection against a lethal influenza virus challenge by immunization with a haemagglutinin-expressing plasmid DNA. *Vaccine* **11,** 957.
8. Wolff, J. A., Malone, R. W., Williams, P., et al. (1990) Direct gene transfer into mouse muscle *in vivo*. *Science* **247,** 1465.
9. Manickan, E., Yu, Z., Rouse, R. J., Wire, W. S., and Rouse, B. T. (1995) Induction of protective immunity against herpes simplex virus with DNA encoding the immediate early protein ICP 27. *Viral Immunol.* **8,** 53.
10. Donnelly, J. J., Martinez, D., Jansen, K. U., Ellis, R. W., Montgomery, D. L., and Liu, M. A. (1996) Protection against papillomavirus with a polynucleotide vaccine. *J. Infect. Dis.* **173,** 314.

11. Yokoyama, M., Zhang, J., and Whitton, J. L. (1995) DNA immunization confers protection against lethal lymphocytic choriomeningitis virus infection. *J. Virol.* **69**, 2684.
12. Phillpotts, R. J., Venugopal, K., and Brooks, T. (1996) Immunisation with DNA polynucleotides protects mice against lethal challenge with St. Louis encephalitis virus. *Arch. Virol.* **141**, 743.
13. Ulmer, J. B., Donnelly, J. J., Parker, S. E., et al. (1993) Heterologous protection against influenza by injection of DNA encoding a viral protein. *Science* **259**, 1745.
14. Huygen, K., Content, J., Denis, O., et al. (1996) Immunogenicity and protective efficacy of a tuberculosis DNA vaccine. *Nat. Med.* **2**, 893.
15. Tascon, R. E., Colston, M. J., Ragno, S., Stavropoulos, E., Gregory, D., and Lowrie, D. B. (1996) Vaccination against tuberculosis by DNA injection. *Nat. Med.* **2**, 888.
16. Xu, D. and Liew, F. Y. (1995) Protection against leishmaniasis by injection of DNA encoding a major surface glycoprotein, gp63, of *L. major*. *Immunology* **84**, 173.
17. Sedegah, M., Hedstrom, R., Hobart, P., and Hoffman, S. L. (1994) Protection against malaria by immunization with plasmid DNA encoding circumsporozoite protein. *Proc. Natl. Acad. Sci. USA* **91**, 9866.
18. Doolan, D. L., Sedegah, M., Hedstrom, R. C., Hobart, P., Charoenvit, Y., and Hoffman, S. L. (1996) Circumventing genetic restriction of protection against malaria with multigene DNA immunization: CD8$^+$ cell-, interferon gamma-, and nitric oxide-dependent immunity. *J. Exp. Med.* **183**, 1739.
19. Niedrig, M., Rabanus, J. P., L'Age Stehr, J., Gelderblom, H. R., and Pauli, G. (1988) Monoclonal antibodies directed against human immunodeficiency virus (HIV) gag proteins with specificity for conserved epitopes in HIV-1, HIV-2 and simian immunodeficiency virus. *J. Gen. Virol.* **69**, 2109.
20. Mitsudomi, T., Viallet, J., Mulshine, J. L., Linnoila, R. I., Minna, J. D., and Gazdar, A. F. (1991) Mutations of ras genes distinguish a subset of non-small-cell lung cancer cell lines from small-cell lung cancer cell lines. *Oncogene* **6**, 1353.
21. Graham, F. L. and Eb, A. J. (1973) A new technique for the assay of infectivity of human adenovirus 5 DNA. *Virology* **52**, 456.
22. Wolf, H., Modrow, S., Soutschek, E., Motz, M., Grunow, R., and Döbl, H. (1990) Production, mapping and biological characterisation of monoclonal antibodies to the core protein (p24) of the human immunodeficiency virus type 1. *AIFO* **1**, 24.
23. Wagner, R., Deml, L., Fliessbach, H., Wanner, G., and Wolf, H. (1994) Assembly and extracellular release of chimeric HIV-1 Pr55gag retrovirus-like particles. *Virology* **200**, 162.
24. Sharp, P. M. and Li, W. H. (1987) The codon adaptation Index—a measure of directional synonymous codon usage bias, and its potential applications. *Nucleic Acids Res.* **15**, 1281.
25. Graf, M., Bojak, A., Deml, L., Bieler, K., Wolf, H., and Wagner, R. (2000) Concerted action of multiple cis-acting sequences is required for Rev dependence of late human immunodeficiency virus type 1 gene expression. *J. Virol.* **74**, 10822.
26. Wagner, R., Graf, M., Bieler, K., et al. (2000) Rev-independent expression of synthetic gag-pol genes of human immunodeficiency virus type 1 and simian immunodeficiency virus: implications for the safety of lentiviral vectors. *Hum. Gene Ther.* **11**, 2403.
27. Haas, J., Park, E. C., and Seed, B. (1996) Codon usage limitation in the expression of HIV-1 envelope glycoprotein. *Curr. Biol.* **6**, 315.
28. Schwartz, S., Campbell, M., Nasioulas, G., Harrison, J., Felber, B. K., and Pavlakis, G. N. (1992): Mutational inactivation of an inhibitory sequence in human immunodeficiency virus type 1 results in Rev-independent gag expression. *J. Virol.* **66**, 7176.

29. Schneider, R., Campbell, M., Nasioulas, G., Felber, B. K., and Pavlakis, G. N. (1997) Inactivation of the human immunodeficiency virus type 1 inhibitory elements allows Rev-independent expression of Gag and Gag/protease and particle formation. *J. Virol.* **71,** 4892.
30. Fynan, E. F., Webster, R. G., Fuller, D. H., Haynes, J. R., Santoro, J. C., and Robinson, H. L. (1993) DNA vaccines: protective immunizations by parenteral, mucosal, and gene-gun inoculations. *Proc. Natl. Acad. Sci. USA* **90,** 11478.
31. Cardoso, A. I., Sixt, N., Vallier, A., Fayolle, J., Buckland, R., and Wild, T. F. (1998) Measles virus DNA vaccination: antibody isotype is determined by the method of immunization and by the nature of both the antigen and the coimmunized antigen. *J. Virol.* **72,** 2516.
32. Feltquate, D. M., Heaney, S., Webster, R. G., and Robinson, H. L. (1997) Different T helper cell types and antibody isotypes generated by saline and gene gun DNA immunization. *J. Immunol.* **158,** 2278.
33. Pertmer, T. M., Roberts, T. R., and Haynes, J. R. (1996) Influenza virus nucleoprotein-specific immunoglobulin G subclass and cytokine responses elicited by DNA vaccination are dependent on the route of vector DNA delivery. *J. Virol.* **70,** 6119.
34. Donnelly, J. J., Ulmer, J. B., and Liu, M. A. (1995) Protective efficacy of intramuscular immunization with naked DNA. *Ann. NY Acad. Sci.* **772,** 40.

V

PURIFICATION, MODIFICATION, AND RENATURATION OF RECOMBINANT PROTEINS

14

Purification and Immunological Characterization of Recombinant Antigens Expressed in the Form of Insoluble Aggregates (Inclusion Bodies)

Udo Reischl

Abstract

This chapter describes a straightforward protein purification strategy for the specific separation of insoluble recombinant proteins (so-called "inclusion bodies") located in the *E. coli* cytoplasm and their subsequent recovery in form of soluble recombinant proteins. Optimization of this technique can overcome in some cases the application of tedious and yield-reducing standard protein purification procedures. Due to the different behavior of individual recombinant proteins during separation, solubilization, and purification, subsequent purification steps may be required to obtain recombinant proteins in such a purity, that they can be used as antigen components of immunological test systems.

Key Words: Recombinant antigens; protein purification; inclusion bodies.

1. Introduction

In the field of clinical diagnosis, the determination of specific antibodies against distinct structural or functional antigenic proteins of a given pathogen is the most commonly used diagnostic tool for the detection of infections. Although most of the established test systems still use natural antigen from different sources, the advent of nucleic acid engineering opens up the possibility of producing proteins of limited natural availability as well as designing novel proteins using in vitro mutagenesis techniques. Recombinant technology has already proved to be an excellent alternative for the production of specific antigens, which are able to improve sensitivity as well as specificity. In general, the production of recombinant antigens for diagnostic purposes is inexpensive compared with the use of purified natural antigens. The major problems associated with the setup of recombinant test systems are of course the identification of those antigens or antigenic determinants that guarantee a safe serological diagnosis and the expression of these fragments with high efficiency.

Although targeted expression of cloned genes is now possible in a variety of prokaryotic and eukaryotic host organisms, the bacterium *Escherichia coli* is usually the first choice for the overexpression of procaryotic as well as eukaryotic antigens.

However, this convenient expression system embodies some intrinsic limitations including the expression of larger proteins or the appearance of non-full-length gene products owing to different codon usage between eukaryotic genes and the heterologous prokaryotic host organism. Furthermore, prokaryotic expression is limited to proteins for which post-transcriptional or post-translational modifications like acetylation, amidation, or glycosylation are not essential for antigenicity. If those modifications are required for antigenicity, recombinant antigens have to be expressed in homologous eucaryotic systems like the baculovirus or the Semliki Forest virus system (*see* Chap. 9 for details).

With a careful choice of host strains, vectors, and growth conditions, most recombinant proteins can be expressed at high levels in *E. coli*. However, some proteins are highly toxic to cells and are expressed poorly. Other proteins are quite unstable and are subject to degradation. Genes can be constructed so that either the foreign proteins specified by them may be located in the *E. coli* cytoplasm or, by incorporating a leader sequence upstream of the coding sequence, the proteins may be secreted through the cell membrane. In general, recombinant polypeptides accumulate to higher levels when they are expressed intracellularly (up to 60% of total cell protein) than when they are secreted (<1% of total cell protein). Because, aside from heat shock proteins, which may act as chaperons, protein-stabilizing influences are unavailable in the bacterial cytoplasm, the overexpressed heterologous proteins often appear in the form of insoluble aggregates known as inclusion bodies. Thus specific solubilization techniques are required for a subsequent protein purification under denaturing conditions.

Unlike the case of minor conformational epitopes, a "native" conformation is not essential for displaying the major sequential epitopes of antigenic polypeptides. In view of the fact that inclusion bodies can be easily separated from the vast amount of soluble bacterial proteins by physical means, it may be advantageous to adjust individual expression parameters in such a way that the recombinant antigen is produced in the form of insoluble aggregates within the bacterial cell.

This chapter concentrates primarily on a straightforward protein purification strategy that has been established for the specific separation of insoluble recombinant proteins located in the *E. coli* cytoplasm and their subsequent recovery in the form of soluble recombinant antigens. Optimization of this technique can overcome in some cases the application of tedious and yield-reducing standard protein purification procedures. Nevertheless, additional purification steps may be necessary to obtain recombinant proteins in such a purity that their antigenic properties can be evaluated by highly sensitive immunological test systems like enzyme-linked immunosorbent assay (ELISA). In our hands, continuous elution electrophoresis proved to be the method of choice to purify diagnostically relevant antigenic recombinant proteins to near homogeneity. Owing to the different behavior of individual recombinant proteins during separation, solubilization, and purification, the gold standard protocol must be determined empirically by adjusting the protocols presented carefully to the proteins of interest.

2. Materials
2.1. Small-Scale Bacterial Expression Cultures

1. Luria-Bertani (LB) medium: 10 g bacto-tryptone, 5 g bacto-yeast extract, 5 g NaCl; adjust with H_2O to a final volume of 1 L. Autoclave and store at room temperature to recognize possible contamination. If the yellow medium is contaminated, it becomes turbid.
2. 100 mM IPTG solution: 1.41 g IPTG (isopropyl-β-D-thiogalactopyranoside); adjust with H_2O to a final volume of 50 mL. Filter and store at –20°C in the dark.
3. 2X sodium dodecyl sulfate polyacrylamide gel electrophoresis (SDS-PAGE) sample buffer: 5% β-mercaptoethanol, 5% SDS, 0.5% bromophenol blue, 50% glycerol. Store at 4°C.
4. 10X TBE buffer: 1 M Tris-HCl, 830 mM boric acid, 10 mM EDTA; adjust the pH to 8.5 with HCl.
5. Electrophoresis unit for vertical polyacrylamide gels and comb bridges with small analytical and broad preparative teeth (Minigel Twin, Biometra, Göttingen, Germany).
6. Constant voltage power supply.
7. Sonication buffer: 50 mM Na-phosphate, pH 7.8, 300 mM NaCl. Adjust pH with NaOH.
8. Sucrose solution: 30 mM Tris-HCl, pH 8.0, 20% sucrose.

2.2. Denaturing Purification of Insoluble Proteins

The chemicals and enzymes used in this section are available from Roche Molecular Biochemicals (Mannheim, Germany).

1. Suspension buffer: 0.1 M Na-phosphate, pH 8.0, 10 mM Tris-HCl, pH 8.0.
2. Sonifying device: Labsonic U (B. Braun, Melsungen, Germany).
3. Continuous elution electrophoresis device: PrepCell model 491 (Bio-Rad, Munich, Germany) (**Fig. 1**).
4. Lysozyme stock solution: dissolve 100 mg lysozyme in 10 mL of H_2O. Store in aliquots at –20°C.
5. PMSF stock solution: 50 mM phenylmethylsulfonyl fluoride in 2-propanol. Prepare freshly before use. **Caution:** PMSF is highly toxic. Wear a mask and gloves when preparing the stock solution.

2.3. Immunological Assays

1. Nitrocellulose membrane: BA85 (Schleicher & Schuell, Munich, Germany).
2. Electroblotting device (e.g., Fastblot B33, Biometra).
3. PBS: phosphate-buffered saline, pH 7.2.
4. PBST solution: PBS containing 0.05% (v/v) Tween-20.
5. Horseradish peroxidase-conjugated anti-human IgG or IgM antibody (Dako Diagnostica, Hamburg, Germany).
6. POD substrate solution: 0.2 mg/mL diaminobenzidine, 0.014% H_2O_2, 0.1 M sodium citrate, and 0.02% thiomersal in H_2O. Prepare freshly and store in the dark.
7. ELISA coating buffer: 0.1 M $NaCO_3$, pH 9.5, 150 mM NaCl.
8. 96-well ELISA plates with medium binding capacity (e.g., ELISA plates 655001, Greiner, Frickenhausen, Germany).
9. ELISA serum dilution buffer: 3% fetal calf serum (FCS), 2% Tween-20 in PBS.
10. ELISA POD substrate solution: 0.1% (w/v) 1,2-phenylenediamine and 0.03% H_2O_2 in PBS. Prepare freshly and store in the dark.
11. Multiple-wavelength microplate OD reader (e.g., SLT, Kreilsheim, Germany).

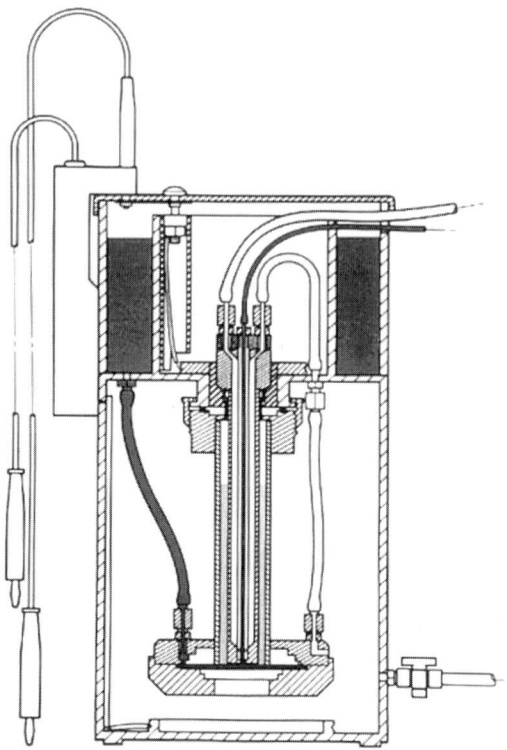

Fig. 1. Schematic representation of the continuous elution electrophoresis device (PrepCell model 531, Bio-Rad, Munich, Germany).

3. Methods

3.1. Rapid Screening of Small-Scale Expression Cultures

The amount of recombinant protein produced by a given recombinant expression construct is mainly dependent on the individual growth and expression conditions and on the type of *E. coli* host strain. To achieve the highest possible expression levels, it is strongly recommended to execute a set of initial optimization experiments by applying different expression parameters to small-scale cultures before large-scale protein production is attempted (see **Table 1** for a listing of relevant parameters). Furthermore, expression levels may vary between different colonies of transformed cells, and preparing a number of mini-cultures allows selection of the optimal expressers.

A basic protocol for the screening of small cultures is described below. As it is performed under denaturation conditions, it is only capable of determining the expression level of individual recombinant proteins and not their intracellular localization or constitution.

1. Inoculate 1.5 mL of prewarmed media (containing appropriate antibiotics) with 500 µL of an overnight culture of the transformant, and grow at 37°C for 30 min, with vigorous

**Table 1
Parameters That Usually Promote the Expression
of Recombinant Proteins in the Form of Inclusion Bodies**

Try different *E. coli* host strains.
Supplement the LB medium with 2% glucose.
Use super medium (*see* **Note 1**).
Grow the cells at an elevated growth temperature (up to 41°C).
Induce the expression of recombinant proteins within the early logarithmic growth phase (at an A_{600} of 0.1–0.4).
Induce the expresssion of recombinant proteins with relatively high IPTG concentrations (up to 10 mM) and grow the induced cultures for up to 5 h.

---

shaking until the A_{600} reaches 0.7–0.9. To analyze the time-course of expression, start with a 20-mL culture and take 1-mL samples at $t = 0, 1, 2, 3$, and 4 h after IPTG induction. Inoculate one extra culture, which serves as an uninduced control. Also inoculate one 1.5-mL culture with a colony transformed with a control plasmid lacking the insert and one 1.5-mL culture with the corresponding *E. coli* host strain.

2. Induce protein expression by adding IPTG to a final concentration of 2 mM. Do not add IPTG to the culture that serves as an uninduced control.
3. Grow the cultures for an additional 3–5 h, and transfer 1 mL to a 1.5-mL reaction tube. Centrifuge in a microfuge at 15,000g for 30 s and discard supernatants. If multiple expression conditions are analyzed, collect the cell pellets and proceed with step 4 or store them at –20°C until all samples are ready for processing.
4. Add 10 μL H$_2$O and 15 μL of 2X PAGE sample buffer to all samples and boil for 5 min at 95°C. Boiled samples may be stored at 4°C for at least 3 d.
5. Prepare an analytical 12.5% SDS-PAGE, centrifuge the boiled samples at 15,000g for 5 min, and apply 10 μL of each viscous supernatant to the gel. Following electrophoresis, visualize proteins by staining with Coomassie blue, and identify the protein of interest with the help of appropriate protein molecular weight markers.

3.2. Checking for Cytoplasmic or Periplasmic Location of the Recombinant Protein

The complex *E. coli* cell wall consists of a lipopolysaccharide-rich outer layer, lined by a membrane bilayer and a peptidoglycan layer, which is tightly associated with the outer cell membrane. The cytoplasmic membrane is separated from the peptidoglycan layer by the periplasmic space. Before deciding on a purification strategy, it is important to determine whether the protein is soluble in the cytoplasm, located in cytoplasmic inclusion bodies, or secreted into the periplasmic space. Many proteins form inclusion bodies when they are expressed at high levels in bacteria, whereas others are well tolerated by the cell and remain in the cytoplasm in their native configuration. To release soluble or insoluble recombinant proteins located in the cytoplasm, both the cell wall and the cytoplasmic membrane must be disrupted. This can be achieved by the enzymatic action of lysozyme in conjugation with detergents or by mechanical techniques like French press or sonication.

1. Grow a 50-mL culture and induce according to the optimized conditions determined in **Subheading 3.1**. Take a 1-mL sample immediately before induction (uninduced control) and before harvesting (induced control). Divide the culture into two aliquots (I and II) and harvest the cells by centrifugation at 5000g for 7 min.
2. Checking for cytosolic localization.
    a. Resuspend pellet (I) in 3 mL of sonication buffer. Freeze sample in dry ice/ethanol, and thaw in cold water.
    b. To lyse the cells, sonicate 10 times on ice with 30-s bursts at 80 W/30 s pause while cooling the reaction tube on ice (*see* **Note 2**). Centrifuge at 10,000g for 5 min, decant the supernatant, and store at 4°C (extract of soluble proteins).
    c. Resuspend the pellet in 3 mL sonication buffer (suspension of the insoluble cell components).
3. Checking for periplasmic localization.
    a. Resuspend pellet (II) in 5 mL sucrose solution. Add EDTA to a final concentration of 1 mM and incubate with occasional shaking for 10 min at room temperature.
    b. Centrifuge at 8000g at 4°C for 10 min and discard the supernatant. Resuspend the pellet in 5 mL ice-cold 5 mM MgSO$_4$ and incubate with shaking or stirring for 10 min in an ice/water bath.
    c. Centrifuge at 8000g at 4°C for 10 min. Collect the supernatant and store at 4°C (periplasmatic extract).
4. Add 5 μL of 2X SDS sample buffer to 5 μL of extracts from steps 2.b. and 2.c. and 10 μL of periplasmic extract, respectively. Boil samples for 5 min at 95°C (along with the uninduced and induced cell samples) and load the supernatants to an analytical SDS polyacrylamide gel. Following electrophoresis, visualize proteins by staining with Coomassie blue and determine the intracellular localization of the protein of interest.

3.3. Lysis of E. coli *by Lysozyme Treatment and Sonication*

In the course of a preparative purification of recombinant proteins from large-scale cultures, a slightly modified lysis protocol turned out to be highly efficient. If the recombinant protein is expected to be present in its native, soluble form, it is best to work quickly and keep the cells at 0–4°C at all times. On the other hand, the procedure can be carried out at room temperature if a localization in inclusion bodies is expected. To inhibit the enzymatic action of cell-borne proteases, specific inhibitors like PMSF should be added to the sonication buffer.

1. Grow and induce a 200 mL culture. Harvest the cells by centrifugation at 4000g for 15 min and resuspend the pellet in 20 mL sonication buffer.
2. Add 0.5 mL of lysozyme solution and 100 μL of a freshly prepared PMSF stock solution. Incubate the suspension at room temperature for 10 min with occasional shaking.
3. Sonicate 10 times on ice: 30-s burst at 80 W/30 s cooling (*see* **Note 2**). If the resulting lysate becomes very viscous, add RNase A and/or DNase I to a final concentration of 10 μg/mL each and incubate on ice for 10 min.
4. Centrifuge at 4000g for 20 min and collect the supernatant carefully (extract of soluble proteins). If the protein of interest is found in the supernatant, proceed with appropriate purification strategies for soluble proteins (*see* Chaps. 6, 11, and 15).
5. Resuspend the pellet in 10 mL sonication buffer, vortex intensely, and centrifuge at 4000g for 10 min. Discard the supernatant. Repeat this step twice (or until a clear supernatant is obtained) to remove most of the contaminating soluble *E. coli* proteins (*see* **Note 3**).

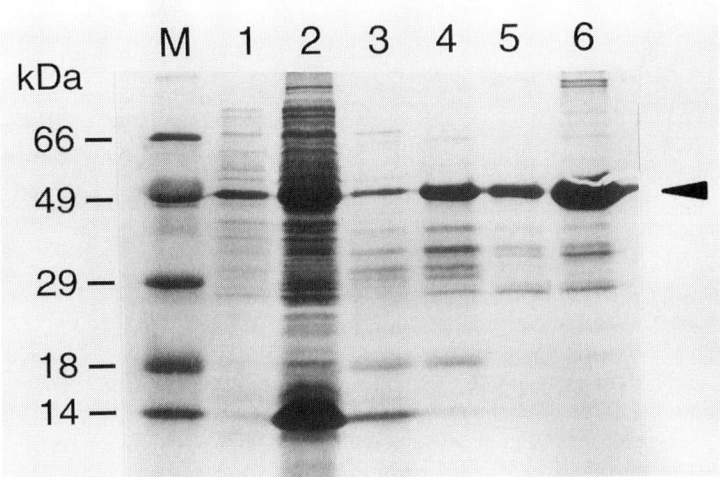

Fig. 2. *E. coli* strain M15[pREP4] was transformed with the vector construct pQE-p23 (which encodes an antigenic protein of the Epstein-Barr virus [EBV]), and protein synthesis was induced with 1 mM IPTG for 4 h. After induction, pQE-p23 led to an intracellular accumulation of the recombinant protein at a very high level, constituting up to at least 50% of the total cell protein, and a preferential localization in inclusion bodies was observed. Following sonication, the soluble and insoluble fractions of the crude cell extract were separated by centrifugation. Purification of the recombinant protein was monitored by Coomassie blue-stained SDS-PAGE (**A**) and immunoblot using a pool of EBV-positive patient sera (**B**). Protein extracts were prepared as described in **Subheading 3.4.** Lanes contain the soluble fraction (lane 1), which is poorly expressed at the recombinant antigen, the insoluble fraction (lane 2) of the crude sonification extract, and the corresponding supernatants after incubation of the insoluble fraction with suspension buffer containing 2 M urea (lane 3), 4 M urea (lane 4), 6 M urea (lane 5), and 8 M urea (lane 6). According to a mixture of marker proteins (lane M), the position of the recombinant antigen is indicated.

6. Collect the remaining pellet, which contains the insoluble cell components. The pellet can be stored frozen at –20°C for subsequent processing. Proceed with **step 1**, **Subheading 3.3.**

3.4. Purification of Insoluble Proteins by Successive Incubation with Solutions Containing Rising Concentrations of Urea

In connection with the high-level expression and subsequent purification of recombinant antigens, it is advantageous to force the intracellular formation of insoluble protein aggregates. This section describes a simple laboratory-proven protein purification strategy that has been established for the specific enrichment of insoluble recombinant proteins located in the *E. coli* cytoplasm and their conversion to a soluble form with the help of denaturants. Briefly, inclusion bodies are separated from the vast amount of soluble bacterial proteins by centrifugation, most of the contaminating *E. coli* proteins are divided off during a successive washing procedure, and the recombinant antigen is solubilized at a particular urea concentration (**Fig. 2**).

Because the solubilization process is strictly dependent on the nature of the individual protein, the minimal urea concentration giving an acceptable level of solubilization has to be determined individually.

1. Resuspend the pellet (consisting of the insoluble protein fraction and cell debris) at room temperature in 5 mL suspension buffer containing 2 M urea by vortexing or vigorous shaking for at least 15 min (*see* **Note 4**).
2. Centrifuge at 4000g for 10 min and collect the supernatant, which may contain the recombinant protein. Take a 15-µL aliquot of the supernatant, add 15 µL of 2X PAGE sample buffer, and boil for 5 min at 95°C. Collect the individual samples and then analyze them together on an SDS-PAGE gel. Store the remaining supernatant at –20°C for subsequent processing; boiled samples may be stored at 4°C for at least 3 d.
3. Resuspend the pellet again in 10 mL suspension buffer containing 2 M urea, vortex for 5 min, and centrifuge at 4000g for 10 min. Discard the supernatant.
4. Repeat **steps 1**, **2**, and **3** for urea concentrations of 4, 6, and 8 M, respectively.
5. Collect the pellet that contains the insoluble cell components. Take a small aliquot, add 10 µL H_2O and 15 µL 2X PAGE sample buffer, vortex, and boil for 10 min at 95°C. The remaining portion of the pellet can be stored frozen at –20°C for subsequent processing.
6. Analyze the aliquots (2, 4, 6, and 8 M urea and pellet) on an SDS-PAGE gel. Following electrophoresis, visualize proteins by staining with Coomassie blue and identify the supernatant/pellet sample that contains the maximum amount of recombinant protein.

3.5. One-Step Purification of Protein Preparations to Near Homogenity with the Help of a Continuous Elution Electrophoresis Device

Although the urea solubilization procedure is capable of separating most of the contaminating *E. coli* proteins, additional purification steps may be necessary to obtain the recombinant protein in such a purity that it could be used as antigen in immunological assay systems. As confirmed by extensive immunoblot analysis, size-dependent fractionation of protein mixtures by continuous elution electrophoresis turned out to be the method of choice for an efficient one-step purification of recombinant antigens at a preparative scale. Since the urea within the applied sample is slowly removed during the electrophoresis procedure, the recovered protein is, in contrast to the applied protein, somehow "renatured" and completely solubilized in standard electrophoresis buffer. Nevertheless, owing to incorrect refolding and thus precipitation of recombinant protein within the electrophoresis column, a significant loss may be observed. We have purified more than 30 recombinant antigens by continuous elution electrophoresis so far and, following the protocol presented, the observed yields of recovered solubilized protein range from 5 to 30% of the total applied protein (**Fig. 3**). Usually, this provides sufficient pure antigen for coating more than 100 standard 96-well ELISA plates for optimal results (*see* **Subheading 3.7.**).

1. To reduce the sample volume, precipitate the "urea-solubilized" proteins by repeated dialysis of the corresponding 10-mL supernatant sample (*see* **Subheading 3.4.**) against 2 × 1 L of 50 mM Tris-HCl, pH 8.0, for at least 8 h.
2. Collect the white precipitate by centrifugation at 4000g for 10 min. Discard the supernatant, add 1.5 mL of 50 mM Tris-HCl, pH 8.0, and 1.5 mL of 2X PAGE sample buffer to

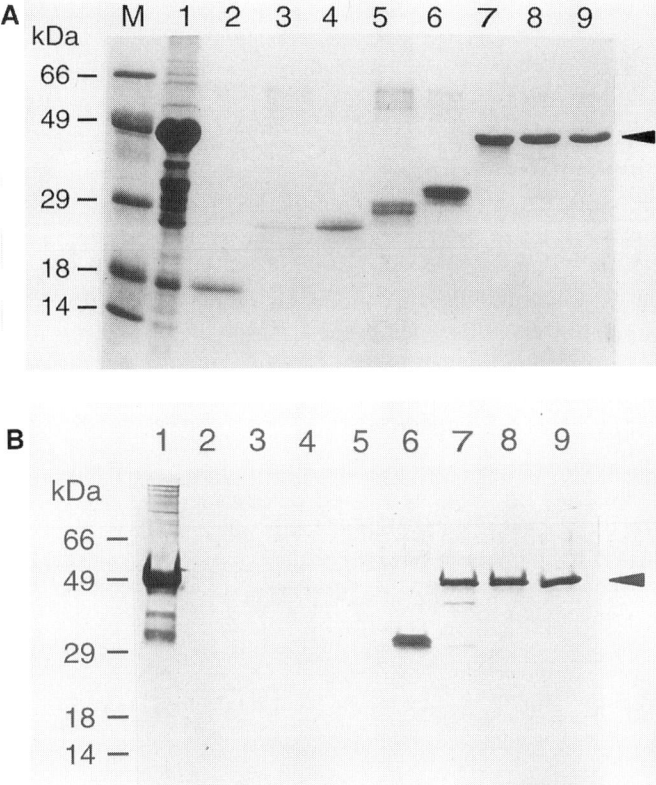

Fig. 3. Purification of the recombinant EBV antigen by quantitative continuous electroelution using the PrepCell device. Aliquots of the eluted fractions (200 µL) were TCA-precipitated and analyzed by Coomassie blue-stained SDS-PAGE (**A**) and immunoblot using a pool of EBV-positive patient sera (**B**). Lanes contain an aliquot of the 8 *M* urea supernatant before application to the electrophoresis device (lane 1) and aliquots of fractions eluted during electrophoresis (lanes 2–9). According to a mixture of marker proteins (lane M), the position of the recombinant antigen is indicated.

the pellet, vortex, and incubate at 95°C for 5 min. Analyze a 20-µL aliquot by SDS-PAGE to estimate the total protein concentration of the sample.

3. Pour a cylindrical SDS-PAGE and assemble the PrepCell device according the manufacturer's instructions. For proteins of less than 50 kDa, a thin collection gel with a polyacrylamide concentration of 2% and a separation gel with a polyacrylamide concentration of 12% is recommended. Let the gel matrix polymerize for at least 4 h, fill the device with 1X TBE buffer, and perform a 20-min prerun at 40 mA.
4. Apply a portion of the sample (≤3 mL), which corresponds to approx 10 µg protein, at the top of the gel column and start electrophoresis at 40 mA (1.5-cm gel diameter) or 70 mA (3-cm gel diameter), respectively. During the 5–8-h run, individual proteins pass directly into the elution chamber of the apparatus and are collected in the form of fractions. Start

the collection of 2-mL fractions when the bromophenol blue band has passed the gel matrix.
5. Analyze 200-µL aliquots of selected fractions (*see* **Note 5**) by Coomassie blue-stained 12.5% SDS-PAGE and/or by immunoblot to monitor the presence and purity of the protein of interest in the different PrepCell fractions.

3.6. Immunoblot Analysis

Once the recombinant antigen is purified to a sufficient extent, immunoblot analysis is usually carried out to both demonstrate the absence of contaminating *E. coli* proteins (*see* **Note 6**) and verify the immunological reactivity/antigenicity of the recombinant protein. Immunoblots are produced by transferring the SDS-PAGE-separated proteins onto nitrocellulose membranes. The coated membrane is incubated with a solution of the corresponding antibody (e.g., human sera) and the protein–antibody complex is pinpointed by incubation with a labeled secondary antibody followed by chromogenic or luminescent detection.

1. Transfer the proteins, which were separated according to their size during SDS-PAGE electrophoresis, to a nitrocellulose membrane via electroblotting (*see* **Note 7**).
2. To visualize the proteins that serve as molecular weight markers, incubate the nitrocellulose membrane in Poinceau-red solution for 5 min with occasional shaking and destain the membrane body subsequently by washing twice with PBS. Mark the protein bands with a black ballpoint pen.
3. Incubate the membrane with 5% nonfat milk in PBS for at least 30 min at room temperature to prevent nonspecific binding of antibodies (*see* **Note 8**).
4. Wash three times with PBST solution and incubate the membrane for at least 1 h at room temperature with the corresponding antibody solution (e.g., human sera diluted 1:100 in 5% nonfat milk/PBS).
5. Wash three times with PBST solution and incubate the membrane with the corresponding horseradish peroxidase-labeled secondary antibody (e.g., horseradish peroxidase-conjugated anti-human IgG antibody diluted 1:1000 in PBS) with occasional shaking.
6. Following a 2-h incubation at room temperature, wash the membrane three times with PBST solution and start color development with POD substrate solution. Stop the chromogenic reaction after 5–10 min by washing several times in H_2O, let the membrane air-dry, and document the protein pattern by photography.

3.7. Enzyme Linked Immunosorbent Assay

Immunoblot analysis is advantageous in certain cases and may be well suited for a preliminary screening of recombinant antigens. For a more extensive screening and to evaluate the diagnostic potential of a particular recombinant antigen in a more systematic manner, an ELISA test format should be established. Usually the principle of indirect ELISA represents a convenient, quantitative, and rapid immunological assay format for the determination of antibody titers against particular recombinant antigens.

1. Dilute the purified recombinant antigen with coating buffer to a final concentration of approx 2 µg/mL and disperse 50-µL aliquots in each well of the ELISA plate. Incubate overnight at 4°C.
2. Discard the antigen solution by tapping the plate upside down on a paper towel and block residual binding sites of the coated ELISA plate by incubating with 2% w/v nonfat dry milk in PBS for 2 h.

3. Disperse 50 µL aliquots of the corresponding antibody solution diluted in ELISA dilution buffer (e.g., human sera, diluted 1:50) in individual wells and incubate for 2 h at 37°C.
4. Discard the remaining antibody solution by tapping the plate upside down onto a paper towel and wash three times with 100 µL ELISA dilution buffer at room temperature.
5. Add 50 µL of the corresponding horseradish peroxidase-labeled secondary antibody (e.g., horseradish peroxidase-conjugated anti-human IgG antibody diluted 1:1000 in ELISA dilution buffer) and incubate for 1 h at 37°C.
6. Wash the plate three times with PBST at room temperature to remove unbound conjugate and start the detection reaction by dispersing 100 µL ELISA POD substrate solution to each well. When a light yellow coloring is visible (5–10 min), stop the color development by adding 100 µL of 1 M H_2SO_4.
7. Put the ELISA plate into a multiwavelength microplate reader and report the data as A_{492} nm–A_{620} nm difference values for the sample wells after subtraction of the blank value (*see* **Note 9**).

4. Notes

1. If a very poor expression of the recombinant protein is observed, the use of "super medium" may be helpful in some cases. This modified enriched medium consists of 25 g bacto-tryptone, 15 g bacto-yeast extract, and 5 g NaCl per liter (*see* **Subheading 2.1.1.**).
2. To achieve an efficient destruction of the bacterial cell wall, it is important to prevent foaming of the bacterial suspension during the sonication procedure. Make sure that the ultrasonic device is immersed completely and is kept at least 1 cm under the surface of the turbid suspension throughout sonication.
3. As inclusion bodies are dense particles, a sophisticated centrifugation step can be employed for their preenrichment. It is generally possible to define empirically a centrifugation speed (usually between 500 and 1200g) and time (5–15 min) that pellets the inclusion bodies, leaving a proportion of other particulate material in suspension together with the soluble host proteins. Optimum separation parameters have to be determined by examination of the relative yield and purity of the recombinant protein in the pellet.
4. For small volumes, inclusion bodies are suspended by vortex mixing; large volumes should be suspended within a 15-mL tube using an overhead homogenizer or overhead shaker for at least 30 min. Take care to minimize foaming of the turbid suspension.
5. If the protein concentration of the samples is very low, it may be necessary to precipitate the proteins before analyzing the samples on SDS-PAGE or immunoblots. Bring up the sample volume to 1 mL with H_2O and add 25 µL of a 2% Na-dodecylsulfate solution. Incubate on ice for 10 min. Add 30 µL of TCA solution (40% trichloroacetic acid in H_2O), vortex, and centrifuge at 15,000g for 10 min. Discard the supernatant, dissolve the white pellet in 50 µL of 1X PAGE sample buffer, and carefully add 0.1 M NaOH solution until the color of the solution switches from yellow to blue. Incubate the neutralized sample at 95°C for 5 min and apply the sample to an SDS-PAGE.
6. Many investigators preincubate their immunoblot sera with sonified extracts of *E. coli* to saturate *E. coli*-specific antibodies and thus reduce the sera's reactivity with *E. coli*-borne proteins. To demonstrate the absence of any contaminating *E. coli* proteins in the final protein preparation, such a preincubation is not recommended.
7. Use a commercially available electroblotting device (e.g., Fastblot B33, Biometra) to achieve optimal results.
8. As in the case of valuable human serum samples, sometimes there are only very small amounts of the corresponding antibody solutions available. Here the use of a multiple-channel immunoblot device (e.g., Miniblotter MN25, Biometra) is recommended.

9. To raise the significance of the ELISA OD values obtained at individual samples, the assay should be performed in duplicate or triplicate for each sample, and each plate should contain appropriate positive and negative controls.
10. For further reading, *see* refs. *4–6*.

References

1. Takacs, B. J. and Gierard, M.-F. (1991) Preparation of clinical grade proteins produced by recombinant DNA technologies. *J. Immunol. Methods* **143,** 231–240.
2. Sambrook, J., Fritsch, E. F., and Maniatis, T. (1989) *Molecular Cloning: A Laboratory Manual*, 2nd ed. Cold Spring Harbor Laboratory Press, Cold Spring Harbor, NY.
3. Reischl, U., Gerdes, C., Motz, M., and Wolf, H. (1996) Expression and purification of a 23-kDa protein encoded by Epstein-Barr virus and characterization of its immunological properties. *J. Virol. Methods* **57,** 71–85.
4. van Grunsven, W. M. J., van Heerde, E. C., de Haard, H. J. W., Spaan, W. J. M., and Middeldorp, J. M. (1993) Gene mapping and expression of two immunodominant Epstein-Barr virus capsid proteins. *J. Virol.* **67,** 3908–3916.
5. Gorgievski-Hrisoho, M., Hinderer, W., Nebel-Schickl, H., et al. (1990) Serodiagnosis of infectious mononucleosis by using recombinant Epstein-Barr virus antigens and enzyme-linked immunosorbent assay technology. *J. Clin. Microbiol.* **28,** 2305–2311.
6. Motz, M., Fan, J., Seibl, R., Jilg, W., and Wolf, H. (1986) Expression of the Epstein-Barr virus 138-kDa early protein in E. coli for the use as antigen diagnostic tests. *Gene* **42,** 303–312.

15

Purification of Recombinant Proteins with High Isoelectric Points

Raffaele A. Calogero and Anna Aulicino

Abstract

The use of recombinant antigens is essential for the construction of robust and sensitive diagnostic assays. A critical step in the preparation of recombinant antigens is protein purification. Purification problems may be very different for related structural proteins expressed in the same host or for the same protein expressed in different hosts, because the biochemical characteristics of a recombinant protein, expressed in a heterologous system, are unique. In this chapter we make a brief introduction to protein purification procedures and we present a quick purification process suitable for the isolation of recombinant protein having high isoelectric points encoding non-conformational epitopes.

Key Words: Protein purification; sequential epitopes; recombinant antigen; *E. coli*.

1. Introduction

The use of recombinant antigens and chemically synthesized peptides is essential for the construction of robust and sensitive diagnostic assays. Moreover, in the field of virology, the use of recombinant antigens eliminates the need to handle highly hazardous material in the preparation of the assay *(1,2)* and allows the assembly of multiepitope polypeptides *(3)*. Protein purification is a critical step in the preparation of recombinant antigens. It is important to point out that the biochemical characteristics of a recombinant protein, expressed in a heterologous system, are unique. Purification problems may be very different for related structural proteins expressed in the same host or for the same protein expressed in different hosts. The designing of an antigen-purification procedure is strongly dependent on the immunodominant epitopes present on its surface. The antibodies recognize chemical groupings, exposed to the solvent, on the surface of an antigen. An immunodominant determinant may be continuous or discontinuous. Moreover, continuous epitopes can be sequential, if they are

Table 1
Recombinant Viral Proteins[a]

	r-HDAg	r-HCaAg	r-p24-NC
Expression system	E. coli	E. coli	E. coli
Expression level	15–25%	5–10%	25–40%
Molecular weight, kDa	30	16	39
Isoelectric point	10.6	12.4	9.1
Mg of 80–90% pure protein/g cells	1–1.5	0.5–0.8	2–4

[a]Three viral proteins having basic isoelectric points are used to evaluate the purification procedure involving lysis in guanidinium chloride, CH_3CN precipitation, and chromatography on a histidil diazobenzil phosphonic acid-agarose column. Recombinant HDAg and recombinant HCcAg are expressed in a soluble form in the *E. coli* cytosol. Recombinant p24-NC is expressed as insoluble aggregates in the *E. coli* cytosol. The described purification procedure (*see* **Subheading 3.**) is also used for the recombinant HCcAg expressed in HCcAg-recombinant baculovirus Sf9 insect cells (data not shown).

only defined by the primary protein sequence, or conformational, if they are associated with secondary structure elements (α-helix, β-sheets, loops). Discontinuous epitopes are, instead, defined by the tertiary structure of the protein. Furthermore, the expression system (bacteria, yeast, insect cells, mammalian cells), the expression condition used *(4)*, the sensitivity of the recombinant antigen to host protease *(5)*, and the number of steps involved in the purification procedure may have relevant effects on the yield and purity of the recombinant antigen. Finally, different purification approaches can produce the same results. For example, a purification procedure can be designed avoiding the use of strong ionic detergent or chaotropic reagents [sodium dodecyl-sulfate (SDS), urea, guanidinium chloride], because the antigen-immunodominant epitopes are structure-dependent (discontinuous or conformational). Alternatively, it is possible to purify the same protein in a denatured form and then reconstitute the antigen-native structure by applying a refolding procedure *(3,6)*.

In this chapter, we describe a procedure for the purification of proteins having a high isoelectric point that consist of organic precipitation combined with ionic exchange chromatography. We applied the procedure to three different recombinant viral antigens: hepatitis Δ antigen (HDAg) *(2)*, hepatitis C core antigen (HCcAg) (R. A. Calogero, unpublished data), and a fragment of HIV-1 Gag polyprotein, p24-NC *(7)* (**Table 1**).

1.1. Heterologous Expression System

Expression systems in *Escherichia coli* are widely available and are used in the biotechnology field for the expression of heterologous proteins. However, *E. coli* does not make protein with post-translational modifications, and, frequently, the heterologous proteins are expressed as insoluble aggregates (inclusion bodies) *(8,9)*. Yeast *(10)*, insect *(11)*, and mammalian cells are also suitable as expression systems, and their use is essential if post-translational modification or glycosylation patterns are antigenically important *(10)*.

1.2. Recombinant Antigens Expressed in E. coli

E. coli allows the production of heterologous proteins at high levels and reduced cost. However, despite many successful examples, there are frequently difficulties in recovering the chains expressed from cloned genes as the correctly folded native proteins. A common outcome is the accumulation of the expressed protein as an aggregate of denatured chains. This problem is generally associated with proteins maturing in the cytoplasm *(12–14)*. Because of the high concentration of the recombinant protein in the reducing *E. coli* environment and the lack of compartmentalization of *E. coli* cytosol, the interchain interactions between the recombinant protein and the host proteins are favored with respect to the intrachain interactions, which are essential for proper protein folding. Moreover, the bacterial outer membrane proteins are involved in the inclusion bodies' formation *(15)*. During the expression of viral membrane proteins (e.g., gp41, HIV-1), inclusion bodies are usually formed, and the interchain interactions are so strong that they can be solubilized efficiently only using strong chaotropic reagents *(16)*. On the other hand, DNA- and RNA-binding proteins, like viral capsid [HCcAg, hepatitis C virus (HCV)] and nucleocapsid proteins (NC, HIV-1), are expressed in soluble form in *E. coli* cytosol and interact strongly with endogenous nucleic acids. Treatment with RNase and DNase cannot completely remove nucleic acids and the oligonucleotides, obtained by nucleic acid degradations, which are still able to interact with the recombinant proteins producing aggregates. Nucleic acids–protein interactions produce changes in the antigen biochemical behavior and interfere with ionic exchange chromatography and gel filtration. Moreover, recombinant proteins complexed with nucleic acids appear to aggregate during freeze/thaw cycles (R. A. Calogero, unpublished observations).

1.2. Bacterial Cell Disruption

Cell disruption is the first step in a purification procedure and it has to be rapid and efficient. It is important to keep the time needed for lysis of the bacterial cells at a minimum, especially if the recombinant protein is sensitive to protease activities. Lab bench lysis techniques, not involving special and expensive apparatus, are sonication, lysozyme treatment followed by osmotic shock, and chaotropic agents for cell disruption. Sonication and lysozyme lysis work well when applied to a small number of cells. In our hands, we found cell disruption mediated by guanidinium chloride at acid pH to be an efficient method for cell lysis and for solubilization of inclusion bodies. Moreover, guanidinium chloride efficiently inhibits cellular proteases and, in conjunction with acid pH, disrupts nucleic acid–protein interactions.

1.3. Fractionation of Protein by Selective Precipitation

The solubility of a protein is dependent on dissolved ion concentration *(17)*. At high salt concentration, most of the solvent molecules are involved in ion solvation, and the numbers of residual molecules are not enough to maintain the protein in solution. Protein salting-out can be achieved by adding substances to the protein solution, which reduces the protein's solubility. Selective protein precipitation, using increasing salt concentration, is routinely used as starting step in the preparation of

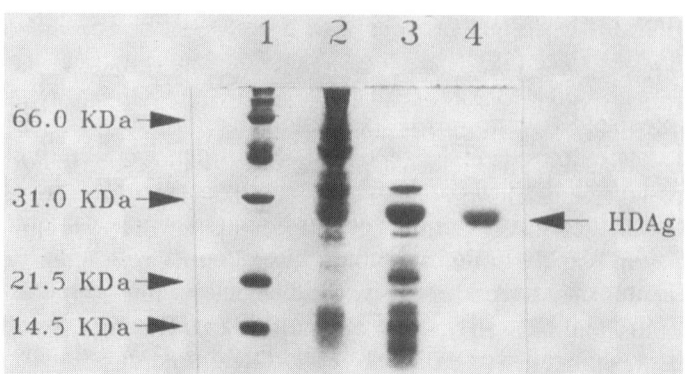

Fig. 1. SDS-PAGE analysis; the recombinant hepatitis Δ antigen (HDAg) is shown by arrow. Lane 1, Bio-Rad low molecular weight markers; lane 2, *E. coli* debris dissolved in solution A; lane 3, protein pattern after CH_3CN precipitation and dialysis against acetic acid; lane 4, ionic exchange chromatography peak fraction pool showing anti-HDV reactivity.

enzymes. Moreover, selective protein precipitation can also be achieved by using organic solvents soluble in water (e.g., alcohols, acetone, acetonitrile) *(2)*. These compounds induce dehydration of the macromolecule and reduce the dielectric constant of the medium, lowering its ability to dissolve polyions, such as proteins. The bacterial lysate treatment with 35% CH_3CN precipitates nucleic acids and high-molecular-weight *E. coli* proteins but does not affect basic recombinant proteins (**Figs. 1–3**).

1.4. Ionic Exchange Chromatography

Ionization of the amino acid side chain functional groups of the protein changes as a function of the dissolution buffer pH. In the ionic exchange process, ions (Ions A^-) bound to an insoluble matrix (Matrix$^+$) are replaced by ions present in solution (Ions B^-):

$$Matrix^+Ions\ A^- + Ions\ B^- \rightleftharpoons Matrix^+Ions\ B^- + Ions\ A^- \tag{1}$$

The protein affinity for an ion exchanger is caused by the net charge of the polypeptide and the dissolved ion concentration, which can compete with the protein in binding to the ion exchanger. During protein purification, pH and salt concentration are selected to allow the protein to bind to the ion exchanger. Subsequently, the protein of interest can be released from the insoluble matrix by modifying the pH or the ionic concentration of the buffer. Ion exchange chromatography is widely used for the preparation of recombinant antigen in native or denaturated conditions, together with other standard protein purification procedures like salting-out, organic precipitation, and gel filtration. The use of different kinds of matrices (cellulose, Sephadex, Sephacryl, silica, polyacrylamide, and others) chemically modified with positively ($-OCH_2CH_2N[C_2H_5]_2$, $OCH_2CH_2N[C_2H_5]_3$) or negatively ($-OCH_2COOH$, $-OPO_3H_2$) charged groups at vari-

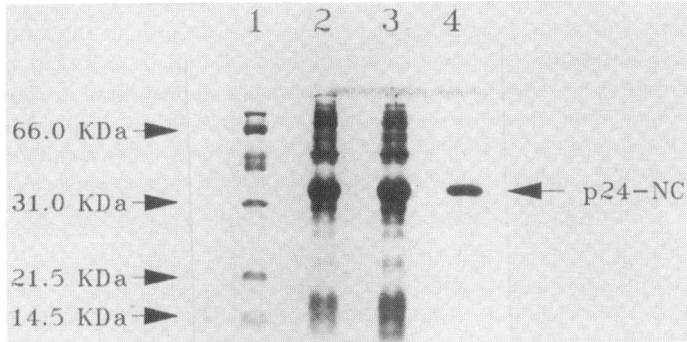

Fig. 2. SDS-PAGE analysis; the recombinant p24-NC is shown by arrow. Lane 1, Bio-Rad low molecular weight markers; lane 2, *E. coli* debris dissolved in solution A; lane 3, protein pattern after CH_3CN precipitation and dialysis against acetic acid; lane 4, ionic exchange chromatography peak fraction pool showing anti-HIV reactivity.

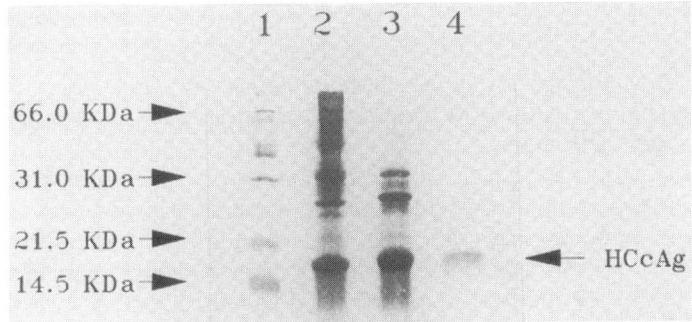

Fig. 3. SDS-PAGE analysis; the recombinant hepatitis C core antigen (HCcAg) is shown by arrow. Lane 1, Bio-Rad low molecular weight markers; lane 2, *E. coli* debris dissolved in solution A; lane 3, protein pattern after CH_3CN precipitation and dialysis against acetic acid; lane 4, ionic exchange chromatography peak fraction pool showing anti-HCV reactivity.

ous pH values and ion concentrations allows the ion exchange purification of virtually any protein. It is convenient to perform batchwise experiments to define the optimal ion exchange chromatography conditions. Samples of the protein mixture, dissolved in buffers at different pH, can be mixed with a small amount of cationic or anionic exchanger. After adsorption, the exchanger can be centrifuged and the supernatant analyzed for the presence of the protein of interest. The ion exchange method, subsequently described, is used to purify the recombinant HCcAg, the recombinant HDAg, and the recombinant p24-NC (**Figs. 4–6**), and it can be easily applied to any basic protein expressed in *E. coli*, as inclusion bodies or in a soluble form.

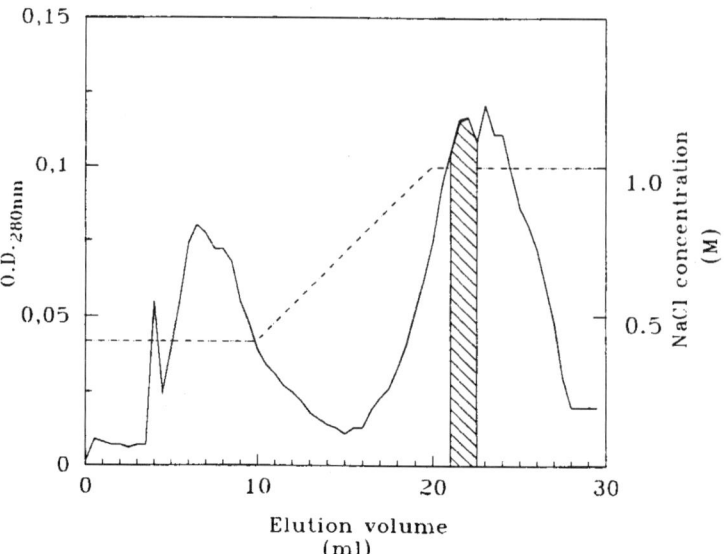

Fig. 4. Purification of the recombinant HDAg; elution profile of the histidil diazobenzil phosphonic acid-agarose column. Four $OD_{280\ nm}$ are loaded on the column, and 1.9 $OD_{280\ nm}$ are eluted in the flow-through. Dashed lines indicate the fractions showing anti-HDV reactivity by Western blot analysis.

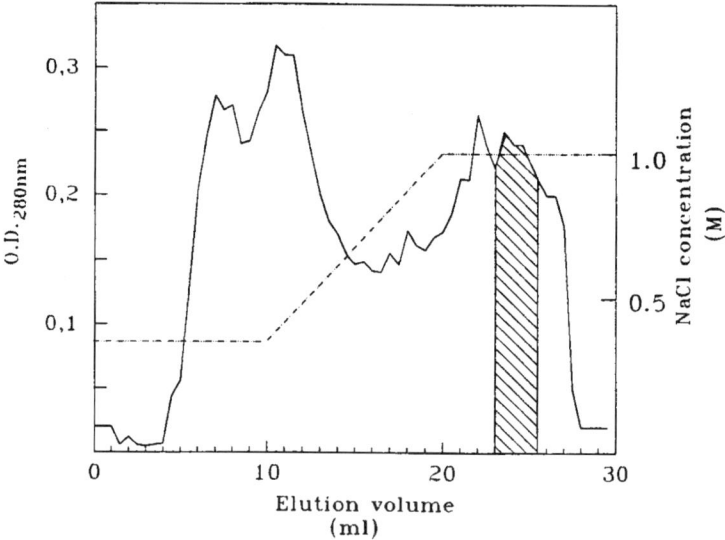

Fig. 5. Purification of the recombinant p24-NC; elution profile of the histidil diazobenzil phosphonic acid-agarose column. Thirty-six $OD_{280\ nm}$ are loaded on the column, and 34 $OD_{280\ nm}$ are eluted in the flow-through. Dashed lines indicate the fractions showing anti-HIV reactivity by Western blot analysis.

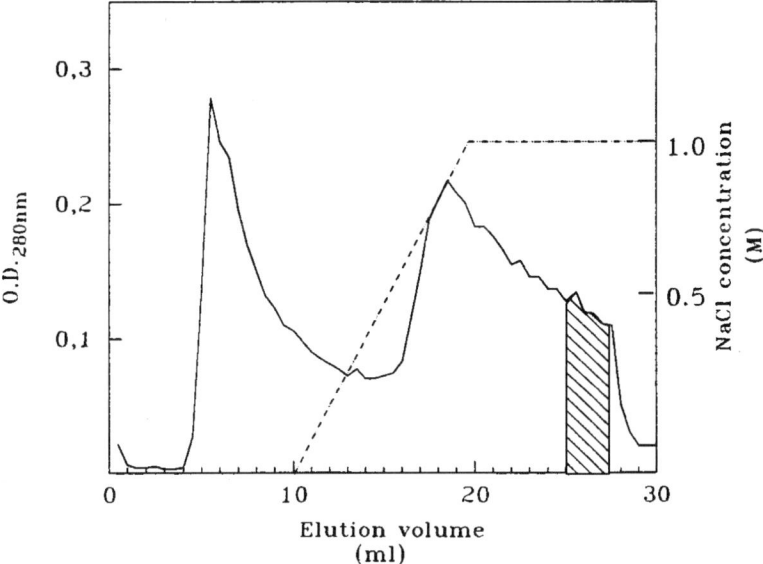

Fig. 6. Purification of the recombinant HCcAg; elution profile of the histidil diazobenzil phosphonic acid-agarose column. Twelve $OD_{280\,nm}$ are loaded on the column, and 5.6 $OD_{280\,nm}$ are eluted in the flow-through. Dashed lines indicate the fractions showing anti-HCV reactivity by Western blot analysis.

1.5. Evaluation of Recombinant Antigen Purity and Specificity

Since their initial development (18,19), noncompetitive solid-phase immunoassays have gained widespread use in the measurement of both antigen and antibodies. Noncompetitive solid-phase immunoassays can also be used as quantitative assays through comparison with a known standard. However, competitive solid-phase immunoassays (see **Fig. 8B**) might be preferred to obtain more specific results. The higher specificity of a competitive immunoassay is caused by the measurement of an unlabeled molecule for binding sites on a limited quantity of solid-phase bound receptor.

The evaluation of recombinant antigen specificity and reliability in a diagnostic assay is a long procedure essential in obtaining indications of the recombinant antigen performance with respect to the native antigen. Moreover, the use of standard sera allows the identification of the optimal antigen coating conditions (pH, antigen concentration) and the best assay format (competitive, noncompetitive, single step, multiple step). In this chapter, we present the evaluation of purified recombinant HCcAg in an indirect enzyme immunoassay (**Fig. 7A**) using commercial characterized anti-HCV sera (BBI). Furthermore, we show the behavior of the purified recombinant HDAg, in a commercial enzyme-linked immunosorbent assay (ELISA) single-step immunoinhibition assay (SORIN, I), in comparison with the natural antigen (**Fig. 8A**).

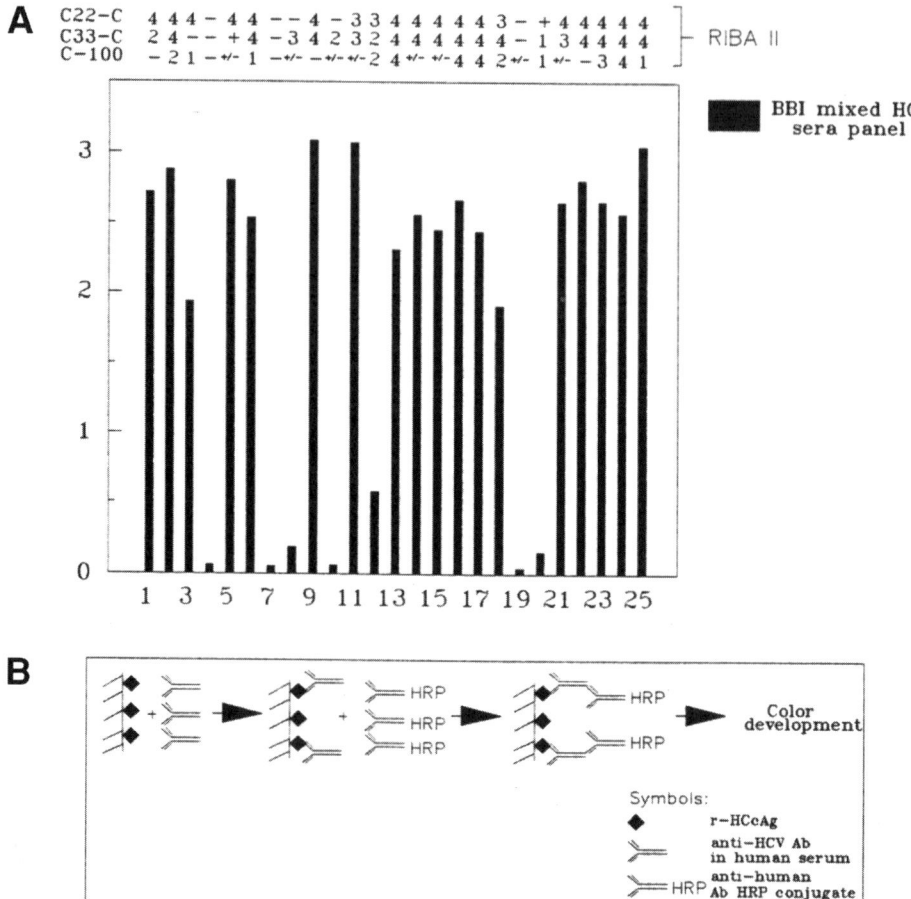

Fig. 7. Immunoreactivity of recombinant hepatitis C core antigen (HCcAg) is evaluated by analyzing its binding to anti-hepatitis C virus (HCV) antibodies using certified anti-HCV-positive and -negative sera panel (BBI). (**A**) Closed bars show the immunoreactivities identified with recombinant HCcAg. Top, values of p22 (C22-c), NS3 (C33-c), and NS4 (C-100) immunoreactivities. These data are obtained by BBI using a RIBA assay for the detection of anti-HCV antibodies (Ortho Diagnostics, Emeryville, CA). Values of 4 to 1 indicate clear positive reactivities, ± indicates reactivities near cutoff, and – indicates absence of reactivity. (**B**) Scheme of indirect noncompetitive ELISA for anti-HCV antibody detection.

2. Materials

The chemicals used in this section are all available from Sigma-Aldrich (St. Louis, MO), unless otherwise indicated.

1. Bicinchonininc acid (BCA) protein concentration determination kit (Pierce, Rockford, IL).
2. Affi-Prep protein A-agarose (Bio-Rad, Hercules, CA).

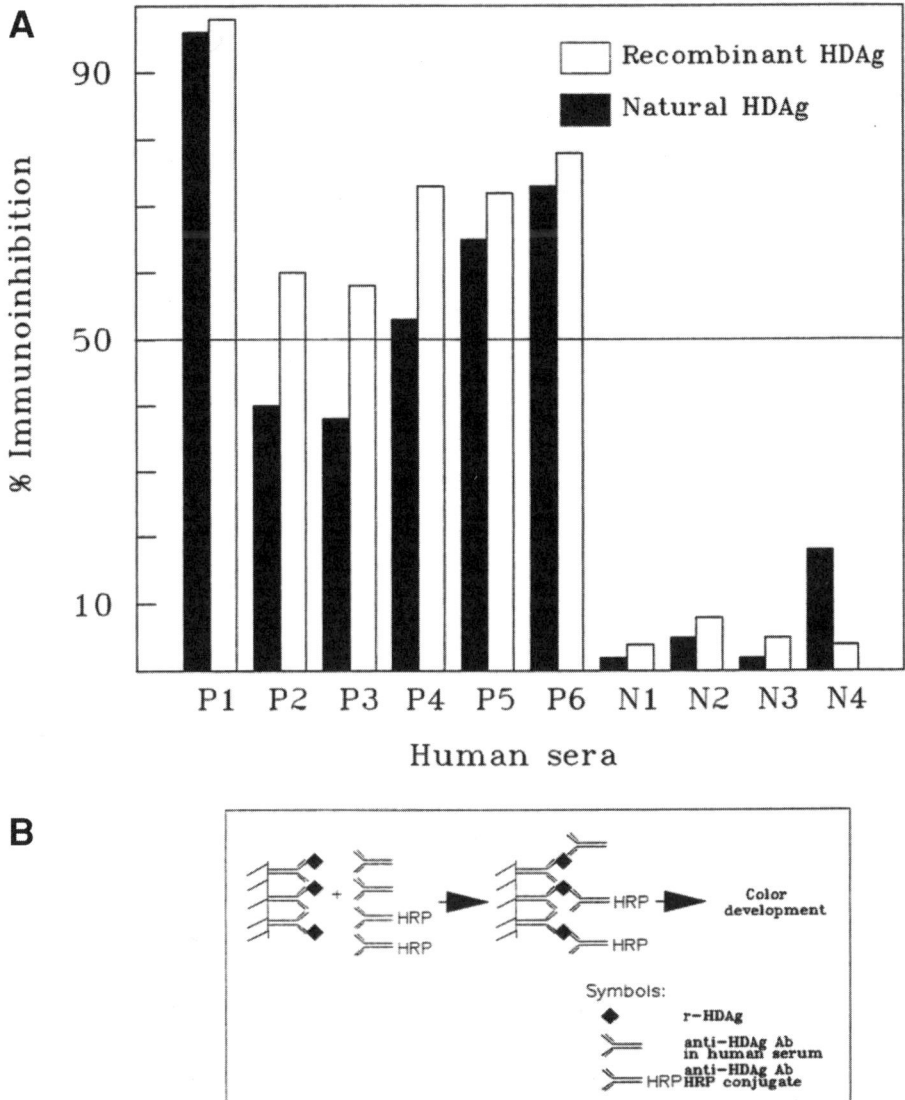

Fig. 8. The purified recombinant hepatitis D antigen (HDAg; closed bars) is tested by the ETI-AB-DELTA-K (SORIN, Saluggia, Italy) assay, in comparison with the natural antigen (open bars). **(A)** Recombinant antigen (closed bars) and natural antigen (open bars), from HDV-infected marmot liver, are used to analyze a group of anti-HDV-positive and -negative sera. The recombinant antigen shows an improved ability to recognize the HDV antibodies with respect to the natural antigen (from liver lysate), which fails to detect P2 and P3 sera. **(B)** Scheme of ELISA single-step immunoinhibition assay.

3. Histidil diazobenzil phosphonic acid agarose (Sigma-Aldrich).
4. Immobilon P nylon membrane filter (Millipore, Bedford, MA).
5. Maxisorb 96-well microtiter plate (Nunc, Kamstrup, Denmark).
6. Peroxidase-conjugated anti-human rabbit polyclonal antibodies (Sigma-Aldrich, D).
7. Certified anti-HCV panel sera (BBI, Boston, MA).
8. ETI-ABDELTA-K, ELISA single-step immunoinhibition kit (SORIN, I).
9. Luria-Bertani (LB) medium: 5 g yeast extract (Sigma-Aldrich), 10 g tryptone (Sigma-Aldrich), 10 g NaCl dissolved in 1 L distilled water, filter-sterilized on a 0.22-µm membrane.
10. Isopropyl thiogalactose (IPTG) solution: 100 mM IPTG in double-distilled water, filter-sterilized on an 0.22-µm membrane.
11. Solution A: 7 M guanidinium chloride, 0.1% trifluoroacetic acid (TFA), freshly prepared. **Caution:** TFA is toxic. In case of contact, immediately flush the skin with water.
12. Solution B: 70% CH_3CN, 29.9% H_2O, 0.1% TFA, freshly prepared. **Caution:** CH_3CN is toxic. In case of contact, immediately flush the skin with water.
13. Solution C: 10 mM Tris-acetate, pH 5.0, 1 mM EDTA, 1 mM phenylmethylsulfonic fluoride (PMSF), 8 M urea. **Caution:** PMSF is extremely destructive to the mucous membranes. It may be fatal if inhaled, swallowed, or absorbed through the skin. In case of contact, immediately flush the skin with water.
14. Starting buffer: 10 mM Tris-acetate, pH 5.0, 1 mM EDTA, 1 mM PMSF.
15. End buffer: 10 mM Tris-acetate, pH 5.0, 1 mM EDTA, 1 mM PMSF, 1 M NaCl.
16. Acetate buffer: 50 mM sodium acetate, pH 4.0.
17. Carbonate buffer: 50 mM sodium carbonate, pH 9.6.
18. Blocking reagent 1X: 3% non fat dry milk in PBS (137 mM NaCl, 2.68 mM KCl, 10.1 Na_2HPO_4, 1.7 mM KH_2PO_4, pH 7.4).
19. Phosphate-buffered saline (PBS)/Tween 20: PBS, 0.05% Tween 20.
20. ELISA chromogen/substrate solution: 0.1 mg/mL 2,2-azino-bis(3-ethylbenzthi-azoline-6-sulfonic acid) (ABTS), 0.014% H_2O_2, 50 mM sodium acetate, pH 4.0, freshly prepared.
21. SDS solution: 20% SDS dissolved in bidistilled water.
22. Affi-prep MAPSII binding buffer (Bio-Rad).
23. Affi-Prep MAPSII elution buffer (Bio-Rad).

3. Methods

The purification procedure, described in this section, is suitable for proteins having an isoelectric point higher than 8.0. We obtain at least 80% pure proteins, with yields ranging between 0.5 and 4 mg/L of culture, associated with guanidinium chloride cell disruption, organic precipitation, and ion exchange chromatography. Here we describe the purification procedures of three viral proteins (**Table 1**): HDAg, HCAg, and a fragment of HIV-1 Gag polyprotein, including all p24 sequence and 59 amino acids of the nucleocapsid protein (p24-NC).

3.1. Recombinant Antigen Expression in E. coli

1. Inoculate 100 µL glycerol stock aliquot of the corresponding recombinant bacterial cells in 10 mL LB supplemented with 50 µg/mL ampicillin and incubate under vigorous shaking at 30°C (*see* **Note 1**).
2. Inoculate 8 mL of the overnight culture in 400 mL LB supplemented with 50 µg/mL ampicillin and incubate under vigorous shaking at 30°C (*see* **Note 2**). When the cell density reaches an A_{590} of 0.5, the culture is transferred into a 42°C water bath (*see* **Note 3**) for 3 h.

3. Cool the bacterial culture to 4°C by immersion in melting ice for 10 min and subsequently centrifuge the culture (JA-14 rotor, 5000g, 30 min, 4°C). Discard the culture medium, weigh the bacterial pellet, and freeze it at –80°C.

3.2. Cells Lysis and Organic Precipitation

1. Dissolve 1 g of the bacterial paste in 4 mL of solution A and incubate at 70°C for 30 min (*see* **Note 4**). Dialyze a 0.2-mL aliquot of the lysis mixture against solution C. Following dialysis, discard the pellet by centrifugation (12,000g, 10 min) and store the supernatant at –20°C (*see* **Note 5**).
2. Add 4 mL of solution B to the lysis mixture and incubate at 4°C for 1 h (*see* **Note 6**).
3. Centrifuge the mixture (8 mL; JA-20 rotor, 10,000g, 1 h, 4°C). Discard the pellet and concentrate the solution to approx 4 mL in a vacuum device (*see* **Note 7**).
4. Dialyze the solution against 1 L of 2% acetic acid (4°C, 3 h) and discard the precipitate by centrifugation (JA-20 rotor, 10,000g, 1 h, 4°C).
5. Dialyze the protein solution (approx 8 mL) against starting buffer (*see* **Note 8**). Dialyze an aliquot of the protein solution (0.1 mL) and store it at –20°C for further evaluations.

3.3. Ionic Exchange Chromatography

1. Suspend 4 mL of the histidil diazophenil phosphonic acid-agarose in 20 mL of starting buffer. Remove the fine particles by changing the buffer three times.
2. Pack the histidil diazophenil phosphonic acid-agarose in a disposable 1.5 × 12-cm Econo-Pac column (*see* **Note 9**).
3. Connect the column to a low-pressure chromatography system and equilibrate the resin with starting buffer (flow rate = 0.5 mL/min; *see* **Note 10**).
4. Load the protein sample (approx 8 mL) on the column at a flow rate of 0.5 mL/min. Wash the column with the starting buffer until the baseline of the UV detector reaches zero.
5. Elute the protein by applying a 20-min linear NaCl gradient (0.4 M → 1.0 M; *see* **Note 11**) in starting buffer, followed by 20-min isocratic elution with end buffer (**Figs. 4–6**). The elution profile is obtained by reading the $A_{280\ nm}$ of the eluted fractions and superimposing their conducibility to the elution profile.
6. Store the fractions at –20°C.

3.4. Evaluation of Recombinant Antigen Purity and Specificity

1. Analyze the column fractions using standard 12% SDS polyacrylamide gel electrophoresis (SDS-PAGE) *(20)*.
2. The proteins are detected by Coomassie blue staining and Western blotting *(21,22)*, using human polyclonal antibodies (*see* **Note 12**), purified from HIV- HDV-, or HCV-positive sera pools, respectively. Pool the fractions containing at least 80% pure recombinant antigen (**Figs. 4–6**, dashed areas) and determine the protein concentration using the BCA assay. Store the corresponding pools at –20°C.
3. Analyze aliquots of the crude lysate (**Figs. 1–3**, lanes 2), of the material loaded on the column (**Figs. 4–6**, lane 3), and of the recombinant antigen-purified pool (**Figs. 1–3**, lane 4) using standard 12% SDS-PAGE. Subsequently, perform a densitometric analysis on the gels to evaluate the level of expression in the crude lysate and the purity of the recombinant antigen, following ion exchange chromatography (**Table 1**).

3.4.1. Purification of Human Polyclonal Antibodies

1. Dilute 10 mL of human sera pools with 20 mL of acetate buffer and adjust the pH to 4.8 with 1 M NaOH.

2. Add dropwise 0.76 mL of caprylic acid to the solution (under vigorous stirring) and incubate the solution at 25°C for 30 min. Subsequently, centrifuge the mixture (JA-20, 10,000g, 4°C, 10 min) and discard the pellet.
3. Dialyze the solution against 1 L PBS, changing the PBS buffer twice.
4. Perform the immunoglobin purification on 4 mL Affi-Prep protein A matrix, packed in a disposable 1.5 × 12-cm Econ-Pac column.
5. Dialyze the immunoglobulin solution against Affi-Prep MAPSII binding buffer and load it onto the column at flow rate of 1 mL/min. Wash the column with the Affi-Prep MAPSII binding buffer until the baseline of the UV detector reaches zero.
6. Elute the bound immunoglobulin with Affi-Prep MAPSII elution buffer and neutralize the eluted peak fraction with 1 M Tris-HCl, pH 8.5.
7. Dialyze the purified immunoglobulins against PBS and determine the protein concentration by the BCA assay. Store the purified immunoglobulins aliquots (10 mg/mL) at –20°C.

3.4.2. Indirect Enzyme Immunoassay

1. Coat a 96-well microtiter plate with purified antigen in carbonate buffer (5 µg/mL, 200 µL/well, 4°C, 18 h; *see* **Note 13**). Remove the coating buffer and add 200 µL 1X blocking reagent per well (37°C, 1 h). wash the microtiter plate 3 times with PBS, let it dry at room temperature, and store at –20°C until needed.
2. Add reference sera to coated microtiter plate (200 µL/well, 37°C, 1 h) and subsequently wash the plate three times with PBS/Tween 20.
3. Add 200 µL peroxidase-conjugated anti-human rabbit polyclonal antibodies (diluted 1:20,000 in 1X blocking reagent) per well and incubate at 37°C for 1 h (*see* **Note 14**). Subsequently, wash the plate three times with PBS/Tween 20.
4. Add 200 µL of ELISA chromogen/substrate solution to each well and incubate at room temperature for 30 min.
5. Block the color reaction by adding 20 µL of SDS solution per well.

4. Notes

1. For strains expressing recombinant HCcAg and recombinant p24-NC, growth is performed at 30°C. The strain expressing recombinant HDAg growth is cultured at 37°C as the promoter controlling the recombinant HDAg expression is not temperature-dependent. Recombinant HDAg expression is induced by adding IPTG at a final concentration of 1 mM.
2. It is important to use a triple-baffled 1-L shack flask to obtain optimal culture aeration.
3. The thermal slope is very important for good expression levels, and it should be very steep. A steep thermal slope can be obtained by incubating the 1-L shake flask in a 80°C water bath for 2–3 min and then transferring the flask to a 42°C water bath.
4. Before incubating the solution at 70°C, the bacterial paste must be dissolved completely by vortexing. During the 70°C incubation, the suspension must be mixed every 10 min for a 30-s period.
5. An aliquot of each purification step is stored to confirm that all passages were performed correctly.
6. After organic precipitation, the mixture can be stored at –80°C. A freeze/thaw cycle improves the precipitation of bacterial proteins and does not affect the recombinant protein.
7. The CH_3CN evaporation is necessary because the organic solvent could be incompatible with the material of the dialysis tubes.

8. In the case of HCcAg, the dialysis is done against starting buffer pH 6.0, because the protein does not bind to the column at pH 5.0.
9. The upper part of the bed resin is covered with a hydrophobic support, preventing the bed from being perturbed by the flow of the buffer and from running dry.
10. For HCcAg purification, the histidil diazophenil phosphonic acid-agarose is equilibrated in starting buffer pH 6.0.
11. For HCcAg the elution is performed by a 20-min linear NaCl gradient ($0\ M \rightarrow 1\ M$) in starting buffer at pH 6.0, followed by 20-min isocratic elution with end buffer pH 6.0.
12. Incubate the blotted filter (Immobilon P) in 20 mL of 1X blocking reagent for 1 h at 37°C and wash subsequently three times with 50 mL PBS/Tween 20 for 5 min at 25°C. Add anti-HCV, anti-HIV, or anti-HDV human polyclonal antibodies (10 µg/mL) to 20 mL of 1X blocking reagent (1 h, 37°C). Wash the filter as previously described. Add 20 mL of peroxidase-conjugated anti-human rabbit polyclonal antibodies (diluted 1:4000 in 1X blocking reagent) to the filter and incubate for 1 h at 25°C. Wash the filter as previously described. Add 50 mL of the substrate (0.006% H_2O_2 in PBS) and the chromogen (30 mg 4-chloronaphthol in 10 mL cold methanol) to the filter. Stop the color development by rinsing the filter in 50 mL of methanol.
13. The optimal antigen-coating concentration has to be determined empirically. The ideal antigen-coating concentration should allow the detection of a minor number of false positives without affecting true-positive detection.
14. The optimal secondary antibody working dilution has to be determined empirically to obtain the best positive/negative (P/N) values.

References

1. Kos, A., Molijn, A., Blauw, B., and Schellekens, H. (1991) Baculovirus-directed high level expression of the hepatitis delta antigen in *Spodoptera frugiperda* cells. *J. Gen. Virol.* **72,** 833–842.
2. Calogero R., Barbieri, U., Borla, M., Osborne, S., Poisson, F., and Bonelli, F. (1993) Purification of recombinant hepatitis delta antigen expressed in *E. coli* cells. *FEBS Lett.* **318,** 322–324.
3. Osborne, S., Cecconato, E., Griva, S., Garetto, F., Calogero, R., and Bonelli, F. (1993) Expression in *E. coli* and purification of a chimeric p22-NS3 recombinant antigen of hepatitis C virus (HCV). *FEBS Lett.* **324,** 253–257.
4. Schein, C. H. and Noteborn, M. H. M. (1998) Formation of soluble recombinant proteins in *E. coli* is favored by lower growth temperature. *Biotechnology* **6,** 291–294.
5. Babbitt, P. C., West B. L., Buechter, D. D., Kuntz, I. D., and Kenyon, G. L. (1990) Removal of a proteolytic activity associated with aggregates formed from expression of creatine kinase in *E. coli* leads to improved recovery of active enzyme. *Biotechnology* **8,** 945–949.
6. Cleland, J. L., Builder, S. E., Swartz, J. R., Winkler, M., Chang, J. Y., and Wang, D. I. (1992) Polyethylene glycol enhanced protein refolding. *Biotechnology* **10,** 1013–1019.
7. Calogero, R., Cecconato, E. Mariani, M., et al. (1989) Expression of HIV-1 gag gene in *E. coli* as a diagnostic test, in *Advanced in Applied Biotechnology Series*, vol. 7 (Papas, T. S., ed.), Gulf Publishing, Houston, TX, pp. 327–336.
8. Georgiou, G. and Bowden, G. A. (1991) Inclusion body formation and the recovery of aggregated recombinant proteins, in *Recombinant DNA Technology and Application* (Ho, C., Prokop, A., and Bajpai, R., eds.), McGraw-Hill, New York, pp. 333–351.

9. Hartley, D. and Kane, J. F. (1988) Properties of inclusion bodies from recombinant *E. coli*. Biochem. Soc. Trans. **16,** 101–102.
10. Cregg, J. M., Tschopp, J. F., Stillman, C., et al. (1987) High-level expression and efficient assembly of hepatitis B surface antigen in the methylothophic yeast, *Pichia pastoris*. *Biotechnology* **5,** 479–485.
11. Fraser, M. J. (1992) The baculovirus-infected insect cell as eukaryotic gene expression system, in *Current Topics in Microbiology and Immunology* (Muzyczka, N., ed.) Springer-Verlag, Berlin, pp. 131–172.
12. Mitraki, A. and King, J. (1989) Protein folding intermediates and inclusion body formation. *Biotechnology* **7,** 690–697.
13. Bowden, G. A., Paredes, A. M., and Georgiou, G. (1991) Structure and morphology of protein inclusion bodies in *E. coli*. *Biotechnology* **9,** 725–730.
14. Haase-Pettingell, C. A. and King, J. (1988) Formation of aggregates from a thermolabile in vivo folding intermediate in P22 tailspike maturation. *J. Biol. Chem.* **263,** 4977–4983.
15. Frankel, S., Sohn, R., and Leinwand, L. (1991) The use of sarkosyl in generating soluble protein after bacterial expression. *Proc. Natl. Acad. Sci. USA* **88,** 1192–1196.
16. Soutschek, E., Hoflacher, B., and Motz, M. (1990) Purification of a recombinantly produced transmembrane protein (gp41) of HIV-1. *J. Chromatogr.* **521,** 267–277.
17. Arakawa, T. and Timasheff, S. N. (1985) Theory of protein solubility. *Methods Enzymol.* **114,** 49–77.
18. Catt, K., Niall, H. D., and Tregear, G. W. (1967) Solid phase radioimmunoassay. *Nature* **213,** 825–827.
19. Catt, K. and Tregear, G. W. (1967) Solid phase radioimmunoassay in antibody-coated tubes. *Science* **158,** 1570–1573.
20. Laemmli, U. K. (1970) Cleavage of structural proteins during the assembly of the head of bacteriophage T4. Nature **227,** 680–685.
21. Towbin, H., Staehelin, T., and Gordon, J. (1979) Electrophoretic transfer of proteins from polyacrylamide gels to nitrocellulose sheets: procedure and some applications. *Proc. Natl. Acad. Sci. USA* **76,** 4350–4354.
22. Talbot, P. V., Knobler, R. L., and Buchmeier, M. (1984) Western blot analysis of viral antigens and antibodies: application to murine hepatitis virus. *J. Immunol. Methods* **73,** 177–188.

16

Refolding of Inclusion Body Proteins

Marcus Mayer and Johannes Buchner

Abstract

Genome sequencing projects have led to the identification of an enormous number of open reading frames that code for unknown proteins. Elucidation of the structure and function of these proteins makes it necessary to produce proteins fast, in high yields and at low cost. The recombinant expression of proteins in bacterial hosts often results in the formation of inclusion bodies. Here, the protein accumulates in large quantities separated from the cellular protein. However, the protein is insoluble and inactive. Thus, it is necessary to establish efficient refolding protocols. Progress has been made recently in this field concerning refolding strategies, the use of low-molecular-weight additives as folding enhancers, and the determination of optimum refolding parameters. Here we present an overview of the refolding technology and give a standard protocol for inclusion body refolding.

Key Words: Recombinant protein expression; protein renaturation; refolding parameters; protein overproduction; protein aggregation.

1. Introduction

The recombinant expression of genes is of great interest for both biotechnology and basic research in the context of protein function and structure. Thousands of DNA targets have been cloned and expressed, but 40% of them were expressed in a insoluble form *(1,2)*. Thus, soluble protein expression seems to be one of the main bottlenecks in this field. The most commonly used expression system is *Escherichia coli*. Protein production in bacteria offers several advantages over other, mostly eukaryotic, hosts such as short doubling times, well-established methods for genetic manipulation, and simple, inexpensive cultivation. However, the expressed proteins often accumulate in insoluble and inactive deposits called inclusion bodies. Inclusion body formation is a quite frequently observed phenomenon upon overexpression in *E. coli (3)*. Especially for eukaryotic proteins or proteins containing disulfide bonds in the native state, inclusion bodies are observed in a host organism like *E. coli*. The different biology of

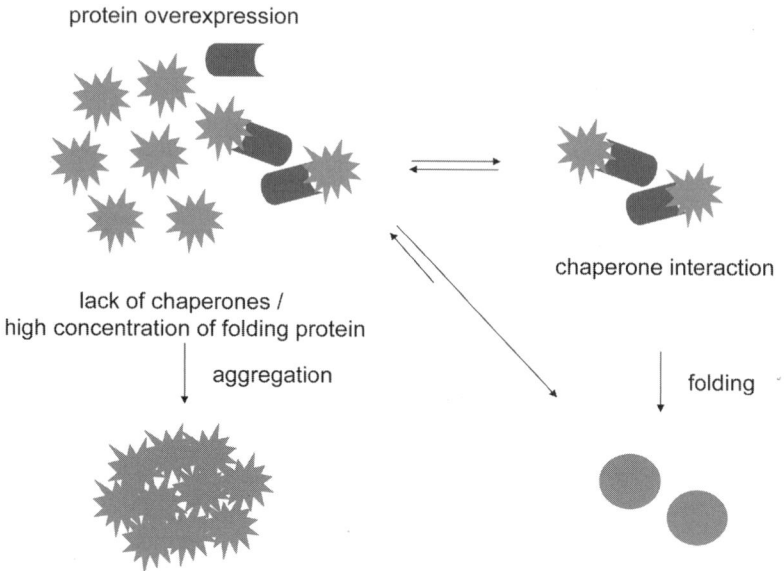

Fig. 1. Protein overexpression leads to protein aggregation owing to the lack of a sufficient amount of chaperones and the high concentration of folding intermediates.

prokaryotic protein production, especially the different codon usage, favors the accumulation of non-native folding intermediates. Furthermore, disulfide bonds cannot be formed in the reducing milieu of the *E. coli* cytoplasm. Therefore, proteins that require disulfide bonds for their stability cannot reach their native states.

During the folding process, starting from the unstructured polypeptide chain, secondary structure elements are formed first. These associate to the tertiary structure and, in the case of oligomeric proteins, the quaternary structure. For many proteins it has been observed, in addition to these intramolecular interactions, that the folding polypeptide chains can undergo unwanted intermolecular interactions that lead to the formation of aggregates. Especially when proteins are produced at unphysiologically high levels, folding competes with aggregation owing to the exposure of hydrophobic surfaces (**Fig. 1**) *(4)*. This is especially true in the case of proteolytically resistant proteins *(5)*. One important factor in the folding process is the protein concentration. It has been shown that proteins tend to aggregate with increasing concentration, whereas low protein concentrations favor the formation of the correctly folded protein *(6)*. This phenomenon has been quantified by a kinetic model with the yield of native protein as a function of the competition between folding and aggregation *(7)*. Thus, a decrease in protein synthesis rate will lead to an increase in the yield of functional protein. This was confirmed experimentally by recombinant protein expression under suboptimal conditions, i.e., the reduction of the cultivation temperature from 37 to 30°C in *E. coli* *(8)*. Inclusion body formation was also observed for *E. coli* proteins upon overexpression in their natural environment *(9)*. Thus, an important aspect for

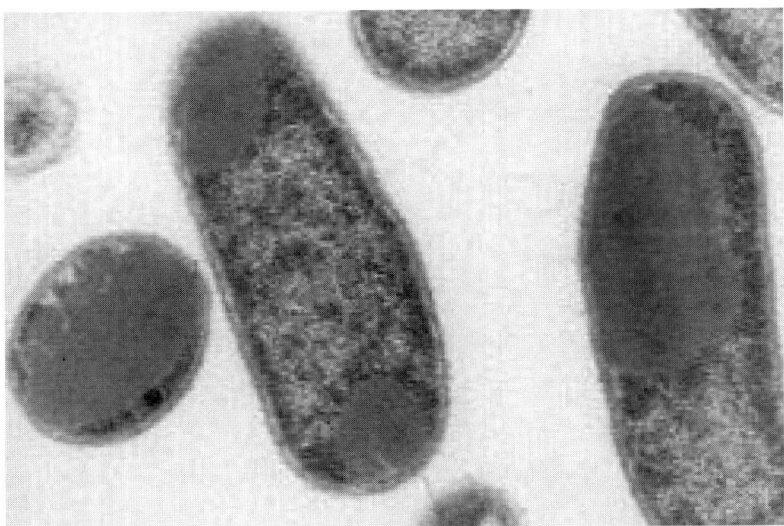

Fig. 2. *E. coli* cells containing inclusion bodies. The recombinant protein deposits are the dark spherical particles in the bacterial cytoplasm. They can amount to up to 50% of the volume of the bacteria.

the aggregation of recombinantly overproduced proteins is the imbalance in the ratio of chaperones and folding helpers versus the folding polypeptide *(10,11)*. In this context, it has been shown that heat shock genes coding for chaperones are overexpressed as a response to protein overproduction *(12,13)*.

At first glance, the deposition of the target protein in insoluble, inactive forms seems to be a failure and a dead end of the expression strategy. However, inclusion bodies also offer some advantages, as the overexpressed protein is often highly enriched and protected from proteolytic degradation *(14)*. In addition, the high-level expression of certain proteins in a soluble form would be toxic to the host organism in some cases.

Since the formation of inclusion bodies was first observed 25 years ago *(15)*, a number of folding protocols for different proteins have been developed. In general, for complex proteins (e.g., multidomain, oligomeric, or disulfide-bonded proteins), several bottlenecks that need to be overcome during the folding process have been identified. Based on these findings, it is possible to develop an efficient refolding strategy for a given protein by optimizing key parameters of the folding reaction.

2. Inclusion Body Morphology and Physiology

Inclusion bodies are very dense particles formed of aggregated protein. Inclusion bodies can reach diameters of up to the micrometer range and can therefore be seen under the light microscope (**Fig. 2**).

Interestingly, in many cases only two large inclusion bodies are observed in an *E. coli* cell. They show an amorphous or paracrystalline structure and contain almost exclusively the overexpressed protein *(16)*. It has been observed that heat shock proteins (Hsps) expressed in response to protein overproduction are tightly coupled to inclusion bodies. In *E. coli*, two inclusion body-associated proteins, termed IbpA and IbpB (for inclusion body protein A and B) have been found to be members of the small Hsp family *(17)*.

The concept of inclusion bodies as irreversibly aggregated proteins may not apply for all forms of aggregated proteins in *E. coli*. In the case of β-galactosidase, inclusion bodies can be redissolved in vivo *(18)*. After protein expression was arrested, inclusion bodies were found to dissociate, leading to the release of soluble and active protein. This implies that at least in this case there is an equilibrium between soluble and native protein and inclusion bodies as protein deposits.

Until now there is little information about the structural properties of inclusion bodies, but it is suggested that they possess at least some amount of secondary structure *(6,19)*. Contaminating proteins in inclusion body preparation are often not part of the inclusion bodies themselves but result from copurification of nonsolubilized membrane proteins. These contaminants can be separated effectively during inclusion body preparation.

3. Inclusion Body Preparation and Purification

The isolation of inclusion bodies is generally performed by centrifugation because of their relative high density (1.3 mg/mL) compared with other cellular components *(20)*. Only intact cells would pellet together with the inclusion bodies. After cell disruption by high-pressure dispersion and a subsequent lysozyme treatment, chromosomal DNA is digested by DNase treatment (10 μg/mL DNase I, 3 mM MgCl2, 30 min, 25°C). Then a detergent such as Triton X-100 at a concentration of 2% (v/v) is added to the lysate to solubilize lipids and membrane proteins *(21)*. The presence of high salt concentrations (e.g., 0.5 M NaCl) is helpful for solubilizing contaminating proteins. After incubation for 30 min at room temperature, the inclusion body pellet can be harvested by centrifugation (25,000g, 60 min, 4°C).

Subsequently, the inclusion body pellet is resuspended and homogenized in a buffer containing 20 mM EDTA to remove contaminants. This washing step is repeated several times, and the resulting pellet is homogenized well between the centrifugation steps to improve the purity of the inclusion body preparation. It should, however, be noted that the presence of contaminating proteins in the final inclusion body preparation does not prevent successful renaturation.

4. Protein Denaturation

Before starting renaturation, the inclusion bodies need to be solubilized by strong denaturants. Either 6 M guanidine hydrochloride (GdmCl) or 8 M urea are normally used. Although there are several different solubilization methods using lower denaturant concentrations in combination with high or low pH values, this method is the most appropriate starting point. The solubilization buffer should also contain a reducing agent, e.g., dithiothreitol (DTT) to keep cysteines reduced, even for proteins lacking disulfide bonds in the native state but containing free cysteines. The time of solubilization depends on the temperature and on the inclusion body concentration used. Normally, 2 h of incubation at room temperature at a protein concentration of 50 mg/mL should be sufficient. Protein concentrations can be determined by the Bradford assay *(22)*. Various buffer substances can be chosen for denaturation. However, for efficient reduction, an alkaline pH (>7.5) is required.

In the case of proteins containing disulfides, it can be useful to modify the denatured protein by creating mixed disulfides with glutathione. To this end, after denaturation, the protein is incubated with an excess of oxidized glutathione (GSSG). This results in the reversible derivatization of the cysteines with glutathione. After dialysis against 6 M GdmCl, the protein is ready for refolding. This procedure has been used successfully in the case of an antibody Fab fragment *(23)*. In this state the protein is more stable, and in the case of pulse renaturation (*see* **Subheading 5.**) the problem of increasing the concentration of reducing agent with every pulse added is avoided.

5. The Refolding Process

Renaturation requires the decrease of the high concentrations of denaturant and reducing agents. This is achieved by either dilution or a buffer-exchange step. The method most commonly used is to dilute the denatured protein in a large volume of refolding buffer (typically 100-fold). It is necessary to provide good mixing during dilution to prevent intermolecular interactions and aggregation.

The optimum protein concentration during refolding has to be determined experimentally for the target protein. In this context, aggregation can be monitored by the increase in absorbance of the solution after addition of the unfolded protein or by determining the soluble and insoluble fraction of the refolding solution by sodium dodecyl sulfate-polyacrylamide gel electrophoresis (SDS-PAGE) after centrifugation (25,000g, 60 min, 4°C). Because the solubilized inclusion body proteins are refolded typically at high dilution (e.g., 10–100 µg/mL), additional steps to increase the concentration after renaturation are required.

Continuous refolding methods have been developed *(24)* whereby the protein is gradually added to the renaturation buffer. Buffer exchange can also be done by diafiltration *(25)* and dialysis *(26)* using ultrafiltration membranes. These methods have the disadvantage that the refolding protein often sticks to membranes or dialysis tubes owing to hydrophobic interactions. Also size exclusion chromatography (SEC) has been used successfully for refolding. This method leads to an effective removal of the denaturant *(27)* including a first purification step, but it is not suitable for large volumes. Because of the limitations outlined above, dilution seems to be the method of choice for most proteins.

Adding the protein in multiple steps to the refolding buffer, known as *pulse renaturation*, has been shown to be successful in many cases *(16,28,29)*. This method allows us to reach higher end concentrations of protein without increasing the concentration of still unfolded and aggregation-prone protein to a level at which aggregation is favored. As outlined above, aggregation is strongly concentration-dependent, whereas folding is concentration-independent. Therefore, after dilution of the protein, aggregation is rapid at high protein concentrations, whereas reactivation is a much slower process. However, aggregation does not occur during the entire time course of

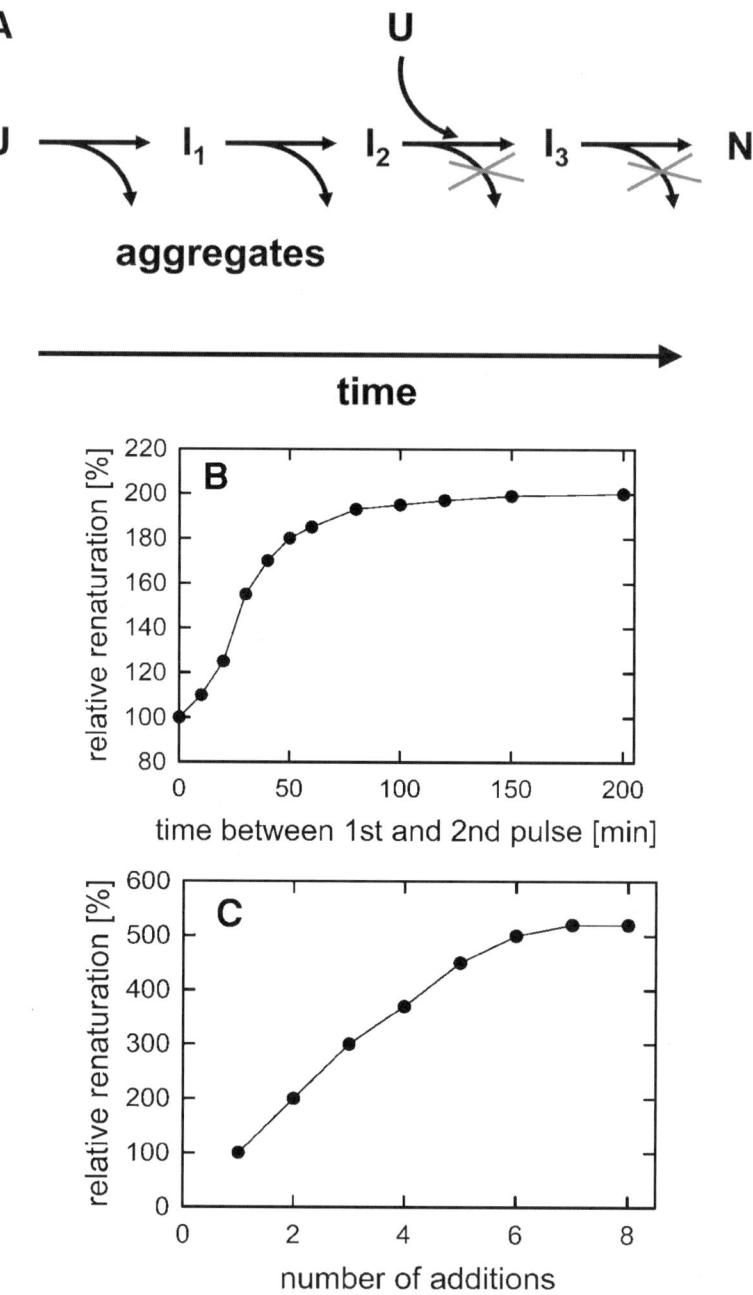

Fig. 3. *Pulse renaturation* enhances the yield of properly folded proteins. (**A**) By stepwise protein addition, the concentration of aggregation-sensitive intermediates (I_1) is kept low. Later intermediates (I_2, I_3) seem to be protected against unwanted interchain interactions and aggregation. (**B**) The time between the different pulses has to be analyzed for optimum

refolding. Aggregation-prone intermediates form early; further intermediates downstream of the folding pathway seem to be protected against interchain interactions leading to aggregation (**Fig. 3A**). Therefore, it is possible to keep the concentration of the aggregation-sensitive intermediate low by a stepwise addition. It is important to choose the right time between the pulses to prevent an accumulation of interchain interacting intermediates (**Fig. 3B**). This procedure allows us to reach high volume yields of the target protein. Of course, here also, an upper limit for the number of productive pulses exists (**Fig. 3C**). However, this number seems to be strongly dependent on the respective protein.

It is important to employ a functional assay for the protein of interest to test the successful regain of the native structure. These assays have to be performed under identical solvent conditions.

6. Optimizing Renaturation Conditions

Temperature is an important parameter that influences folding. In most cases, a higher refolding yield and less aggregation can be observed at lower temperatures *(23)*. This decreases the folding speed and hydrophobic interaction, whereas high temperatures favor aggregation *(30)*. The optimum refolding temperature must be determined experimentally for each target protein (**Fig. 4A**).

For the refolding of disulfide-bonded proteins, the presence of an adequate redox system has been shown to be essential for functional refolding (**Fig. 4B**). In many cases the formation of the correct disulfide bond is the limiting step during the folding process. This process can be influenced by the use of a suitable redox system consisting of oxidized and reduced thiol reagents, which allows disulfide reshuffling, i.e., wrong disulfide bonds will be reduced again because they are not protected by the correct structural context *(23,31–34)*. One of the most frequently used redox couples is oxidized and reduced glutathione (GSSG/GSH). However, in some cases, smaller redox couples seem to be more useful. For example, during renaturation of human proinsulin cystein/cystine seems to be the most favorable redox system *(35)*. The di-thiol reagent Vectrase™-P (BioVectra) also allows us to achieve high renaturation yields *(36)*.

To suppress aggregation, small chemical compounds have been used successfully for different target proteins. A very common additive is L-arginine (**Fig. 4C**). Although the effect of arginine in refolding proteins is not yet understood in detail, it is assumed that arginine destabilizes non-native conformations *(37)*. Provided that the correctly structured species is not affected, the addition of high concentrations of arginine allows

---

refolding. In this example, the maximum renaturation yield for one sample is set to 100%. Doubling of the amount of unfolded protein does not result in an increase in refolding. However, when the protein is added at later time points, an increase in refolding is observed. After 100 min, the refolding yields correspond to those obtained in two independent reactions performed at low protein concentrations. (**C**) First pulses can be done with nearly no loss of renaturation yield. (Refolding yield of first pulse is normalized to 100%.) Later on, *pulse renaturation* also reaches its limits.

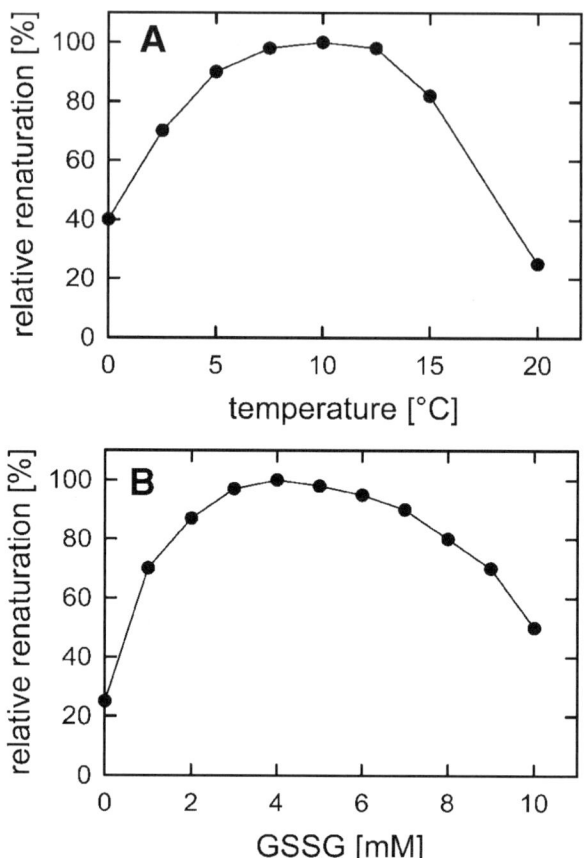

Fig. 4. Optimum folding conditions must be determined experimentally. The optimum curves for four important folding parameters, the temperature (**A**), the redox conditions, i.e., the concentration of [oxidized glutathione (GSSG; reduced glutathione GSH) concentration kept constant at 5 mM] (**B**), the concentration of L-arginine (**C**), and the pH (**D**) are shown for a certain target protein.

reshuffling of molecules trapped in nonproductive side reactions *(37)*. Other low-molecular-weight additives are denaturants such as urea or GdmCl and detergents like SDS, 3-[(3-Cholamidopropyl)dimethylammonio]-1-propane sulfonate (CHAPS) or Triton X-100. Their effects on refolding and the concentration range in which they can be used have been reviewed recently *(29)*. In addition, nondetergent sulfobetaines prevent aggregation and improve refolding effectively *(38)*. A large number of new nondetergent sulfobetaines have been designed that are efficient in improving renaturation. In the refolding of the human p53 tumor suppressor protein, the sulfobetain 3-(1-pyridinio)-1-propanesulfonate (PPS) has been used successfully *(39)*.

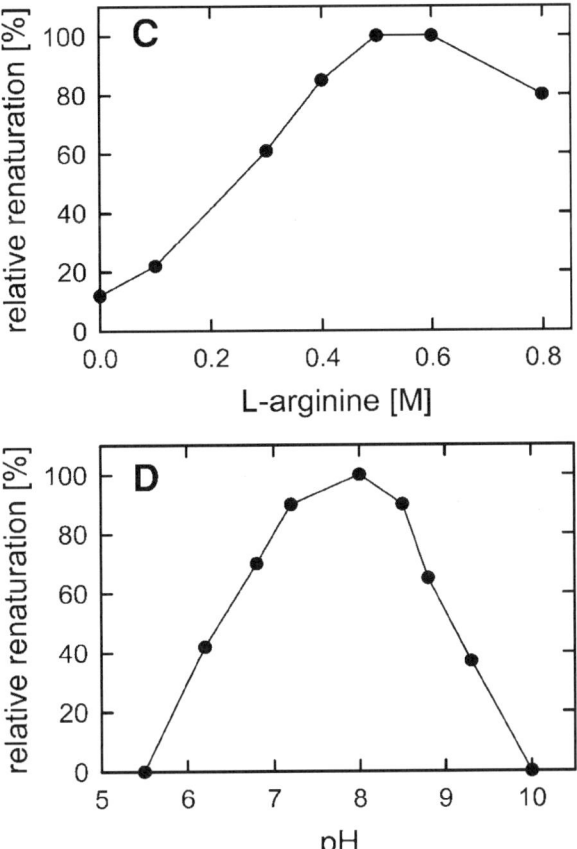

Fig. 4. *(continued).*

The ionic liquid ethylammonium nitrate (EAN) has been shown to have a positive effect on the refolding of denatured lysozyme, with renaturation yields reaching 90%. The ethyl group of EAN interacts with the hydrophobic portion of the protein and protects it from intermolecular association, whereas the charged portion of the salt stabilizes the electrostatic interactions of its secondary structure *(40)*.

The pH of the solution is also an important factor. All proteins have a characteristic pH range at which they can fold efficiently and reach their active state (**Fig. 4D**). For proteins containing disulfide bonds, an alkaline pH is necessary to allow formation and reshuffling of disulfide bonds.

It should be noted here that it is not possible to give a general recipe or a set of parameters optimal for any protein. Finding out the best conditions for refolding of a specific protein can be complicated and laborious. If one wanted to vary 12 parameters systematically, which is not necessary for many proteins, this would require 4096 experiments. Therefore, factorial matrix screens for determining the optimum renaturation conditions have been developed *(41,42)*. Using only a fraction of the full factorial is sufficient to determine multifactor interactions *(43)*. Having found the parameters that have the greatest impact on the refolding yield, the folding conditions can be optimized using a subset of the original folding conditions. Although interactions of different folding parameters are neglected, and it is assumed that the refolding yield depends linearly on the change of the folding factor, which must not be true in all cases, the use of such a fractional factorial design for determining the optimum folding conditions is a helpful tool.

7. Additional Strategies

Refolding of denatured proteins on a column may be a useful option. Typically, this strategy is used for fusion proteins containing suitable tags for affinity binding to a column. Binding occurs under denaturing conditions, and refolding is initiated by going from denaturing to native conditions, using flat gradients and slow flow rates. After refolding, the protein is eluted. Matrix-assisted refolding offers some advantages, i.e., aggregation is prevented by the binding to a column, the target protein already undergoes a first purification step, and the protein can be concentrated by elution.

The most common fusion tag still binding under denaturing conditions is the hexahistidine tag for metal chelate affinity chromatography using a nickel-nitrilotriacetic acid (Ni^{2+}-NTA) resin. Matrix-assisted refolding has been shown to be the method of choice, for example, in the case of polyhydroxyalkanoate synthase from *Pseudomonas aeruginosa (44)*. Other tags have also been used successfully. Using the cellulose binding domain from *Clostridium thermocellum* as a fusion tag, which retains its specific cellulose binding capability up to 6 M urea, a single-chain Fv antibody has been refolded efficiently *(45)*. Polyionic peptides are used as fusion tags for the purification and refolding of proteins *(46)*. For the purification of a viral coat protein, a tag containing eight glutamic acid residues was used that allowed efficient refolding and purification by ion-exchange chromatography *(46,47)*.

A new approach to protein refolding is folding proteins inside a micelle. Micelles provide a separate folding volume for the protein in which intermolecular interactions and aggregation are suppressed and therefore higher protein concentrations during folding can be achieved. As a model, denatured RNase A was refolded in reversed micelles formulated with sodium di-2-ethylhexyl sulfosuccinate in isooctane *(48)*. The denatured RNase A was dissolved into the reversed micellar solution under ultrasonic irradiation. With this method, a high concentration of protein during refolding could be achieved. At 4.8 mg/mL RNase A, a renaturation yield of 100% was achieved, and after recovery from the micelles almost 60% of active protein was obtained.

8. Molecular Chaperones and Folding Helpers

In vivo, molecular chaperones, protein disulfide isomerases, and peptidyl prolyl isomerases are known to prevent protein aggregation and enhance the correct folding and oxidation. This has to be a highly effective process, considering the very unfavorable folding conditions in vivo (e.g., high protein concentrations, high temperature) and the severe impact of misfolding and aggregation on the metabolism. The principal property of molecular chaperones is their ability to bind unfolded or partially folded polypeptides and therefore to suppress aggregation effectively. The low specificity of the hydrophobic interaction and the conformational flexibility of folding intermediates ensures that chaperones bind to a large variety of polypeptides *(49)*. The in vivo coexpression of chaperones increased the solubility of recombinantly expressed proteins *(50–53)*. The overproduction of GroEL/ES and DnaK in particular has proved to be helpful *(54)*. Protein disulfide isomerase (PDI), when it was coexpressed, had an positive effect on the expression of an antibody Fab fragment *(55)* and bovine pancreatic trypsin inhibitor *(56)*. However, in a number of cases coexpression of a restricted number of chaperones was not successful. For proper folding, a whole network of chaperones and other folding helpers as it is present and active in vivo could be necessary. Nevertheless, it is tempting to test molecular chaperones for optimizing the refolding of proteins in vitro. In the refolding of an antibody fragment, it had been shown that the ER Hsp70 homolog BiP and PDI have an cooperative effect *(57)*. Because of high costs, the use of these proteins will be restricted to small analytical refolding experiments and will probably not be suitable for large-scale purposes if the chaperone components cannot be recycled. In principle, however, it seems possible to reuse immobilized chaperone compounds for refolding in vitro *(58)*.

9. Purification of Refolded Proteins

In most cases studied, the renaturation yield does not depend on the purity of the denatured protein. Therefore, it is not necessary to purify the protein under denaturing conditions prior to refolding except in special cases, e.g., when degradation during refolding is observed.

After refolding, protein purification is performed under native conditions using conventional techniques such as ion exchange, size exclusion, or hydrophobic interaction chromatography. Often, the isolated and refolded protein from inclusion body preparations is quite pure so that further purification may be limited to one or two steps. Incorrectly folded species can be easily separated from the native protein because they differ in the exposure of hydrophobic side chains and the charge distribution, among others.

10. A Standard Protocol for Protein Refolding

Importantly, the critical parameters of the refolding process have to be varied for each unfolded protein. Taking this into account and knowing the effects of the different parameters discussed above, the following procedure can be seen as a basic protocol and a guideline for optimizing the renaturation process. **Figure 5** shows a schematic overview of the different steps involved.

In most cases dilution either in a single step or in a pulse renaturation seems to be the most efficient and reproducible way to dilute the denaturant and establish refolding conditions. The protein concentration is normally kept low (10–100 µg/mL). Renaturation conditions must be carefully optimized regarding solvent parameters like temperature, pH, and ionic strength. Refolding is performed at temperatures between

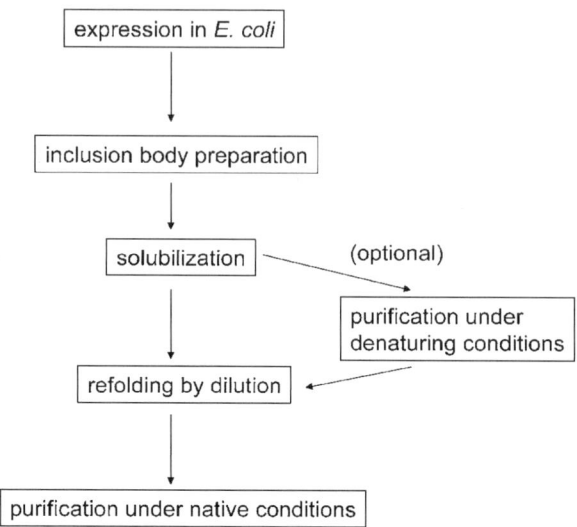

Fig. 5. Flowchart of a standard protocol for inclusion body preparation and refolding.

5 and 15°C. It is helpful to perform small-scale screening experiments to find the optimum buffer system for refolding.

For proteins containing disulfide bonds, a redox system should be present, because oxidation by air even in the presence of a metal catalysator is not efficient enough in most cases. Low-molecular-weight thiol reagents like glutathione or cysteine are suitable for this purpose. The ratios of reduced to oxidized compounds should be between 3:1 and 1:1, and total thiol concentrations of 5–15 mM were found to be optimal *(23,29,59,60)*. The addition of low-molecular-weight compounds like the popular amino acid L-arginine can suppress aggregation *(58)*. In some cases, the presence of specific cofactors can be required. For example, refolding of the human p53 tumor suppressor protein requires the presence of Zn^{2+} for effective renaturation *(39)*.

In the case of pulse renaturation, it must be considered that the final concentration of denaturant increases with every pulse added. The increase in denaturant concentration is the main limitation in achieving higher protein concentrations with this method. The concentration of reducing agent is also increased with every pulse added. Therefore, GSSG must be added with every pulse to keep the redox conditions equal. In some cases it may be advantageous to overcome the problem by producing mixed disulfides prior to renaturation, as outlined above.

11. Perspectives

The main obstacles to expressing recombinant proteins in a prokaryotic host like *E. coli* are the inability to perform post-translational modifications, the lack of chaperones in the case of protein expression at unphysiological high amounts, and the redox conditions in the bacterial cytoplasm, which are unfavorable for forming disulfide bonds.

Several strategies are emerging that may allow us to avoid inclusion body formation. There have been successful attempts to overcome the problem of the reducing environment of the cytoplasm by using *E. coli* knockout strains lacking parts of the glutharedoxin or thioredoxin system *(61,62)*. These knockout strains show a less reducing milieu in the cytoplasm and allow the formation of structurally important disulfide bonds in some model proteins tested. It remains to be seen whether these strains, combined with appropriate chaperones, will allow us to achieve the protein yields so far obtained with expression in inclusion bodies.

The coexpression of chaperones seems to be a helpful tool in achieving soluble protein expression. Different experiments have been performed with coexpressing chaperones and other folding helpers that led to an increased yield in soluble and active protein *(11,55,56,63)*. Nevertheless, recombinant expression of proteins resulting in inclusion bodies and subsequent refolding will still remain an important method, especially for large-scale applications. Although the optimum conditions for protein renaturation differ from protein to protein, it is possible to find suitable conditions for the target protein by the strategies outlined above and by considering the growing knowledge of the principles of protein folding and aggregation. Ongoing studies in this field will further improve existing refolding strategies. Thus, recombinant expression of proteins in inclusion bodies offers an efficient means of producing protein for structural therapeutic and industrial applications.

References

1. Service, R. F. (2002) Structural genomics. Tapping DNA for structures produces a trickle. *Science* **298,** 948–950.
2. Christendat, D., Yee, A., Dharamsi, A., et al. (2000) Structural proteomics of an archaeon. *Nat. Struct. Biol.* **7,** 903–909.
3. Marston, F. A. (1986) The purification of eukaryotic polypeptides synthesized in *Escherichia coli. Biochem. J.* **240,** 1–12.
4. King, J., Haase-Pettingell, C., Robinson, A. S., Speed, M., and Mitraki, A. (1996) Thermolabile folding intermediates: inclusion body precursors and chaperonin substrates. *FASEB J.* **10,** 57–66.
5. Corchero, J. L., Viaplana, E., Benito, A., and Villaverde, A. (1996) The position of the heterologous domain can influence the solubility and proteolysis of beta-galactosidase fusion proteins in *E. coli. J. Biotechnol.* **48,** 191–200.
6. Zettlmeissl, G., Rudolph, R., and Jaenicke, R. (1979) Reconstitution of lactic dehydrogenase. Noncovalent aggregation vs. reactivation. 1. Physical properties and kinetics of aggregation. *Biochemistry* **18,** 5567–5571.
7. Kiefhaber, T., Rudolph, R., Kohler, H. H., and Buchner, J. (1991) Protein aggregation in vitro and in vivo: a quantitative model of the kinetic competition between folding and aggregation. *Biotechnology (NY)* **9,** 825–829.
8. Schein, C. H. and Noteborn, M. H. M. (1988) Formation of soluble recombinant proteins in *Escherichia coli* is favored by lower growth temperature. *Biotechnology* **6,** 291–294.
9. Ceciliani, F., Caramori, T., Ronchi, S., Tedeschi, G., Mortarino, M., and Galizzi, A. (2000) Cloning, overexpression, and purification of *Escherichia coli* quinolinate synthetase. *Protein Expr. Purif.* **18,** 64–70.
10. Lorimer, G. H. (1996) A quantitative assessment of the role of the chaperonin proteins in protein folding in vivo. *FASEB J.* **10,** 5–9.

11. Thomas, J. G. and Baneyx, F. (1996) Protein misfolding and inclusion body formation in recombinant *Escherichia coli* cells overexpressing heat-shock proteins. *J. Biol. Chem.* **271**, 11,141–11,147.
12. Goff, S. A. and Goldberg, A. L. (1985) Production of abnormal proteins in *E. coli* stimulates transcription of lon and other heat shock genes. *Cell* **41**, 587–595.
13. Jurgen, B., Lin, H. Y., Riemschneider, S., et al. (2000) Monitoring of genes that respond to overproduction of an insoluble recombinant protein in *Escherichia coli* glucose-limited fed-batch fermentations. *Biotechnol. Bioeng.* **70**, 217–224.
14. Cheng, Y. S., Kwoh, D. Y., Kwoh, T. J., Soltvedt, B. C., and Zipser, D. (1981) Stabilization of a degradable protein by its overexpression in *Escherichia coli*. *Gene* **14**, 121–130.
15. Prouty, W. F., Karnovsky, M. J., and Goldberg, A. L. (1975) Degradation of abnormal proteins in *Escherichia coli*. Formation of protein inclusions in cells exposed to amino acid analogs. *J. Biol. Chem.* **250**, 1112–1122.
16. Lilie, H., Schwarz, E., and Rudolph, R. (1998) Advances in refolding of proteins produced in *E. coli*. *Curr. Opin. Biotechnol.* **9**, 497–501.
17. Allen, S. P., Polazzi, J. O., Gierse, J. K., and Easton, A. M. (1992) Two novel heat shock genes encoding proteins produced in response to heterologous protein expression in *Escherichia coli*. *J. Bacteriol.* **174**, 6938–6947.
18. Carrio, M. M. and Villaverde, A. (2001) Protein aggregation as bacterial inclusion bodies is reversible. *FEBS Lett.* **489**, 29–33.
19. Oberg, K., Chrunyk, B. A., Wetzel, R., and Fink, A. L. (1994) Nativelike secondary structure in interleukin-1 beta inclusion bodies by attenuated total reflectance FTIR. *Biochemistry* **33**, 2628–2634.
20. Mukhopadhyay, A. (1997) Inclusion bodies and purification of proteins in biologically active forms. *Adv. Biochem. Eng. Biotechnol.* **56**, 61–109.
21. Rudolph, R., Böhm, G., Lilie, H., and Jaenicke, R. (1997) Folding proteins, in *Protein Function, a Practical Approach* (Creighton, T. E., ed.), IRL Press, Oxford, pp. 57–99.
22. Bradford, M. M. (1976) A rapid and sensitive method for the quantitation of microgram quantities of protein utilizing the principle of protein-dye binding. *Anal. Biochem.* **72**, 248–254.
23. Buchner, J. and Rudolph, R. (1991) Renaturation, purification and characterization of recombinant Fab-fragments produced in *Escherichia coli*. *Biotechnology (NY)* **9**, 157–162.
24. Katoh, S. and Katoh, Y. (2000) Continuous refolding of lysozyme with fed-batch addition of denatured protein solution. *Process Biochem.* **35**, 1119–1124.
25. Varnerin, J. P., Smith, T., Rosenblum, C. I., et al. (1998) Production of leptin in *Escherichia coli*: a comparison of methods. *Protein Expr. Purif.* **14**, 335–342.
26. West, S. M., Chaudhuri, J. B., and Howell, J. A. (1998) Improved protein refolding using hollow-fibre membrane dialysis. *Biotechnol. Bioeng.* **57**, 590–599.
27. Fahey, E. M., Chaudhuri, J. B., and Binding, P. (2000) Refolding and purification of a urokinase plasminogen activator fragment by chromatography. *J. Chromatogr. B Biomed. Sci. Appl.* **737**, 225–235.
28. Buchner, J., Pastan, I., and Brinkmann, U. (1992) A method for increasing the yield of properly folded recombinant fusion proteins: single-chain immunotoxins from renaturation of bacterial inclusion bodies. *Anal. Biochem.* **205**, 263–270.
29. De Bernardez, C. E., Schwarz, E., and Rudolph, R. (1999) Inhibition of aggregation side reactions during in vitro protein folding. *Methods Enzymol.* **309**, 217–236.
30. Chalmers, J. J., Kim, E., Telford, J. N., et al. (1990) Effects of temperature on *Escherichia coli* overproducing beta-lactamase or human epidermal growth factor. *Appl. Environ. Microbiol.* **56**, 104–111.

31. Rajesh Singh, R. and Appu Rao, A. G. (2002) Reductive unfolding and oxidative refolding of a Bowman-Birk inhibitor from horsegram seeds (*Dolichos biflorus*): evidence for "hyperreactive" disulfide bonds and rate-limiting nature of disulfide isomerization in folding. *Biochim. Biophys. Acta* **1597,** 280–291.
32. Thies, M. J., Talamo, F., Mayer, M., et al. (2002) Folding and oxidation of the antibody domain C(H)3. *J. Mol. Biol.* **319,** 1267–1277.
33. Chatrenet, B. and Chang, J. Y. (1992) The folding of hirudin adopts a mechanism of trial and error. *J. Biol. Chem.* **267,** 3038–3043.
34. Wetlaufer, D. B., Branca, P. A., and Chen, G. X. (1987) The oxidative folding of proteins by disulfide plus thiol does not correlate with redox potential. *Protein Eng.* **1,** 141–146.
35. Winter, J., Klappa, P., Freedman, R. B., Lilie, H., and Rudolph, R. (2002) Catalytic activity and chaperone function of human protein-disulfide isomerase are required for the efficient refolding of proinsulin. *J. Biol. Chem.* **277,** 310–317.
36. Woycechowsky, K. J., Wittrup, K. D., and Raines, R. T. (1999) A small-molecule catalyst of protein folding in vitro and in vivo. *Chem. Biol.* **6,** 871–879.
37. Jaenicke, R. and Rudolph, R. (1989) Folding proteins, in *Protein Structure, a Practical Approach* (Creighthon, T. E., ed.), IRL Press, Oxford, pp. 191–223.
38. Goldberg, M. E., Expert-Bezancon, N., Vuillard, L., and Rabilloud, T. (1996) Non-detergent sulphobetaines: a new class of molecules that facilitate in vitro protein renaturation. *Fold. Des* **1,** 21–27.
39. Bell, S., Hansen, S., and Buchner, J. (2002) Refolding and structural characterization of the human p53 tumor suppressor protein. *Biophys. Chem.* **96,** 243–257.
40. Summers, C. A. and Flowers, R. A. (2000) Protein renaturation by the liquid organic salt ethylammonium nitrate. *Protein Sci.* **9,** 2001–2008.
41. Armstrong, N., de Lencastre, A., and Gouaux, E. (1999) A new protein folding screen: application to the ligand binding domains of a glutamate and kainate receptor and to lysozyme and carbonic anhydrase. *Protein Sci.* **8,** 1475–1483.
42. Chen, G. Q. and Gouaux, E. (1997) Overexpression of a glutamate receptor (GluR2) ligand binding domain in *Escherichia coli*: application of a novel protein folding screen. *Proc. Natl. Acad. Sci. USA* **94,** 13431–13436.
43. Box, G. E. P., Hunter, W. G., and Hunter, J. S. (1978) *Statistics for Experimenters*. John Wiley & Sons, New York.
44. Rehm, B. H., Qi, Q., Beermann, B. B., Hinz, H. J., and Steinbuchel, A. (2001) Matrix-assisted in vitro refolding of *Pseudomonas aeruginosa* class II polyhydroxyalkanoate synthase from inclusion bodies produced in recombinant *Escherichia coli*. *Biochem. J.* **358,** 263–268.
45. Berdichevsky, Y., Lamed, R., Frenkel, D., et al. (1999) Matrix-assisted refolding of single-chain Fv-cellulose binding domain fusion proteins. *Protein Expr. Purif.* **17,** 249–259.
46. Stempfer, G., Holl-Neugebauer, B., Kopetzki, E., and Rudolph, R. (1996) A fusion protein designed for noncovalent immobilization: stability, enzymatic activity, and use in an enzyme reactor. *Nat. Biotechnol.* **14,** 481–484.
47. Stubenrauch, K., Bachmann, A., Rudolph, R., and Lilie, H. (2000) Purification of a viral coat protein by an engineered polyionic sequence. *J. Chromatogr. B Biomed. Sci. Appl.* **737,** 77–84.
48. Goto, M., Hashimoto, Y., Fujita, T., Ono, T., and Furusaki, S. (2000) Important parameters affecting efficiency of protein refolding by reversed micelles. *Biotechnol. Prog.* **16,** 1079–1085.

49. Walter, S. and Buchner, J. (2002) Molecular chaperones—cellular machines for protein folding. *Angew. Chem. Int. Ed.* **41,** 1098–1113.
50. Levy, R., Weiss, R., Chen, G., Iverson, B. L., and Georgiou, G. (2001) Production of correctly folded Fab antibody fragment in the cytoplasm of *Escherichia coli* trxB gor mutants via the coexpression of molecular chaperones. *Protein Expr. Purif.* **23,** 338–347.
51. Nishihara, K., Kanemori, M., Kitagawa, M., Yanagi, H., and Yura, T. (1998) Chaperone coexpression plasmids: differential and synergistic roles of DnaK-DnaJ-GrpE and GroEL-GroES in assisting folding of an allergen of Japanese cedar pollen, Cryj2, in *Escherichia coli*. *Appl. Environ. Microbiol.* **64,** 1694–1699.
52. Nishihara, K., Kanemori, M., Yanagi, H., and Yura, T. (2000) Overexpression of trigger factor prevents aggregation of recombinant proteins in *Escherichia coli*. *Appl. Environ. Microbiol.* **66,** 884–889.
53. Goloubinoff, P., Gatenby, A. A., and Lorimer, G. H. (1989) GroE heat-shock proteins promote assembly of foreign prokaryotic ribulose bisphosphate carboxylase oligomers in *Escherichia coli*. *Nature* **337,** 44–47.
54. Thomas, J. G., Ayling, A., and Baneyx, F. (1997) Molecular chaperones, folding catalysts, and the recovery of active recombinant proteins from *E. coli*. To fold or to refold. *Appl. Biochem. Biotechnol.* **66,** 197–238.
55. Humphreys, D. P., Weir, N., Lawson, A., Mountain, A., and Lund, P. A. (1996) Co-expression of human protein disulphide isomerase (PDI) can increase the yield of an antibody Fab' fragment expressed in *Escherichia coli*. *FEBS Lett.* **380,** 194–197.
56. Ostermeier, M., De Sutter, K., and Georgiou, G. (1996) Eukaryotic protein disulfide isomerase complements *Escherichia coli* dsbA mutants and increases the yield of a heterologous secreted protein with disulfide bonds. *J. Biol. Chem.* **271,** 10,616–10,622.
57. Mayer, M., Kies, U., Kammermeier, R., and Buchner, J. (2000) BiP and PDI cooperate in the oxidative folding of antibodies in vitro. *J. Biol. Chem.* **275,** 29421–29425.
58. Buchner, J., Brinkmann, U., and Pastan, I. (1992) Renaturation of a single-chain immunotoxin facilitated by chaperones and protein disulfide isomerase. *Biotechnology (NY)* **10,** 682–685.
59. Tran-Moseman, A., Schauer, N., and De Bernardez, C. E. (1999) Renaturation of *Escherichia coli*-derived recombinant human macrophage colony-stimulating factor. *Protein Expr. Purif.* **16,** 181–189.
60. Wetlaufer, D. B. and Xie, Y. (1995) Control of aggregation in protein refolding: a variety of surfactants promote renaturation of carbonic anhydrase II. *Protein Sci.* **4,** 1535–1543.
61. Prinz, W. A., Aslund, F., Holmgren, A., and Beckwith, J. (1997) The role of the thioredoxin and glutaredoxin pathways in reducing protein disulfide bonds in the *Escherichia coli* cytoplasm. *J. Biol. Chem.* **272,** 15,661–15,667.
62. Stewart, E. J., Aslund, F., and Beckwith, J. (1998) Disulfide bond formation in the *Escherichia coli* cytoplasm: an in vivo role reversal for the thioredoxins. *EMBO J.* **17,** 5543–5550.
63. Cole, P. A. (1996) Chaperone-assisted protein expression. *Structure* **4,** 239–242.

17

Small-Molecule–Protein Conjugation Procedures

Stephen Thompson

Abstract

Small-molecule–protein conjugates are often required to act as immunogeneic complexes in the production of both monoclonal and polyclonal antibodies against small antigens. When antibodies have been obtained, they (and/or the small antigens) need to be labeled to facilitate their use in diagnostic assays. It is often impossible or extremely expensive to obtain the required conjugates. This chapter therefore discusses the common procedures used to couple small molecules to proteins and the analysis of the resulting conjugates. Practical guidance is given on the coupling of small molecule carboxyl, hydroxyl, and amine residues to amine and sulfhydryl residues on proteins using linkage techniques in which the author has extensive experience. Although a comprehensive list and analysis of every available linker is not given, the practical advice should enable the reader to use any commercially available linker productively to its optimum potential.

Key Words: Proteins, labeling; coupling; conjugation; enzymes; alkaline phosphatase; biotin; complexes; small antigens.

1. Introduction

The conjugation of small molecules (SM) to proteins is required in many areas of medical diagnostics and research. Many molecules are simply too small to elucidate an immune response, and they have to be coupled to a carrier protein such as bovine serum albumin (BSA) or keyhole limpet hemocyanin (KLH) in order to form antigenic conjugates. Even after polyclonal (or monoclonal) antibodies have been raised to the small molecule, similar SM–enzyme conjugates are required for competitive immunoassays in which the antibodies are used to coat an enzyme-linked immunosorbent assay (ELISA) plate. Alternatively, the SM–protein conjugates can be used to coat the ELISA plate, but then the detecting antibody has to be labeled. Indeed, labeled antibodies and antigens are required in the vast majority of the diagnostic techniques discussed in this book.

It is often impossible or extremely expensive to obtain the required conjugates. This chapter therefore concentrates on the general procedures (**Subheading 3.**) that can be used to couple SM carboxyl (**Subheading 3.1.1.**), hydroxyl (**Subheading 3.1.2.**), and amine (**Subheading 3.1.3.**) residues to proteins and the analysis of the resulting conjugates. The coupling of biotin to proteins is specifically demonstrated in **Subheading 3.1.1.** Biotinylated antigens and antibodies are often used in medical diagnosis, as biotin can easily be further complexed with commercial avidin and streptavidin conjugates.

Protein–protein conjugation is sometimes required to elucidate an effective immune response. A large diagnostic antibody, marker, or protein may not be immunogeneic, as it is a natural product. However, KLH can be conjugated to it to increase its immunogenicity. Protein–protein coupling is also required to make antibody–enzyme conjugates and bispecific antibodies. A second section is therefore included that discusses antibody–KLH (**Subheading 3.2.1.**), antibody–enzyme (**Subheading 3.2.2.**), and antibody–antibody (**Subheading 3.2.3.**) conjugations for the sake of completeness. Although many of the principles discussed in **Subheadings 3.1.1.–3.1.3.** are equally applicable to the formation of protein–protein conjugates, all such protein–protein conjugations have to be carried out in aqueous solutions to prevent denaturization/ deactivation of the proteins. This second section is therefore also highly relevant in the conjugation of SM, which are only soluble in aqueous solutions to proteins.

2. Materials

1. Dry organic solvents: dioxan, dimethylformamide (DMF), and dimethyl sulfoxide (DMSO). These are all bought over molecular sieves to minimize their water content.
2. Proteins: BSA, alkaline phosphatase (AP), KLH. (mc-KLH from Pierce is both very soluble and has a well-defined molecular weight.)
3. 1-Ethyl-3-(3-dimethylamino-propyl)carbodiimide (EDC).
4. N-hydroxysuccinamide (NHS).
5. 1,3-Dicyclohexylcarbodiimide (DCC).
6. Biotin.
7. Carbonyldiimidazole (CDI).
8. Phosgene (Fluka).
9. p-Maleimidophenyl isocyanate (PMPI).
10. P10 desalting column (Pharmacia).
11. N-(maleimidocaproyloxy)succinimide ester (EMCS).
12. Vitamin B_{12}.
13. Sodium dodecyl sulfate (SDS).
14. S-acetylthioglycolic acid NHS-ester (SATA).
15. 3-(2-Pyridyldithio)propionic acid NHS-ester (SPDP).

3. Methods
3.1. Small Molecule Coupling to Proteins
3.1.1. Small Molecule Carboxyl Residues

Carboxylic acid groups are normally coupled to protein amine groups using carbodiimide condensation reagents. These react with carboxylic acids to form highly reactive mixed anhydride and isourea esters *(1,2)*. There are two major practical prob-

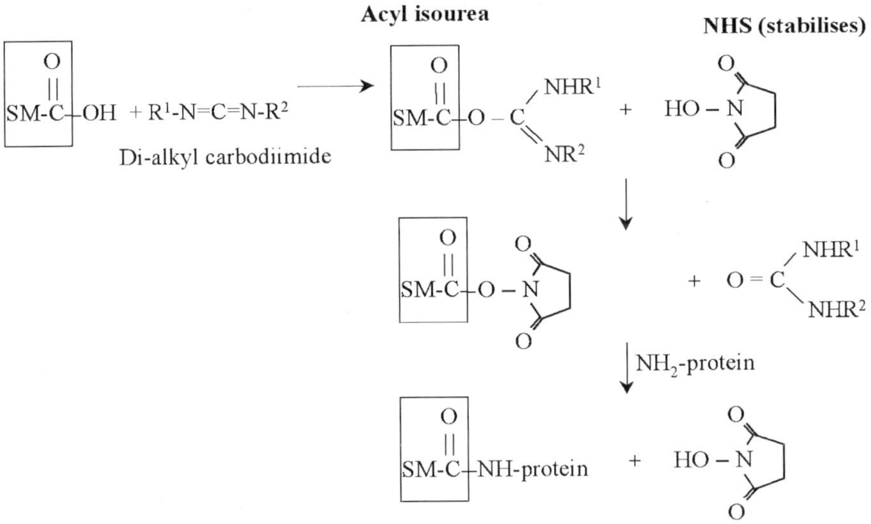

Protein Conjugation Procedures 257

Fig. 1. The reactions involved in coupling an SM–carboxylic acid residue to a protein.

lems. The first is that carbodiimides are quite unstable, especially in the presence of water. The second is that the activated derivatives they form with acid groups (*1,2*) are even less stable, hydrolyzing in fractions of a second (*3*), often before you have time to add the activated SM to the protein. Luckily, most SMs (vitamins, dyes, and so on) and carbodiimides are soluble in dry organic solvents, which gets us round the first problem. The second problem can be overcome by adding NHS to the SM before the carbodiimide is added. This reacts with the SM–isourea acid derivatives as they are formed to give rise to activated NHS esters. These intermediates are stable indefinitely in dry conditions and have a half-life of approx 40 min in aqueous solutions (*3*). These activated esters can then be added to proteins in aqueous solutions and react with protein amine groups to form SM–protein amide bonds and free NHS (**Fig. 1**). A typical example of this conjugation procedure is given below using biotin, BSA, and two different carbodiimides.

3.1.1.1. Biotin-BSA Conjugation

1. Suspend 10 mg biotin in 1 mL of dry dioxan or DMF.
2. Divide into two halves and add 250 µL of NHS (10 mg/mL) to each 500-µL aliquot.
3. Add 250 µL of either EDC or DCC (both at 16 mg/mL) to each aliquot and leave for 1 h to react. As the NHS and carbodiimides are dissolved in the same dry organic solvent, this solution of activated biotin–NHS esters is stable indefinitely if kept in dry solution (*see* **Note 1**).
4. Add varying amounts of either the EDC or DCC NHS-activated biotin solutions (20, 50, or 100 µL) to 1-mL aliquots of BSA (10 mg/mL in 0.1 *M* sodium bicarbonate) and leave for 16 h to react (*see* **Note 2**).
5. Dialyze the biotin-BSA conjugates exhaustively against phosphate-buffered saline to remove uncoupled reaction products.

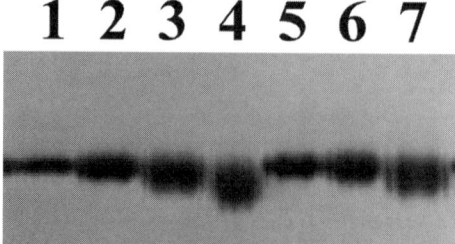

Fig. 2. The migration of BSA-biotin conjugates in an 8% polyacrylamide gel. All lanes contain 5 μg BSA. Lane 1 contains unconjugated BSA. Lanes 2–4 contain BSA-biotin conjugates made when 20, 50, and 100 μL of EDC-activated biotin-NHS were added to BSA. Lanes 5–7 contain BSA-biotin conjugates made by the addition of the same amounts of DCC-activated biotin.

The formation of biotin–BSA conjugates can be confirmed by nondenaturing electrophoresis of the conjugates in 8% polyacrylamide gels (lane 4, **Fig. 2**) using the discontinuous buffer system of Laemmli *(5)* without SDS. As more biotin is bound, the negative charge on the BSA increases (owing to loss of the positively charged NH_2 groups), and the conjugates migrate faster. Slightly more conjugation can be obtained with EDC than DCC with approx one, four, and eight residues of biotin bound to BSA in each conjugate.

It is normally fairly easy to determine how many SM residues are conjugated to a protein by comparing the OD of the substituted protein with the OD of the unsubstituted protein (**Fig. 3**). Unfortunately, biotin has no easily measured absorbance value. This is why the formation of conjugates was verified by electrophoresis. ELISA assays with labeled avidin were also used to confirm the presence of conjugates.

It is very easy to couple an SM to a protein residue via a spacer arm using this technique (*see* **Note 3**).

3.1.2. Small-Molecule Hydroxyl Residues

3.1.2.1. USING CARBONYLDIIMIDAZOLE

It is generally accepted that carbonyldiimidazole (CDI) can be used to couple hydroxyl residues to proteins in a manner similar to that discussed above for carboxylic acid residues. The CDI is added as above at an equimolar concentration to the SM in dry solvent for 2 h, and then aliquots are added to protein in 0.1 *M* sodium bicarbonate for up to 16 h. For example, dioxigenin can be bound to BSA using CDI, and it is possible to couple the light-sensitive protecting group 2-nitrobenzylalcohol to BSA *(6)*. However, the secondary alcohol, 2-nitrobenzyl ethanol, cannot be coupled to proteins by this method *(6)*. As this is a simple and cheap procedure, it should be attempted, but it will not always work.

3.1.2.2. PRE-DERIVITIZATION OF THE HYDROXYL RESIDUE

It is much easier to conjugate a hydroxyl group to proteins if it is first chemically modified prior to conjugation. This can easily be done by converting the hydroxyl

Protein Conjugation Procedures

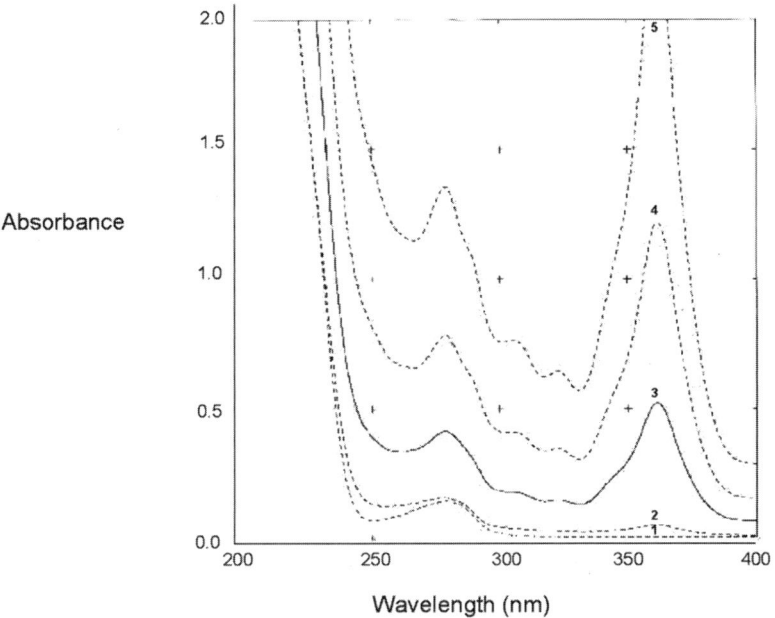

Fig. 3. Spectrophotometric scans (1–5) of vitamin B_{12}-BSA conjugates (at 0.2 mg/mL) made by the addition of increasing amounts of vitamin B_{12}-chloroformate (0, 1, 8, 25, and 50 mg) to 10 mg BSA.

groups to highly reactive chloroformates by treating them with equimolar amounts of diphosgene in dry dioxan in the presence of pyridine as a catalyst. The resulting chloroformates are highly reactive and spontaneously couple to protein amine groups at pH >7.5 (6). Although the chloroformates are highly reactive toward protein amine groups, they also hydrolyze rapidly in aqueous solutions. Therefore, they are normally added in large excess to the protein. A typical example of this procedure is given below in which vitamin B_{12} is coupled to BSA:

1. Suspend 84 mg of vitamin B_{12} in 250 µL of dry dioxan with 6 µL of pyridine and 8 µL of diphosgene. The vitamin immediately changes color from red to purple (see **Note 4**).
2. Leave the mixture for 15 min to react.
3. Evaporate unreacted materials.
4. Resuspend the chloroformates in 250 µL of dry solvent.
5. Add approx 1, 8, 25, and 50 mg of the B_{12}–chloroformate to 2 mL aliquots of BSA (5 mg/mL in 0.1 M bicarbonate) and leave for 4 h to react.
6. Dialyze the B_{12}–BSA conjugates for 2 d against PBS to remove uncoupled B_{12} and unwanted reaction products.

The pink solutions can then be measured by scanning spectrophotometry (**Fig. 3**), and the amount of vitamin coupled can be measured by the increase in absorbance at

280 nm, or more accurately by measuring the absorbance at 360 nm that was only due to B_{12}. In this case, the four conjugates were found to contain 0.1, 1.1, 2.8, and 5.2 mg of B_{12}, which represented approx 0.5, 7, 22, and 30 molecules of B_{12} bound to each BSA molecule. The covalent bonding of the vitamin groups can be confirmed by using the conjugates in competitive ELISA assays.

3.1.2.3. UTILIZING A COMMERCIAL LINKER CONTAINING A HYDROXYL-SPECIFIC ISOCYANATE GROUP

There is one commercial linker, PMPI, which can be used to couple hydroxyl residues to protein sulfhydryl groups *(7)*. One end of the linker consists of an isocyanate group, which is normally first coupled at equimolar concentrations to the SM hydroxyl residue in dry organic solvents. The other end is a maleimide residue, which couples to free protein sulfhydryl groups in aqueous buffers between pH 6.5 and 7.5.

3.1.3. Small Molecule Amine Residues

SMs containing an amine residue can be coupled to proteins by simply reversing the above carbodiimide reaction. NHS is added to the protein, and it is then treated with a large excess of the water-soluble carbodiimide (EDC). This procedure is discussed in much greater practical detail in **Subheading 3.2.1**. The SM (dissolved in 0.1 *M* bicarbonate) is then added and left to couple for 2–16 h. After extensive dialysis, conjugation can be confirmed by spectrophotometry or non-denaturing electrophoresis. Here, however, conjugated proteins would migrate more slowly due to their loss of carboxylic acid residues.

If the SM is insoluble in water, then it can be coupled to a commercial crosslinker, which contains an NHS-activated ester at one end. This is done by simply adding the SM to the linker at an equimolar ratio for 24 h in dry DMF or DMSO. This type of linkage is also required when the protein (antibody or enzyme) loses its activity after it is treated with EDC. I have found this to happen with AP. This type of crosslinker can also be used to create a spacer arm between the SM and the protein, but as they are more commonly used to couple antibodies and enzymes together, their practical use is described below in **Subheading 3.2.2.** and **3.2.3**.

3.2. The Formation of Protein-Protein Conjugates

Protein–protein conjugates can be made by most of the procedures already discussed. However, there is one critical difference. Proteins have to be coupled in aqueous solutions in order to prevent their denaturization, precipitation, and loss of functional activity. In practical terms, this means that the organic solvents often needed to dissolve the carbodiimides and commercial linkers should never exceed 10% of the final reaction volumes.

3.2.1. Antibody–KLH Conjugates

Antibody–KLH conjugates can be used to increase the murine–murine immune response to an antibody above and beyond that obtained solely by changing the mouse strains *(8)*. The simplest way to prepare such conjugates is to use the water-soluble carbodiimide EDC at a large molar excess to allow for the rapid hydrolysis of

Protein Conjugation Procedures

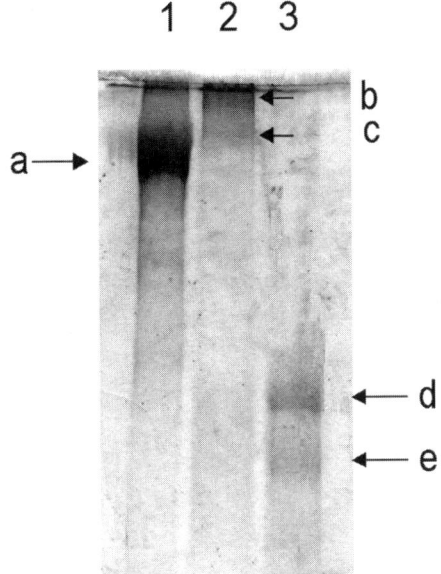

Fig. 4. The conjugation of antibody to KLH. Lanes 1, 2, and 3 contain 15 µg mc-KLH, 20 µg of antibody-KLH conjugates, and 5 µg of unconjugated antibody, respectively.

both the EDC and its reactive intermediates. It is now absolutely essential to include NHS in reaction mixtures prior to the addition of the carbodiimide *(3)*. I normally treat the antibody with the carbodiimide to activate its carboxylic acid groups and then add the activated antibody to the KLH to enable it to bind to the KLH amine groups as described below.

1. Dialyze 1 mg (0.5–3 mL) of monoclonal antibody against distilled water for at least 4 h at 0°C. This is done to remove all preservatives (e.g., azide), and buffers containing carboxylic acid and amine groups.
2. Prepare one vial of mcKLH (20 mg in 5 mL) by dialysis against 0.1 M bicarbonate for the same length of time.
3. Add 5 mg of NHS to the antibody followed by 2 mg EDC. Leave for 15 min to activate carboxylic acid groups.
4. Add 800 µL (approx 3 mg) of the KLH solution to the above and leave this antibody/KLH mixture to conjugate overnight. This should result in a roughly 1:1 complex given the molecular weights of IgG and KLH as 165 and 480 K, respectively.
5. Dialyze the conjugate against 0.9% saline for 24 h.

Conjugation can be confirmed by electrophoresis in SDS in 5% polyacrylamide gels (**Fig. 4**). The complexes (b and c) are at a much higher molecular weight than the KLH (a) and the monoclonal antibody (d and e; note: probably two antibodies!) starting components. Given the low intensity of staining of b and c, most of the complexes do not even penetrate the gel.

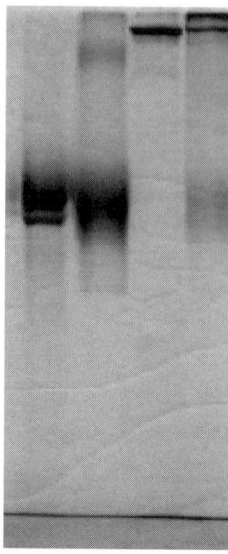

Fig. 5. The conjugation of a monoclonal antibody to AP using EMCS. Lane 1 contains AP, lane 2 contains AP treated with EMCS, lane 3 contains reduced antibody, and lane 4 contains the antibody-AP conjugates crosslinked by the EMCS.

3.2.2. Antibody–Enzyme Conjugates

It is possible to conjugate antibodies to enzymes exactly as is given for KLH in **Subheading 3.2.1.** However, the carbodiimide-derivatized antibody can lose much of its activity. The complexes can contain several crosslinked molecules of each component! The only thing you can control is the ratio of each component to another compound. Most antibody–enzyme conjugates are therefore produced in more precise conditions using well-defined crosslinkers. The most popular consist of an NHS-ester linked to a maleimide group via a spacer arm *(9)*. The spacer can be a benzene ring [*m*-maleimidobenzoyl-NHS ester (MBS)], cyclohexane [succinimidyl 4-(*n*-maleimidomethyl)-cyclohexane-1-carboxylate (SMCC)], or various lengths of nonaromatic carbon chain (EMCS). The selection of an appropriate spacer is very important. Rigid cyclic and aromatic spacers are themselves highly immunogeneic *(9)*. Nonrigid carbon spacers have very little immunogenicity *(9)*, and their flexibility can greatly enhance the sensitivity of competitive enzyme-linked immunosorbent assay (ELISAs) *(10)* when they are used to link SM to AP or horseradish peroxidase. NHS-esters link to primary amines at pH 7–9, whereas the maleimide groups react with sulfhydryl groups at pH 6.5–7.5. An antibody–AP conjugation procedure is given below as an example.

1. Dissolve 1–2 mg of the IgG in 1 mL 50 mM phosphate buffer, pH 7.0, containing 10 mM EDTA (*see* **Note 5**).

2. Reduce specifically at the hinge region by the addition of 2-mercaptoethylamine (6 mg in 100 µL buffer) for 90 min at 37°C.
3. Separate the reduced antibody from the reducing agent on a P10 desalting column using the same phosphate buffer containing EDTA. If 1-mL fractions are eluted, the reduced antibody normally elutes in the fourth and fifth tubes and is stable for several hours in the presence of EDTA.
4. Dissolve one vial of AP (10,000 units, approx 4 mg) in 1 mL distilled water and prepare for coupling by dialyzing against 50 mM phosphate, pH 7.5, to remove amine-containing buffers and preservatives (*see* **Note 6**).
5. Dissolve 1 mg of the crosslinker (EMCS) in 100 µL DMSO and add immediately to the dialyzed AP (*see* **Note 7**).
6. Leave 30 min (no longer!) for the amine groups of the AP to couple to the NHS-ester end of the linker.
7. Remove excess unreacted linker on a P10 column.
8. Add the AP-EMCS fraction immediately to the reduced antibody and leave the final mixture for 2–16 h for the reduced antibody to react with the maleimide end of the EMCS–AP complex.
9. Dialyze to remove the EDTA from the antibody–EMCS–AP conjugate.

The antibody–EMCS–AP conjugate is checked by SDS electrophoresis in an 8% polyacrylamide gel (**Fig. 5**). Very little unreacted antibody and AP can be seen in lane 4.

3.2.3. Antibody–Antibody Conjugates

Antibody–antibody conjugates are prepared in exactly the same way as for antibody–AP conjugates, which have just been described. Equal amounts of each antibody are used. The only problem to be considered is that one of the antibodies has to be linked via its amine groups. This can inactivate an antibody when the antibody has a susceptible amine residue in its binding region. However, a highly susceptible antibody is always reduced and linked via its hinge sulfhydryl residue to the maleimide group. It is unlikely that both antibodies would be susceptible to a loss in activity when they are linked through their amine residues. Some workers prefer to use the thiolating groups SATA *(11)*, SPDP *(12)*, and Trauts' reagent [iminothilane *(13)*] to introduce sulfhydryl groups into antibodies and enzymes rather than use a reducing agent. All these agents react spontaneously with protein amine residues in aqueous solutions. Two reagents, SATA and SPDP, produce protected sulfhydryl residues, which can be released when required for crosslinking; the third directly incorporates a free SH group, but this can be unstable under certain conditions *(14)*.

4. Notes
1. This is by far the most commonly used coupling procedure. Always use SM and carbodiimides in dry organic solvents at 1:1 molar ratios to activate the carboxyls. If the SM is only soluble in water, a large excess of EDC (around 50X) is required, and the stabilizing NHS has to be added before you add the carbodiimide.
2. The NHS-activated SM is always added in a large molar excess (20–50-fold) to the protein, and the extent of coupling required is discovered by trial and error. A pH > 7.0 is required to enable the activated ester to couple to the protein amine groups, as they have to be de-protonated to be able to react. At 0.1 M, sodium bicarbonate has a pH of 8.3 and is a convenient and simple coupling solution.

3. It is very easy to couple an SM to a protein residue via a spacer arm using this technique. The NHS-activated SM in dry organic solvent is added to an equimolar solution of 6-aminocaproic acid (also in dry solvent) and left overnight. A peptide bond is formed with the amine residue on the caproic acid just as it would be with proteins, but here it is still in dry organic solvent. The SM acid group is now coupled via a five-carbon chain to another carboxylic acid. SM-COOH + $_2$HN-(CH$_2$)$_5$-COOH in the presence of DCC and NHS results in SM-CONH-(CH$_2$)$_5$-COOH. Another cycle of NHS and carbodiimide is added, and the SM-(CH$_2$)$_5$-CO-NHS-activated ester group is then coupled to the protein via the spacer. Alternatively, as this linker is formed in dry organic solutions, several cycles can be performed, and thus any length linker can be made, prior to the SM being coupled to the protein (10).
4. Although this is a quick and reliable method of derivatizing hydroxyl residues, diphosgene dissociates into phosgene gas, which is very dangerous. Handle very carefully in a fume hood!
5. It is very important that EDTA be present in the solutions used to reduce the antibody and in the P10 elution of the reduced antibody or it will reoxidize.
6. **It is also essential that the NHS-ester amine coupling to the enzyme be carried out at pH 7.5.** If a higher pH is used (e.g., the more normal 0.1 M bicarbonate buffer, pH 8.3 for 1–2 h), the maleimide group at the other end of the linker rapidly hydrolyzes. It is then impossible to bind the maleimide end of the linker to the reduced antibody in the second stage. The crosslinker maleimide group only has a half-life of 30 min at pH > 7.5. A large excess of linker at pH 7.5 and a short reaction time (20–30 min) followed by a quick separation (of unreacted linker) in a P10 column is used to reduce this problem. Lower temperatures can also be used to reduce the rate of maleimide hydrolysis. Maleimide hydrolysis while the first protein/enzyme is coupling via its amine groups is the major problem associated with maleimide-spacer-NHS-ester crosslinkers. A five-carbon straight chain linker was used in this example (EMCS), but three other linkers [MBS, SMCC, and N-(γ-maleimidobutyryloxy) succinimide ester (GMBS)] have successfully been used in an identical procedure.
7. More expensive sulfo-NHS derivatives of the above crosslinkers can be used if necessary. Their use is promoted because they are soluble in water. There are usually no problems using the normal linkers predissolved in a small amount of DMSO. However, some proteins and enzymes are extremely sensitive to the presence of organic solvents, and the sulpho-crosslinker derivatives could then be directly substituted into the conjugation procedure.

References

1. Carraway, K. L. and Koshland, D. E. (1972) Carbodiimide modification of proteins. *Methods Enzymol.* **25,** 616–623.
2. Rebek, J. and Feitler, D. (1973) An improved method for the study of reaction intermediates. The mechanism of peptide synthesis mediated by carbodiimides. *J. Am. Chem. Soc.* **95,** 4052–4053.
3. Grabarek, Z. and Gergely, J. (1990) Zero length crosslinking procedure with the use of active esters. *Anal. Biochem.* **185,** 131–135.
4. Thompson, S., Wong, E., Cantwell, B. M. J., and Turner, G. A. (1990) Serum alpha-1-protease inhibitor with abnormal properties in ovarian cancer. *Clin. Chim. Acta* **193,** 13–26.
5. Laemmli, U. K. (1970) Cleavage of structural proteins during the assembly of the bacteriophage T4. *Nature [Lond.]* **227,** 680–685.

6. Thompson, S., Spoors, J. A., Fawcett, M-C., and Self, C. H. (1994) Photocleavable nitrobenyl-protein conjugates, *Biochem. Biophys. Res. Comm.* **201,** 1213–1219.
7. Annunziato, M. E., Patel, U. S., Ranade, M., and Palumbo, P. S. (1993) p-Maleimidophenyl isocyanate: A novel heterobifunctional linker for hydroxyl to thiol coupling. *Bioconjugate Chem.* **4,** 212–218.
8. Maruyama, M., Sperlagh, M., Zaloudik, J., et al. (2002) Immunization procedures for antiidiotypic antibody production in mice and rats. *J. Immunol. Methods* **264,** 121–133.
9. Peeters, J. M., Hazendonk, T. G., Beuvery, E. C., and Tesser, G. I. (1989) Comparison of four bifunctional reagents for coupling peptides to proteins and the effect of the three moieties on the immunogenicity of the conjugates. *J. Immunol. Methods* **120,** 133–143.
10. Bieniarz, C., Husain, M., Barnes, G., King, C. A., and Welch, C. J. (1996) Extended length heterobifunctional coupling agents for protein conjugations. *Bioconjugate Chem.* **7,** 88–95.
11. Julian, R., Duncan, S., Weston, P. D., and Wrigglesworth, R. (1983) A new reagent which may be used to introduce sulphydryl groups into proteins, and its use in the preparation of conjugates for immunoassay. *Anal. Biochem.* **132,** 68–73.
12. Carlsson, J., Drevin, H., and Axen, R. (1978) Protein thiolation and reversible protein-protein conjugation. *Biochem. J.* **173,** 723–737.
13. Jue, R., Lambert, J. M. Pierce, L. R., and Traut, R. R. (1978) Addition of sulphydryl groups to *Escherichia coli* ribosomes by protein modification with 2-iminothiolane. *Biochemistry* **17,** 5399–5406.
14. Singh, R., Kats, L., Blatter, W. A., and Lambert, J. M. (1996) Formation of N-substituted 2-iminothiolanes when amino groups in proteins and peptides are modified by 2-iminothiolane. *Anal. Biochem.* **236,** 114–125.

VI

CHARACTERIZATION OF RECOMBINANT PROTEINS

18

Structural Characterization of Proteins and Peptides

Rainer Deutzmann

Abstract

The primary structure of proteins is nowadays determined by DNA sequencing, and a variety of genomes are already known. Nevertheless, protein sequencing/identification is still indispensable to analyze the proteins expressed in a cell, to identify specific proteins, and to determine posttranslational modifications. Proteins of interest are typically available in low microgram amounts or even less. The separation method of choice is gel electrophoresis, followed by blotting to PVDF membrane for N-terminal sequencing or by in-gel digestion to generate peptides that can be separated by HPLC. Structural analysis can be done by Edman degradation or mass spectrometry (MS). Edman degradation is the older method based on successive removal of N-terminal amino acids by chemical methods. Sequencing of a peptide requires many hours, the sensitivity is in the range of 2–5 pmol of a purified peptide. Nevertheless, Edman degradation is still the workhorse in the lab for routine work such as identification of blotted proteins. It is also the method of choice for sequencing unknown proteins/peptides and modified peptides. MS has routinely been used with peptides in the range of 100 fmol or even less. In contrast to Edman degradation, complex mixtures such as tryptic digests can be analyzed, making HPLC separation of peptides unnecessary. MS is a very fast method that can be automated. It is the method of choice for sensitive analysis and large-scale applications (proteomics). Two different ionization methods are commonly used to generate peptide/protein ions for MS analysis. These are MALDI (matrix assisted laser desorption and ionization) and ESI (electrospray ionization). They can be combined with a variety of mass analyzers (TOF, quadrupole, ion trap). Proteins are either identified by searching databases with the masses of proteolytic peptides (peptide mass fingerprinting) or using fragmentation data (raw MS/MS spectra or sequence tags). This approach requires that the protein is known and listed in the database. *De novo* sequencing by MS of peptides is possible, but very time consuming and not a routine application, in contrast to Edman degradation. The aim of this chapter is to introduce to basic theory, practical applications and limitations of the various methods, to enable the non-expert scientist to decide which method is best suited for his project and which kind of sample preparation is necessary.

Key Words: In-gel digestion; Edman degradation; MALDI; ESI; MS/MS techniques; peptide mass fingerprinting; sequence-tag.

1. Introduction

For many years protein sequencing was the only method for elucidating the primary structure of proteins, whereas nowadays DNA sequencing can do this much more conveniently. Nevertheless, structural analysis of proteins is still a great challenge. The genomes of many organisms are known, and researchers want to know which proteins are expressed in different cells and how they interact. In many cases this requires study of the composition of protein complexes, such as the transcription factor complexes. Moreover, biological effects can often be mapped to specific proteins by biochemical and cell biological methods, and researchers want to know whether these proteins are already listed in the database or whether to clone them using polymerase chain reaction (PCR) methods. Last, but not least, proteins often require extensive modification, such as phosphorylation or glycosylation, to be biologically active. Such information can only be obtained at the protein level.

Many excellent textbooks and reviews have covered all aspects of modern structural analysis of proteins (e.g., *1–7*). A comprehensive treatise is beyond the scope of this chapter. A few topics will be explored here that are important to the nonexpert scientist who needs to characterize a protein or set of proteins found in the course of biological or biochemical investigations. Usually the amount of protein available is at the lower limit. Therefore, it is essential to know which protein isolation protocols work reliably with proteins available in the low microgram range and how to prepare peptides for sequencing. This can be done in most labs, whereas ensuing structural analysis by classical Edman degradation or modern mass spectrometry methods requires special equipment. Nevertheless, some background in these techniques is required to decide which method is best suited for the current project and which kind of sample preparation is necessary.

2. Purification of Proteins and Preparation of Peptides for Sequence Analysis

2.1. General Strategy

A variety of powerful column separation techniques are available for isolation of proteins, including affinity, ion exchange, molecular sieve, and reverse-phase chromatography. Whenever possible, high-performance liquid chromatography (HPLC) methods should be used because of their higher resolution power, which results in purer samples *(8,9)*. Otherwise contamination caused by column bleeding and washout of impurities might interfere with the subsequent analysis.

Combinations of these techniques are quite efficient but require rather large amounts of starting material. It is generally not a good idea to try protein isolation involving several purification steps when only a few micrograms of proteins are available. Losses increase more than proportionally with decreasing amounts of protein.

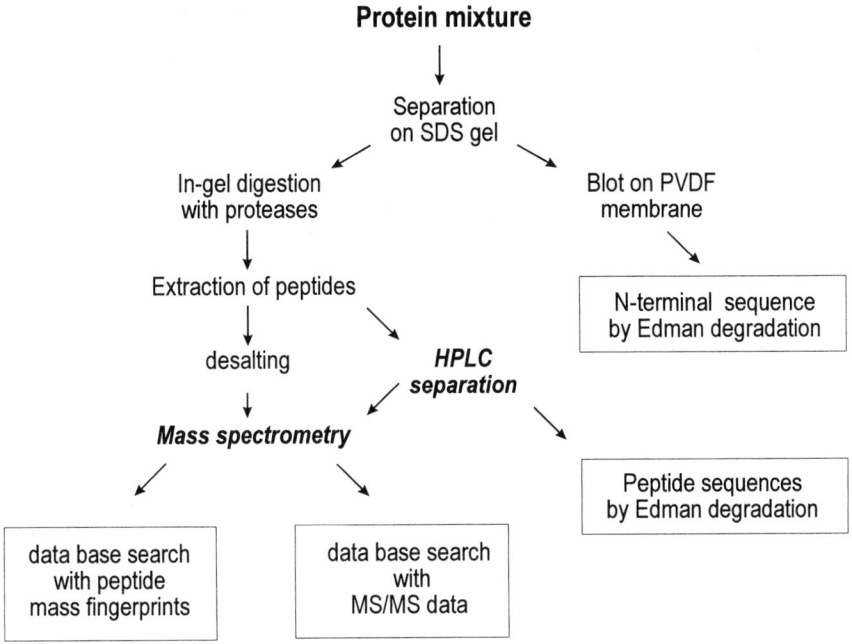

Fig. 1. Characterization of gel-separated proteins by Edman degradation and mass spectrometry (MS). HPLC, high-performance liquid chromatography; PVDF, polyvinyldifluoride.

Even pipeting from one Eppendorf tube to another can result in complete loss of protein at levels below about 1 μg. All handling should be restricted to a minimum.

Fortunately, it is sufficient to enrich a protein to a stage at which it can be visualized as a discrete band or spot on a sodium dodecyl sulfate (SDS) gel or a two-dimensional (2D) gel. A variety of sensitive methods are available to gain structural information from gel-separated proteins. These are discussed below and are summarized in **Fig. 1**.

The strategies are different for simple protein identification or for determination of N-terminal sequence data. The best, if not the only, way to find the N-terminal sequence is to do Edman degradation of the blotted protein. Unfortunately, about 50% of all eukaryotic proteins are N-terminally blocked, pyroglutamic acid or acetylated amino acids being very common (**Fig. 2**). In such cases N-terminal sequencing is not possible. Deblocking can be tried, but results are variable. Removal of pyroglutamic acid by pyroglutamase is usually quite successful *(6,7,10)*, whereas yields of deacetylation are poor *(11,12)*. Moreover, the kind of N-terminal modification is not known in advance, and recovery of proteins for deblocking is generally not successful when the proteins have already been subjected to a few cycles of Edman degradation. Usually fresh material has to be prepared. If the N-terminus is blocked internal sequence data from proteolytic peptides must be generated. This is also recommended if cloning of the gene is intended, since the chance of obtaining suitable oligonucle-

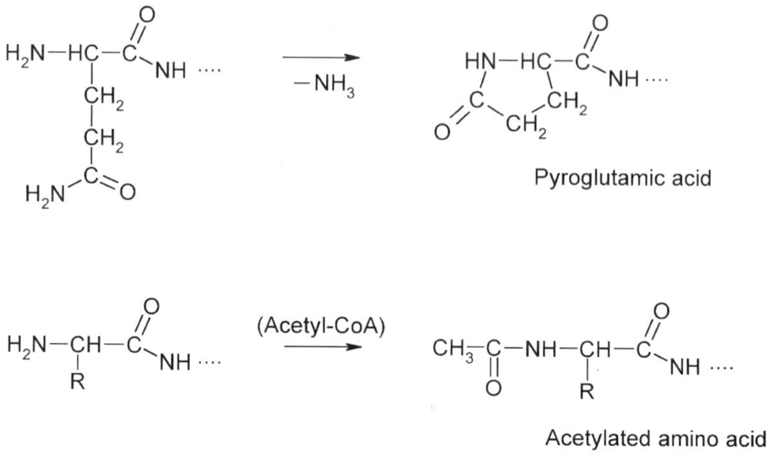

Fig. 2. N-terminal blocking of eukaryotic proteins by acetylation and cyclization of glutamine to pyroglutamic acid.

otides increases with the number of peptide sequences. For preparation of peptides, proteolytic cleavage of the proteins in an SDS gel is probably the best method. The peptides can then be subjected to Edman degradation after HPLC separation and/or analyzed by mass spectrometry. Depending on the method used, even less material might be required than for N-terminal sequencing.

2.2. Blotting of Proteins for N-Terminal Sequence Analysis

Electroblotting is the preferred method of obtaining pure proteins for N-terminal sequencing, either out of a mixture or when the sample contains too much detergent or buffers. Blotting for sequencing is performed in much the same way as for Western blotting. The semidry blot procedure is the most widely used (13,14). The gel, covered with the blotting membrane of the same size, is sandwiched between three layers of Whatman paper soaked with buffer. This assembly is placed between the electrodes (**Fig. 3**). After the current is switched on, the proteins, on their way from the gel toward the anode, are electrophoretically transferred to the membrane, where they are fixed by hydrophobic interactions. A procedure is outlined in **Protocol 1** in the Appendix.

Different materials are used for the electrodes (graphite, platinum-coated metal sheets, carbon fiberglass). Graphite electrodes have the disadvantage that they are rapidly oxidized by nascent oxygen formed at the anode. For blotting, polyvinylidene (PVDF) membrane is most frequently used (15), such as Bio-Rad's PVDF, Applied Biosystems' ProBlott, or Millipore's PVDF PSQ or PVDF-P. Glass fiber-based membranes can also be applied (for an overview of suitable membranes, see ref. 14). The popular nitrocellulose and nylon membranes are not suitable, since they are not resistant to the reagents used in the protein sequencer! The blotted proteins can be visualized on the PVDF membrane by staining with Coomassie Blue, which does not

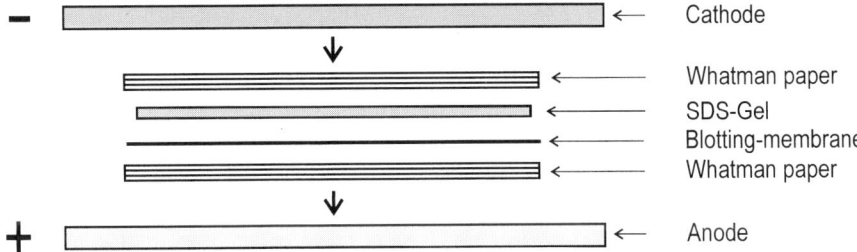

Fig. 3. Semidry blotting of sodium dodecyl sulfate (SDS) gel-separated proteins; the schematic drawing shows the assembly of the blotting sandwich.

interfere with sequencing. Afterward the blot should be shaken for at least 1 h in pure water to remove traces of buffer bound to the membrane that would otherwise interfere with Edman degradation.

Optimal transfer of proteins would require the same buffer as that used for electrophoresis, but under these conditions the proteins would not bind efficiently to the blot membrane, as SDS interferes with binding. Usually SDS-free buffers are preferred, but low amounts (approx 0.5%) can be added to elute hydrophobic proteins. Bound SDS molecules help to drag the protein out of the gel toward the anode. During blotting, SDS migrates faster than the protein, resulting in dissociation of the complex. Adsorption of protein to the blot membrane is facilitated by the addition of methanol to the blot buffer (20%), promoting the dissociation of SDS and protein. A variety of buffer systems are reported in the literature, e.g., 50 mM boric acid buffered to pH 9.0 with NaOH *(16)*, 10 mM CAPS (3[cyclohexylamino]-1-propane sulfonic acid, pH 11) *(15)*, various Tris/glycine buffers *(17,18)*, and others (summarized in ref. *13*). These buffers should contain methanol in the range of 10% for larger proteins and 20% for smaller ones (<50 kDa).

Electroblotting is a straightforward procedure, but some pitfalls should be avoided:

1. For Edman degradation the protein should be visible by Coomassie Blue staining; otherwise the amount of protein is too low. (Weakly stained bands correspond to 50–200 ng protein; 200 ng of a 20-kDa protein correspond to 10 pmol, and 200 ng of a 200-kDa protein correspond to 1 pmol only.)
2. The protein should be as concentrated as possible. Sometimes diluted samples are loaded onto many slots. Better sequencing results are obtained when the sample is concentrated before putting on the gel.
3. In case of poor blotting, it is not recommended to prolong elution times. This does not improve the yield significantly but might lead to unnecessary destruction of blotted proteins by reactive agents formed by electrolysis at the electrodes. The current should not exceed 5 mA/cm^2, in our hands about 1.5 mA for 2–3 h being optimal. However, many people prefer higher amps and shorter times. The best conditions have to be checked for each system.
4. Often very poor sequence data are obtained even though the protein is known not to be N-terminally blocked naturally. In these cases artificial blocking has occurred during electrophoresis or sample preparation due to modification of the N-terminal amino group, thus preventing reaction with the Edman reagent.

The N-terminal amino acid is susceptible to oxidation or derivatization with all reagents that react with free amino groups, such as aldehydes. Using fresh, high-quality reagents can minimize this problem. Immediately after casting, SDS gels contain residual radicals (from TEMED and ammonium persulfate) as well as unpolymerized acryl amide that can react with free amino groups. The gel should be allowed to polymerize overnight. Furthermore, the wells should be rinsed after removing the comb to remove nonpolymerized material. Addition to the sample loading buffer of scavengers, such as 1 mM mercaptopropionate or thioglycolate or small peptides/proteins, running ahead of the sample, can also improve results. It is also a good idea to reserve one lane for a control protein known to have a free N-terminus.

Care should also be taken during sample preparation. Detergents like Triton often contain peroxides that can destroy the N-terminus. Whenever possible, urea should be avoided. In aqueous solution urea is in equilibrium with ammonium cyanate:

$$H_2N-\overset{\overset{O}{\|}}{C}-NH_2 \rightleftharpoons NH_4^+ \; NCO^-$$

This compound can carbamylate proteins, probably involving addition of the free amino groups of the proteins to the unionized cyanic acid HN=C=O (correctly named isocyanic acid). If it is necessary to include urea in the buffer, freshly deionized urea or freshly dissolved high-quality reagent-grade urea should be used. In general, at acidic pH the N-terminal amino group is quite resistant to modification because in its protonated form it lacks the free electron pair for attack by oxidants or aldehydes.

2.3. Other Applications with Blotted Proteins

It is possible to digest proteins on the PVDF membrane after blocking reactive sites with PVP-40 (polyvinylpyrolidone) (according to refs. *19* and *20*), but for production of peptides, digestion in the SDS gel is generally superior (see below) and avoids blotting that is not always quantitative. In addition, the extent of digestion is highly dependent on the kind of blot membrane.

Proteins can be eluted from the membrane for further analysis, but yields using volatile reagents, such as 50–70% isopropanol containing 0.5–5% trifluoroacetic acid (TFA), are usually very low. In our laboratory, blotted *E. coli* lysates could be eluted nearly quantitatively with a mixture of 5% Triton X-100/2% SDS *(21)*. Either detergent alone was not as effective.

2.4. Preparing Peptides from SDS Gel-Separated Proteins for Sequence Analysis

The best method for generation of peptides is digestion of the protein in the SDS gel. This method is superior to digestion of blotted proteins as long as the protease used is able to diffuse into the gel pores. It is also possible to elute the protein from the gel prior to digestion *(22)*. However, the proteins recovered in this way are contaminated by salts, detergents, and soluble gel components that interfere with subsequent analysis and have to be removed by microdialysis, precipitation of protein, HPLC, or other methods. This can lead to substantial losses of material. In addition, the extraction is not always quantitative. Therefore, this method has no advantage over in-gel digestion for samples present in low microgram amounts.

2.4.1. Sample Preparation

Digestion requires denaturing the proteins to make all susceptible bonds available for the protease. Frequently used denaturing agents are 8 M urea and 6 M guanidinium chloride. Guanidinium chloride forms a white precipitate with SDS and has to be removed prior to SDS gel electrophoresis, which can be problematic with low amounts of protein, as mentioned above. The simplest way to denature proteins is heating samples in SDS containing sample buffer prior to electrophoresis. However, proteins containing disulfide bonds are not fully denatured by this treatment. S–S bridges have to be reduced by mercaptoethanol, DTT (dithiothreitol, Cleland's reagent) or its isomer DTE (dithioerythrol). The easiest way to do this is to add mercaptoethanol to the sample loading buffer. However, the resulting free cysteines are not detected by Edman degradation, and mass spectroscopy will mainly detect the product resulting from addition of acrylamide. Therefore, this method can only be recommended if just a few partial sequences are required. Otherwise the cysteine residues have to be chemically modified. This will also inhibit reoxidation by atmospheric oxygen. Frequently used blocking agents are vinylpyridine or iodoacetate/iodoacetamide:

Prot—SH + N⟨pyridine⟩—CH—CH=CH$_2$ → N⟨pyridine⟩—CH—CH=CH—S—Prot

Prot—SH + I—CH$_2$—C(=O)OH (-NH$_2$) $\xrightarrow{- HI}$ Prot—S—CH$_2$—C(=O)OH (-NH$_2$)

A common procedure is to incubate the protein for 1 h in 0.02–0.1 M Tris buffers containing denaturing agents and 2 mM DTT (under nitrogen), at a pH between 7.5 (vinylpyridine) and 8.5 (iodoacetate/amide). Then a slight excess (4.5–5 mM) of the blocking agent is added, and incubation continues for another hour at room temperature. Finally, mercaptoethanol is added to a concentration of 1% *(1,6,7)*. Excess reagents are removed by dialysis by desalting on Sephadex G10/Biogel P10, or by reverse-phase chromatography. This is not necessary if the sample is immediately mixed with sample loading buffer and loaded onto the SDS gel. Alternatively, reduction can be done in the sample loading buffer containing DTT (5 mM) followed by addition of the alkylating agent. It is also possible to modify the peptides directly on the filter disk used for Edman degradation *(23)* or after electrophoresis in the excised band *(24)*.

Even though the procedure is simple, it is not without problems. At high pH (>9) or with the use of excess reagents, other amino acids including the N-terminus can be alkylated. During the reaction of iodoacetate/amide with cysteine, the strong acid HI will be formed, which has to be neutralized by the buffer. Vinylpyridine has the advantage of yielding a PTH amino acid that can easily be detected by Edman degradation. A major disadvantage of vinylpyridine is the increased hydrophobicity of the derivatized peptide. In contrast, carboxymethylation with iodoacetate makes the peptide more hydrophilic. The iodoacetate/amide used must be colorless. Any iodine present, revealed by a yellow color, causes rapid oxidation of thiol groups and may modify tyrosine residues.

2.4.2. In-Gel Digestion

A recipe for in-gel digestion *(25)* as used in our lab is given in **Protocol 2** in the Appendix. The method is straightforward and simple, but some details should be noted. As in blotting, the gel lanes should be overloaded to the maximum extent possible. The band of interest should be excised and cut into small cubes with a sharp scalpel or razor blade. Crushing of the gel piece must be avoided, even though this seems to be logical to enhance the accessible surface area for attack of the protease. Many unwanted contaminants would be set free by this treatment that would interfere with the detection of peptides in subsequent HPLC separation. Prior to digestion, the gel pieces are thoroughly washed with solutions of increasing acetonitrile concentrations to remove SDS, buffer, and soluble gel components. During this procedure the gel size decreases to a small volume. After the last extraction with pure acetonitrile, the gel pieces are air-dried and then rehydrated with buffer containing the protease. It is sufficient to rehydrate the gel pieces, they should not be immersed in buffer, since digestion occurs only in the gel, not in the supernatant. After cleavage, the diffusible proteolytic peptides are extracted and after lyophilization they are ready for HPLC separation.

In-gel digestion is also possible with silver-stained proteins, but a special protocol for staining is required *(24)*. Use of glutaraldehyde or formaldehyde, which result in greatly reduced sensitivity, has to be avoided.

2.4.3. Proteases for Digestion

Digestion of proteins is nowadays performed with specific proteases. Chemical methods are employed in special cases only. For instance, cleavage by BrCN is used if long peptides are required, but this often has the great disadvantage of being incomplete and yielding fragments that are difficult to separate via HPLC due to their length and hydrophobicity. Several useful proteases for generation of peptides from denatured proteins are listed in **Table 1**. As a first trial, trypsin or endoproteinase Lys-C are good choices. Nonspecific proteases will yield too many small peptides and should only be used if the more specific ones fail. This can happen in rare cases when bulky modifications sterically block access of proteases. Nonspecific enzymes might also be useful for obtaining fragments from nondenatured proteins. Details about the various proteases can be found on the web (www.expasy.org/enzyme). Some commercially available enzymes are contaminated by other proteins or have low specific activity. Therefore, the quality should be checked before use with precious samples. If the quality of a protease is questionable, use of sequencing-grade proteases is recommended (available from Roche or Promega in the case of modified trypsin).

Table 1
Frequently Used Enzymes for Fragmentation of Proteins

Enzymes (major sources)	EC number	Specificity (X-↓-Y)	Optimum pH
Enzymes of high specificity			
Endoproteinase Arg-C	3.4.22.8	Arg-↓-Y	7–8
(*Clostridium histolyticum*)			
Endoproteinase Lys-C	3.4.21.50	Lys-↓-Y	8–9
(*Achromobacter lyticus,*			
Lysobacter enzymogenes)			
Trypsin	3.4.21.4	Lys-↓-Y, Arg-↓-Y	7–9
(*Bos taurus*)			
Endoproteinase Glu-C	3.4.21.19	Glu-↓-Y, (Asp-↓-Y)^a	8 (4)
(*Staphylococcus aureus*)			
Endoproteinase Asp-N	3.4.24.33	X-↓-Asp	7–8
(*Pseudomonas fragi*)			

Enzymes (major sources)	EC number	Major cleaved bonds	Optimum pH
Enzymes of lower specificity			
Chymotrypsin	3.4.21.1	X = Tyr, Phe, Trp, (Leu)	7–9
(*Bos taurus*)			
Pepsin	3.4.23.1	X,Y = aromatic,	1.8–2.5
(*Sus scrofa*)		large aliphatic aa	
Elastase	3.4.21.36	X = Ala, uncharged,	8–9
(*Sus scrofa*)		nonaromatic aa,	
Thermolysin	3.4.24.27	Y = L, F, and other aromatic	8
(*Bacillus thermoproteolyticus*)		or large hydrophobic aa	

Abbreviation: aa, amino acid.
^{a}Glu-↓, ammonium bicarbonate, pH 7.8, ammonium acetate, pH 4.0; Glu-↓, Asp-↓, phosphate buffer, pH 7.8 (cleavage at Asp often slow).

2.5. Separation of Peptides

The amount of protein used for in-gel digestion is often so low that it is not possible to optimize the separation. Therefore, a reliable and robust method is required that should work with all kinds of samples. The method of choice is reverse-phase chromatography *(8,9,26,27)*. The peptides are separated on the basis of their hydrophobicity by distribution between an apolar stationary phase and a polar mobile phase. An example for a separation by reverse-phase chromatography is shown in **Fig. 4**. Most reverse-phase column packings are based on porous silica particles that contain covalently bonded alky chains of different length, depending on the column type. These packings are typically made by reaction of silica particles with chlorosilanes carrying organic substituents, such as octadecyl (C_{18}) groups:

$$\left(\text{Si}-\text{OH}\right) + \text{Cl}-\underset{\underset{R}{|}}{\overset{\overset{R}{|}}{\text{Si}}}-(CH_2)_{17}CH_3 \xrightarrow{-HCl} \left(\text{Si}-\text{O}-\underset{\underset{R}{|}}{\overset{\overset{R}{|}}{\text{Si}}}-(CH_2)_{17}CH_3\right)$$

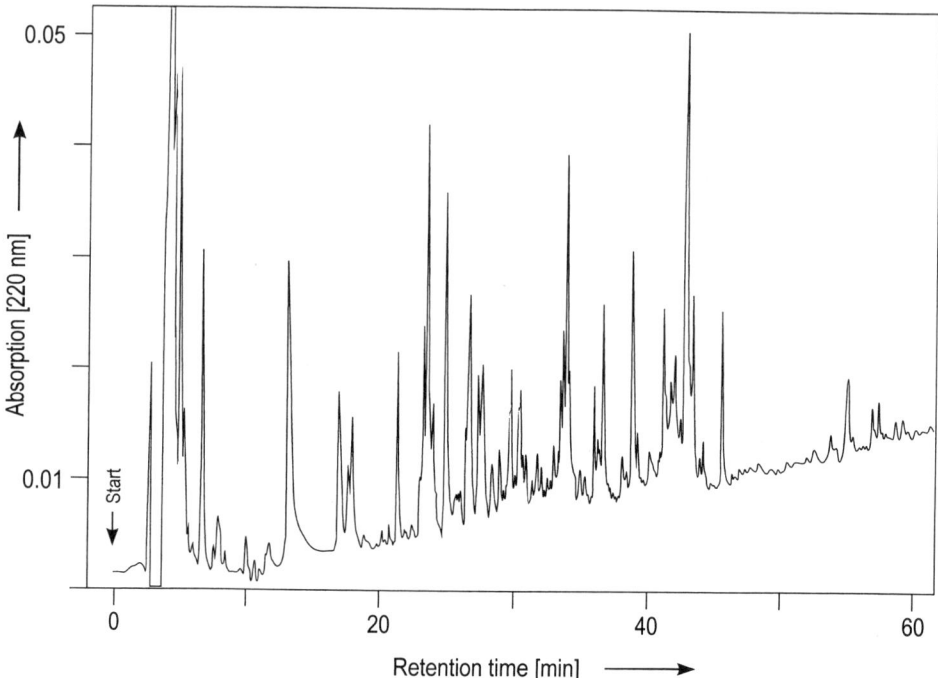

Fig. 4. Separation of tryptic peptides of reduced and carboxymethylated ferritin (100 pmol) by reverse-phase HPLC. The separation was achieved on a Vydac-C18 column (2.1 × 250 mm) using a 60-min gradient of 5–70% acetonitrile in 0.1% aqueous trifluoroacetic acid.

Other commonly used moieties are octyl (C_8) and butyl (C_4) groups. Steric hindrance inhibits quantitative derivatization, thus leaving residual silanol groups that interfere with reverse-phase chromatography by forming a polar surface with weak ion exchange properties. Free silanol groups tend to adsorb proteins and peptides very strongly. To avoid these complications, column manufacturers usually block these groups by a second reaction with a small reagent like trimethylchlorosilane ("end-capping").

Hydrophobicity of the column increases with increasing chain length of the bound moieties. Concomitantly, the retention times of proteins and peptides increase due to stronger hydrophobic interactions. Therefore, C_4 columns are used mainly for separation of proteins and large hydrophobic peptides (>20–30 amino acids), whereas C_{18} columns are best suited for separation of smaller peptides such as those obtained by tryptic cleavage.

Other important parameters for the selection of the right column include column length and diameter and pore and particle size of the packing. Efficiency of the column increases with decreasing particle diameter, but unfortunately the back pressure also increases. For analytical separations, spherical particles with diameters of 3–5 µm should be used having back pressures on the order of 50–100 atm. Particles of

10 μm are not recommended. The pore size for separation of peptides and small proteins should be 150–300 Å to allow penetration into the matrix. Otherwise only the surface is available for separation, resulting in loss of column capacity.

Until recently, the standard reverse-phase column had a length of 25 cm and an inner diameter of 4.6 mm. These dimensions are definitely too large for separation of peptides in the microgram range. All columns will irreversibly bind a certain amount of the sample, and this nonspecific binding is proportional to column size. Therefore, columns with inner diameters of 2 or even 1 mm should be used. The recommended flow rates are about 200 and 50 μL/min, respectively. Even narrower columns are available (down to the 100-μm range), but these columns are not recommended for two reasons. First, they are difficult to pack, and good packing is a prerequisite for good separations. Second, the small-bore columns require a special instrument design. For instance, the pumps must be able to perform reproducible gradients at flow rates of a few microliters per minute. Because of the reduced volumes, dispersion of the solute during passage through the tubing and connections becomes a serious problem. Even with a 2-mm column, connecting the column outlet with the inlet of a UV detector by 20-cm-long steel tubing with a 0.5-mm inner diameter will abolish your separation. This of course is even more critical with smaller columns. All tubing must be kept as short as possible, and the inner diameter should be 0.1 mm (or approx 0.005"), and zero dead-volume fittings should be used. Also, the volume of the flow cell of the UV detector should be matched (0.5–5 μL).

Reducing the length of the column is another choice to minimize nonspecific sample loss. For separation of proteins, the length can be reduced to 5 cm, but for separation of complex peptide mixtures the length of the column is more critical. Using 15-cm instead of 25-cm-long columns can already result in loss of resolution.

The standard solvent system for elution of peptides is water/acetonitrile containing 0.05–0.1% trifluoroacetic acid (TFA). Acetonitrile gives much better peak shapes than alcohols of similar polarity. Proteolytical peptides have to be separated by gradient elution. Typically, gradients from 5 to 70% acetonitrile are used. The steepness can be adjusted according to the complexity of the peptide mixture. More complex mixtures require a decrease in gradient rates. As a rule, peptide digests from a protein such as albumin can be separated using a gradient of 5–70% acetonitrile in about 1 h. Elevation of temperature to 50°C results in shorter retention times and gives sharper peaks.

Other solvent systems are of minor importance. Highly hydrophobic peptides can be eluted with less polar solvents like propanol, often in mixtures with acetonitrile. In some cases, especially when a peak has to be further purified, separation can be done at pH 7 using ammonium acetate (10–50 mM) instead of TFA. This solvent is also volatile upon lyophilization.

The use of acetonitrile and TFA, which are both available as HPLC-grade reagents, permits detection of peptides at wavelengths of 210–220 nm, at which all peptides absorb. (At 280 nm, tyrosine and tryptophane will be detected, and at 254 nm phenylalanine will be detected). At 220 nm and below, a rise in baseline will be observed in gradient separations. This rise can be counterbalanced by increasing the percentage of TFA in buffer A to 0.12% (or lowering it in buffer B to 0.88%).

Last but not least, collection of the separated peaks should be done with some care. Quite often beautiful chromatograms are seen, but the collected fractions do not contain peptides for various reasons such as neglecting the delay between the UV detector-outlet or that of the screen display.

The collected peaks can be directly applied to the protein sequencer; no further sample preparation is necessary. Therefore, if possible, reverse-phase chromatography using acetonitrile/TFA should be the last purification step in a series of separations.

Suitable columns can be bought from a variety of suppliers such as Vydac, Merck, Macherey-Nagel, and Waters, to mention a few. In addition to silica, other materials based on organic polymers or aluminum oxide are available. These materials are for special purposes, and up to now most of them have been inferior in resolution. Polymer-based columns can be used if separations above pH 7 are necessary, because silica-based columns are not stable for a long time at neutral or basic pH.

3. Identification/Sequencing of Proteins

Two methods are available for determination of peptide sequences. The older one is Edman degradation, which had long been the only way to determine complete sequences of proteins. Today sequencing the genes is much faster, but sequencing of proteins is still important for characterization of isolated proteins. In the last few years, Edman degradation has been complemented and partially replaced by mass spectrometry methods, which also allow identification of post-translational modifications.

In the following section the theoretical background of these methods will be discussed briefly, followed by a comparison of their strengths and weaknesses for different applications.

3.1. Edman Degradation

Edman degradation is a cyclic process; in each cycle one amino acid is cleaved off from the N-terminal side and identified by reverse-phase chromatography. Each cycle can be divided into three parts: coupling, cleavage, and conversion *(6,28)*. The reaction scheme is outlined in **Fig. 5**. In the first step phenylisothiocyanate (PITC) is coupled to the N-terminal amino acid under slightly alkaline conditions. After excess reagents and byproducts have been washed away, the N-terminal amino acid is cleaved off with TFA as the thiazolinone derivative. This is done with anhydrous TFA to avoid hydrolytic cleavage of peptide bonds. The thiazolinones are not stable; for HPLC analysis they have to be converted to the more stable phenylthiohydantoin (PTH) amino acids, using aqueous TFA. After cleavage, the next amino acid is the N-terminal one, and the next cycle can be started.

Edman degradation, including HPLC analysis, has been fully automated. Modern instruments are gas-phase or pulsed-liquid sequencers *(29)*. The reaction chamber of these instruments has a very simple design (**Fig. 6**). It consists of two glass blocks between which a glass fiber filter disk is placed. The protein/peptide solution is applied to the filter disk and evaporated to dryness in a stream of argon. This will immobilize the protein on the disk. Immobilization is enhanced by coating the filter disk with polybrene (a polymer of tetramethyl-1,6-hexanediamine and 1,3-dibromopropane),

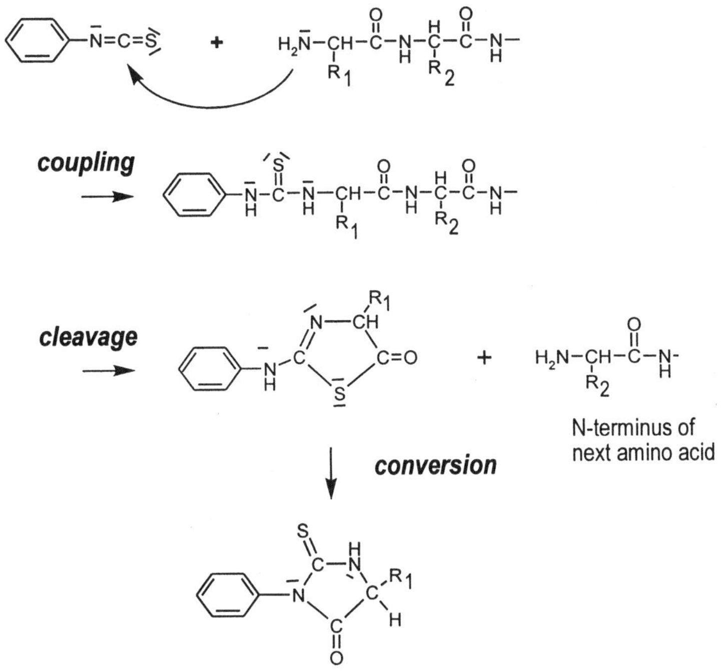

Fig. 5. Sequential steps of Edman degradation.

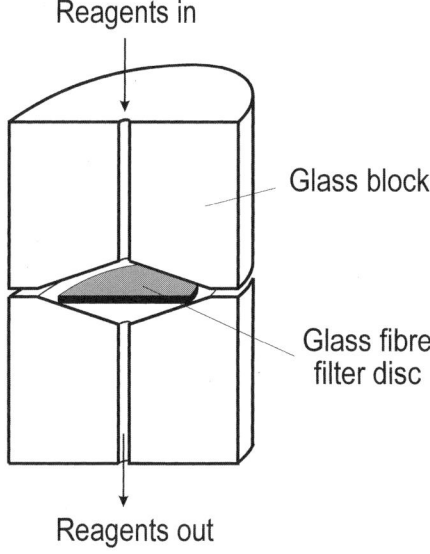

Fig. 6. Reaction cartridge of modern gas-phase type protein sequencers.

which binds tightly to the glass fiber as well as to the protein/peptide. Alternatively, a protein blotted on PVDF membrane can be placed on top of the filter. All reaction steps take place on the filter, the reagents being applied through a narrow channel in the center of the cylindrical glass blocks. The cleaved thiazolinone is extracted with butylchloride and converted to the PTH–amino acid in a separate vessel.

Edman degradation needs a free N-terminal amino group, otherwise coupling cannot occur. The samples must be clean; impurities like salts, detergents, and buffers interfere with the reaction. In particular, chemicals containing free amino groups such as Tris/glycine buffers cannot be tolerated because these compounds react with PITC forming products that interfere with the detection of amino acids. Peptides obtained by reverse-phase separation are best suited for sequencing because acetonitrile/TFA is volatile and the HPLC grade reagents are of highest quality. Routinely about 30–50 amino acids can be sequenced in one run. This requires a repetitive yield of better than 95%. Otherwise the overlap of the preceding amino acids increases very rapidly, making the identification of the actual amino acid very difficult. Proteolytic peptides can usually be sequenced until the C-terminal amino acid.

3.2. Mass Spectrometry

3.2.1. Principles

Mass spectrometry has long been known as a powerful technique for structural determination of organic compounds. Unfortunately *only charged molecules in the gas phase can be analyzed*, because all mass determinations are based on the properties of charged molecules in electromagnetic fields. For instance, two different masses with the same charge and kinetic energy will have different flight paths in a magnetic field (magnetic sector mass spectrometers). This physical property allows calculation of their masses, provided the path is not disturbed by collisions with air molecules, explaining the requirement of high vacuum. The deflection of an ion is dependent on its *m/z* ratio, that is, the mass divided by its charge. Actually, all kinds of mass spectrometers determine *m/z* values, because it is this value that determines the behavior of ions in electromagnetic fields. From the *m/z* ratio the mass can be calculated if the charge is known.

Transfer of peptides and proteins from aqueous solutions to the gas phase had naturally been a great problem, but in recent years several powerful methods have been developed. Two of them, electrospray ionization (ESI) and matrix-assisted laser desorption ionization (MALDI), have gained practical importance and are today indispensable tools for protein analysis *(30,31)*.

3.2.1.1. MALDI

In MALDI an analyte present in the solid state is transferred directly into the gas phase *(32,33)*. The sample is cocrystallized with organic matrix molecules and irradiated in high vacuum with a short laser pulse, leading to evaporation of the organic matrix and concomitantly of the protein. At the point of impact the laser beam produces a small hole that looks like a meteorite crater. Astonishingly, the proteins are not destroyed under these harsh conditions but can be desorbed intact. This is possible

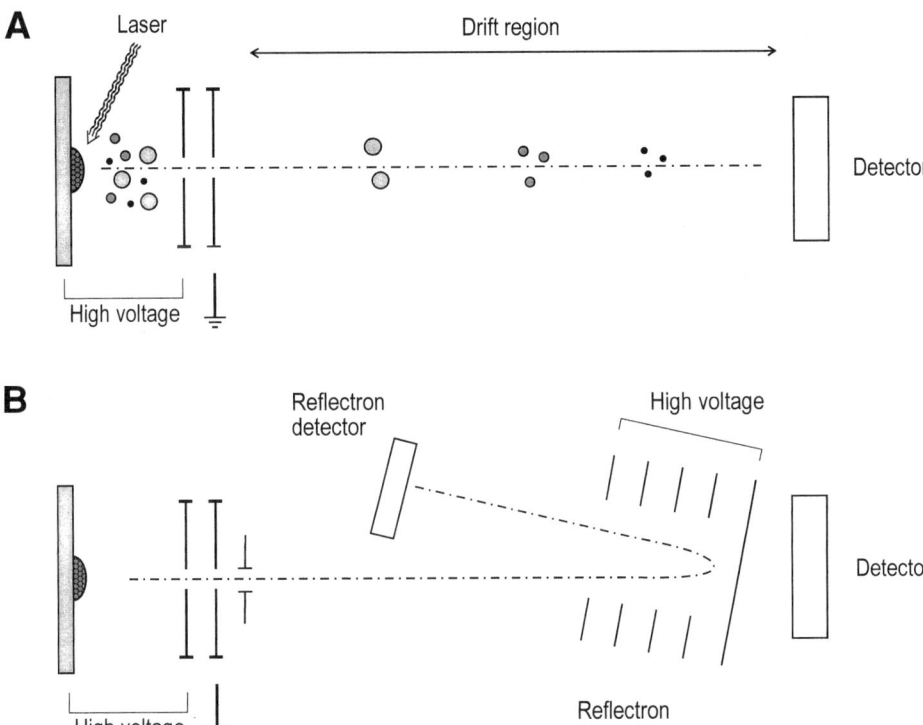

Fig. 7. Basic design of matrix-assisted laser desorption/ionization mass spectrometers (MALDI-MS). (**A**) MALDI-MS. (**B**) MALDI-MS with reflectron to enhance mass resolution.

because the energy of the laser is matched with the absorption properties of the matrix. UV lasers are normally used and as matrix molecules weak organic acids like α-cyano-4-hydroxycinnamic acid, sinapic acid (3,5-dimethoxy-4-hydroxy-cinnamic acid), 2,5-dihydroxybenzoic acid, and others. The energy of the laser is almost exclusively absorbed by these molecules, which are in large excess. During evaporation of the matrix and coevaporation of the analyte molecules, a proton is transferred from the weak organic acid to basic groups of the analyte. In this way charged species are formed that can then be detected by mass spectrometry. In MALDI, preferentially singly charged molecules are detected. Molecules with higher charges are probably also formed but are lost by reactions in the cloud of gaseous ions.

Values of m/z are determined by measuring the time the ions need to travel a given distance in high vacuum [time of flight (TOF)]. The principle is shown in **Fig. 7**. Exiting the source, the ions are accelerated by a potential V. Traversing in the direction of the electric field, they gain the same kinetic energy that is equal to their potential energy ze V in the field (z = number of units of charge, e = fundamental unit of charge, 1.6×10^{-19} Coulomb). The larger the molecule, the longer it takes to reach the

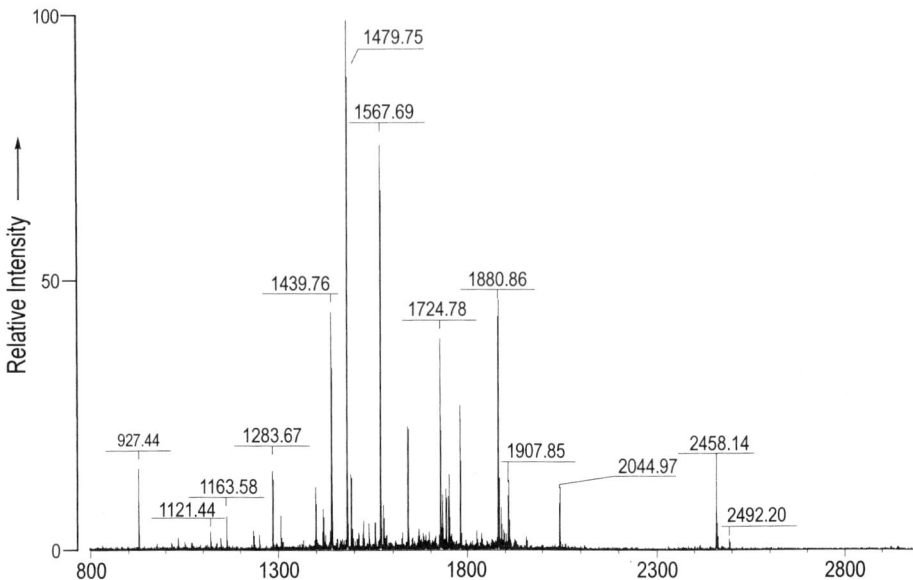

Fig. 8. MALDI mass spectrum of tryptic peptides of bovine serum albumin. (Kindly provided by Bruker Daltonics, Bremen.)

detector. Drift times are typically in the microsecond range, and time measurement is triggered by the laser pulse, the duration of which is a few nanoseconds.

Such a simple device has a rather poor mass resolution. The resolution can be dramatically improved by two techniques, usually applied in combination. One method uses a modified procedure of ion extraction (*delayed extraction*; *34,35*). The other method uses a different device for TOF measurement (reflectron instruments; *see* **Fig. 7**): In the cloud of ions and molecules formed by absorption of the laser pulse, the analyte molecules have different initial kinetic energies (in part due to collisions), superimposed on the kinetic energy gained in the electric field. The molecules can be energy-focused by entering an electrostatic mirror that consists of a series of electrical lenses. The ions are slowed down until they come to rest; their direction of movement is reversed, and they are accelerated in the reverse direction. The faster moving ions penetrate deeper into the reflectron. Thus they travel a longer distance before hitting the detector, which compensates for the shorter flight time due to their higher initial velocity. A typical MALDI mass spectrum of an unseparated tryptic digest of bovine serum albumin is shown in **Fig. 8**.

3.2.1.2. ESI

ESI is a soft ionization method (*36,37*). In contrast to MALDI, the analyte molecules are transferred not from the solid, but from the liquid state into the gas phase. The principle of ESI is shown in **Fig. 9**. A solvent containing the analyte flows continuously at a rate of a few microliters per minute through a narrow capillary, the tip of

Structural Characterization of Proteins and Peptides

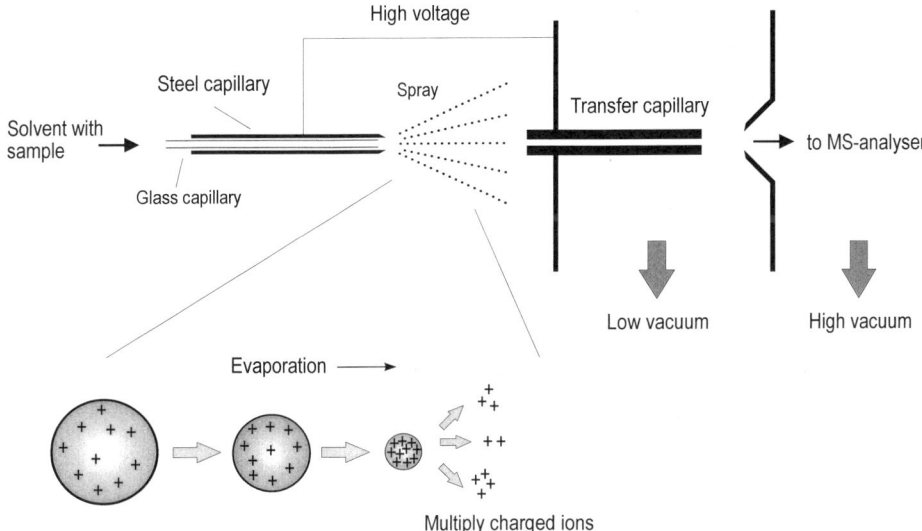

Fig. 9. Ion formation by electrospray ionization (ESI). The schematic drawing shows an ESI source that uses a heated capillary for transmission of ions from atmospheric pressure to the vacuum system. Alternatively, ions can be transmitted through a small orifice. For details, see text. MS, mass spectrometry.

which is held at a high potential up to about 5000 V. Because of this high voltage, the solvent does not exit the capillary as droplets but rather as a fine spray. Spray formation is facilitated by adding methanol or acetonitrile to the aqueous solution (final concentration approx 50%). Weak organic acids (<5%) help in ionization of the solute by transferring protons to basic amino acid residues. A stream of heated nitrogen gas evaporates the droplets on their way to the counterelectrode. With shrinking droplet size, the charge density increases until it becomes so high that the droplets explode due to Coulomb repulsion. In this way smaller droplets are formed that continue to be evaporated until disruption due to Coulomb repulsion, and so on. Finally, the solvent-free analyte ions are released. Spraying and most desolvation occur at atmospheric pressure, whereas mass spectrometry requires high vacuum, as mentioned earlier. Thus, an interface is needed to introduce the ions into the mass spectrometer. In some instruments the ions are allowed to enter through a narrow orifice (about 100 µm in diameter), whereas others use a transfer capillary. The ion beam first traverses a chamber with low vacuum before being guided by electronic lenses into the high vacuum chamber for mass analysis.

A recent fascinating development is the *nanospray technique*, which is able to produce flow rates of a few *nanoliters* per minute *(38,39)*. This allows sensitive analysis of complex mixtures (required, for instance, for sequencing peptides in an unfractionated digest; see below). A few microliters (typically 2–5 µL) of sample are sufficient for measurements over a time period of 0.5–1 h.

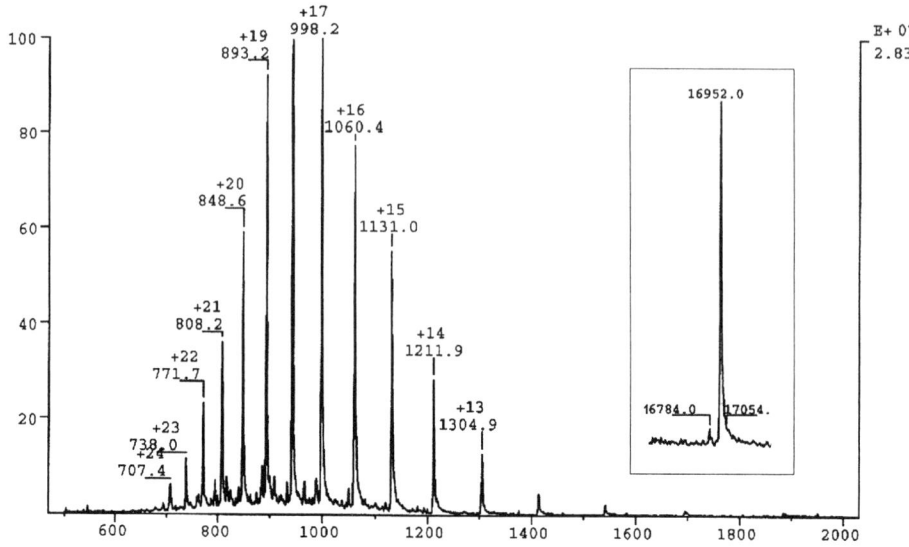

Fig. 10. ESI mass spectrum of myoglobin. Each peak represents a different charge state.

Figure 10 shows the ESI spectrum of pure myoglobin. Surprisingly, a whole series of peaks with different m/z ratios are detected, the neighboring ones differing by one charge unit. It is typical for ESI to produce protein/peptide ions with multiple charges, depending on the number of ionizable groups. In the case of myoglobin, a large number of basic residues (lysines and arginines) are protonated by the weak acid present in the solvent. The protonation is a statistical process: it is improbable that only one or a few basic residues are protonated, but it is also improbable that all accessible residues are protonated; thus the observed gaussian-like distribution is formed. When one knows that neighboring peaks differ by only one charge, the molecular weight of the protein can easily be calculated from the various m/z values. This can be done manually or more easily by a simple program (deconvolution; *37*). Tryptic peptides usually form singly or doubly charged species by protonation of the N-terminal amino group and/or the C-terminal lysine/arginine residue.

Analysis of ions produced by ESI is usually performed with *quadrupole analysers* or *ion traps* (**Fig. 11**).

Quadrupole instruments are the most frequently used mass spectrometers in a biochemical lab *(40)*. A quadrupole functions as a mass filter. It consists of four parallel metal rods. A DC voltage is applied to the rods superimposed by a radiofrequency potential. Each pair of opposite rods is electrically connected and has the same polarity of the DC voltage and the same phase of the AC voltage. The neighboring rods have the opposite polarity, and the phase is shifted by 180°. It is possible to choose values for the AC frequency and the AC and DC amplitudes such that only ions with a certain m/z value can traverse the quadrupole field on stable trajectories, whereas all other ions, with a different m/z ratio, are diverted from their linear path and are stopped

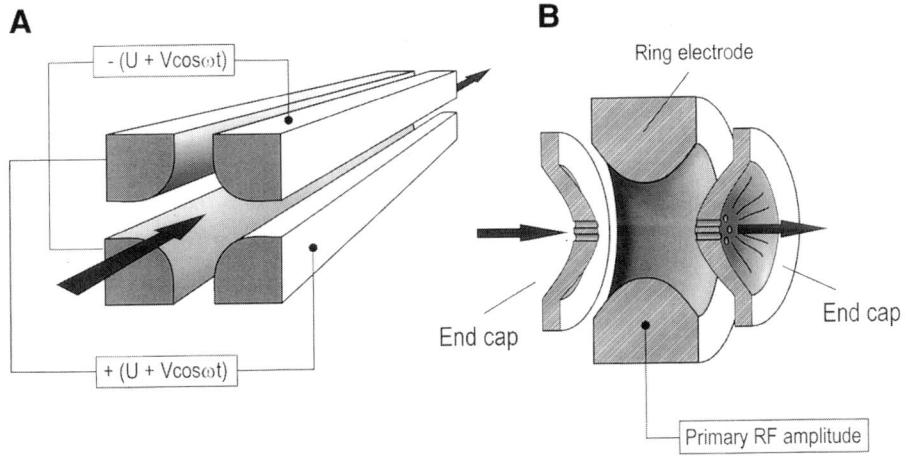

Fig. 11. Schematic drawing of (**A**) quadrupole and (**B**) ion trap mass analyzers. For details, see text. RF, radiofrequency.

by collision with the metal rods. If the applied voltages are varied over time in the right way, ions with increasing m/z ratios can pass the quadrupole filter in succession (ion scan). By recording the number of exiting ions as a function of the m/z ratio, a mass spectrum of the sample is obtained.

Quadrupole analyzers are technically quite simple and relatively cheap. In contrast to MALDI, the mass range that can be scanned is rather limited for technical reasons. Most quadrupole machines can only measure m/z ratios below 2500. Nevertheless, in combination with ESI, the molecular weight of proteins (up to about 100 kDa) can be measured since multiply charged species are formed.

Ion traps are mass spectrometers consisting of a doughnut-shaped ring electrode and two end caps with a hyperbolic cross-section, which have small openings for inlet and ejection of ions *(41)*. By applying appropriate AC and DC voltages to the electrodes, ions of a wide range of m/z values can be forced to move on stable trajectories in the cavity between the electrodes. This allows for accumulation and storage ("trapping") of ions for several milliseconds prior to analysis. Then, by changing the voltages, ions of successive m/z values can be ejected to record the mass spectrum. The function of the ion trap essentially follows the same principles as the quadrupole. (The mathematical treatment of the ion trajectories is almost the same.) This similarity can be seen looking at the cross-section, which resembles a two-dimensional slice through a quadrupole. One advantage of the ion trap is its high sensitivity because ions are accumulated for several milliseconds. Another advantage is the possibility of performing MS^n measurements (see below).

3.2.2. Instrumentation for Sequencing (MS/MS Techniques)

To obtain information about the amino acid sequence of a peptide, it is not sufficient to measure its mass; the peptide also has to be fragmented. The resulting pattern

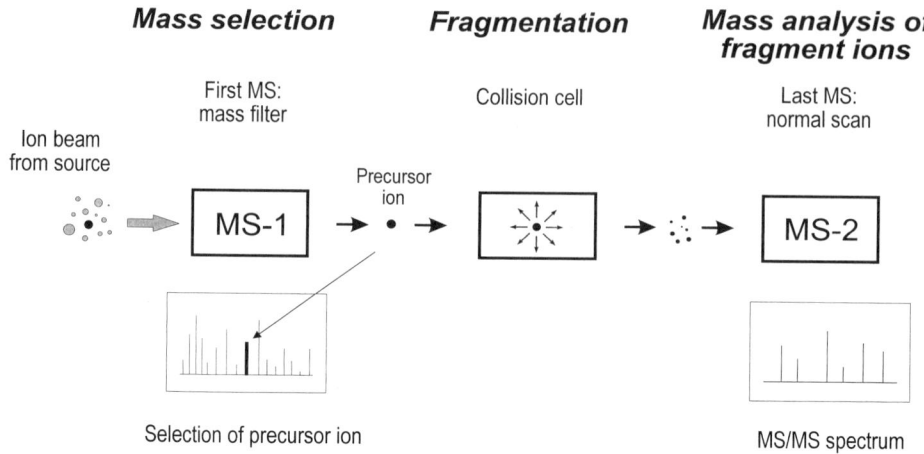

Fig. 12. Basic concepts of tandem mass spectrometry (MS).

of fragment ions is characteristic for a given peptide and thus contains information about its primary structure. This task can be performed by MS/MS techniques, that is, the coupling of two mass spectrometers, sometimes also called *tandem mass spectrometry* (**Fig. 12**):

1. The first mass spectrometer (MS1) selects the ion of interest (*precursor ion*). This step is necessary if the samples contain several peptides and is also recommended with pure peptides to reduce background.
2. The precursor ion is fragmented in the collision cell. Fragmentation can be induced by acceleration of the precursor ion to enhance its kinetic energy, followed by collisions with neutral gas atoms/molecules (argon, nitrogen, helium). The strength of the fragmentation is dependant on the acceleration voltage and the gas pressure in the collision cell.
3. The fragment ions are analyzed in a second mass spectrometer (MS2).
4. The mass spectrum is evaluated to obtain structural information (see below).

A variety of combinations of mass analyzers are commercially available. Triple quadrupole analyzers, hybrid mass spectrometers, and ion traps are widely used.

Triple quadrupole instruments use three quadrupoles, as the name indicates, arranged in tandem. The first quadrupole Q1 selects the ion of interest. The second quadrupole Q2 serves as a collision cell. This quadrupole does not select ions as usual, but rather parameters for the radiofrequency field are chosen that allow for focusing all fragment ions within the boundaries of Q2 and to prevent losses through the rods. The third quadrupole serves again as a conventional mass filter.

Triple quadrupole instruments have been the standard technique for many years, but in recent years *hybrid mass spectrometers* have been developed that use two different mass analysis principles, thereby profiting from the best performance features of both types of mass analyzers. Very popular is the combination of a quadrupole as the first analyzer and a TOF instrument as the second mass analyzer (*Q-TOF*). This device is more sensitive and has a better mass resolution than triple quadrupoles.

The most recent development is *MALDI-TOF-TOF*, combining two TOF analyzers. This type of spectrometer offers very high mass resolution and sensitivity. Another advantage is the fact that in the first step, protein identification on the basis of peptide fingerprints can be done as with normal MALDI (*see* below). If this information is not sufficient, in a second step structure information can be obtained by fragmentation of peptide ions using the same sample target (*see* below).

Ion traps represent a special group of instruments. They are not really tandem mass spectrometers because MS/MS measurements are performed within the same cell. This is possible because values for AC and DC voltages can be adjusted either to store a wide range of ions or to entrap single ions only: (1) First a normal mass spectrum is recorded, then the desired precursor ion is isolated by ejecting all other ions from the trap. (2) The selected ion is accelerated and fragmented by collisions with helium gas atoms in the trap. (3) The fragment ions are accumulated in the trap. (4) The fragment ions are successively ejected to record the mass spectrum. This principle makes it possible to perform "MS^n": In the next round, a fragment ion can be selected, fragmented once more followed by analysis of its fragment ions, and so on. This feature can be useful for structure determination and is not available with other types of spectrometers.

3.2.3. Identification of Proteins by Mass Spectrometry

Even though it is possible to determine complete sequences of peptides (*de novo* sequencing), even for experts this is not a trivial task *(42)*. However, mass spectrometry is a powerful and widely used technique for identification of proteins, either based on peptide masses alone or in combination with fragmentation data (for short overviews, *see* refs. *43–45*). Essentially three techniques are commonly used, as outlined below.

3.2.3.1. SEARCHING DATABANKS BY PEPTIDE MASSES (PEPTIDE MASS FINGERPRINTING)

This is the simplest method, and is most often performed in combination with MALDI. (The prevalence of singly charged peptides results in simpler spectra than with ESI.) The technique is based on the concept that the peptide pattern (peptide masses) obtained by cleavage with a specific protease like trypsin is characteristic for a given protein. Algorithms have been developed to identify proteins by correlating experimental peptide masses, determined by mass spectrometry, with calculated masses of theoretical proteolytic peptides derived from protein sequences in the database. The database is searched with a set of masses, with the bias that a user-defined minimum number must match for a protein to be counted as a positive hit. A score is then calculated to provide a measure of fit between experimentally derived and calculated peptide masses. This approach requires that the complete sequence of a protein be in the database and is sensitive to factors that cause a discrepancy between observed and calculated masses. These can be caused, for instance, by post-translational modifications such as glycosylation, or errors in the database entries (not a rare event). The method will also fail if the sample digested is mixture of proteins.

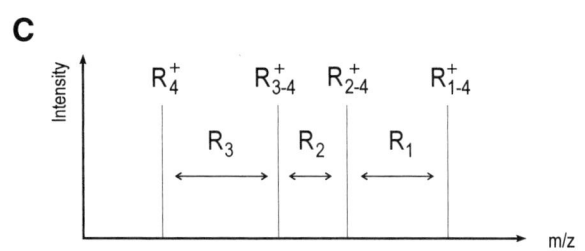

Fig. 13. Principles of sequencing by mass spectrometry, demonstrated for the Y-series (see **Fig. 14**) of a tryptic tetrapeptide. (**A**) Statistical cleavage of the chain's peptide bonds by collision with neutral gas atoms. (**B**) Resulting fragment series. (**C**) Neighboring mass peaks differ by just one amino acid residue.

3.2.3.2. SEARCHING DATABASES WITH UNINTERPRETED MASS SPECTRA

This approach uses information from the fragment ion spectrum of peptides to identify the corresponding protein in the database. The technique takes advantage of the fact that peptides fragment nonrandomly, yielding distinct series of fragments. To understand the principle, let us assume that a tryptic peptide carries the positive charge at the C-terminal Arg/Lys residue and that only peptide bonds are broken during fragmentation (**Fig. 13**). With a high probability, fragmentation occurs statistically at any peptide bond (the bonds are energetically almost equivalent), yielding a ladder of fragments differing by exactly one amino acid residue, from which the sequence can easily be deduced. In fact, this kind of ion series represents a major fragmentation series of tryptic peptides and is called the Y-series. In practice, the observed fragmentation pattern is more complex, because cleavage occurs at other bonds of the peptide backbone as well, and the charge can reside at the N-termi-

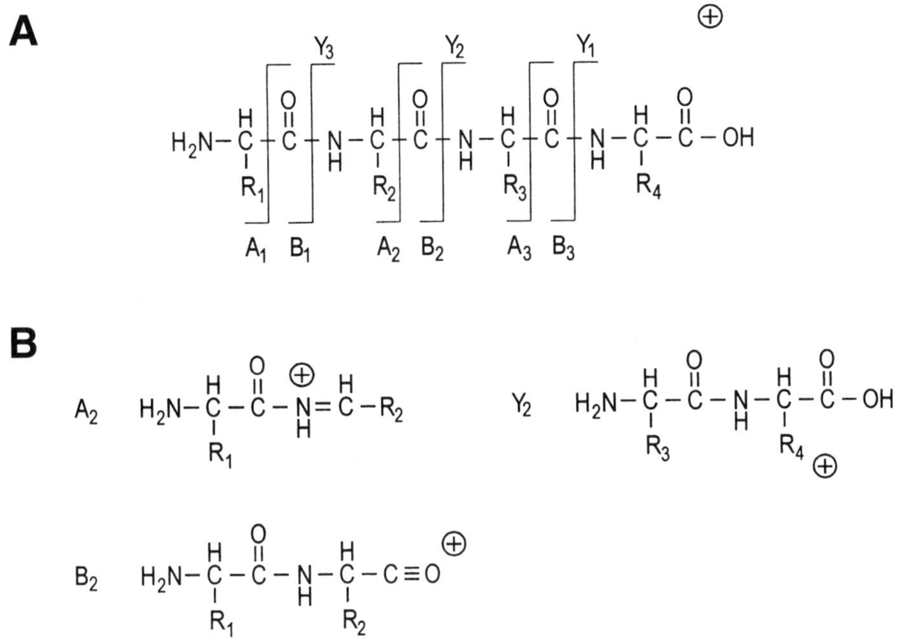

Structural Characterization of Proteins and Peptides

A

[Figure showing peptide backbone with fragmentation sites labeled Y_3, Y_2, Y_1 and $A_1 B_1$, $A_2 B_2$, $A_3 B_3$, with residues R_1, R_2, R_3, R_4]

B

A_2: H$_2$N–CH(R$_1$)–C(=O)–N$^+$H=CH–R$_2$

Y_2: H$_2$N–CH(R$_3$)–C(=O)–NH–CH(R$_4$)–C(=O)–OH (charged)

B_2: H$_2$N–CH(R$_1$)–C(=O)–NH–CH(R$_2$)–C≡O$^+$

Fig. 14. Most common C- and N-terminal ion series. (**A**) Y-series represents ions with the charge at the C-terminal amino acid, whereas A- and B-series carry charges at the N-terminal fragment ions. (**B**) Selected examples of the three ion series.

nal or C-terminal fragments. Some major ion series are shown in **Fig. 14** using the nomenclature given by Roepstorf and Fohlmann *(46)*. Cleavage can also occur within the side chains, making interpretation even more difficult.

Nevertheless, it is possible to calculate theoretical fragmentation spectra. Programs have been developed, such as *Sequest* and *Mascot*, that can identify a protein by using an experimental MS/MS scan to search a protein database for matches with patterns of fragment ions that are calculated for theoretical proteolytic peptides of the same mass as the selected peptide ion. After specification of the enzyme to be used for digestion, theoretical fragment patterns are generated for all peptides of proteins in the database having the specified mass. The program produces a list with the most probable proteins that has to be evaluated critically. This method is especially suited for ESI MS, because the degree of fragmentation can be better controlled.

3.2.3.3. SEQUENCE-TAG METHOD

Crucial for this method is the experimental determination of a partial sequence of a peptide; three or four amino acids are usually sufficient. Such short stretches of contiguous amino acids can often be recognized in the fragmentation patterns without problems. With knowledge of the mass and a partial sequence of a proteolytic peptide and with specification of the type of protease used for digestion, the database can be searched for matches. The technique is very reliable and tolerant to nonexact mass determinations, but it requires manual interpretation of the sequence tag.

4. Comparison of MS Methods and Edman Degradation for Sequencing/Identification of Proteins and Peptides

Each method has distinct advantages and disadvantages that the experimenter should know to be able to select the method best suited for solution of the particular problem.

Edman degradation is still the only practical method to obtain N-terminal sequence data and is also unsurpassed in obtaining longer (>15–25 amino acids) and continuous stretches of amino acids, whereas *de novo* sequencing with mass spectrometry is still a problem. To derive the complete sequence of a peptide, much experience in interpreting the fragmentation patterns of MS/MS spectra is required, in addition to special instrumentation *(42)*. Several chemical modification steps are often used to help determine the type of ion series. Thus Edman degradation is the first choice for generation of partial sequence data from unknown proteins, e.g., for construction of oligonucleotides for PCR cloning.

Identification of modified amino acids by the usual HPLC separation of the PTH amino acids can be a serious problem. Some modifications, such as methylation of amino acids, cause PTH amino acids to elute at unusual positions; others, like glycosylation, result in PTH amino acids that do not show up at all in the chromatogram. However, coupling and cleavage generally proceed normally, so sequencing is still possible, with some "holes" in the sequence. This information is usually sufficient for identification of the protein in the database, whereas MS methods often fail in these cases.

The sensitivity of Edman degradation is about 5 pmol of peptide; with a specially designed protein sequencer (CLC sequencer from Applied Biosystems), sensitivity can be extended to the high femtomole range. Unfortunately, Edman degradation requires pure peptides. Sequencing mixtures of peptides present in similar amounts is not possible due to ambiguities in assignment of the PTH amino acid to the respective sequence. The samples for Edman degradation must be of high purity and be free of salts, detergents, and nonvolatile additives such as urea.

In contrast to Edman degradation, mass spectrometry is ideally suited for analysis of complex peptide mixtures such as unfractionated protein digests, and peptide mass fingerprinting by MALDI takes advantage of this feature. If fingerprints yield ambiguous results, additional sequence information by MS/MS methods is required to allow database searching with uninterpreted MS/MS spectra or sequence tags, as shown above. Up to now the best technique for this application has been nanoESI-MS/MS, which permits the researcher to work with small amounts of an unfractionated digest for about 30 min. This is sufficient time to select a large number of peptides out of a digest and to optimize the conditions for fragmentation. ESI techniques can also be coupled with HPLC (LC/MS-coupling), although sensitivity drops dramatically using this technique. When one works in the low pmole range, substantial loss of material has to be expected, in part owing to irreversible adsorption to the column material.

Mass spectrometry techniques are fast and sensitive. A single MS analysis can be completed in a few minutes, whereas sequencing of a 15-amino-acids-long peptide by Edman degradation needs half a day. Therefore, mass spectrometry is the method of choice in proteomics. Identification of all spots from a 2D gel by Edman degradation is impossible. MALDI permits processing of multiple samples; peptide fingerprinting of several hundred samples or even more can be done automatically.

For recording of a single mass scan, only a few tens of attomols of sample are consumed, with MALDI being more sensitive than ESI. These numbers are a bit misleading, however, because the spectra are usually averaged over 5–10 scans for a better signal-to-noise ratio, and only a small fraction of the sample contributes to the signal: in MALDI, only a part of the target is evaporated into the gas phase, and in ESI only part of the spray enters the capillary/orifice to the mass analyzer. In routine applications, MALDI needs sample concentrations of about 0.1–1 pmol/µL; ESI requires somewhat higher concentrations.

In the literature, examples of subfemtomole peptide sequencing are reported (e.g., ref. *47*) using nanoHPLC/nanoESI or capillary electrophoresis/ESI techniques. However, this level of perfection is hardly achieved with routine samples. First, very small sample volumes are required (in the range of 1 µL or even less). Handling of such small volumes is not easy and requires some practice. Second, and most important, high-quality samples are required. Sample preparation often makes the difference between a failed analysis and good results. Pure samples can be obtained quite easily by dilution of standard proteins/peptides to the desired concentration, but high purity is very difficult to achieve with real samples. The popular purification techniques on classical separation media like Sephadex and Dowex are not suitable (as the last step), even though many people do not believe this. The background signals can be so high that it is impossible to measure samples below 20–50 pmol/mL, apart from pollution of the instrument, which is a major concern of the mass spectroscopist. A single dirty sample can block the instrument for several days. Proteolytic digests have to be carefully desalted for MS measurements, for instance, using pipet tips/glass capillaries containing a small bed volume of reverse-phase beads (*39*). Samples for mass spectrometry should also be of high purity, MALDI being somewhat more tolerant than ESI. To protect the samples from being contaminated by keratin from your fingers you should wear gloves, and the Eppendorf tubes should be rinsed with reagent-grade methanol or acetonitrile before use.

Mass spectrometry has evolved into a powerful method, with the major weak point of *de novo* sequencing and analysis of highly modified peptides. Its application is not restricted to protein identification; it is also the method of choice for analysis of post-translational modifications. Comparison of the mass expected on the basis of the gene sequence with the experimentally observed mass permits conclusions about the kind of modification. However, as a daily workhorse for the laboratory, Edman degradation still has its importance. Many samples to be analyzed are just blotted proteins to check whether the right protein had been isolated.

Even though Edman degradation could probably be further miniaturized employing new technologies for delivery and handling of small amounts of reagents, improvements is this technology have been rare in recent years. Mass spectrometry, on the other hand, has been a rapidly developing technique. Sensitivity, mass accuracy, mass resolution, sample throughput, and handling of samples have been improved continuously, as has new software. We can expect many new developments over the next few years to facilitate the characterization of the molecular design of life.

References

1. Walker, J. M., ed. (2002) *The Protein Protocols Handbook*. Humana, Totowa, NJ.
2. Howard, G. C. and Brown, W. E., eds. (2002) *Modern Protein Chemistry: Practical Aspects*. CRC Press, Boca Raton.
3. Smith, B. J., ed. (1997) Protein sequencing protocols, in *Methods in Molecular Biology*, vol. 64 (Walker, J. M., ed.), Humana, Totowa, NJ.
4. Kellner, R., Lottspeich, F., and Meyer, H. E., eds. (1994) *Microcharacterization of Proteins*. VCH, Weinheim.
5. Matsudaira, P., ed. (1993) *A Practical Guide to Protein and Peptide Purification for Microsequencing*. Academic, San Diego.
6. Allen, G. (1989) Sequencing of proteins and peptides, in *Laboratory Techniques in Biochemistry and Molecular Biology*, vol. 9 (Burdon, R. H. and van Knippenberg, P. H., eds.), Elsevier, Amsterdam.
7. Darbre, A. (1986) *Practical Protein Chemistry—A Handbook*, John Wiley & Sons, New York.
8. Gooding, K. M. and Regnier, F. E., eds. (2002) *HPLC of Biological Macromolecules*, 2nd ed., in *Chromatographic Science Series*, vol. 87, Marcel Dekker, New York.
9. Oliver, R. W. A., ed. (1998) *HPLC of Macromolecules*, in *The Practical Approach Series* (Rickwood, D. and Hames, B. D., eds.), IRL, Oxford.
10. Podell, D. N. and Abraham, G. N. (1978) A technique for the removal of pyroglutamic acid from the amino terminus of proteins using calf liver pyrogutamate amino peptidase. *Biochem. Biophys. Res. Commun.* **81,** 176–185.
11. Hirano, H., Komatsu, S., and Tsunagawa, S. (1997) On-membrane deblocking of proteins, in *Protein Sequencing Protocols* (Smith, B. J., ed.), Humana, Totowa, NJ, pp. 285–292.
12. Bergmann, T., Gheorghe, M. T., Hjelmqvist, L., and Jörnvall, H. (1996) Alcoholytic deblocking of N-terminally acetylated peptides and proteins for sequence analysis. *FEBS Lett.* **390,** 199–202, and references therein.
13. Gravel, P. (2002) Protein blotting by the semidry method, in *The Protein Protocols Handbook* (Walker, J. M., ed.), Humana, Totowa, NJ, pp. 321–334.
14. Eckerskorn, C. (1994) Electroblotting, in *Microcharacterization of Proteins* (Kellner, R., Lottspeich, F., and Meyer, H. E., eds.), VCH, Weinheim, pp. 75–92.
15. Matsudaira, P. (1987) Sequence from picomole quantities of proteins electroblotted onto polyvinylidene difluoride membranes. *J. Biol. Chem.* **262,** 10,035–10,038.
16. Eckerskorn, C., Mewes, W., Goretzki, H. W., and Lottspeich, F. (1988) A new siliconized-glass fiber as a support for protein chemical analysis of electroblotted proteins. *Eur. J. Biochem.* **176,** 509–519.
17. Bjerrum, O. J. and Schafer-Nielson, C. (1986) Buffer systems and transfer parameters for semidry electroblotting with a horizontal apparatus, in *Electrophoresis 1986* (Dunn, M. J., ed.), VCH, Weinheim, pp. 315–327.

18. Towbin, H., Staehlin, T., and Gordon, J. (1979) Electrophoretic transfer of proteins from polyacrylamide gels to nitrocellulose sheets: procedure and some applications. *Proc. Natl. Acad. Sci. USA* **76**, 4350–4354.
19. Aebersold, R. H., Leavitt, J., Saavedra, R. A., Hood, L. E., and Kent, S. B. (1987) Internal amino acid sequence analysis of proteins separated by one- or two-dimensional gel electrophoresis after in situ protease digestion on nitrocellulose. *Proc. Natl. Acad. Sci. USA* **84**, 6970–6974.
20. Fernandez, J. and Mische, S. M. (2002) Enzymatic digestion of membrane-bound proteins for peptide mapping and internal sequence analysis, in *The Protein Protocols Handbook* (Walker J. M., ed.), Humana, Totowa, NJ, pp. 523–532.
21. Szewczyk, B. and Summers, D. (1988) Preparative elution of proteins blotted to immobilon membranes. *Anal. Biochem.* **168**, 48–53.
22. Jenö, P. and Horst, M. (2002) Electroelution of proteins from polyacrylamide gels, in *The Protein Protocols Handbook* (Walker, J. M., ed.), Humana, Totowa, NJ, pp. 299–305.
23. Amons, R. (1987) Vapor-phase modification of sulfhydryl groups in proteins. *FEBS Lett.* **212**, 68–72.
24. Shevchenko, A., Wilm, M., Vorm, O., and Mann, M. (1996) Mass spectrometric sequencing of proteins from silver-stained polyacrylamide gels. *Anal. Chem.* **68**, 850–858.
25. Rosenfeld, J., Capdevielle, J., Guillemot, J. C., and Ferrara, P. (1992) In-gel digestion of proteins for internal sequence analysis after one- or two-dimensional gel electrophoresis. *Anal. Biochem.* **203**, 173–179.
26. Aguilar, M.-I. and Hearn, M. (1996) High resolution reverse-phase high-performance liquid chromatography of peptides and proteins. *Methods Enzymol.* **270**, 3–26.
27. Stone, K. L. and Williams, K. R. (2002) Reverse-phase HPLC separation of enzymatic digests of proteins, in *The Protein Protocols Handbook* (Walker, J. M., ed.), Humana, Totowa, NJ, pp. 533–540.
28. Lottspeich, F., Houthave, T., and Kellner, R. (1994) The Edman degradation, in *Microcharacterization of Proteins* (Kellner, R., Lottspeich, F., and Meyer, H. E., eds.), VCH, Weinheim, pp. 117–130.
29. Hunkapiller, M. W., Hewick, R. M., Dreyer, W. J., and Hood, L. E. (1983) High-sensitivity sequencing with a gas-phase sequenator. *Methods Enzymol.* **91**, 399–413.
30. Chapman, J. R., ed. (1996) Protein and peptide analysis by mass spectrometry, in *Methods in Molecular Biology* (Walker, J. M., ed.), Humana, Totowa, NJ.
31. Dass, C. (2001) *Principles and Practice of Biological Mass Spectrometry*, John Wiley & Sons, New York.
32. Karas, M. and Hillenkamp, F. (1988) Laser desorption ionisation of proteins with molecular masses exceeding 10000 D. *Anal. Chem.* **60**, 2299–2301.
33. Beavis, J. F. and Chait, B. T. (1996) Matrix-assisted laser desorption ionization mass-spectrometry of proteins. *Methods Enzymol.* **270**, 519–551.
34. Vestal, M. L., Juhasz, P., and Martin, S. A. (1995) Delayed extraction matrix-assisted laser desorption time-of-flight mass spectrometry. *Rapid Commun. Mass Spectrom.* **9**, 1044–1050.
35. Brown, R. S. and Lenon, J. J., (1995) Mass resolution improvement by incorporation of pulsed ion extraction in a matrix-assisted laser desorption/ionization linear time-of-flight mass spectrometer. *Anal. Chem.* **67**, 1998–2002.
36. Fenn, J. B., Mann, M., Meng, C. K., Wong, S. F., and Whitehouse, C. M. (1989) Electrospray ionization for the mass spectrometry of large biomolecules. *Science* **246**, 64–71.

37. Banks, J. F. and Whitehouse, C. M. (1996) Electrospray ionization mass spectrometry. *Methods Enzymol.* **270**, 486–518.
38. Wilm, M. and Mann, M. (1996) Analytical properties of the nano electrospray ion source. *Anal. Chem.* **66**, 1–8.
39. Jensen, O. N. and Wilm, M. (2002) Peptide sequencing by nanospray tandem mass spectrometry. in *The Protein Protocols Handbook* (Walker, J. M., ed.), Humana, Totowa, NJ, pp. 693–710.
40. Dawson, P. H. (1995) *Quadrupole Mass Spectrometry and Its Applications*, AIP Press, Woodbury, NY.
41. Jonscher, K. R. and Yates III, J. R. (1997) The quadrupole ion trap mass spectrometer—A small solution to a big challenge. *Anal. Biochem.* **244**, 1–15.
42. Shevchenko, A., Chernushevich, I., Wilm, M., and Mann, M. (2000) De novo peptide sequencing by nanoelectrospray tandem mass spectrometry using triple quadrupole and quadrupole/time-of-flight instruments, in *Mass Spectrometry of Proteins and Peptides* (Chapman, J. R., ed.), Humana, Totowa, NJ, pp. 1–16.
43. Lamond, A. I. and Mann, M. (1997) Cell biology and the genome projects—a concerted strategy for characterizing multiprotein complexes by using mass spectrometry. *Trends Cell Biol.* **7**, 139–142.
44. Patterson, S. D. (1997) Identification of low to subpicomolar quantities of electrophoretically separated proteins: towards protein chemistry in the post-genome era. *Biochem. Soc. Trans.* **25**, 255–262.
45. Fenyö, D. (2000) Identifying the proteome: software tools. *Curr. Opin. Biotechnol.* **11**, 391–395.
46. Roepstorff, P. and Fohlmann, J. (1984) Proposal for a common nomenclature for sequence ions in mass spectra of peptides. *Biomed. Mass Spectrom.* **11**, 601–611.
47. Martin, S. E., Shabanowitz, J., Hunt, D. F., and Marto, J. A. (2000) Subfemtomol MS and MS/MS peptide sequence analysis using nano-HPLC micro-ESI Fourier transform ion cyclotron resonance mass spectrometry. *Anal. Chem.* **72**, 4266–4276.

Appendix
Protocol 1: Electroblot

1. Prepare transfer buffer by titrating 50 mM boric acid to pH 9. The buffer should contain methanol to a concentration of 10–20% (depending on the molecular weight of the protein to be blotted; see text).
2. Cut blotting paper (Whatman 3MM) and PVDF membrane to the same size as the gel to be blotted and equilibrate for 30 min in transfer buffer, changing the buffer at least once. The PVDF membrane *has to be wetted with methanol* before soaking with transfer buffer (1–2 min; instead of methanol, 5% Triton might be used). The PVDF membrane is highly hydrophobic. If you omit immersion in methanol, the membrane will not be wetted by the transfer buffer, resulting in a lack of protein binding. Once wetted, the membrane must stay immersed in buffer and not be allowed to dry.
3. After electrophoresis, cut out the part of the SDS gel (mini-gel, 1-mm thick) to be blotted with a scalpel and equilibrate for 15 min in transfer buffer.
4. Assemble the sandwich, consisting of three layers of blotting paper, gel, and membrane and another three layers of blotting paper on top of the bottom electrode. The membrane must be located toward the anode. *Avoid air bubbles between the layers!* Cover the assembly with the top electrode. A soft pressure can be applied by putting a weight of about 1 kg on top.

5. Perform electroelution for 2–3 h at a current of 1.5 mA/cm^2.
6. Stain the membrane with Coomassie Blue (0.1% in a solution containing 50% methanol and 10% acetic acid) for 5–10 min, followed by destaining using a mixture of methanol/water/acetic acid (5:4:1) until the background is clear. The stain does not interfere with sequencing. Afterward the blot *should be shaken for at least 1 h in pure water* to remove traces of buffer bound to membrane. Otherwise the background tends to be so high that the first amino acids cannot be detected in Edman degradation. To check the efficiency of the blotting, the gel should also be stained.
7. Cut out the bands of interest from the moist blot with a scalpel; they are now ready for sequencing. Store at –20°C, but sending air-dried spots by mail is possible.
Caution: Always wear gloves to avoid contamination of your sample.

Protocol 2: In-Gel Digestion

1. Run SDS gel. (The pore size of the gel should be sufficiently large; 10% gels are very good.)
2. Stain the gel with Coomassie Blue as usual, for 30 min.
3. Destain the gel with 7% acetic acid with or without 30% methanol (destaining without methanol takes longer but gives more intense bands) for 2–3 h.
4. Rinse the gel in water for 1 h or overnight.
5. Cut out bands of interest, following the contours of the stained band as closely as possible. Estimate the volume of the excised band. Cut the band with a sharp scalpel into small cubes of about 1-mm side length.
6. Wash the gel pieces with gentle shaking in a 2-mL Eppendorf tube successively with 0.2 M NH$_4$HCO$_3$, 0.2 M NH$_4$HCO$_3$/25% acetonitrile, 25% acetonitrile in water, 50% acetonitrile, 100% acetonitrile. During washing (30 min per step) the gel pieces will shrink to a small percentage of the original volume.
7. Air-dry washed gel pieces at room temperature. (They should no longer stick to the walls of the cup.)
8. Digest overnight at 37°C with trypsin in 0.2 M NH$_4$HCO$_3$ (final concentration: 2 µg trypsin/100 µL gel); the total volume of buffer added should exceed the original gel volume by about 50% because in NH$_4$HCO$_3$ the gel pieces swell to a larger volume compared with water. The buffer is added in two portions of equal size:
    a. Add buffer containing the protease at a concentration of 4 µg/100 µL, and wait 5–10 min until the buffer has been soaked up by the gel. Gently turn over the gel pieces with a spatula to wet the whole surface.
    b. Add the same volume of buffer without protease, gently turning over the gel pieces for even wetting.
9. Extract peptides twice with 5% TFA, followed by 5% TFA/acetonitrile (1:1, optional), 30 min to 1 h each. The combined extracts are lyophilized.
10. Dissolve the peptides in 5% TFA (100 µL) for HPLC separation.
Caution: Always wear gloves to avoid contamination of your sample.

19

Determination of Kinetic Data Using Surface Plasmon Resonance Biosensors

Claudia Hahnefeld, Stephan Drewianka, and Friedrich W. Herberg

Abstract

The use of biosensors employing surface plasmon resonance (SPR) provides excellent instrumentation for a label-free, real-time investigation of biomolecular interactions. A broad range of biological applications including antibody–antigen interactions can be analyzed. One major advantage of kinetic analysis using SPR-based biosensors is the option of determining separately distinct association and dissociation rate constants exceeding the classical steady-state analysis of biomolecules. Based on these data new possibilities for drug design, characterizing human pathogens, and the development of therapeutic antibodies can be achieved. The hardware of commercially available systems is described, practical step by step procedures are given, and possibilities and limitations of the technology are discussed.

Key Words: Biacore; biomolecular interaction analysis; antibody analysis; surface plasmon resonance; kinetics; biosensor.

1. Introduction

Biomolecular interaction analysis (BIA) is nowadays used to describe in detail interactions among small molecules, proteins, peptides, nucleotides, sugars, and other biomolecules. Applications also cover the characterization of antibodies and intra- and extracellular components like infectious agents. The outcome of such detailed investigations may be the knowledge of potential agonists, antagonists, or diagnostic targets. Furthermore, there is an urgent demand for advances and development of novel methods in protein analysis with high throughput, high sensitivity, and analytical flexibility.

Although a vast variety of technologies exists to determine the interaction of clinically relevant biomolecules in vivo and in vitro, only a few techniques allow the direct determination of separate association and dissociation rate constants. However, only the knowledge of these distinct kinetic parameters allows the detailed characterization of several classes of important molecules, for example, antibodies. These could be polyclonal as well as monoclonal antibodies, fragments of antibodies, or genetically engineered gene products.

The tremendous amount of antibodies produced by both hybridoma and recombinant technologies requires solid and accurate methods for the process of selecting and characterizing suitable candidates. These selections include determination of the exact specificity of an antibody (epitope mapping), as well as knowledge of the affinity of an antibody–antigen complex and of the stability of the complex. Although a number of classical methods exists to determine the interaction of an antibody with an antigen, methods like enzyme-linked immunosorbent assays (ELISA) and radioimmunoassays (RIA) represent equilibrium binding technologies with the advantages of being rather fast, easily automated, and allowing a large number of samples to be analyzed. However, these methods do not distinguish between binding patterns based on specific association and dissociation phases. Additionally, the interaction by those methods cannot be determined directly, making it necessary to add labels (fluorescent labels, radioactive markers) or enzymatic reactions to make those interactions visible. Most methods require precipitation of the antibody–antigen complex, making quantification more difficult and less reproducible and thereby requiring repetitive assays to achieve an accurate result. The analysis of binding patterns is especially important for the characterization of antibodies. One demand could be that the antibody should bind fast (to achieve short assay times) and bind tightly, i.e., display a slow dissociation rate (to allow highly sensitive reactions with a low amount of antibody).

1.1. Instrumentation

Generally biosensors are composed of three parts: a sensor device employing different physical and/or optical principles, a sample delivery system, and a sensor surface whereby one of the interaction partners is immobilized. Interaction of a binding partner from the soluble phase is then monitored in real-time. Several biosensor systems employ surface plasmon resonance (SPR) as the physical principle of detection. Based on sales and on the amount of scientific literature published, SPR devices, for example, the Biacore instruments, are the most commonly used *(1)*.

Biacore (Uppsala, Sweden) introduced the first optical biosensor at the beginning of the 1990s, and several different systems are now available. For research purposes, the Biacore 1000, 2000, and 3000 as high-end instruments with four separate flow cells and the Biacore X and J in the lower price segment were developed. For the food industry, Biacore Q allows the qualitative or quantitative determination of analytes in food, for example, water-soluble vitamins or veterinary residues. More recently, the validated Biacore C system has allowed measurements under good laboratory practice (GLP) and good manufacturing practice (GMP) conditions important for clinically applied antibodies. A complete new platform is the S-series of instruments; the Biacore S51, introduced at the end of 2001, is geared to drug screens and the analysis of low-molecular-weight compounds with increased sensitivity and throughput. Future advances are aimed at the development of parallelization (eight-channel instruments) and multispot arrays to achieve increased throughput, a prerequisite for the analysis of the vast number of available antibody products and their prospective antigens.

Other instruments based on resonant mirrors [for example, the IAsys system by Thermo Finnigan, San Jose, CA) and the instrument of biomolecular interaction sensing (IBIS), Windsor Scientific, Slough, UK *(2)*] are available (for a more complete overview, *see* ref. *3*). In addition to the method of detection, there are principle differences in sample delivery, i.e., cuvet-based systems and systems with constant flow (*see* also **Note 7** for emerging technologies).

1.1.1. Basis of Detection

In BIA the binding of a molecule in the soluble phase (the "analyte") is directly measured to a "ligand" molecule immobilized on a sensor surface. In the sensor device the binding of the ligand is monitored by an optical phenomenon termed surface plasmon resonance (SPR).

The sensor device of the Biacore system consists of a LED-emitting near infrared light, a glass prism fixed to a sensor microchip, and a position-sensitive diode array detector. Binding events cause changes in the refractive index at the surface layer, which are detected as changes in the SPR signal (*see* below). In general, the refractive index change for a given change in mass concentration at the surface layer of a sensor chip is practically the same for all proteins and peptides *(4)*; however, it is slightly different for glycoproteins, lipids, and nucleic acids. This value is plotted as response units (RUs): for a general carboxymethylated dextran-coated chip (CM5), 1000 RU correspond to 1 ng protein/mm^2 sensor surface *(4)*.

SPR arises when light illuminates thin conducting films (gold is used in the case of Biacore instruments) under specific conditions. The resonance is a result of the interaction between electromagnetic vectors in the incident light and free electron clouds, called plasmons, in the conductor. SPR can arise because of a resonant coupling between the incident light energy and surface plasmons in the conducting film at a specific angle of incident light. Absorption of the light energy causes a characteristic drop in the reflected light intensity at that specific angle (**Fig. 1**).

The resonance angle θ is sensitive to a number of factors, including the wavelength of the incident light, the nature and thickness of the conducting film, and the temperature. Most important for this technology, the angle depends on the refractive index of the medium opposite to the incident light. When other factors are kept constant, the resonance angle is a direct measure of the refractive index of the medium. Only the angle on which SPR occurs is altered and detected with the diode array detector; the intensity of the "shadow" in the reflected light is unchanged. One thousand RU correspond to a 0.1° arc in the SPR angle.

The signal exponentially decays with the distance from the interface between a high refractive (prism with conducting film) and a low refractive index medium (interaction chamber/flow cell) *(5)*. In consequence, SPR only detects changes in refractive index very close to the surface, i.e., in the Biacore system about 300 nm. Therefore, the setup has to be optimized for interactions close to the detection surface in a biocompatible environment. Dextrans of different lengths have been proved to be

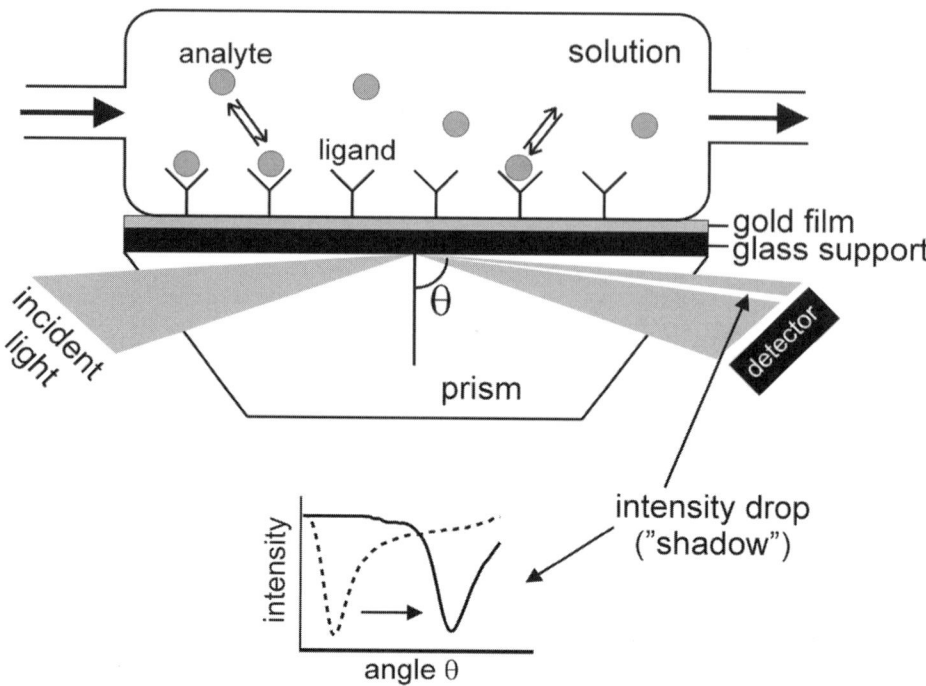

Fig. 1. Schematic view of a surface plasmon resonance (SPR) detector as utilized in a Biacore system. SPR arises when light is totally internally reflected from a metal-coated interface between two media of different refractive index (a glass prism and solution). If the incident light is focused on the surface in a wedge, the drop in intensity at the resonance angle appears as a "shadow" in the reflected light wedge, which is detected by a position-sensitive diode array detector. When an interaction between an immobilized ligand (e.g., an antibody, Y) and an analyte in solution (filled circles) occurs, the "shadow" is shifted on the detector, i.e., the angle θ changes.

excellent for interaction studies allowing for high surface densities and low nonspecific binding. However, for detailed kinetic analysis, the ligand molecule has to be immobilized in a biologically active conformation to a suitable sensor chip (*see* **Subheading 3.1.** and **Note 1**).

1.1.2. Sensor Chips

A wide range of functionalized sensor surfaces developed for covalent immobilization and noncovalent capturing of biomolecules including peptides, proteins, oligonucleotides, and membrane-bound receptors is available. All sensor chips are designed for chemical stability, stable baselines, high sensitivity, and reproducibility. They should allow extensive regeneration procedures, a prerequisite for multiple reuse, and should display low nonspecific binding. The sensor chip surface consists of a glass support with a gold layer. For most applications the gold film is linked to a non-

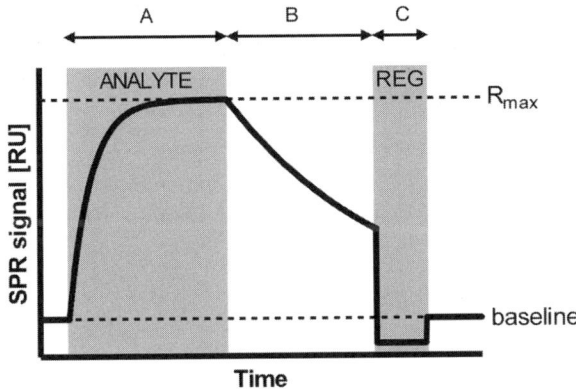

Fig. 2. Basic parts of a sensorgram. A typical sensorgram consists of three phases. **(A)** In the *association phase*, the analyte is injected over the immobilized ligand on the surface (ANALYTE, gray bar). With increasing interaction of analyte and ligand, an increasing response is detected [displayed in response units (RU)]. The maximal binding is specified as R_{max}. **(B)** The injection of the analyte is stopped by switching the system back to buffer (*dissociation phase*). In many cases the dissociation of the analyte is not complete after a reasonably long time. **(C)** Therefore an injection with an appropriate regeneration solution (REG, gray bar) is performed. After this *regeneration phase*, the baseline response level should be reached. If this is not the case, the regeneration can be repeated with either the same or another solution, keeping in mind the maintenance of the biological activity of the surfaces.

crosslinked carboxymethylated dextran hydrogel bound via an alkyl thiol layer (for a complete list of available sensor chips, *see* **Note 1**). Covalent coupling is performed using primary amines, aldehydes, or reactive thiols. High-affinity capture is based on streptavidin interactions, interactions between antibodies and specific fusion tags, or ligand-specific interactions.

Once the ligand surface shows a stable baseline, the analyte is injected. Association and dissociation phases are now monitored separately and plotted in form of a sensorgram (*see* **Fig. 2** and legend for details). From those data the association rate (k_a) and dissociation rate (k_d) constants are determined. With a known concentration of analyte, apparent equilibrium binding constants (K_D or K_A) are calculated. Furthermore, median effective concentration (EC_{50}) values for competitors are determined by solution competition or surface competition. For experiments in the solution competition assay format, analytes bind either to immobilized ligands (which is detected) or to competitors in solution. In contrast, surface competition occurs when a competitor competes with analytes for the same binding site of an immobilized ligand *(6)*.

1.2. Applications of Biomolecular Interaction Analysis

A broad range of biosensor data have been published *(1)*. Networks within the cells are mapped in the signal transduction field; multimolecular complexes containing proteins as well as nucleotides or lipid components are analyzed; and the role of posttranslational modifications in the kinetics of interaction patterns is elucidated, or even

1.2.1. Analysis of Human Pathogens

SPR has been employed for the analysis of human pathogens. In this context the complement system as part of the innate human immune defence is a major research topic. Complement proteins are efficient in the direct neutralization of microbes and at the same time enhance subsequent immune mechanisms of the acquired immune defense. Complement regulators protect the host cells against the activation of the complement proteins in response to pathogens. With the SPR technology, specific complement reactions are followed *in situ* in real time *(7,8)*. This leads to the identification of microbial surface structures of different organisms such as *Streptococcus pneumoniae (9)*, *Borrelia burgdorferi (10)*, and *Candida albicans (11)*. Kinetic analysis helps to identify the most important ligands for the proposed pathogenic mechanisms. In many cases pathogenic microorganisms utilize the same binding sites as for the binding of complement regulators, which are also important for the physiological functions of the regulators.

The human pulmonary surfactant protein A (hSP-A) is thought to play a key defensive role against airborne invading pulmonary pathogens like *Mycobacterium tuberculosis*. Sidobre et al. *(12)* developed an SPR assay to analyze the molecular basis for the recognition of a mycobacterial cell wall lipoglycan (ManLAM) by hSP-A.

Furthermore, Biacore instruments have been used to monitor *Salmonella* infections employing serological methods *(13)*, providing quality control and safety measures in food processing. Thus, microorganisms (such as *Salmonella*, *Campylobacter*, *E. coli O157*, and *Mycobacterium avium paratuberculosis*), vitamins, toxins, and residues of veterinary drugs are detected. Systems like the Biacore Q and the not yet available Biacore TAS (= *t*hroughput *a*nalysis *s*ystem) with eight flow cells are developed for this kind of analysis.

SPR-based systems have also been employed to quantify virus titers. Usually the determination of serum potentially infected with human immunodeficiency virus (HIV) is a two-step process. First serum samples are screened using a standard ELISA assay on microtiter plates or immunobeads or by latex aggregation. If positive results show up, an additional test must be performed to confirm these seropositives finally. Using Western blot analysis, each of the HIV proteins is then identified individually by probing specific antibodies. Both analytical methods are needed to obtain an accurate judgment for HIV-infected serum. Recently Hifumi et al. *(14)* described an assay on a custom-made SPR-based system that allows simultaneous analyses of screening (ELISA and others) and confirmation steps (Western blotting). By removing serum components causing nonspecific reactions, an antibody for p24 in human serum sample could be detected in a range from 1 to 20 µg/mL.

1.2.2. Therapeutic Antibodies

Recent reports describe the use of therapeutic antibodies in cancer, infectious diseases, thrombosis, or even neurogenerative diseases. McLaurin et al. *(15)* immunized transgenic Alzheimer mice using amyloid-beta peptide (Abeta) and could demonstrate a reduction of both the Alzheimer's disease-like neuropathology and the spatial memory impairments of these mice. These findings provide the basis for the development of improved immunizing antigens as well as attempts to design small-molecule mimics for alternative therapies.

Several problems are associated with the use of therapeutic antibodies for applications in human infectious diseases or cancer. Murine monoclonal antibodies show a high degree of immunogenicity and thus are not compatible for use in humans. Therefore, chimeric antibodies with a human constant region combined with the variable regions from the light and heavy chains of the therapeutic murine antibodies are developed using genetic engineering technologies. Further modifications are made when variable framework regions (in addition to the constant regions) are replaced with human sequences, leaving from the murine antibody only the small complementarity-determining regions (CDRs) essential for epitope mapping. The resulting humanized antibodies will cause no immunogenic response in patients but should retain the binding characteristics of the parent murine antibodies. Antibodies with strong neutralizing activity are designed containing only minimal murine CDRs with optimal binding kinetics for the epitope to select suitable candidates for therapeutic applications *(16)*.

SPR technology is used to optimize such chimeric or humanized antibodies by comparing the kinetics of the respective antigen–antibody interactions. Therefore, the antigen is immobilized on a chip surface (for example, employing the CM5 sensor chip; *see* **Note 1**) via primary amines, resulting in a ligand population rather heterogeneously immobilized. A site-directed coupling is achieved by engineering a cysteine residue into the antibody and subsequently immobilizing this antibody using thiol coupling (*see* **Subheading 2.1.**).

There are several ways to analyze antigen–antibody interactions. For accurate measurements of the association rate constant (k_a), the exact analyte concentration is needed, a parameter that is not easily available for antibodies. Ranking of different antibodies is more useful by the dissociation rate constant (k_d), because the analyte concentration is not required for calculation. For further optimization, humanized antibodies are chosen with high k_a values and very low k_d values.

Another application for SPR technology in the field of therapeutic antibodies is the detection and quantification of antibody markers during diagnosis and therapy of various immune-mediated diseases, classically carried out by ELISA screening. SPR analyses on high-density surfaces are performed by injecting diluted serum samples in buffer containing 1 mg/mL carboxymethyl (CM) dextran to minimize nonspecific binding of serum proteins to the CM–dextran matrix. A reference surface (containing either nonrelated antibody of the same type or an activated/deactivated CM-dextran matrix) is subtracted. Linearity of the assay should be tested by a series of different sample dilutions. Further enhancement of specificity and sensitivity of the assay is achieved by injection of secondary antibodies.

Alaedini et al. *(17)* describe the use of SPR in the diagnosis of patients and in accurate monitoring of serum antibody levels in response to treatment of infectious or autoimmune diseases.

2. Materials

The Biacore system allows one to perform direct measurements between an immobilized ligand and an analyte in the liquid phase. Because the ligand needs to be maintained in an almost biologically active state, extreme care in the preparation of the ligand surface is needed. Therefore, before starting an experiment, careful consideration regarding the immobilization strategy has to be taken. Depending on the coupling chemistry, covalent or noncovalent (site-specific) immobilization strategies, coupling chemicals as well as chip surfaces, have to be selected (for an overview, see **Table 1** and **Note 1**).

2.1. Coupling

Basically, two different covalent coupling chemistries were developed for Biacore systems: coupling via primary amines using NHS/EDC or coupling via thiols, either by surface or ligand thiol coupling.

2.1.1. Amine Coupling

When performing an *amine coupling* (*see* also **Fig. 3**), a ligand with a primary amine function (for example, free N-terminus or lysine residue) is needed.

1. Appropriate immobilization buffer with low ionic strength ranging from pH 3 to 6; this buffer should contain *no primary amines*, e.g., *do not use Tris*, use HBS-EP (10 mM HEPES, pH 7.4, 150 mM NaCl, 3.4 mM EDTA, 0,005% surfactant P20, filtered and degassed).
2. 100 mM N-hydroxysuccinimide (NHS).
3. 400 mM N-ethyl-N'-(dimethylaminopropyl)-carbodiimide (EDC).
4. 1 M ethanolamine hydrochloride, pH 8.5.
5. Ligand solution: 1–100 µg/mL ligand in an appropriate immobilization buffer.

2.1.2. Thiol Coupling

Thiol coupling provides an alternative to amine coupling and is recommended for ligands when amine coupling cannot be used or is unsatisfactory, e.g., for acidic proteins or peptides and other small ligands. Generally thiol coupling is performed by two different approaches: coupling via intrinsic thiol groups in the ligand (e.g., cysteines, ligand thiol procedure) or coupling via thiol groups introduced into carboxyl or amino groups of the ligand (e.g., engineered cysteine residues, surface thiol procedure).

2.1.2.1. Intrinsic Ligand Thiol Coupling

1. Ligand solution: 10–200 µg/mL ligand in an appropriate immobilization buffer.
2. 80 mM 2-(2-pyridinyldithio)-ethaneamine hydrochloride (PDEA) in 0.1 M borate buffer, pH 8.5; freshly prepared.
3. 50 mM L-cysteine, 1 M NaCl in 0.1 M formate buffer, pH 4.3 (cysteine/NaCl); freshly prepared.

2.1.2.2. Surface Thiol Coupling

1. Ligand solution: 1 mg/mL in 0.1 M MES buffer, pH 5.0.
2. Fast desalting column (NAP10 column, Amersham Biosciences, or equivalent).
3. 40 mM cystamine dihydrochloride in 0.1 M borate buffer, pH 8.5.
4. 0.1 M dithiothreitol (DTT) or dithrioerythrol (DTE) in 0.1 M borate buffer, pH 8.5.
5. 20 mM PDEA, 1 M NaCl in 0.1 M sodium formate buffer, pH 4.3 (PDEA/NaCl); freshly prepared.

Table 1
Overview of the Available Sensor Chips for the Biacore Systems as Distributed by Biacore[a]

Type	Surface characteristics	General applications
Sensor Chip CM5	CM-dextran	Standard surface, suitable for most applications
Sensor Chip SA	CM-dextran + streptavidin	Capture of biotinylated ligands
Sensor Chip NTA	CM-dextran + nitrilotriacetic acid (NTA)	Capture of poly His-tagged proteins via chelated nickel ions
Sensor Chip HPA	Thioalkane-covered gold surface	Creation of lipid monolayers from liposomes
Pioneer Chip B1	CM-dextran with lower degree of carboxymethylation	Lower immobilization capacity, reduces nonspecific binding
Pioneer Chip C1	Carboxylated surface without dextran matrix	If dextran matrix interferes with the interaction being studied; for binding particles (e.g., cells) too large to enter the dextran matrix
Pioneer Chip F1	CM-dextran with shorter dextran matrix	Lower immobilization capacity
Pioneer Chip J1	Plain gold surface	Build your own sensor surface inside the instrument!
Pioneer Chip L1	CM-dextran + lipophilic groups	Direct capture of liposomes
SIA Kit Au	Plain gold surface	Build your own sensor surface outside the instrument!

[a]Note that specific surfaces could be produced on most of the available sensor chips. For further details, *see* **Note 1**.

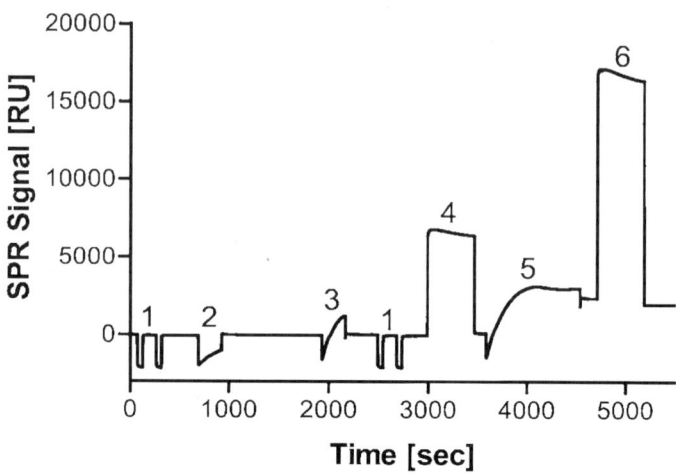

Fig. 3. Immobilization of a murine IgG type antibody on a Sensor Chip CM5. All steps were performed at a flow rate of 5 μL/min at 25°C in 20 mM MOPS, pH 7.0, 150 mM NaCl, 0.005% surfactant P20 on a Biacore 3000. First the sensor surface was treated with two injections of 10 mM NaOH for 1 min (**1**). Now the antibody has to be attracted to the carboxymethylated (CM) matrix via electrostatic interaction. Therefore the pH optimum for immobilization has to be optimized, since, depending on the antibody, the pH optimum for immobilization can vary from pH 4.5 to 5.8 10 mM sodium acetate. It is recommended to perform preconcentration runs—either short injections of varying pH or using different antibody concentrations (**2** and **3**)—to obtain an appropriate immobilization level. Check that baseline levels are reached after changing to running buffer to make sure no nonspecific binding occurs. After treating the surface again with 10 mM NaOH (**1**), the immobilization is started with the injection of NHS/EDC (*see* **Subheading 2.**) for 8 min (activation of chip surface, **4**). The antibody solution (5 μg/mL) is then injected under optimized conditions for 15 min (immobilization, **5**). In the example shown, above a saturation of the sensor surface was already achieved after 8 min. Ethanolamine hydrochloride 1 M, pH 8.5, is used for deactivation of residual reactive groups (**6**). RU, response units; SPR, surface plasmon resonance.

2.1.3. Noncovalent Coupling

Noncovalent coupling is performed using fusion tags employing biotinylated components or by generating lipid-containing sensor surfaces. Fusion proteins are captured via site-specific antibodies against the fusion tag, i.e., anti-GST or anti-poly His antibodies (*see* **Figs. 4–6**) or in case of poly His fusion proteins by using patented nickel/nitrilotriacetic acid (Ni/NTA) surfaces:

For interaction with Ni-NTA surfaces, the following materials are needed:

1. Running buffer: 10 mM HEPES, pH 7.4, 150 mM NaCl, 50 μM EDTA, 0.005% surfactant P20, filtered and degassed; in using only low EDTA, contaminating metal ions are neutralized; on the other hand, the nickel is not stripped from the surface.
2. Nickel solution: 500 μM NiCl$_2$ in running buffer.

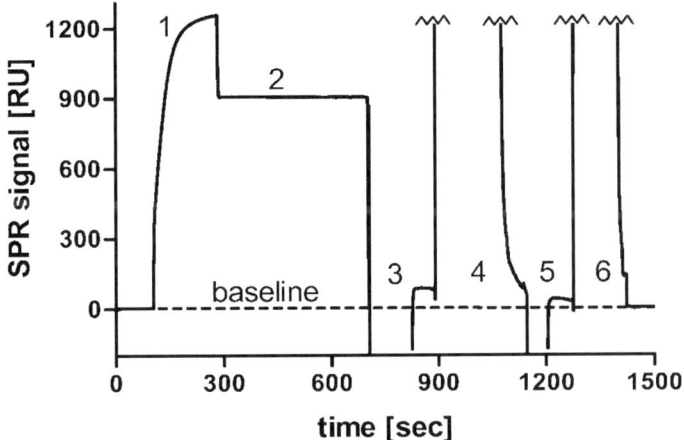

Fig. 4. Interaction and regeneration of a His-tagged protein with an anti RGS-His antibody immobilized via amine coupling on a CM5 sensor chip surface. All experiments were performed on a Biacore 3000 with a flow rate of 20 µL/min at 25°C and a running buffer of 20 mM MOPS, 150 mM NaCl, 0.005% P20. Note that the data presented are not corrected by subtracting a reference surface. The His-tagged protein was injected for 3 min (**1**) followed by a 5-min dissociation phase with running buffer using the kinject command (**2**). Note a 400 response unit (RU) bulk shift at the beginning and the end of the association phase. This shift—clearly only noticeable in the dissociation phase—is due to differences in the buffer composition. The regeneration of the chip surface required rather harsh conditions [1-min injection of 10 mM glycine, pH 2.2 (**3**), injection of 3 M guanidinium hydrochloride for 1.5 min (**4**), 10 mM glycine, pH 2.2, for 0.5 min again (**5**), and finally a 1-min injection of 3 M guanidinium hydrochloride (**6**)]. Note that full regeneration back to the baseline was not achieved until the last regeneration step was performed. RGS-His, arginine, glycine, serine sequence followed by poly histidine; SPR, surface plasmon resonance.

3. Ligand solution: be careful not to use EDTA and bivalent metal ions in the buffer; nonspecific binding is prevented by varying ionic strength and pH; additionally 10–20 mM imidazole can be advantageous.
4. Regeneration solution: 10 mM HEPES, pH 8.3, 150 mM NaCl, 350 mM EDTA, 0.005% surfactant P20.
5. Dispensor buffer: 10 mM HEPES, pH 7.4, 150 mM NaCl, 3 mM EDTA, 0.005% surfactant P20.

For the different surfaces an appropriate regeneration solution has to be chosen. Herberg and Zimmermann *(18)* give an overview of Biacore-compatible solutions [e.g., urea, guanidinium hydrochloride, sodium dodecyl sulfate (SDS), NaOH].

3. Methods

The following methods are described in detail for Biacore systems, but the principles can easily be transferred to other biosensor devices.

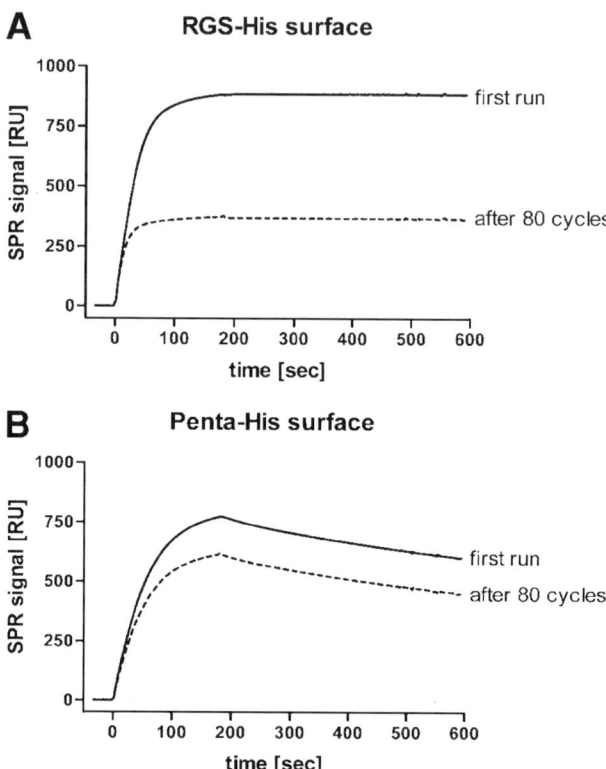

Fig. 5. Loss of binding capacity of immobilized antibody during regeneration. The antibody surface can be damaged in the performance of harsh regeneration procedures as described in the legend to **Fig. 4**. (**A**) Comparison of the first run of an injection series of an RGS-His-tagged protein with an injection of the same protein after 80 regeneration cycles reveals a loss in binding capacity of about 60% for the RGS-His antibody surface. (**B**) The Penta-His antibody on the same chip showed a loss of only 20% in binding activity during the same series of experiments. Note that different antibodies can exhibit different stability to the same kind of regeneration protocols. In both experiments blank runs were subtracted using an activated/deactivated sensor surface. RU, response units; SPR, surface plasmon resonance.

3.1. Immobilization: Step-by-Step Procedure for Coupling an Antibody via Primary Amines

1. Let the chip reach ambient temperature while still enclosed in the nitrogen atmosphere.
2. Insert the chip (in the cover slide) into the Biacore instrument.
3. Run *Prime* (Tools–Working Tools) to fill the syringes and compartments with the immobilization buffer.
4. Start a sensorgram (*Run Sensorgram*). Decide on flow cells for blank subtraction (*see* **Note 2**).
5. Wait until a stable baseline is reached (preferably at a flow rate between 50 and 100 µL/min). If no baseline is reached within 10 min, try an injection of 10 mM NaOH for 30 s.

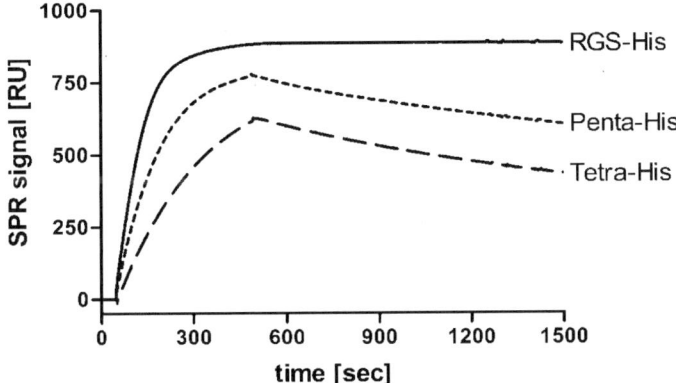

Fig. 6. Antibody characterization of three different antibodies against a His-tag. Three different antibodies against the RGS-His, Penta-His and Tetra-His epitope were coupled on the same CM5 sensor chip via amine coupling. Special care was taken to immobilize always the same amount of antibody to the three flow cells. Then an RGS-His-tagged protein was injected for 8 min at a flow rate of 20 µL/min in running buffer (20 mM MOPS, pH 7.0, 150 mM NaCl, 0.005% P20) with the kinject command monitoring the dissociation phase for 16 min. The RGS-His antibody showed very slow dissociation of the bound RGS-His-tagged recombinant protein, whereas antibodies against the penta-His or the tetra-His epitope showed increased dissociation rates and subsequently lower equilibrium binding levels. Blank runs were subtracted to remove bulk shifts. Note that the RGS-His curve is identical to the raw data curve shown in **Fig. 4**. RU, response units; SPR, surface plasmon resonance.

6. Switch to the flow cell where the immobilization should take place (if this has not been chosen at the start of the sensorgram). The flow rate should be set to 5 µL/min. Again, make sure that the baseline is stable.
7. Perform preconcentration runs as described in detail in **Fig. 3** and the figure legend.
8. Mix the thawed solutions of NHS and EDC in a 1:1 ratio (50 µL each; see **Subheading 2.**).
9. *Quickinject* 40 µL of the mixture (corresponds to 8 min) to activate the CM surface. Add the command *Extraclean* to wash the IFC.
10. *Quickinject* 50 µL of the interaction partner (e.g., anti-GST antibody).
11. *Quickinject* 40 µL of 1 M ethanolamine hydrochloride, pH 8.5, to deactivate excess reactive groups and to remove noncovalently bound material from the surface. Again perform *Extraclean* (for **steps 7–11**, see also **Fig. 3**).
12. Repeat from **step 6** until each flow cell has been treated (*see* **Note 3**).
13. Wash the surface(s) with a washing solution, i.e., regeneration solution that is tolerated by the ligand.
14. The sensor chip is either used directly or stored at suitable conditions (*see* **Note 4**). Sometimes it is recommended to run a sensor chip overnight in buffer to ensure a stable baseline for the following interaction analysis.

3.2. The Kinetic Experiment

1. Insert sensor chip with immobilized ligand into the Biacore instrument. Take care that both sides of the chip contain no salt deposits or storage solution. (You can carefully rinse the chip surfaces with deionized water and soak excess water off the sensor surfaces with a precision wipe placed to one edge on the surface; do not touch the center of the surface with the immobilized ligand!)
2. Run the *Prime* (Tools–Working Tools) procedure or start directly a new sensorgram if the chip was still in the instrument, as described in **Subheading 3.1., steps 3–5**.
3. Choose desired flow cell pathways for the injected analyte (*see* **Note 2**).
4. Inject an appropriate dilution of analyte using the *Quickinject* command. For kinetic analyses, we prefer the *Kinject* command, which consumes more analyte but monitors the dissociation phase without disturbing peaks caused by needle movements. Note that potential shifts and bulk effects may occur at the beginning and the end of the injection if buffer composition of the analyte solution differs from the running buffer. The refractive index of solutions containing even small additional amounts of glycerol, sucrose, detergents, or other buffer components changes dramatically when switching between buffers [for an example of a 400 response unit (RU) buffer shift, *see* **Fig. 4**]. These effects are minimized by subtracting sensorgrams, i.e., flow cell 2-1, 3-1, or 4-1. It is always recommended to match running buffer and analyte buffer as closely as possible, for example, by the use of buffer exchange columns like PD10 or NAP5 (Amersham Biosciences) during analyte preparation.
5. Following the association and dissociation phase, an appropriate regeneration of the sensor chip surfaces has to be developed to disrupt the analyte–ligand interaction without damaging the biological function of immobilized ligand. A thirty second injection of 10 mM glycine, pH 2.2, is suitable for immobilized antibodies. In case no stable baseline down to the previous response level is achieved, try longer injections, lower the pH of the 10 mM glycine regeneration buffer carefully by using 0.1-pH steps down to pH 1.9, or alternatively use 0.05% SDS. The use of SDS results in a drifting baseline, and you have to wash the surface longer with running buffer, water, or Biacore desorb solution 2 (50 mM glycine, pH 9.5). *See* **Fig. 4** for an optimized regeneration procedure also using chaotrophic salts like 3 M guanidinium hydrochloride.
6. Perform a second injection with the same analyte solution at identical conditions to control for stability of the immobilized ligand during the regeneration procedure. No loss of binding activity should be detected. Note that sometimes during a first regeneration step, noncovalently immobilized ligand may dissociate off the chip surface. Thus, if you observe a loss of binding capacity after the first regeneration test, a third injection of analyte has to be performed. Sometimes antibodies displaying favorite binding kinetics (fast association and very slow dissociation) require rather harsh conditions for regeneration. **Figure 5** shows binding kinetics of an identical analyte to two different antibodies immobilized at the start of an experiment and after 80 cycles displaying a significant drop in binding capacity for one of them.
7. Once these technical details have been established, a series of experiments with several cycles are started. Biacore systems offer the possibility of writing methods for automatization; additionally, a wizard function is available.

3.3. Data Processing

Raw data need to be processed before the sensorgrams are evaluated. Accurate data processing is essential for subsequent evaluation. Two types of data are derived from advanced Biacore systems; besides raw data, an on-line reference module provides

Determination of Kinetic Data Using SPR Biosensors 313

data where a control surface was subtracted. The BIAevaluation software (Biacore AB) is then used to calculate rate and equilibrium binding constants from the original data for Biacore systems. The following list summarizes a guideline for the necessary data processing steps (*see* also ref. *19*).

1. Zeroing
   a. *y*-axis: zero the response just prior to the start of the association phase.
   b. *x*-axis: set the start of the injection on the time axis.
2. Reference subtraction. For elimination of refractive index changes and subtraction of nonspecific binding. If the binding curves contain bulk shifts, it is difficult to fit the data (compare **Figs. 4** and **6** for data processing). A software routine is available to detect and subtract bulk refractive index changes; however, you should not trust those data manipulations (*see* **Subheading 3.2., step 4**).
3. Overlay. All curves of one dataset should be overlaid (after **steps 1** and **2** have been performed), i.e., injection of different concentrations of analyte over the same sensor surface. Another option is to overlay data derived from three different ligand surfaces on a chip simultaneously and subtract an additional reference surface. **Figure 6** shows an overlay plot: an RGS-His-tagged protein was injected over three different antibody surfaces.

3.4. Evaluation of Kinetic Data

Preprocessed data are now evaluated. Several kinetic modules are available in the BIAevaluation software. Potential models for data evaluation are also discussed in refs. *19* and *20*. BIAevaluation supports three principal means of data evaluation. A global fit module allows one to fit an entire set of association and dissociation curves with one set of rate constants, improving the robustness of the fitting procedure. Separate fitting of the association and dissociation phases, is another option. Furthermore, transient kinetics are fit with equilibrium binding analysis according to a Scatchard analysis.

A 1:1 Langmuir fit model should be applied as a first try (*see* **Note 6**). However, it is important to consider the biological system when deciding on the fit model. More complex models of interaction are also available. As the complexity of those models increases, the ability to fit the equations to an experiment will improve automatically! This is simply because there is more scope for varying parameters to generate a close fit. Therefore, assumptions about the mechanism of interaction should be decided on before applying a more complex model. Complex systems are extremely difficult to interpret, and even sophisticated evaluation software cannot substitute for careful experimental design.

3.4.1. Data Evaluation with BIAevaluation 3.X

The following protocol describes a global fit analysis. For details regarding different models, *see* **Note 6**.

1. Open overlaid plot of the processed data from one dataset (*see* **Subheading 3.3.**).
2. Choose *Fit kinetics simultaneous* k_a/k_d (= global fit).
3. Select the injection start and end points as well as the area for the association and dissociation phases. This is narrowed down with the option *split view* (*see* **Note 5**).
4. Next enter the concentrations for each curve, choose the appropriate model, and press *Fit* (for picking a model, *see* **Note 6**).

A global fit in which one set of rate constants is used for the approximation to the association and dissociation phases should be performed to test the reaction model of choice.

The binding of an analyte to a ligand under constant flow is regarded as a pseudo-first-order reaction, because the concentration of the analyte is constant in the flow cell. This is not absolutely true, especially when a cuvet system is used (21,22); the depletion of analyte may have a significant effect on the analyte concentration. The same might also be true for flow systems; owing to mass transfer limitations, the concentration of analyte might be reduced close to the dextran matrix, where interaction with the immobilized ligand takes place (23). This inherent problem may produce the same kind of deviations from pseudo-first-order binding processes. Therefore, global fitting may potentially result in conclusions as doubtful as those derived from conventional linear analysis of data (24).

4. Notes

1. Sensor chips. The following detailed description of the available sensor chips for the Biacore systems should help to choose the appropriate surface for a specific application (*see also* **Table 1**). At the end of each paragraph, references are given in which the particular sensor chip has been employed.
    a. Sensor chip CM5. This is the most versatile sensor chip surface for general purposes, with a very stable carboxymethylated (CM)-dextran matrix with high binding capacity allowing a flexible choice of well-defined covalent immobilization chemistries. It provides a hydrophilic environment for biomolecular interaction studies including kinetic measurements, concentration analysis, and direct immobilization of large and small molecules. High-density customized surfaces are prepared for ligand capture (e.g., with covalently immobilized anti-mouse Ig antibodies for capturing monoclonal antibodies) or small-molecule studies down to 180 Daltons (e.g., searching for inhibitors of HIV protease; characterizing low-affinity antibodies against maltose or vitamin content in food preparations). High-capacity CM5 sensor surfaces also provide the basis for the combination of Biacore analysis with mass spectrometry. This technology—also termed BIA-MS—allows one to identify analytes screened for in proteomic studies in very small volumes directly eluted or by removing the sensor chip from the Biacore instrument. Direct identification on the sensor surface using matrix-assisted laser desorption ionization (MALDI) or electrospray ionization (ESI) technology is performed for ligand fishing in bacterial lysates (13,25,26).
    b. Pioneer chip B1. This is similar to sensor chip CM5 but with a lower degree of carboxymethylation to reduce nonspecific binding of highly positively charged molecules (e.g., cell culture supernatants, cell homogenates). Because most DNA binding proteins are positively charged, this chip is useful in protein–DNA binding studies. Note that this chip has a lower immobilization capacity than sensor chip CM5, but immobilization and regeneration techniques developed with sensor chip CM5 can be used identically with the Pioneer chip B1 (27,28).
    c. Pioneer chip F1. This chip has the same degree of carboxymethylation as sensor chip CM5 but a shorter dextran matrix, resulting in a lower binding capacity, bringing the interactants closer to the detection surface and thereby increasing the signal. If you intend to work with large molecules and particles, use the Pioneer chip F1.

This applies, e.g., to whole cells or virus particles, as in some cases even these large compartments are immobilized to the surface to study the way molecules bind to these particles. Examples include interaction of immobilized antibody with protein A on staphylococci, direct analysis and selection of recombinant proteins in phage display libraries, and studies of platelet adhesion mechanisms. All immobilization and regeneration procedures of sensor chip CM5 may be used directly with the Pioneer chip F1 *(13,29)*.

d. Sensor chip SA. Chip with CM–dextran matrix with preimmobilized streptavidin to capture biotinylated ligands such as peptides, proteins, nucleic acids (matrix is optimized for high-capacity binding of long DNA or RNA fragments), and carbohydrates but also to capture lipid vesicles. On-chip reactions of protein–nucleic acid as well as nucleic acid–nucleic acid interactions (for example, ligation, cleavage, strand separation, hybridization, and polymerization) may be followed. It is recommended to use high salt concentrations (up to 0.5 M NaCl) to reduce electrostatic repulsion effects and to increase the efficiency of DNA binding to the surface *(30)*.

e. Sensor chip NTA. The CM–dextran matrix is derivatized with NTA for capturing poly histidine-tagged recombinant proteins via free coordination sites of chelated metal ions such as nickel. Optimal side exposure of the ligand is achieved by control of steric orientation of the complex. Regeneration of the surface is performed with EDTA or other strong chelating agents, stripping nickel ions from the surface to release both the his-tagged ligand and the bound analyte. The regenerated surface is then treated with nickel chloride solution to capture another his-tagged ligand *(31)*.

f. Pioneer chip L1. Lipophilic groups are attached on a CM–dextran matrix for rapid and reproducible capture of liposomes containing the ligand of interest without the need to incorporate anchor molecules. Intact liposomes or a flat lipid bilayer are captured depending on the unaltered conditions and lipid composition already in use for liposome formation. These captured liposomes as well as bilayers mimic biological membranes for quantitative binding studies of ligands that are normally embedded in membranes and of transmembrane components (e.g., investigation of antibiotic binding to membrane-anchored peptides, development of lipid-modified proteins for signal transduction studies; elucidation of membrane aggregation mechanisms). Anchored liposomes on the Pioneer chip L1 are extremely stable under a wide range of salt and pH conditions in detergent-free buffers. Regeneration is achieved by removing the analyte from the captured liposomes or by stripping the liposomes from the surface with detergent *(32)*.

g. Sensor chip HPA: The thioalkane-covered flat hydrophobic gold surface can adsorb lipid monolayers. For studying membrane-associated interactions, the sensor chip HPA provides an alternative to solubilization techniques. The adsorbed lipid monolayer is highly stable, withstanding typical regeneration conditions of the embedded or associated ligands. Note, however, that this surface demands a very clean instrument. [*Desorb* and *Sanitize* (Tools—Working Tools) have to be run and the instrument should be left on standby with water overnight.] Use the Pioneer chip L1 instead, if the native configuration/structure of the embedded protein is important *(33)*.

h. Pioneer chip C1. The C1 chip has a carboxymethylated surface that supports the same immobilization chemistry as CM–dextran surfaces but lacks a dextran matrix. The less hydrophilic character of this chip makes it useful when the dextran matrix interferes with the interaction under study. This chip may also improve experimental setups in which proteins tend to bind to anionic surface, e.g., working with particles, viruses, cell culture supernatants, or cell homogenates. The lower degree of carboxymethylation results in a much lower immobilization capacity than CM–dextran chips. The Pioneer chip C1 is the best choice for ligand fishing in culture media because of the low nonspecific binding and is also useful for protein–DNA binding studies since most DNA binding proteins are positively charged. All immobilization and regeneration techniques developed for the sensor chip CM5 may be directly applied to Pioneer chip C1 *(13,29,34)*.

i. Pioneer chip J1. For the study of interaction between surface materials and biomolecules or for the design of customized surface chemistries using self-assembled monolayers or other modifications, for example, spin-coating, the Pioneer chip J1 offers a plain gold sensor surface *(35)*.

j. SIA Kit Au. This kit is similar to Pioneer chip J1 with a plain gold surfaces. Any chemistry may be applied on this sensor surface but is supplied separately from a chip carrier (e.g., spin coating of thin polymeric film from dilute solutions, coating with organic self-assembled monolayers, polymer adsorption via different functional groups). Studies of the properties of surface materials are important for applications like solid-phase immunoassay, artificial biomaterial implantation, biochemical separation, or support for cell growth. Polymeric adsorption and protein adhesion is viewed as a dynamic process *(36)*.

2. Flow cells and predefined flow cell subtraction. In Biacore 3000 you have the choice of using either one of the four flow cells or several flow cells at once (flow cells 1 and 2; 1, 2, and 3; 1, 2, 3, and 4; 3 and 4). At the beginning of the sensorgram you will be asked to select the flow cell(s) needed; however, when the sensorgram is running, you can redefine your choice. If an on-line blank subtraction is needed (this should be done routinely), it is important to note that not every flow cell may be subtracted from another, i.e., flow cell 2-1, 3-1, 4-1, 4-3. In most cases it is convenient to use flow cell 1 as the reference surface. Reference surfaces are an NHS/EDC-activated/ethanolamine hydrochloride-deactivated surface or, in case of antibody analysis, a surface with an unrelated antibody of the same class and with the same immobilization level. In some cases (for example, Ni-NTA sensor chips), it is difficult to subtract blank surfaces due to unspecific binding or ionic interference.

3. Activation of the chip surface. Immobilization of ligand does not necessarily have to be performed on all four flow cells at one time. Therefore, only the flow cell(s) intended for use is activated and used for coupling. In other cases the user wants to immobilize ligands on each flow cell of the sensor chip simultaneously. Although it is possible to run the activation solution over all flow cells at the same time, experience shows that the activation level of each flow cell might not be the same, probably due to dilution effects. However, you can deactivate all flow cells together at the end of the immobilization procedures.

4. Storage conditions. Once a covalent immobilization has been performed, the sensor chip with bound ligand may be taken out of the Biacore instrument and put into a 50-mL Falcon tube filled with approx 35 mL buffer. (The sensor surface should be covered with liquid.) It is not recommended to cover the chip completely, otherwise the buffer may be contaminated when the chip is taken out and/or the labeling may come off, further contaminating the buffer.

5. Evaluation with *split view*. When evaluating a curve set, it is important to know which area should be selected for implementing the fit. The BIAevaluation software offers a *split view* function whereby the plot window is split into two panels with the original curves on

the top panel and derivative functions on the bottom panel. Depending on the part of the sensorgram that should be analyzed, the user has the option of choosing between several mathematical operations. For the *dissociation phase*: $\ln(dR_0/R_t)$ vs. time [$= \ln(Y_0/Y)$ in the program]. For the *association phase* $\ln(dR/dt)$ vs. time [$= \ln(abs(dY/dX))$ in the program]. This helps to judge whether the model and the parts of the sensorgram selected are appropriate for data evaluation. The functions $\ln(dR/dt)$ and $\ln(R_0/R_t)$ are linear for 1:1 interactions, constant for mass-transfer-limited interactions, and curved for more complex systems. It is easier to judge curves in *split view* when the overlay function is turned off. Do not forget to perform the overlay again before proceeding to the next step in the evaluation procedure.

6. 1:1 (Langmuir) binding.

$$A + L \underset{k_d}{\overset{k_a}{\rightleftharpoons}} AL$$

This model displays the simplest situation of an interaction between analyte (*A*) and immobilized ligand (*L*). It is equivalent to the Langmuir isotherm for adsorption to a surface. The Langmuir isotherm was developed by Irving Langmuir in 1916 to describe the dependence of the surface coverage of an adsorbed gas on the pressure of the gas above the surface at a fixed temperature *(37,38)*. The equilibrium that exists between gas adsorbed on a surface and molecules in the gas phase is a dynamic state, i.e., the equilibrium represents a state in which the rate of adsorption of molecules onto the surface is exactly counterbalanced by the rate of desorption of molecules back into the gas phase. It should therefore be possible to derive an isotherm for the adsorption process simply by considering and equating the rates for these two processes. These considerations are assigned to the SPR detection system.

The 1:1 Langmuir module also allows for deviation in the raw data. Sometimes the baseline shows a slight drift that is largely eliminated by the use of a reference cell. However, in analysis with low surface binding capacity (R_{max} of 100 RU or less), a model including linear drift may be appropriate (1:1 binding with drifting baseline).

A third 1:1 binding model considering mass transfer limitations is also included in the BIAevaluation software. Thus, kinetic data are produced even though the interaction analyzed is mass transfer limited. However, it is recommended to perform the experiment in such a way as to avoid mass transfer limitations, e.g., use higher flow rates and lower surface densities. For mass transfer limitations and rebinding refer to ref. *18*.

Alternative models for more complex interaction patterns are available such as bivalent analyte, heterogeneous analyte (competing reactions), heterogeneous ligand (parallel reactions), and two-state reaction (conformation change). Refer to the Biaevaluation (3.0 or later) manual for details.

It is recommended to use the global fit module for data evaluation. However, for some datasets, it is necessary to perform a separate k_a/k_d determination, e.g., if one of the phases is obscured by bulk shifts or if different conditions apply during association and dissociation phases. BIAevaluation includes a module to fit the association and dissociation phases separately.

Additionally, a general fit module including four-parameter equation, linear fit, solution affinity, and steady-state affinity is available. Finally, additional models may be imported into BIAevaluation software.

7. Emerging technologies. Apart from SPR-based detections, several other techniques are evolving for use in interaction studies. An interferometric evanescent waveguide is being developed (Fraunhofer-Institut für Physikalische Messtechnik, Freiburg, Germany) combining high resolution with respect to surface mass coverage and a low sensitivity toward undesired external influences. Furthermore, with only few optical components needed, the production costs are low.

The BIAffinity system (Analytik Jena, Germany) is based on reflectometric interference spectroscopy (RIfS) and is already commercially available. Like the Biacore instrument, it is intended for the determination of biomolecular interactions as well as drug screens, epitope mapping, quality control, and on-line control of biotechnological processes. For a comparison between RfIS, Biacore, and IAsys refer to ref. *39*.

Quartz crystal microbalances detect frequency changes in response to the binding of biomolecules. The kinetic properties of the molecular interaction influence the dynamic of the frequency. The Fraunhofer Institute has developed a quartz crystal microbalance (QCM) device [Fraunhofer-Institut für Festkörpertechnologie, Munich *(40)*] for antibody development and quality control as potential applications. A multichannel unit is planned. A second device is the Affinix Q quartz balance (distributed in Europe by Probior, Munich, Germany). First applications are described in ref. *41*.

Acknowledgments

We thank Bastian Zimmermann and Oliver Diekmann for valuable input. This work was supported by the Bundesministerium für Bildung und Forschung (BMBF, 031U102F) and the Deutsche Forschungsgemeinschaft (He 1818/4).

References

1. Rich, R. L. and Myszka, D. G. (2001) Survey of the year 2000 commercial optical biosensor literature. *J. Mol. Recognit.* **14,** 273–294.
2. Wink, T., de Beer, J., Hennink, W. E., Bult, A., and van Bennekom, W. P. (1999) Interaction between plasmid DNA and cationic polymers studied by surface plasmon resonance spectrometry. *Anal. Chem.* **71,** 801–805.
3. Baird, C. L. and Myszka, D. G. (2001) Current and emerging commercial optical biosensors. *J. Mol. Recognit.* **14,** 261–268.
4. Stenberg, E., Persson, B., Roos, H., and Urbaniczky, C. (1991) Quantitative determination of surface concentration of protein with surface plasmon resonance using radiolabeled proteins. *J. Colloid Interface Sci.* **143,** 513–526.
5. Kovacs, G. D. (1982) In *Electromagnetic Surface Modes* (Boardman, A. D., ed.), Wiley, New York, p. 143.
6. Zimmermann, B., Hahnefeld, C., and Herberg, F. W. (2002) Applications of biomolecular interaction analysis in drug development. *TARGETS* **1,** 66–73.
7. Jokiranta, T. S., Westin, J., Nilsson, U. R., et al. (2001) Complement C3b interactions studied with surface plasmon resonance technique. *Int. Immunopharmacol.* **1,** 495–506.
8. Jokiranta, T. S., Hellwage, J., Koistinen, V., Zipfel, P. F., and Meri, S. (2000) Each of the three binding sites on complement factor H interacts with a distinct site on C3b. *J. Biol. Chem.* **275,** 27,657–27,662.
9. Jarva, H., Janulczyk, R., Hellwage, J., Zipfel, P. F., Bjorck, L., and Meri, S. (2002) *Streptococcus pneumoniae* evades complement attack and opsonophagocytosis by expressing the pspC locus-encoded Hic protein that binds to short consensus repeats 8-11 of factor H. *J. Immunol.* **168,** 1886–1894.

10. Hellwage, J., Meri, T., Heikkila, T., et al. (2001) The complement regulator factor H binds to the surface protein OspE of *Borrelia burgdorferi*. *J. Biol. Chem.* **276**, 8427–8435.
11. Meri, T., Hartmann, A., Lenk, D., et al. (2002) The yeast *Candida albicans* binds complement regulators factor H and FHL-1. *Infect. Immun.* **70**, 5185–5192.
12. Sidobre, S., Puzo, G., and Riviere, M. (2002) Lipid-restricted recognition of mycobacterial lipoglycans by human pulmonary surfactant protein A: a surface-plasmon-resonance study. *Biochem. J.* **365**, 89–97.
13. Jongerius-Gortemaker, B. G., Goverde, R. L., van Knapen, F., and Bergwerff, A. A. (2002) Surface plasmon resonance (BIACORE) detection of serum antibodies against *Salmonella enteritidis* and *Salmonella typhimurium*. *J. Immunol. Methods* **266**, 33–44.
14. Hifumi, E., Kubota, N., Niimi, Y., Shimizu, K., Egashira, N., and Uda, T. (2002) Elimination of ingredients effect to improve the detection of anti HIV-1 p24 antibody in human serum using SPR apparatus. *Anal. Sci.* **18**, 863–867.
15. McLaurin, J., Cecal, R., Kierstead, M. E., et al. (2002) Therapeutically effective antibodies against amyloid-beta peptide target amyloid-beta residues 4-10 and inhibit cytotoxicity and fibrillogenesis. *Nat. Med.* **8**, 1263–1269.
16. Gonzales, N. R., Schuck, P., Schlom, J., and Kashmiri, S. V. (2002) Surface plasmon resonance-based competition assay to assess the sera reactivity of variants of humanized antibodies. *J. Immunol. Methods* **268**, 197–210.
17. Alaedini, A. and Latov, N. (2001) A surface plasmon resonance biosensor assay for measurement of anti-GM(1) antibodies in neuropathy. *Neurology* **56**, 855–860.
18. Herberg, F. W. and Zimmermann, B. (1999) Analysis of protein kinase interactions using biomolecular interaction analysis, in *Protein Phosphorylation—A Practical Approach* (Hardie, D. G., ed.), vol. 2, Oxford University Press, Oxford, pp. 335–371.
19. Myszka, D. G. (2000) Kinetic, equilibrium, and thermodynamic analysis of macromolecular interactions with BIACORE. *Methods Enzymol.* **323**, 325–340.
20. Karlsson, R. and Falt, A. (1997) Experimental design for kinetic analysis of protein-protein interactions with surface plasmon resonance biosensors. *J. Immunol. Methods* **200**, 121–133.
21. Hall, D. R., Gorgani, N. N., Altin, J. G., and Winzor, D. J. (1997) Theoretical and experimental considerations of the pseudo-first-order approximation in conventional kinetic analysis of IAsys biosensor data. *Anal. Biochem.* **253**, 145–155.
22. O'Shannessy, D. J. and Winzor, D. J. (1996) Interpretation of deviations from pseudo-first-order kinetic behavior in the characterization of ligand binding by biosensor technology. *Anal. Biochem.* **236**, 275–283.
23. Hall, D. R., Cann, J. R., and Winzor, D. J. (1996) Demonstration of an upper limit to the range of association rate constants amenable to study by biosensor technology based on surface plasmon resonance. *Anal. Biochem.* **235**, 175–184.
24. Schuck, P. and Minton, A. P. (1996) Analysis of mass transport-limited binding kinetics in evanescent wave biosensors. *Anal. Biochem.* **240**, 262–272.
25. Natsume, T., Nakayama, H., and Isobe, T. (2001) BIA-MS-MS: biomolecular interaction analysis for functional proteomics. *Trends Biotechnol.* **19(10 suppl)**, S28–33.
26. Cain, K. D., Jones, D. R., and Raison, R. L. (2002) Antibody-antigen kinetics following immunization of rainbow trout (*Oncorhynchus mykiss*) with a T-cell dependent antigen. *Dev. Comp. Immunol.* **26**, 181–190.
27. Ciolkowski, M. L., Fang, M. M., and Lund, M. E. (2000) A surface plasmon resonance method for detecting multiple modes of DNA-ligand interactions. *J. Pharm. Biomed. Anal.* **22**, 1037–1045.

28. Schindler, J. F., Godbey, A., Hood, W. F., et al. (2002) Examination of the kinetic mechanism of mitogen-activated protein kinase activated protein kinase-2. *Biochim. Biophys. Acta* **1598,** 88–97.
29. Hoffman, T. L., Canziani, G., Jia, L., Rucker, J., and Doms, R. W. (2000) A biosensor assay for studying ligand-membrane receptor interactions: binding of antibodies and HIV-1 Env to chemokine receptors. *Proc. Natl. Acad. Sci. USA* **97,** 11,215–11,220.
30. Hwang, J., Fauzi, H., Fukuda, K., et al. (2000) The RNA aptamer-binding site of hepatitis C virus NS3 protease. *Biochem. Biophys. Res. Commun.* **279,** 557–562.
31. Kamionka, A. and Dahl, M. K. (2001) *Bacillus subtilis* contains a cyclodextrin-binding protein which is part of a putative ABC-transporter. *FEMS Microbiol. Lett.* **204,** 55–60.
32. Baird, C. L., Courtenay, E. S., and Myszka, D. G. (2002) Surface plasmon resonance characterization of drug/liposome interactions. *Anal. Biochem.* **310,** 93–99.
33. Satoh, A., Hazuki, M., Kojima, K., Hirabayashi, J., and Matsumoto, I. (2000) Ligand-binding properties of annexin from *Caenorhabditis elegans* (annexin XVI, Nex-1). *J. Biochem. (Tokyo)* **128,** 377–381.
34. Schlecht, U., Nomura, Y., Bachmann, T., and Karube, I. (2002) Reversible surface thiol immobilization of carboxyl group containing haptens to a BIAcore biosensor chip enabling repeated usage of a single sensor surface. *Bioconjug. Chem.* **13,** 188–193.
35. Gau, J. J., Lan, E. H., Dunn, B., Ho, C. M., and Woo, J. C. (2001) A MEMS based amperometric detector for *E. coli* bacteria using self-assembled monolayers. *Biosens. Bioelectron.* **16,** 745–755.
36. Cooper, M. A., Fiorini, M. T., Abell, C., and Williams, D. H. (2000) Binding of vancomycin group antibiotics to D-alanine and D-lactate presenting self-assembled monolayers. *Bioorg. Med. Chem.* **8,** 2609–2616.
37. Langmuir, I. (1916) The constitution and fundamental properties of solids and liquids. Part I. Solids. *J. Am. Chem. Soc.* **38,** 2221–2295.
38. Langmuir, I. (1918) The adsorption of gases on plane surfaces of glass, mica and platinium. *J. Am. Chem. Soc.* **40,** 1361–1404.
39. Hänel, C. and Gauglitz, G. (2002) Comparison of reflectometric interference spectroscopy with other instruments for label-free optical detection. *Anal. Bioanal. Chem.* **372,** 91–100.
40. Hengerer, A., Decker, J., Prohaska, E., Hauck, S., Kosslinger, C., and Wolf, H. (1999) Quartz crystal microbalance (QCM) as a device for the screening of phage libraries. *Biosens. Bioelectron.* **14,** 139–144.
41. Matsuno, H., Niikura, K., and Okahata, Y. (2001) Direct monitoring kinetic studies of DNA polymerase reactions on a DNA-immobilized quartz-crystal microbalance. *Chemistry* **7,** 3305–3312.

20

Affinity Measurements of Biological Molecules by a Quartz Crystal Microbalance (QCM) Biosensor

Uwe Schaible, Michael Liss, Elke Prohaska,
Jochen Decker, Karin Stadtherr, and Hans Wolf

Abstract

We present a immunosensing system based on a quartz crystal microbalance (QCM) that allows detection and evaluation of a broad range of molecules relevant to medicine and biology. Soluble analyte is applied by injection into a flow-through chamber where receptor molecules are immobilized, thus enabling real-time measurement of the specific intermolecular interactions. This system can be used for detection purposes, e.g., screening for antibodies, and for determination of affinity constants including affinity evaluation of newly selected or designed aptamers, as k_d and k_a rates can be derived from the sensograms.

Key Words: Affinity constants; aptamers; immunosensor; quartz crystal microbalance; specific intermolecular interactions.

1. Introduction

The quartz crystal microbalance (QCM) is an acoustic immunosensor that provides a method for label-free measurement of specific interactions between immobilized molecules and analytes in solution. Binding of a soluble analyte causes a frequency shift in the resonance frequency of the quartz and can thus be detected in real time. The time-course of this frequency change can be evaluated to determine kinetic and equilibrium constants of the binding reaction.

A broad range of biological molecules, e.g., antibodies, proteins, or peptides, are suitable as receptor molecules. Among the applications QCM has successfully been used for are immunoassays, e.g., screening for HIV-specific antibodies *(1)*, screening of combinatorial phage display gene libraries *(2–5)*, determination of affinity constants *(2,6)*, and detection of bacterial toxin. It has also recently provided a suitable way to evaluate aptamers *(7)*.

From: *Methods in Molecular Medicine, vol. 94: Molecular Diagnosis of Infectious Diseases, 2/e*
Edited by: J. Decker and U. Reischl © Humana Press Inc., Totowa, NJ

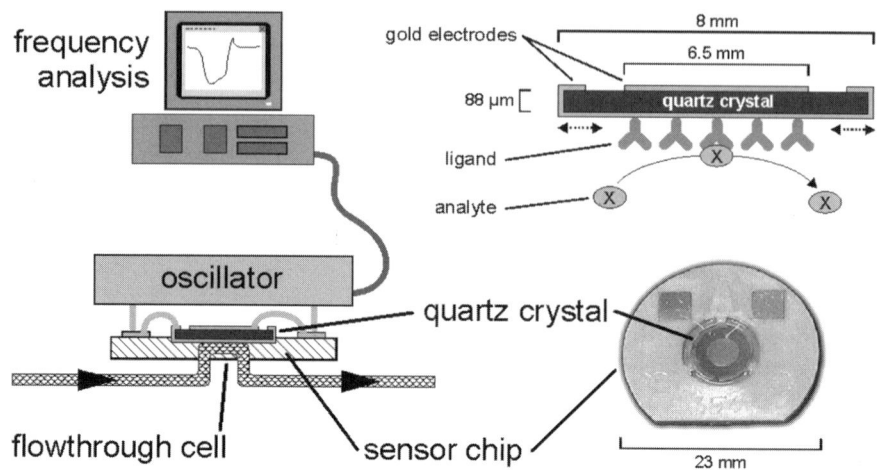

Fig. 1. Schematic view of the applied quartz crystal microbalance (QCM) (left) and the quartz crystal (top right). The quartz crystal mounted on the sensor chip is shown at bottom right.

In this chapter, we present the detection of a 28-kDa protein by an IgG receptor layer and the detection of human IgE using aptamer receptor molecules as exemplary applications.

1.1. Basic Principles

The core of the QCM biosensor is a quartz crystal with a diameter of 8 mm, serving as a piezoelectric transducer. The crystal is coated on both sides with gold layers, which provide the electrical contact to the oscillator as well as support for ligand immobilization. Receptor molecules are immobilized on the gold layer facing a flow-through chamber where a constant flow of buffer is maintained and where the analyte is injected. The oscillator and flow-through injection system can easily be controlled by any ordinary PC (**Fig. 1**). The quartz crystal chip unit is exchangeable and inexpensive.

The detection of analyte binding is based on the mass dependency of the resonance frequency of the quartz; the measurable increase of its mass due to the binding of an analyte can be calculated from the frequency change according to the formula of Sauerbrey (8):

$$\Delta f_r = -c_f \cdot \Delta m / A \tag{1}$$

with

$$c_f = -[(f_R^2)/(v_q \rho_q)] \tag{2}$$

where Δf_r (Hz) is the change of the resonance frequency, Δm (ng) is the change in mass on the quartz surface, A (cm^2) is the size of the coated quartz surface, and c_f is the sensitivity constant. The constant c_f is calculated from f_r, the original resonance frequency of the quartz (here, 20 MHz), v_q, is the velocity of sound in the quartz crystal (3340 m s^{-1}); and ρ_q, is the quartz density (2648 kg m^{-3}). The sensitivity constant c_f is approx 0.9 cm² Hz ng⁻¹. With a sensor surface of 0.33 cm², this results in a frequency change of -27 Hz/10 ng. Thus, the quartz resonance frequency is in direct linear correlation with the mass surface density at the immobilized receptor layer.

2. Materials

1. Biosensor device: the biosensor device was manufactured by the Fraunhofer Institut für Mikroelektronische Schaltungen und Systeme (IMS), München, Germany *(9)*.
    a. Transducer: diameter 8 mm, thickness 84 µm, accessible surface 33 mm², crystal cut AT, resonant frequency 20 MHz, electrodes of nickel/gold bilayer (Quarzkeramik, Gauting, Germany) with diameter 8 mm and thickness 65 nm.
    b. Flow-through cell: height 90 µm, length 30 mm, width 8 mm, total volume 4 µL, top and bottom of polymethylmethacrylate, spacer of polyvinylchloride foil, transparent, thickness 90 µm (Alkor, Gräfeling/Munich, Germany).
    c. Housing/contacting: silicone adhesive: Elastosil N10, transparent (Wacker, Munich, Germany), bonding wires gold, diameter about 5 µm, contacting glue Elekolite 489 (Bürklin, Munich, Germany), contacting pins of stainless steel, diameter 0.8 mm. Temperature regulation by Peltier module.
    f. Signal processing: oscillator manufactured in-house by the IMS, frequency counter Kontron model 6030 (Kontron Messtechnik, Eching, Germany), used in combination with appropriate computer software for data registration and analysis.
    e. Fluid handling: injection valve six-port standard high-performance liquid chromatography (HPLC) valve (e.g., VICI, Switzerland), tubings 1/16 in. Teflon capillaries with an inner diameter of 0.5 mm, pump 1000-µL double-syringe pump, manufactured in-house by the IMS.
2. Micropipet (10–200 µL).
3. 250-µL Teflon-free syringe (Hamilton).
4. 250-µL syringe (Hamilton).
5. 30% H_2O_2 (Merck).
6. 95–97% H_2SO_4 (Merck).
7. DSP [3,3-dithio-dipropionic acid-bis-(*N*-succinimidylester; Fluka].
8. *N*,*N*-dimethylacetamide (DMA) (Sigma).
9. Phosphate-buffered saline (PBS).
10. PBS+: PBS containing 1 m*M* $MgCl_2$.
11. Streptavidin (Sigma): 1 mg/mL in PBS.
12. Bovine serum albumin (BSA): 0.025% in PBS.
13. Piranha: 1 vol 30% H_2O_2, 4 vol concentrated H_2SO_4.
14. 5'- or 3'-biotin-labeled aptamer solutions: 0.2–2 pmol/µL in PBS+.

3. Method

3.1. Immobilization

3.1.1. Basic Approach

1. Clean and activate gold surfaces by two injections with 200 µL of ice-cold piranha solution followed by an immediate wash with 500 µL of cold H_2O after each injection.
2. Wash the quartz three times with cold H_2O and dry for 1 h at 100°C.
3. Inject 10 µL of a 4-mg/mL solution of the homobifunctional crosslinking reagent DSP in water-free DMA.
4. Seal with adhesive tape and incubate at room temperature for 15 min.

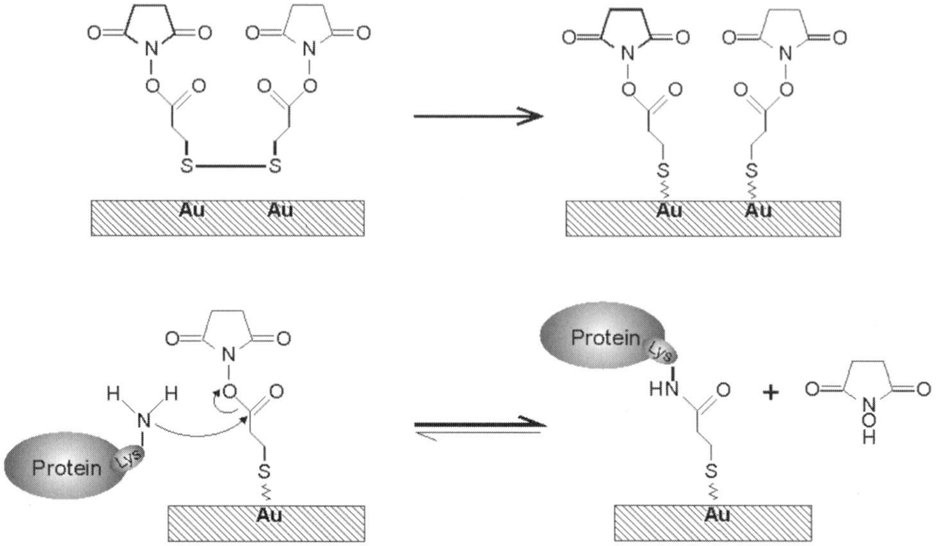

Fig. 2. Immobilization of proteins to a gold surface using 3,3'-dithiodipropionic acid-di(N-succinimidylester) (DSP).

DSP binds to gold by a cleavable disulfide bridge and forms a peptide bond to free amino groups of lysine residues (**Fig. 2**). Further procedure depends on the aspirated receptor layer, protein/antigen, or aptamer (*see* **Note 1**).

3.1.2. Proteins and Antibodies

Immobilization via DSP is an easy and fast method to achieve a stable protein layer on the quartz crystal.

1. For direct immobilization of protein, wash the surfaces after the DSP incubation three times with 150 µL of PBS.
2. Immediately inject 10 µL of a protein dilution of choice in PBS (*see* **Note 2**).
3. Seal with tape and incubate at 4°C overnight.
4. After protein immobilization, rinse the flow-through chambers five times with 100 µL of PBS.
5. Use immediately or seal with adhesive tape and store at 4°C. Use the quartz within 1 wk.

However, if directly bound to DSP, the receptor molecules are immobilized in random orientation. In contrast, site-specifically biotinylated molecules fixed to immobilized streptavidin build up an ordered quasi-monolayer structure. Biotinylation of desired receptor molecules can be performed using commercially available biotinylation kits or by recombinant methods (*see* **Note 3**). Whether biotinylation is desirable has to be evaluated in each individual case, as it is often not necessary to perform biotinylation of the receptor molecules for satisfactory results.

1. For immobilization via streptavidin wash the surfaces with 150 µL of PBS.
2. Immediately inject 10 µL of streptavidin solution.
3. Seal with tape and incubate at 4°C overnight.
4. Rinse the surface with 100 µL PBS.
5. Apply the biotinylated receptor molecule. Incubate for 1 h at room temperature.
6. After protein immobilization, rinse the flow-through chambers five times with 100 µL of PBS.
7. Use immediately or seal with adhesive tape and store at 4°C. Use the quartz within 1 wk.

3.1.3. Aptamers

1. Coat DSP-treated quartzes with streptavidin as described above. Important: use PBS+ instead of PBS.
2. Heat 5'- or 3'-biotin-labeled aptamer solutions to 95°C for 3 min and chill on ice to ensure correct intramolecular folding.
3. Inject 10 µL of aptamer solution into the quartz crystal chamber.
4. Seal and incubate at room temperature for 1 h.
5. Use aptamer-coated chips either immediately or cover with a 0.02% sodium azide solution in PBS+ and store at 4°C for <1 mo.

3.2. Measurement

The measurement basically consists of the following steps:

1. Rinse the prepared quartz crystal with 100 µL of PBS and insert into the flow-injection analysis system.
2. Apply a constant flow of buffer to establish an equilibrium resonance frequency. Environmental conditions, especially temperature, have some influence on the equilibrium resonance frequency and should therefore be kept stable (*9–11*). Although the flow-through chamber is thermostated to prevent effects from different temperatures of running buffer and samples, changes in room temperature affecting the running buffer could lead to drift of the signal.
3. Inject BSA to block uncovered gold sites (*see* **Notes 4** and **5**).
4. Inject the sample and sample data. The sample should be diluted in the running buffer (*see* **Note 6**). The quartz resonance frequency decreases when the sample is injected, because of increasing mass surface density by association of binding molecules as well as because of the usually higher viscosity of the sample (*see* **Note 7**). After the sample goes through the flow-through chamber and is replaced again by buffer, the viscosity effect vanishes and a new—lower—equilibrium resonance frequency is established because of the mass of the binding molecules. To distinguish a signal from background noise, a minimum frequency shift of at least −2 Hz is required (*see* **Note 8**). Dissociation starts depending on the kinetic constants of the binding reaction (**Fig. 3**). The size of the sample loop is changeable and is chosen depending on the association kinetics. Our standard as used in these examples is 100 µL, but slow kinetics may require larger sample loops to increase the incubation time for the association reaction. Flow rates are typically 10–30 µL/min.
5. Inject appropriate negative controls to exclude nonspecific binding. Sample data as described above. There is no nonspecific binding if no permanent frequency decrease is detectable.

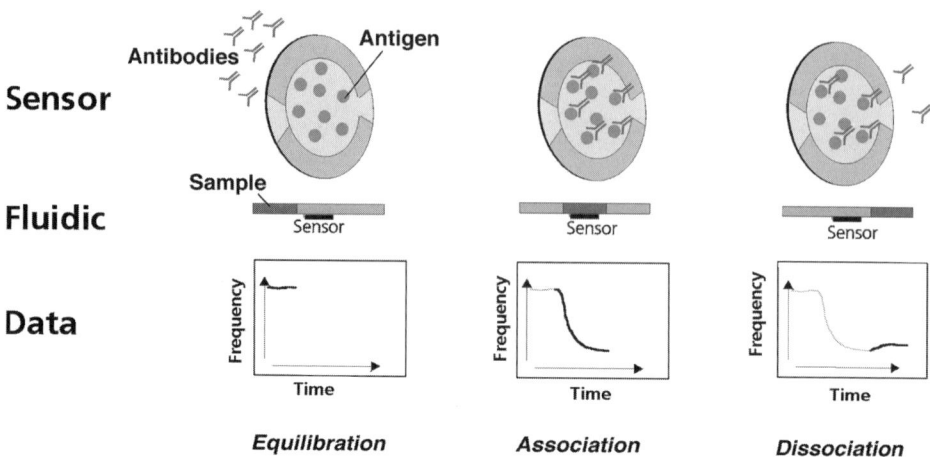

Fig. 3. Connections among flow-through of the sample, binding reaction, and signal output. In this example, antigen is immobilized as receptor molecule, and the sample contains antibodies as soluble analyte. (Courtesy of S. Hauck.)

Our first example sensogram (**Fig. 4**) shows the detection of a 28-kDa protein, performed at a continuous flow rate of 10 µL/min (running buffer PBS) and a sample size of 100 µL in each case. The receptor layer consists of a IgG with specific affinity to the analyte protein (*see* **Note 9**).

The injection of negative controls (bacteria and bacteriophages) at comparatively high concentrations proves that no nonspecific binding occurs, as there is no permanent frequency decrease. The injection of the analyte at a concentration of 50 ng protein/mL results in a frequency shift far above the required minimum for a positive signal. Further injections of the analyte at higher concentrations lead only to smaller frequency shifts because many binding sites of the receptor layer are already occupied after the first injection. At the end of the measurement, saturation is already reached.

In our second example, the detection of human IgE via aptamer receptor molecules is shown. Aptamers are nucleic acids that have been selected from large randomized oligonucleotide libraries. Because of their three-dimensional structure, they are capable of binding to target proteins with high affinity. Their small size allows the generation of denser receptor layers, meaning that for a given receptor affinity, the sensitivity of these layers can be increased. Their binding affinity, specificity, and stability can easily be manipulated and improved by rational design or by techniques of molecular evolution. They can be modified with functional groups or tags to allow covalent, directed, or indirect immobilization on biochips, resulting in highly ordered receptor layers *(7)*. Thus, the QCM biosensor can be used for affinity evaluation of newly selected or designed aptamers. Based on the association and dissociation kinetics, the binding constants K_d, K_a, and K can be determined *(2,4–7)* (*see* **Notes 10** and **11**).

Here, a single-stranded DNA aptamer that was either 5'- or 3'-biotinylated was immobilized. With this receptor layer, the detection of a minimum IgE concentration

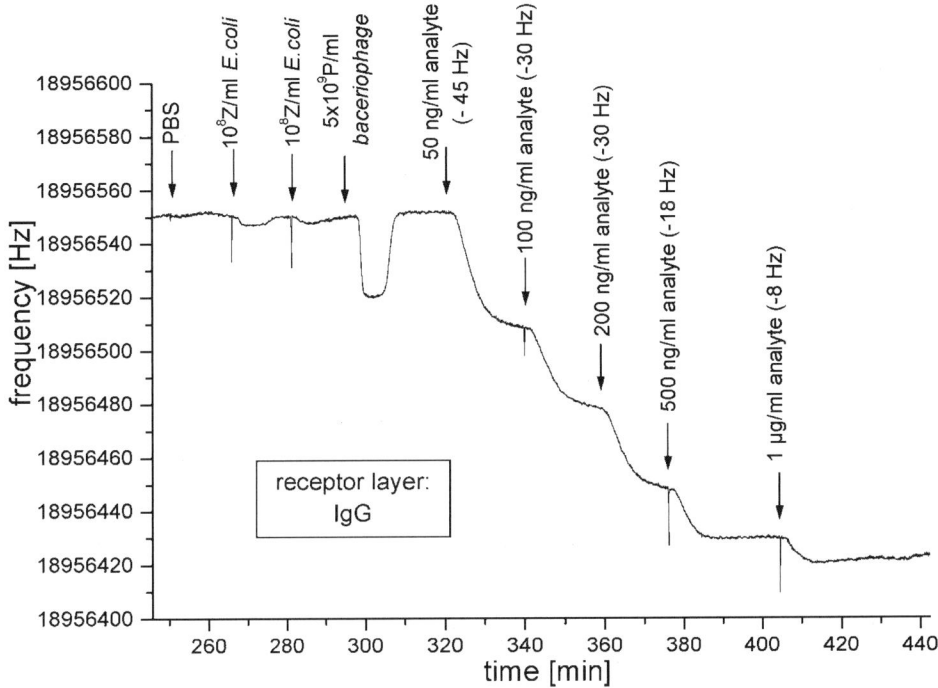

Fig. 4. Sensogram of the detection of a 28-kDa protein. Receptor: specific IgG. The injections of bacteria (two different serotypes of *E. coli*) and bacteriophages as negative controls at the beginning of the measurement show a temporary viscosity effect but no permanent frequency decrease.

of 100 ng/µL was possible (data not shown). The sensogram shows 10 cycles of measurement and regeneration of the receptor layer. First, the absence of nonspecific binding is proved by injection of BSA and an IgG molecule to which the aptamer has no affinity. The cycles show that the sensitivity is only very slightly affected by the regeneration procedure (**Fig. 5**).

Binding of IgE to the aptamers is an almost completely reversible process. Injection of 50 m*M* EDTA leads to dissociation of the analyte, the receptor layer being completely reconstituted simply by returning to original buffer conditions (PBS+) (*see* **Note 12**). This is because most aptamers need bivalent metal ions to assemble into their three-dimensional structure and unfold in the presence of EDTA. They refold very rapidly when EDTA is withdrawn and replaced by Mg^{2+}-containing buffer. This effective regeneration procedure can be applied to all aptamers requiring Mg^{2+} for their tertiary structure. It is possible to regenerate protein receptor layers as well (e.g., with 0.2 *M* glycine-HCl), but this usually goes with a loss of sensitivity with each regeneration cycle. Sensor layers based on aptamer receptors can be regenerated more easily than antibody-based layers and are more resistant to denaturation and degradation because of their simple structure (*see* **Note 13**).

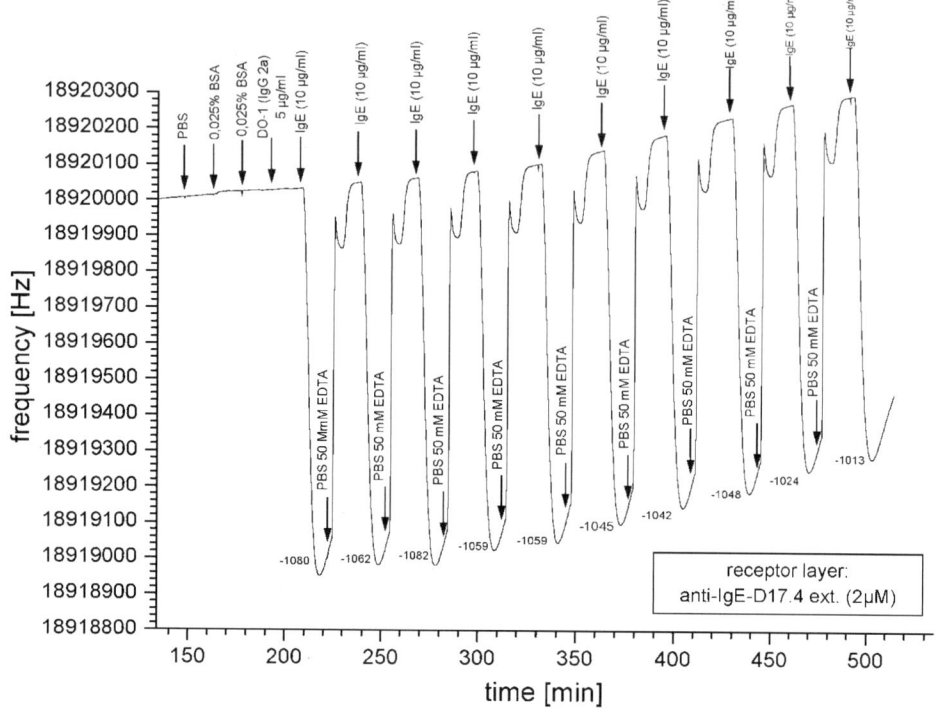

Fig. 5. Sensogram of repetitive measurements and regenerations of the receptor layer. Receptor: anti-IgE aptamer; regeneration with 50 m*M* EDTA.

4. Notes

1. In principle, proteins can be fixed noncovalently onto gold by physical adsorption *(12)*. However, these surfaces are not stable during prolonged buffer rinses.
2. For best results concerning sensibility of the receptor layer, protein concentrations should be about 2 mg/mL. However, determination of affinity constants might require lower surface concentration (*see* **Note 9**).
3. Site-specific in vivo biotinylation can also be achieved by cloning the gene of interest into a Pinpoint vector downstream to a biotinylation signal. As a consequence, the expressed antigen molecules are modified with exactly one biotin within the respective signal sequence *(13)*. These biotinylated antigens will bind unidirectionally to a streptavidin layer.
4. Uncovered gold sites may be a source of nonspecific binding *(12)*. Therefore, we recommend testing for nonspecific binding before starting the measurements by injection of BSA. If a frequency shift occurs because of nonspecific binding, the injection of BSA can be repeated several times until the uncovered gold sites are blocked by BSA. However, frequency shifts above –50 Hz indicate an unsuccessful immobilization, and the quartz should be dismissed.

5. As the injection of BSA can lead to a slight decrease in the sensibility of the receptor layer, testing for nonspecific binding can alternatively be performed by injection of analytes different from the one supposed to bind.
6. Analytes should be diluted in the same buffer that is used as running buffer. We routinely use PBS as running buffer. Any other buffer can be used as well, or a solution can be used as running buffer, respectively. In any case, differences in viscosity between running buffer and sample cannot be avoided completely, as even little differences in temperature lead to different viscosity, and in some cases (e.g., resuspended phage) suspension of the analyte alone is sufficient to increase the viscosity of the buffer.
7. Frequency decreases obtained with whole bacteria or viruses as soluble analytes are usually below the value estimated by the Sauerbrey equation. Obviously, owing to some wobble effect within the particle body, comparatively large microorganisms cannot be taken into account with all their mass. However, the achievable frequency shift can be improved when these analytes are reduced to small pieces by appropriate methods such as ultrasonic disruption.
8. Small constant drifts (<2 Hz/min) can be tolerated especially if one is working with fast kinetics. They can be corrected mathematically by multiplication of the whole curve with the slope of the drift. Larger drifts indicate a malfunction of the fluid system. In this case the whole system should be checked for leakage and air bubbles and cleaned with water (optionally with 0.2 M glycine-HCl, pH 2.2, or urea).
9. In the case of receptor layers consisting of mouse or rabbit IgG, the value for the minimum detectable concentration can also be improved via Protein A. This 42-kDa protein from *Staphylococcus aureus* has a very high affinity to the F_c-region of mouse and rabbit IgG ($K_D = 10^{-8}\,\text{mol}^{-1}$) and can be immobilized to DSP like any other protein. After rinsing with 150 µL PBS, the antibody is injected and incubated for at least 1 h at room temperature.
10. Dissociation constants are determined by nonlinear regression of the data by applying Origin software. The rate constant of dissociation k_d is assessed as follows:

$$F(t) = F_0 \cdot \exp(k_d \cdot [t - t_0]) \tag{3}$$

The kinetics of association is fitted to **Eq. 4**:

$$F(t) = (k_a \cdot C \cdot \Delta F_{max})/(k_a \cdot C + k_d) \cdot \{\exp([k_a \cdot C + k_d] \cdot [t - t_0]) - 1\} + F_0 \tag{4}$$

The dissociation constant is the quotient of k_d and k_a:

$$K_D = k_d/k_a \tag{5}$$

$F_{(t)}$, frequency (Hz); k_a, rate constant of association [$s^{-1}\,M^{-1}$]; k_d, rate constant of dissociation (s^{-1}); C, concentration (M); ΔF_{max}, maximal frequency shift for completely occupied surface (Hz); t, time (s); t_0, start time for fitting procedure (s); F_0, frequency at t_0. First, the dissociation reaction should be analyzed, as **Eq. 3** depends only on k_d. k_d can then be inserted into **Eq. 4** to determine k_a by nonlinear regression analysis. K_D can be calculated by using **Eq. 5**.

For an accurate determination of dissociation kinetics, care has to be taken that rebinding of dissociated molecules cannot take place. It is therefore important to work with low surface concentrations of the antigen on the gold electrode, achievable by protein concentrations in the range of 10 to 100 µg/mL during the immobilization procedure.

11. If one is working with analytes that have more than one potential binding site (e.g., outer cell membrane fragments of Gram-negative bacteria, phages) it should be noted that the

affinity between receptor molecule and one single binding site is not determined, but rather the (biological much more interesting) avidity.
12. Aptamer receptor layers can alternatively be regenerated with acidic (0.2 M glycine-HCl, pH 2.2) or chaotropic (6 M urea) buffers with no significant loss of sensitivity.
13. Many details of the highly elaborate (genome-based) DNA microarray technologies are easily transferred to aptamer chips; therefore these low-molecular-weight receptors could become the perfect complement for proteome-based analyses.

References

1. Aberl, F., Wolf, H., Kösslinger, C., Drost, S., Woias, P., and Koch, S. (1994) HIV serology using piezoelectric immunosensors. *Sensors Actuators* **B18–19,** 271–275.
2. Decker, J., Weinberger, K., Prohaska, E., Hauck, S., Kösslinger, C., Wolf, H., and Hengerer, A. (2000) Characterization of a human pancreatic secretory trypsin inhibitor mutant binding to *Legionella pneumophila* as determined by a quartz crystal microbalance. *J. Immunol. Methods* **233,** 159–165.
3. Hengerer, A., Decker, J., Prohaska, E., Hauck, S., Kösslinger, C., and Wolf, H. (1999) Quartz crystal microbalance (QCM) as a device for the screening of phage libraries. *Biosensors Bioelectronics* **14,** 139–144.
4. Hengerer, A., Kösslinger, C., Decker, J., et al. (1999) Determination of phage antibody affinities to antigen by a microbalance sensor system. *Biotechniques* **26,** 956–964.
5. Prohaska, E., Kösslinger, C., Hengerer, A., Decker, J., Hauck, S., and Dübel, S. (2001) Affinity measurements of antibody fragments on phage by quartz crystal microbalance (QCM). *Antibody Engineering* (Dübel, S., ed.), Springer, Berlin, pp. 397–406.
6. Dorsch, S., Kaufmann, B., Schaible, U., Prohaska, E., Wolf, H., and Modrow, S. (2001) The VP1-unique region of parvovirus B19: amino acid variability and antigenic stability. *J. Gen. Virol.* **82,** 191–199.
7. Liss, M., Petersen, B., Wolf, H., and Prohaska, E. (2002) An aptamer-based quartz crystal protein biosensor. *Anal. Chem.* **74,** 4488–4495.
8. Sauerbrey, G. (1959) Verwendung von Schwingquarzen zur Wägung dünner Schichten und zur Mikrowägung. *Z. Physik* **155,** 206.
9. Uttenthaler, E. (2002) Hochempfindliche akustische Sensorelemente für die Flüssigkeits- und Biosensorik; thesis, Universität der Bundeswehr, München.
10. Kanazawa, K. and Gordon, J. (1985) The oscillation frequency of a quartz resonator in contact with a liquid. *Anal. Chim. Acta* **175,** 99–106.
11. Muramatsu, H., Tamiya, E., and Karube, I. (1988) Computation of equivalent circuit parameters of quartz crystals in contact with liquids and study of liquid properties. *Anal. Chem.* **60,** 2142–2146.
12. Janshoff, A., Galla, H., and Steinem, C. (2000) Piezoelectric mass-sensing devices as biosensors—an alternative to optical biosensors? *Angew. Chem. Int. Ed* **39,** 4004–4032.
13. Schatz, P. J. (1993) Use of peptide libraries to map the substrate specificity of a peptide-modifying enzyme: a 13 residue consensus peptide specifies biotinylation in *Escherichia coli*. *Biotechnology* **11,** 1138.

VII

EVALUATION OF RECOMBINANT PROTEINS IN IMMUNOLOGICAL TEST SYSTEMS

though not necessarily unique

21

Solid Supports in Enzyme-Linked Immunosorbent Assay and Other Solid-Phase Immunoassays

John E. Butler

Abstract

Most modern immunoassays involve the use of synthetic solid phases to immobilize one of the reactants, often by simple adsorption. These solid-phase immunoassays (SPIs) involve ligand–receptor interactions that occur within a reaction volume close to the solution/solid-phase interface. As a consequence, the immunochemistry/biochemistry of these ligand–receptor interactions differ from their counterparts in solution. Nevertheless, mass law equations can be derived for measuring the antigen capture of solid-phase antibodies, for determining the affinity of solid phases for protein adsorption, and for estimating antibody affinity.

Many proteins adsorbed on polystyrene or silicone suffer adsorption-induced conformational changes (ACC) and are partially or largely denatured. Alternative methods for immobilizing proteins and virus, while preserving antigenicity, may yield only a modest increase in functional reactant concentration. Peptides and small recombinant proteins appear to benefit especially from nonadsorptive immobilization. Not all solid phases commonly used in SPIs have the same properties, the same capacity for reactant immobilization, cause the same level of denaturation, or experience the same level of nonspecific binding. Empiricism, adherence to a few practical rules of thumb, and avoidance of certain "old wives tales" can be valuable in the successful development of SPIs.

Key Words: Solid-phase immunoassay (SPI); mass transfer; affinity; protein denaturation; non-specific binding; protein adsorption.

1. Immunochemical Aspects of Receptor–Ligand Interactions at Solid/Solution Interfaces

1.1. General Principles

Ligand–receptor interactions that occur at a solid/solution interface, as in solid-phase immunoassays (SPIs), obey general mass law principles while displaying unique

From: *Methods in Molecular Medicine, vol. 94: Molecular Diagnosis of Infectious Diseases, 2/e*
Edited by: J. Decker and U. Reischl © Humana Press Inc., Totowa, NJ

characteristics that differ from those of ligand–receptor interactions that occur in solution. These differences may complicate data interpretation, can lead to certain misconceptions regarding SPIs, and may contribute to less than optimal assay protocols. These characteristics include: (1) reaction kinetics, (2) reaction volumes, (3) "functional" versus total reactant concentration, and (4) the molecular configuration of immobilized reactants.

1.2. Diffusion Dependence

The time required for equilibrium to be reached in most SPIs, except for some involving microparticles or surface plasmon resonance (SPR), is longer than for solution-phase interactions and increases in proportion to the ratio of the total volume occupied by the fluid phase to that occupied by the interfacial (solid-phase) receptor. Interfacial reaction kinetics display a pronounced diffusion dependence when conducted in microtiter wells (**Fig. 1A**). This can be reduced by vortex agitation, which compresses the liquid to a small area in contact with the receptor (reactant)-coated interface *(1)* (**Fig. 1B**). Forcing the reactive solution phase into the small volume occupied by the solid-phase reactant can also be done using an inert plunger and by concentrating the liquid phase by vacuum or centrifugation on a porous matrix with a large surface such as nitrocellulose. Using flow-through capillary channels as in SPR or using microparticles as the solid phase (*see* **Note 1** and **Fig. 1C**) also reduces diffusion dependence. Plastic tubes with internal "fins" to increase surface area have been proposed *(2)*, and microtiter "starwells" are available based on this principle (Starwells; NUNC, Roskilde, Denmark). These wells increase surface area, reduce the ratio of total volume to reaction volume, and thus shorten diffusion distances. The ability to greatly reduce the time required for the establishment of equilibrium using the various procedures and configurations described above supports the view that diffusion dependence is directly correlated with the ratio of the solution-phase volume to the volume of the reactive interface (*see* **Note 2**).

The diffusion dependence of SPI and the mass transfer issues involved raise questions about the true reaction volume (*see* **Note 2**) in SPI. This is difficult to determine and may be much less than the total fluid volume of the reaction vessel in which the assay is being performed. One must conceptualize that solid-phase antigen–antibody reactions, much like intracellular interactions, occur in "microenvironments," i.e., at the fluid–solid-phase interface. Interactions probably occur within the attraction distance of the strongest primary bonds, a distance that is less than 100 Å and probably closer to 10Å. Diffusion or mass transfer is needed to move reactants into this true interfacial reaction volume. The reactive surface area of microparticles (**Fig. 1C**) is much larger than that of microtiter wells and constitutes a correspondingly larger proportion of the total volume. The true reactant concentration depends on the true reaction volume, which has been estimated but not precisely calculated. The lack of such data creates uncertainty in the calculation of the equilibrium constant (K_{eq}) using common graphic forms of the mass law, thus favoring calculations based on relative and proportional measurements such as "extent of reaction" equations or equations derived by us for evaluating capture antibody (CAb) performance

Solid Supports in ELISA and Other Assays 335

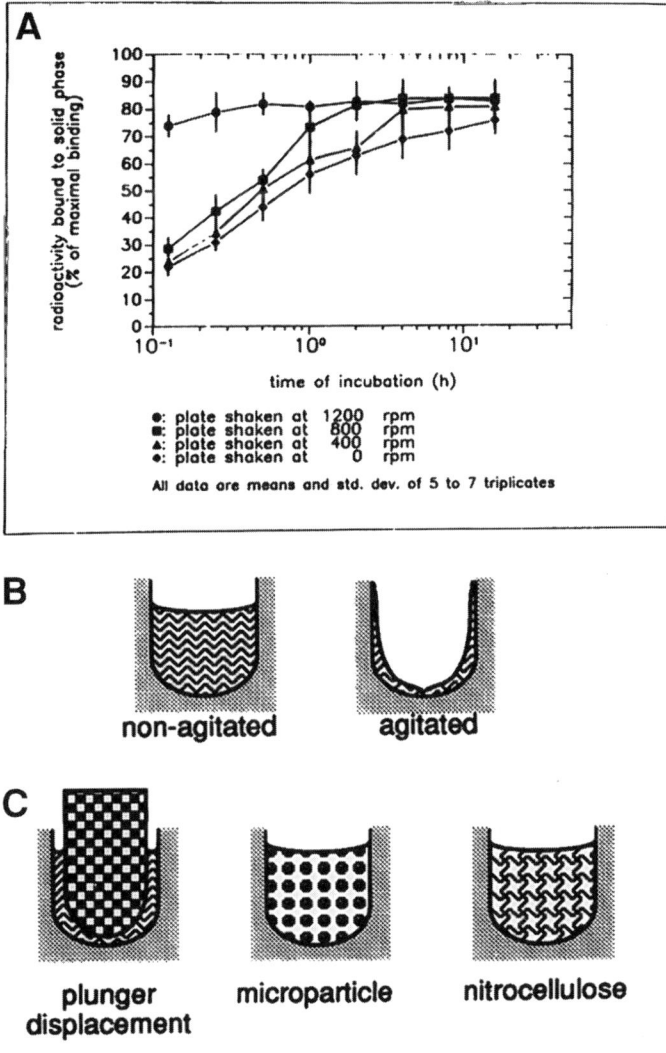

Fig. 1. The diffusion dependence of solid-phase immunoassay and methods used to reduce its influence. (**A**) The effect of vortexing (shaking) microtiters wells on establishment of equilibrium (from ref. *13*). (**B**) Illustration of the physical effect of vortexing microtiter wells (rotary agitation) on the distribution of the fluid phase relative to the solid phase. The fluid phase is depicted by wavy lines. (**C**) Alternative methods of confining the reaction volume to within close proximity to the solid phase bearing the immobilized reactant.

(3) and protein adsorption *(4)*. These features of SPIs mean that values reported for solid-phase antigen-antibody reactions may be incomparable to those obtained for fluid-phase reactions *(5)*.

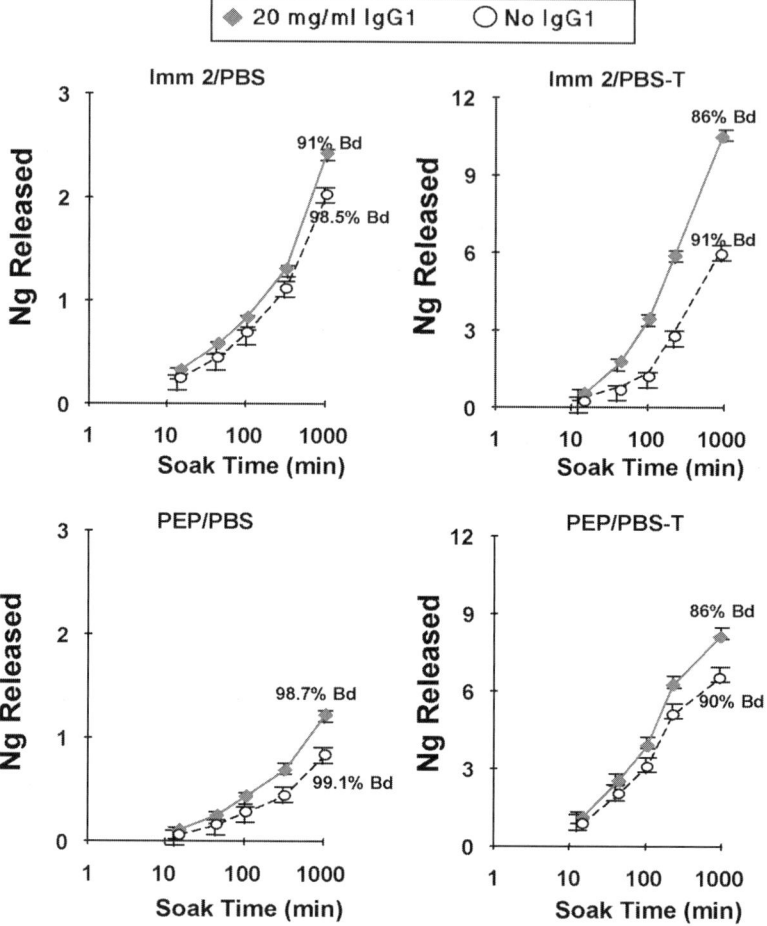

Fig. 2. Effect of protein and detergent on desorption of bovine IgG1. Data are expressed as the nanograms of IgG1 released over time in the presence or absence of excess ligand (20 μg/mL IgG1) in PBS and in the presence of a non-ionic detergent Tween-20 (PBS-T). The percentage value given is the amount remaining bound at the conclusion of the study (16 h). From Butler et al. *(4)*.

1.3. Reaction Rates of Interfacial Interactions

Dissociation rates for interfacial reactions, including those on cell surfaces, may be two orders of magnitude lower than those occurring in solution. The similarity between the slow dissociation rates observed for SPI and for those occurring on cell surfaces is interesting for a number of reasons:

1. Many interactions on cell surfaces may involve aggregation of the cell surface receptors to form "reactive islands" that result in a corresponding increase in avidity due to reduced dissociation of multivalent complexes *(6)* (*see* **Subheading 1.5.**). Not surprisingly,

we and others have shown that passively adsorbed antigens and antibodies also appear to be clustered or aggregated *(7,8)*, i.e., the active interfacial reactant may be "the cluster" or "reactive island."
2. The greater energy needed for dissociation of most antigen-antibody bonds compared with that needed to prevent their association (hysteresis) indicates that secondary bonds have formed after formation of the initial bond, and many of these bonds may be a consequence of aggregation. "Aggregation hysteresis" may account for the synergistic behavior of monoclonal antibodies (MAbs), which bind different epitopes on the same antigen (*see* **Subheading 1.5.**, below). It has been reported that as more MAbs bind, the avidity of their interaction with the ligand increases *(9)*. This suggests that in SPIs and cell surface reactions, secondary bond formation is important.
3. A very high solid-phase reactant concentration, especially as might occur in clusters, and resulting from a confined interfacial reaction volume (see above), could facilitate more rapid reassociation of dissociated reactants than might occur in fluid-phase systems. This might account for the higher K_{eq} of antibodies when tested in SPI versus in solution *(10)*.

Much of this remains speculation since the major purpose of SPI is to quantify biological parameters, not to determine the biological mechanisms involved. Fortunately for the user of the technology, it is the very slow "off-rate" that makes enzyme-linked immunosorbent assay (ELISA) and SPI technology very forgiving in that most of the primary receptor–ligand interaction remains associated throughout the repeated washing steps that characterize many of these assays (**Fig. 2**).

1.4. Do Solid-Phase and Solution-Phase Ligand–Receptor Interactions Differ in Affinity?

All immunoassays are affinity-dependent, a fact that is self-evident from examination of the mass law principles governing these interactions. The relevant question is whether SPIs and ELISAs are particularly affinity-dependent. For example, microtiter assays performed using adsorbed complex antigens may display apparent affinity dependence compared with their solution-phase counterparts *(11)*. This seems paradoxical in light of the stability of solid-phase receptor–ligand interactions until one observes the small proportion of available antibody that is captured by a several log excess of adsorbed antigen *(12,13)*. Perhaps this effect could be due to: (1) the small proportion, and thus low functional concentration, of antigenic epitopes on adsorbed antigens that survive the passive adsorption process; (2) steric hindrance of epitopes when molecules are crowded into the monolayer or into aggregates; or (3) alteration but not loss of native epitopes, which results in a much lower affinity for their paratopes.

The stable adsorption of 50–80% of most protein antigens when optimally adsorbed (**Fig. 3C–F**) *(14) should not* result in a deficiency of total antigen. If 500–800 ng of antigen is stably adsorbed on a microtiter well (when a 200-µL sample is used), a 1/10,000 dilution of serum containing 500 µg antibody/mL should still provide a more than 10-fold excess of antigen. Because many tests use only 100 µL, the amount of solid-phase antigen and serum antibody are reduced by half. Considering an average K_{eq} for the interaction, a limitation in *total adsorbed antigen* does not seem to explain the affinity dependence of SPI, rather, loss, alteration, or steric unavailability of

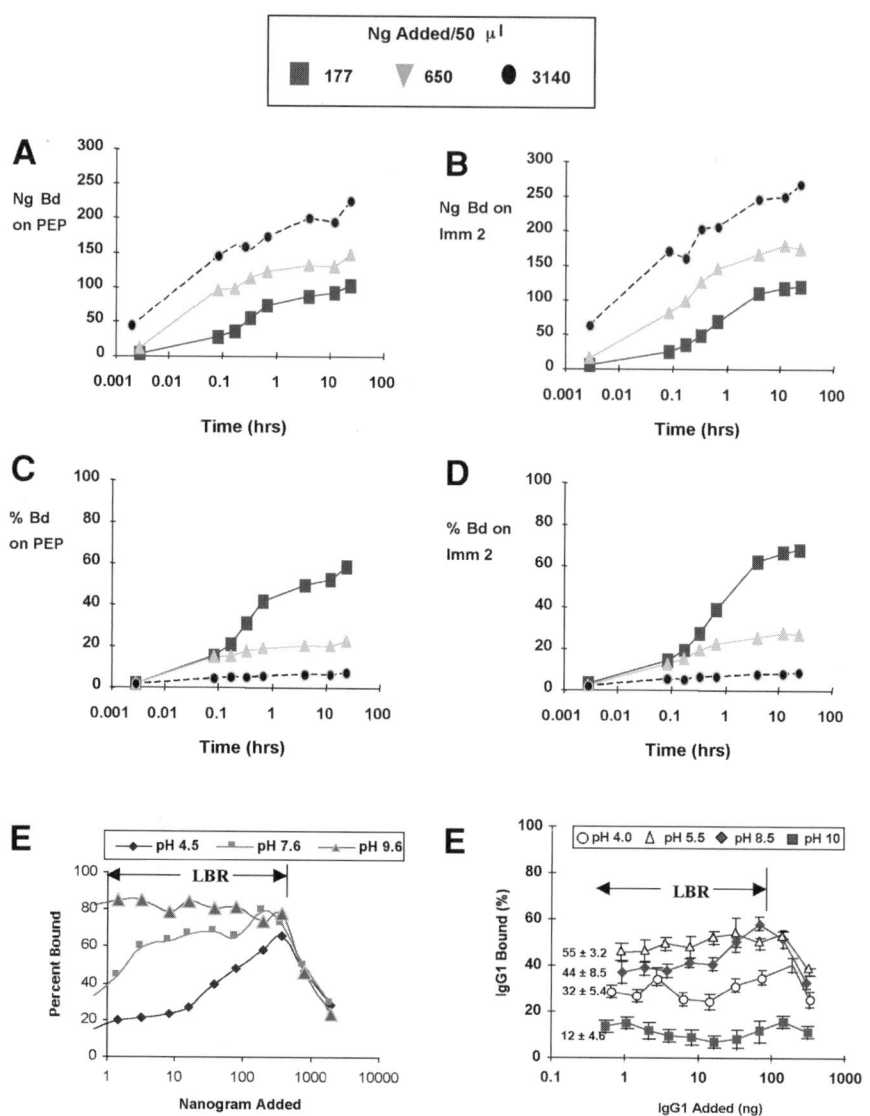

Fig. 3. Characteristics of adsorption of IgG on hydrophobic surfaces. Kinetics of adsorption of human IgG on polydimethylsiloxane elaster (PEP) (**A,C**) and Immulon 2 (Imm 2) (**B,D**). Adsorption was studied at three different concentrations, the lowest (177 ng/well) approximates the upper limits of the LBR. Data are presented both in terms of nanograms bound (**A, B**) and proportion adsorbed (**C, D**). From Butler et al. *(4)* (**E**) The influence of pH on the adsorption of rabbit IgG on Imm 2 expressed as percentage bound. From Butler et al. *(15)*. Only at alkaline pH is a region of constant percent adsorption, i.e., an LBR, observed. All IgGs studied to date behave similarly. (**F**) The effect of pH on the adsorption of bovine IgG1 (pI = 5.5) on PEP, expressed as percentage adsorption. The mean percentage bound in the LBR ± SD is given to the left of plots. Note that adsorption is best at the pI of IgG1. From Butler et al. *(4)*.

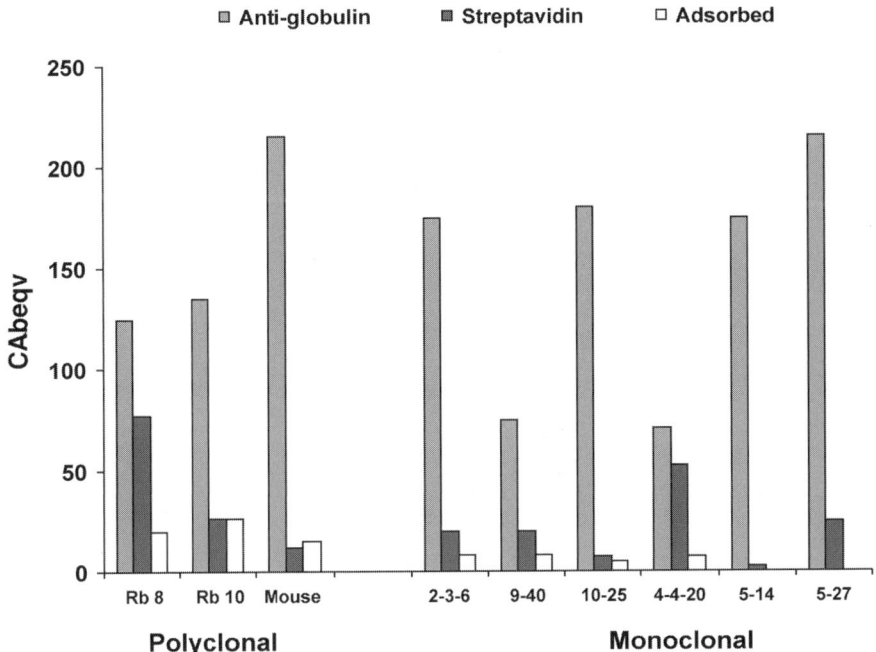

Fig. 4. The proportion of functional capture antibody equivalent (Cabeqv) after immobilization of capture antibodies (CAbs) specific for fluorescein, by different methods. The three CAbs on the left are polyclonals whereas the six CAbs on the right-hand side are monoclonals. CAbeqv is the equivalent of an antibody with two functional sites. Because the standard for this assay was an adsorbed Cab with only one functional site, a value of "200 percent functional" means that 100% of the antibodies are functional with two binding sites. The immobilization procedures included use of a primary antiglobulin, a streptavidin–biotin linkage *(44)*, and direct adsorption. Equal amount of immobilized CAbs were established using iodinated CAbs. From Butler et al. *(15)*.

epitopes seems to be more plausible. Thus, "functional" reactant concentrations in SPIs and ELISA may be much lower than total reactant concentrations *(15,16)*. This has been quantitatively demonstrated for adsorbed antibodies (Abs) (**Fig. 4**) and antigen (Ag) (**Figs. 5** and **6**). The possibility that immobilized reactants may not be conformationally displayed in the same manner as in solution is supported by a variety of experiments. This is further addressed below.

Steric factors are also operative on solid phases. Antigenic epitopes may become buried during immobilization, either at the solid-phase protein interface or at the protein–protein interface of a tightly packed monolayer. As there is probably some orientation ordering of molecules in the packed monolayer, not all epitopes are likely to be equally affected, and not all solid phases behave in the same manner (*see* **Subheading 2.2.**). Steric hindrance may also occur when the specific antibody bound to its epitope cannot be recognized by, e.g., the antibody–enzyme conjugate used in an ELISA *(17)*.

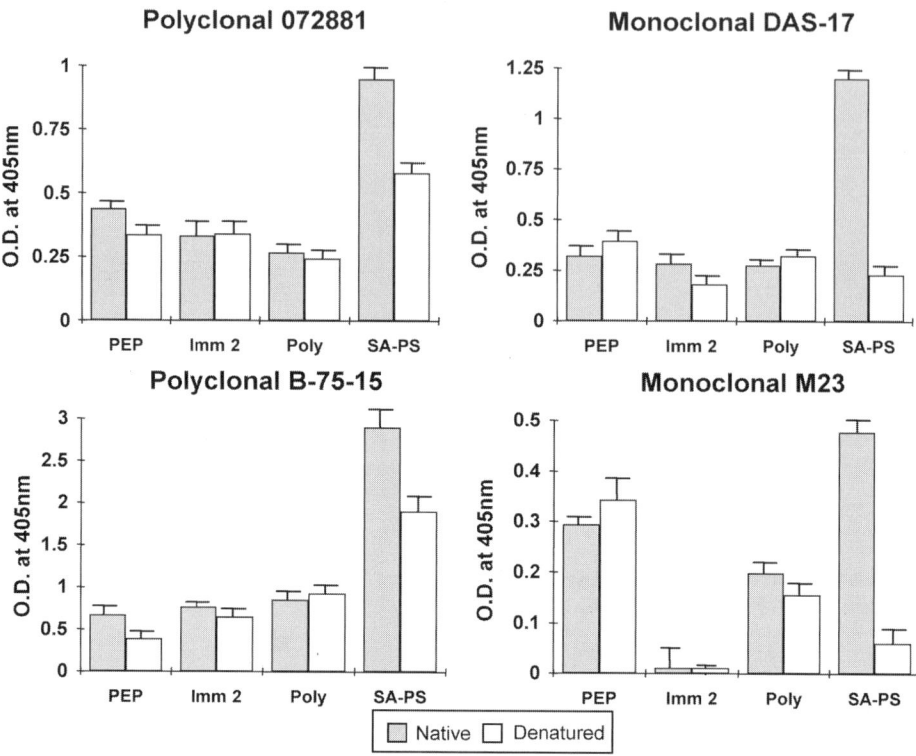

Fig. 5. The specificity of two pAbs and two mAbs for bovine IgG1 immobilized on various surfaces by adsorption or via a streptavidin–biotin linkage (SA-PS) with or without prior denaturation in 6 M guanidine-HCl. Histograms were constructed using OD_{405} values calculated from triplicate log-log titration plots. From Butler et al. *(16)*.

1.5. Hysteresis and the "Joy of Aggregation" in SPI

It is well recognized that aggregation of receptor molecules on cell surfaces is a normal event during receptor-mediated stimulation of cells. This was suggested from initial observations of "capping" on lymphocytes. Although such aggregation may be necessary to assemble the various kinases and adapter molecules with the receptor and its signaling molecules to transduce signals to the nucleus, such aggregation is likely to increase the avidity of the ligand–receptor interactions and produce a hysteresis effect. Although it is recognized that receptor molecules can move in the fluid matrix of the cell membrane lipid bilayer, it is less well recognized that lateral diffusion also occurs during the adsorption of proteins on hydrophobic polymers *(18)*. Such movement may contribute to the formation of the "clusters" or "reactive islands" of adsorbed proteins that have been reported *(7,8,19)* and also the concentration dependence of adsorption efficiency at certain pH values (**Fig. 3E**).

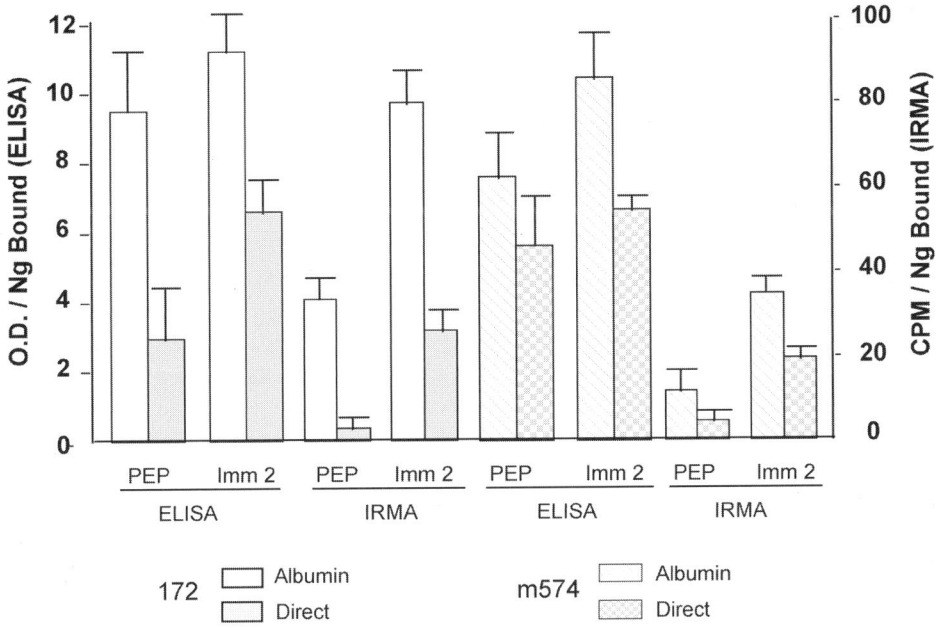

Fig. 6. Comparative detection of IgG2[b] when directly or secondarily adsorbed on silicone (PEP) or polystyrene (Imm 2). Data are given for detection by an ELISA and an immunoradiometric assay (IRMA) to test whether steric hindrance of the enzyme–antibody complex affected detectability. 172 (left side) is a polyclonal rabbit antibody and m574 (right side) is a mouse monoclonal antibody. From Butler et al. *(16)*.

It is also quite likely that aggregation results in multilayering of proteins within clusters such that the most superficial layers are the least denatured and therefore retain most of the properties of the native antigen or antibody. This is demonstrated by adsorbing IgG to the surface of an already existing surface protein layer (**Fig. 6**) and when capture antibodies are immobilized indirectly by a surface-adsorbed antiglobulin (**Fig. 4**).

The concept of aggregation or clustering of immobilized reactants on solid phase gives rise to a modified version of the mass law equation:

$$\text{Ag}_{SLD} + \text{Ab}_{SOL} \underset{k_2 D}{\overset{k_1 D}{\rightleftharpoons}} \text{Ag}^*_{SLD}\text{- - - -Ab}_{SOL} \underset{k_2 R}{\overset{k_1 R}{\rightleftharpoons}} \text{Ag}_{SLD}\text{–Ab}_{SOL} + \text{Ag}_{SLD}\text{–Ab}_{SOL} \underset{k_2 A}{\overset{k_1 A}{\rightleftharpoons}} (\text{Ag}_{SLD}\text{Ab}_{SOL})_n \quad (1)$$

The dashed bond indicates that both Ag_{SDL} and Ab_{SOL} are present within the true reaction volume but not yet combined.

Equation 1 describes an interfacial interaction using an immobilized antigen (Ag_{SLD}). The equilibrium reaction D governs the diffusion-dependent phase of the reaction, which moves the soluble reactant (Ab_{SOL}) to the interface. Reaction R, governed by rate constants k_1R and k_2R, describes the interaction of Ag_{SLD} and Ab_{SOL} within the true interfacial reaction volume. Whereas the average forward rate constant (k_1) for solution-phase interaction is $10^7\,M^{-1}s^{-1}$, those for reactions on synthetic solid phase and cell surfaces are two to four orders of magnitude slower *(20)*. **Equation 1** assumes that this overall slower k_1 is not the consequence of the kinetics of the interaction within the true interfacial reaction volume, i.e., reaction R, but the result of slower diffusion and lower mass transfer of reactants to the site of interaction (reaction D). Hence, **Eq. 1** distinguishes between D and R, and, as described above, k_1D can be greatly increased by a number of means (*see* **Subheading 1.2.** and **Fig. 1B** and **C**).

It is also known that the overall dissociation rate (k_2), from synthetic or cellular interfaces is on the order of 10^{-5} to $10^{-4}\,s^{-1}$, i.e., up to two orders of magnitude slower than for solution-phase systems *(21–23)*. Polyvalent interactions in solution also have higher K_{eq}s values, which are believed or have been shown to result from lower k_2s *(23,24)*. This hysteresis effect, combined with the apparent clustering of solid-phase reactants *(6,7,19)*, suggests that multivalent interactions involve secondary bonding to other proteins or to the solid phase itself. Extensive crosslinking at the interface may also be associated with surface coagulation or aggregation via translational diffusion *(18)*. This could also increase surface contact and further increase hysteresis and lower k_2. **Equation 1** treats this secondary aggregation phase as a third distinct interaction, i.e., reaction A, that is governed by k_1A and k_2A.

2. Immobilization of Reactants on Synthetic Solid Phases for Use in Immunoassay

2.1. General

Modern immunoassay takes its origin from the historic work of Berson and Yalow *(25)*, who used the specificity and sensitivity of the antigen–antibody reaction to quantify biomolecules on the basis of their antigenicity. Their contribution, known as radioimmunoassay (RIA), was based on competitive inhibition and is therefore not dissimilar in principle to complement fixation, albeit considerably more direct. Many so-called blocking ELISAs that are used to measure antibacterial or antiviral antibodies depend on a similar principle *(26–28)*. Despite its popularity through the 1960s and 1970s, traditional RIA was technically troubled by the difficulty in separating the bound reactant from the free reactant, hence the observation by, e.g., Catt and Tregear *(29)*, that proteins spontaneously adsorbed to plastic surfaces allowed the simple plastic test tube to be not only a reaction vessel but also to provide a solid phase for immobilizing the solid-phase reactant, thus providing a convenient means of separating bound reactants from free reactants. For example, once the receptor is adsorbed on the plastic tube and is allowed to bind its ligand, bound ligand and free ligand can be quickly separated by merely inverting the tube. Since the pioneer development of ELISAs and other SPIs in the 1970s, plastic test tubes have given way to microtiter plates, nitrocellulose and nylon membranes, beads of polystyrene and methylmethacylate, and, especially today, microparticles. In any case, what constituted the "slowest ship in the RIA convoy (separation of bound and free ligand)" was

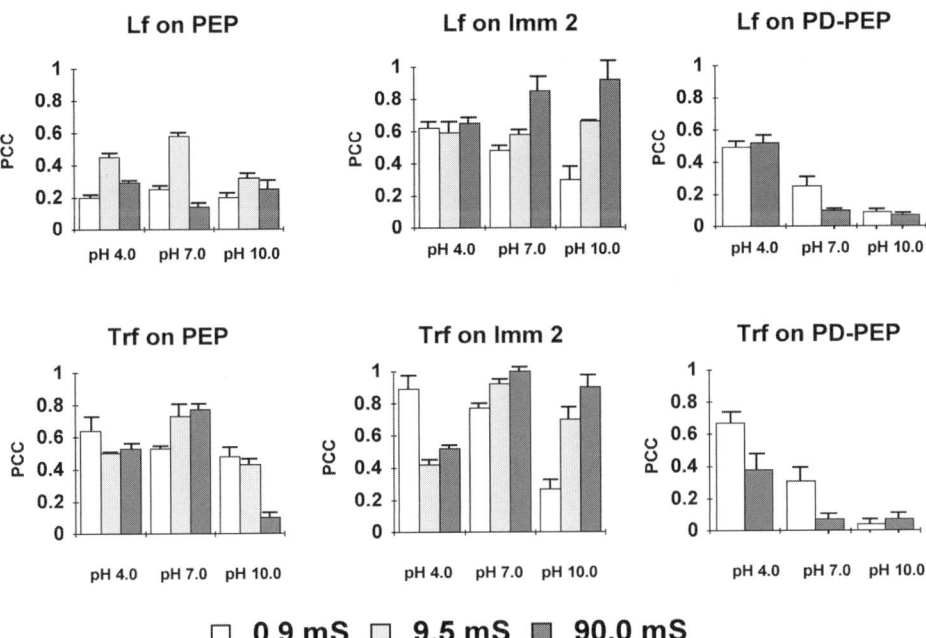

Fig. 7. Summary of the protein capture capacity (PCC) of Imm 2 and PEP for lactoferrin (Lf) (pI = 7.9) and transferrin (Trf) (pI = 5.3) at various pHs and ionic strengths indicated in milli Siemens (mS). Error bars depict standard deviations. PD-PEP = plasma discharge treated PEP. From Butler et al. *(4)*.

revolutionarily overcome by the introduction of a solid phase that held the immobilized receptor firmly.

2.2. The Diversity of Solid Phases Used in SPI

The popularity of a new principle like SPI often results in its widespread usage on all types of solid phase without regard to the chemical differences among them. It is usually only later that investigations reveal that all solid phases are not "created equal" and thus do not behave in the same manner. This is quantitatively demonstrated in **Figs. 2** and **7**.

Differences in volume-to-surface ratios of the various solid-phase surfaces affect both the kinetics and dynamic range of SPI. Immobilization chemistries also differ among solid phases and specific applications. **Table 1** summarizes the characteristics of the more commonly used solid phases with respect to parameters that influence their performance in SPI. Solid phases can be grouped into at least three broad categories, with plastic labware such as tubes or microtiter or tissue culture plates constituting a rather physically and chemically homogeneous group. Not surprisingly, these materials generally perform similarly, regardless of manufacturer or type (*see* **Note 3**). The 96-well microtiter plate is ubiquitous in laboratories where, e.g., ELISAs, are performed. Nearly a half-dozen companies manufacture plate readers, a smaller

Table 1
Characteristics of the More Commonly Used Solid Phases

Solid phase	Bonding force	Relative surface area	Performance characteristics
Plastic labware			
Polystyrene	Hydrophobic	Modest	
Polystyrene-irradiated	Hydrophobic, hydrophilic, and covalent	Modest	Low background, reproducible
Surface-functionalized	Hydrophobic and polystyrene	Modest	Readily adapted to automation
Bead			
Polystyrene (PS) beads	Hydrophobic	Moderate	Yield assays with broad dynamic ranges
Derivatized PS beads	Covalent, hydrophobic, and hydrophilic	High	Less convenient to use than labware; more difficult to automate
Beaded agarose and derivatives	Hydrophilic and covalent	High	Minimal protein denaturation; high background and difficult to automate
Microparticles	Hydrophobic and covalent	Very high	"Solution-phase performance" owing to colloidal nature; wide dynamic range; magnetized variants make them automatable
Membranes			
Nitrocellulose (NC)	Hydrophobic and hydrophilic	Very high	Desorption and background problems hinder their use in quantitative assays
Nylon	Hydrophobic	Very high	Serious background problems reduce signal:noise ratio
Charge-modified nylon	Hydrophilic, covalent, and hydrophobic	Very high	Problems similar to nylon but perhaps less denaturation and less desorption
Functionalized nitrocellulose	Hydrophobic, covalent, and hydrophilic	Very high	Similar to NC but less desorption
PVDF (Immobilon P)	Hydrophobic	Very high	Very high and stable binding; may be best for immunoblotting

From ref. *3*.

number supply automatic washers and diluters designed for such microtiter plates, and some offer full-robotic systems for SPIs in microtiter plates.

The most commonly used material for microtiter plates is polystyrene, with irradiated "plasma-treated" (*see* **Note 4**) forms having somewhat greater capacity to adsorb protein, probably due to covalent bonds made possible by the free radicals that generate reactive carboxyls during the irradiation. Polystyrene has a relatively flat surface with a characteristic windrow pattern (**Fig. 8**). Both Costar (Cobind) and NUNC (Covalink) offer functionalized polystyrene and thus covalent attachment. However, the overwhelming influence of the surrounding hydrophobic polystyrene surface in such wells means that most immobilization might still be a consequence of protein–plastic bonding. Thus, the performance or such functionalized surfaces is only slightly different than on nonfunctionalized surfaces. However, functionalized plates may differ substantially in performance from nonfunctionized plates when used with small molecules, especially peptides (*see* **Subheading 2.9.**).

Polystyrene beads function much like plates, although they provide greater surface area. Their relative inconvenience in use has limited their popularity, whereas microparticle assays have become popular. Microparticles differ from beads in behaving as a colloid, thus greatly increasing surface area and greatly reducing the diffusion dependence of SPI performed using the more traditional solid phases (**Fig. 1**). Magnetic microparticles readily facilitate the separation of the solid phase and its bound reactant from the fluid phase containing the free reactants after the ligand–receptor interaction has occurred. Magnetic beads are also popular in cell separations, affinity purification of proteins, and preparation of mRNAs.

Beaded materials composed of carbohydrates are unique on the list provided in **Table 1** in both chemistry and performance. Binding to their hydrophilic surface is usually covalent and proteins immobilized in this manner are only minimally altered (*see* below). However, hydrophilic beads have never achieved popularity in immunoassay and have been largely restricted to affinity chromatography.

The third group of solid phases is membranous materials, and they form the basis of ELISA-based immunoblotting. Their adsorptive surface areas are 100–1000 times greater than those of plastic, presumably due to their immense internal surfaces. Also, the porous nature of membranes allows flow-through technology to be employed in immunoassay, which is another means of reducing the diffusion-dependent phase of SPI (*30*; **Eq. 1**; and *see* **Subheading 1.2.**). A variety of types exist, including those composed of cellulose nitrate ester [nitrocellulose (NC)], nylon, and polyvinylidene difluoride (PVDF). NC and PVDF are preferred for proteins, whereas nylon is popular for nucleic acids. Membranes, especially nylon, have a high propensity toward nonspecific binding (NSB) of proteins due in part to their large surface area. Thus, they require "blocking agents" (*see* **Subheading 3.3.**) and extensive washing when used in protein immunoassays. The use of nylon for nucleic acid research is a different matter and for such applications is generally superior. Millipore offers NC in a 96-well format (the well bottoms are NC), and several companies offer assemblies for clamping NC sheets into templates for 96-well systems (e.g., Pierce Chemicals, Rockford, IL). The principal chemical bonds between proteins and nonmodified membranes are

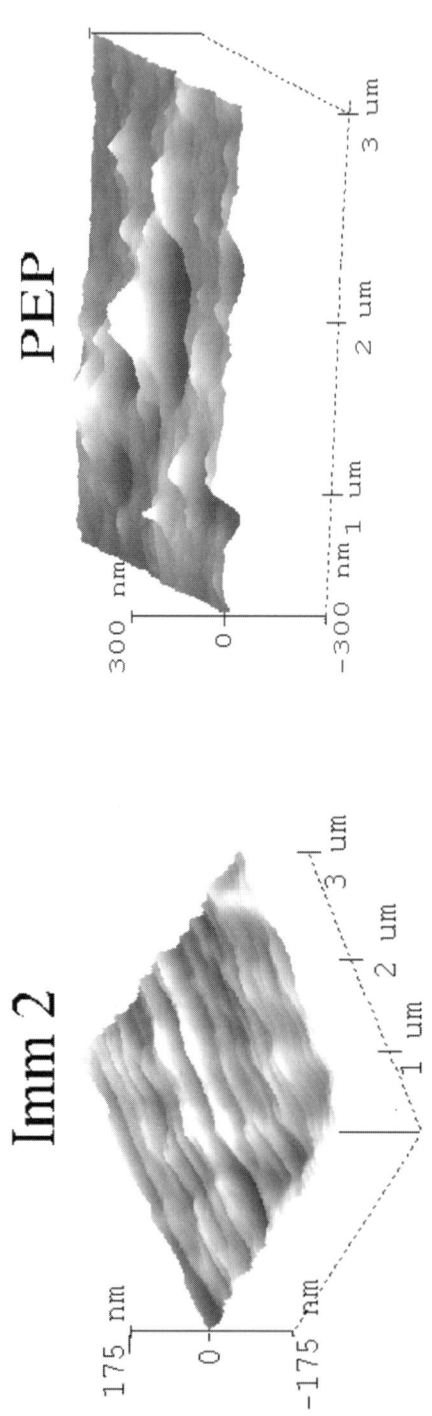

Fig. 8. The surface topography of PEP (right) and Imm 2 (left). Images are for 3 μm squares obtained by atomic force microscopy in the Tapping Mode. Note that the maximum surface height depicted (Z axis) is 175 nm for Imm 2 but 300 nm for PEP. From Butler et al. (*4*).

hydrophobic, although evidence exists for some hydrophilic bonding (*see* **Note 4**). It is possible that a certain amount of the latter represents protein–protein bonds, which dissociate under conditions designed to break hydrophilic bonds (*see* **Note 5**).

Overall, synthetic surfaces used in immunoassay should not be thought of as "flat," but as having highly contoured surfaces (**Fig. 8**). Membranous surfaces are the most extreme, but even the superficially smooth surface of a silicone elastomer is highly contoured compared with that of polystyrene (**Fig. 8**). Differences in surface contour can significantly affect estimates of planar surface and can also influence the effect of washing procedures *(4,16)*. In fact, surface "roughness" may be important in immobilization by adsorption *(31)*.

2.3. General Methods Used for Reactant Immobilization

Solid-phase reactants can be immobilized by four general procedures: (1) direct adsorption, (2) covalent attachment to functionalized solid phases, (3) immunochemical immobilization, and (4) other nonadsorbent, noncovalent methods of attachment. The last group includes immobilization via streptavidin–biotin linkages and the use of bacterial Ig-binding proteins, e.g., Protein-A. Covalent attachment is most common for hydrophilic beads (agarose) or heavily functionalized polystyrene beads. Direct adsorption is most widely used for ELISAs on microtiter plates and for immunoblotting on NC.

2.4. Immobilization of Proteins by Direct Adsorption

The adsorption of proteins on synthetic surfaces, such as polystyrene latex and glass, was studied as early as 1956 *(32)*, although its popular application to immunoassay dates to Catt and Tregear *(29)*. **Figure 3** illustrates the kinetics and characteristics of protein adsorption on two different hydrophobic surfaces, polystyrene and polydimethylsiloxane. Percentage bound plots (*see* details below) generate values that are independent of assumptions regarding the true reaction volume, a controversial issue in any discussion of solid-phase interactions (*see* **Subheading 1.3.** and ref. *13*). Since the mean percentage bound in the linear binding region (LBR; *14*; **Fig. 3E** and **F**) is a measure of the ligand capture capacity by a receptor attached to a solid phase, one can adopt the term protein capture capacity (PCC) to express the ability of a solid phase to adsorb (capture) proteins. We have explained a similar approach for measuring the antigen capture capacity (AgCC) of capture antibodies and an appropriately derived equation *(3)*. In studies on protein adsorption, the solid phase is not an antibody but the surface itself and an equation (Eq. 2) derived from those of Karush *(33)* for describing the surface "affinity," i.e., the PCC of solid phase is

$$\text{PCC} = \frac{N_m \cdot K}{1 + N_m \cdot K} \quad (2)$$

where K is the equilibrium constant of the monolayer adsorption reaction and N_m is the maximum value for n, i.e., the number of sites available for adsorption on the solid phase. N_m can be calculated from Langmuir plots of $(1/b) \times (1/c)$ in which the

Y-intercept is $1/N_m$. According to **Eq. 2**, PCC cannot exceed 1, e.g., 100% bound, and PCC approaches 1 as a $N_m \cdot K \gg 1$.

In percentage bound plots, the relative length of the LBR (**Fig. 3E and F**) is proportional to N_m, and PCC becomes a measure of the relative K when N_m is constant. One might assume that for a given surface area and a molecule of fixed mass, N_m would be constant if indeed monolayer formation is surface-area-dependent.

The application of equations derived for antibody–antigen reactions to surface adsorption of macromolecules assumes that surface adsorption follows mass action principles, i.e., that adsorption is a reversible, equilibrium interaction. We have verified this assumption experimentally *(4)* (**Fig. 2**).

Most soluble proteins have their highest PCC for noncharged surfaces at neutral pH and physiological ionic strength (**Figs. 3E and F, 7**). When the kinetics of adsorption for a typical protein like polyclonal IgG for polystyrene or silicone (**Fig. 3A–D**) is studied under these conditions, several phenomena are noteworthy. First, almost no additional adsorption is seen after 4 h regardless of the initial concentration used. Second, the proportion adsorbed is inversely correlated to the amount added. Finally, when 177 ng/50 µL (corresponding to approx 800 ng/200 µL) is added, 50–60% of the protein becomes adsorbed in 4 h. When adsorption is plotted as the proportion bound versus the amount added, a so-called linear binding region (LBR) is seen (**Fig. 3F**) in which the proportion stably bound is constant so when data are plotted as amount bound versus amount added, a straight line is obtained *(7)*. As described in **Eq. 2**, the value for PCC (a measure of avidity of the interaction) is calculated over the entire LBR.

We have calculated by different methods, including Langmuir plots, that the upper limit of the LBR corresponds to the formation of a protein monolayer *(4)*. The results in **Fig. 3A–C** suggest that at concentrations of more than 177 ng/50 µL, the IgG was applied at a supersaturating concentration or above the LBR. Nevertheless, as more protein is added above this level, progressively more protein becomes adsorbed (**Fig. 3A and B**), leading to the formation of so-called protein multilayers *(34)*. Secondary layers are less stable *(4)* but also appear to represent less denatured proteins *(16)*. Secondary layers or multilayers are believed to form through protein–protein interactions and are likely to involve hydrophilic interactions. This may explain why they are less stable than protein adsorbed directly on hydrophobic surfaces. Therefore, comparisons of different surfaces for their ability to adsorb a protein or for different proteins to become adsorbed to the same surface should be done in the LBR and expressed as PCC. The results of such a study are presented in **Fig. 7**, in which the effect of pH and ionic strength on the adsorption of two similar proteins with very different pI values was studied. As indicated, PCC for both lactoferrin (Lf) and transferrin (Trf) was optimal for adsorption to polydimethylsiloxane elaster (PEP) at pH 7.0 and physiological ionic strength (9.5 mS). Raising the pH to 10 reduced the PCC of PEP for both proteins even though at pH 10 Lf would be only slightly above its pI (negatively charged). On polystyrene, raising both the pH and ionic strength increases the PCC of Lf and Trf regardless of the difference in their pI values. **Figure 7** serves to indicate that: (1) neutral pH and physiological ionic strength are good conditions for the adsorption of two proteins of very different pI; and (2) the effect of pH and ionic strength is not a reliable predictor of PCC even when the protein pI is known. We have also shown that the PCC of proteins like IgG is surprisingly little affected by pH or ionic strength even when the mean pIs of the IgG differ *(4)*.

When the PCC of Lf and Trf for plasma-treated silicone was studied (making the surface highly hydrophilic due to free radical generation), PCC was markedly reduced above pH 4.0, especially as ionic strength was increased (**Fig. 7**). In other words, creating a hydrophilic surface significantly decreases stable (hydrophobic) adsorption. Thus, hydrophilic surfaces have not been popular for the adsorption of proteins used in solid-phase immunoassays.

Although the PCC of proteins is difficult to predict based on charge, there is a general correlation between PCC and the molecular weight for proteins in the same family. The PCC of IgM (1000 kDa) is significantly greater than that of IgA (400 kDa), which is in turn slightly greater than IgG (150 kDa) *(14)*. IgG has a higher PCC than small globular proteins like lactoferrin, α-lactalbumin, and so on. Consistent with monolayer formation on polystyrene, fewer large molecules are needed for the monolayer than small molecules *(13)*. Others have also described modifications of the Langmuir equation to study protein adsorption. Stevens and Kelso *(35)* estimated that the protein capacity for a 100-μL well was 250–325 ng, dependent on the PCC. These values are consistent with our data on adsorption in the lBR (**Fig. 3E**).

In summary: (1) protein adsorption as measured by PCC is biased to favor hydrophobic surfaces, (2) the effect of pH and ionic strength is not always predictable even if the pI of the protein is known, (3) ionic strength-dependent increases in PCC (expected for hydrophobic reactions) are observed for polystyrene but not hydrophobic surfaces in general, (4) PCC is proportional to molecular size for proteins of the same family, and (5) PCC is an empirical value rather than predictable value. The last statement takes on greater significance when adsorption to hydrophobic membranes like nylon and NC are considered.

NC and other membranes show: (1) a much higher capacity for adsorption per planar surface area than plastic and (2) considerable adsorption heterogeneity among membranes and proteins. In contrast, adsorption of IgG on polystyrene microtiter wells from various manufacturers at alkaline pH differs only subtly. However, we and others have observed that adsorption pH can affect the activity of adsorbed MAb *(36)*, and preincubation of Immulon 2 (Imm 2) or NUNC Maxisorp plates with buffers of different pH can alter subsequent adsorption, even when adsorption is done at another pH (unpublished data). A change in surface chemistry is suggested, perhaps equivalent to the effect seen after γ-irradiation or photo oxidation (*see* below).

2.5. Stability of Adsorbed Proteins

Macromolecules adsorbed on polystyrene and silicone, even within the LBR, dissociate at a regular rate, but this seldom exceeds 15% over a 16-h period even when both nonionic detergent and blocking proteins are present in the incubation solution (**Fig. 2**). When exchangeable protein is absent in the incubation solution, desorption drops to less than 10%, and when no detergent is used, only 1–2% of adsorbed protein is released (**Fig. 2**). The ionic detergent sodium dodecyl sulfate (SDS) can increase desorption, especially from silicone. When proteins are adsorbed above the LBR, they form less stable multilayer aggregates, as evidenced by their greater dissociation when secondary antibodies and conjugates are added to the system *(16)*. As much as one-third of the adsorbed protein is released after incubation with secondary antibodies and conjugates if initial adsorption occurred above the LBR, whereas desorption after such treatment in the LBR is no greater than that which occurs with PBS-T alone *(16)*. Ionic detergent also preferentially causes dissociation of protein–protein interactions that form on the surface above saturation. The relatively stable monolayer of protein adsorbed on polystyrene, at least in short-term experiments, is such that a urea/SDS/2-mercaptoethanol cocktail or dilute alkali can be used to remove the immunochemically bound reactant after completion of an ELISA without completely dislodging the adsorbed layer. This allows the same antigen-coated wells to be reused in subsequent tests. This can result in a major reduction in cost and labware waste but should not be indiscriminately used until empirically tested in one's own assay system.

2.6. The Biological Integrity of Adsorbed Reactants

Various investigators have studied the consequence of passive adsorption of proteins on polystyrene, and their findings are summarized in **Table 2**. These investigations reached the conclusion that passive adsorption results in the loss or alteration of antigenic epitopes, the loss of enzymatic activity, the generation of new epitopes, demonstrable physical chemical changes, and the loss of CAb activity *(15)*. We have generically referred to this as adsorption-induced conformation change (ACC) *(16)*. As described below, losses in CAb activity can exceed 90% (**Fig. 4**), and if losses of the same magnitude occur with antigen (**Fig. 5**), it can explain the discrepancy between total serum antibody measured by quantitative precipitation and that measured by ELISA *(12,13)*. At least for polystyrene and silicone, there appears to be sufficient evidence that passive adsorption involves conformational changes as proteins unfold to permit internal hydrophobic side chains to form strong hydrophobic bonds with the solid phase.

The loss or alterations of antigenicity have recently been demonstrated by comparing the ability of monoclonal and polyclonal antibodies to recognize immunoglobulin IgG1 that had been (1) adsorbed on polystyrene or silicone within the LBR, (2) first denatured in 6 M guanidine hydrochloride and then adsorbed, or (3) immobilized nonadsorptively in a native or denatured form using a streptavidin linkage. The results of this study are summarized in **Fig. 5**. The results confirm that (1) IgG1 not adsorbed on a solid phase is more antigenic than IgG1 adsorbed on any surface, (2) significantly larger differences in antigenicity are observed between native and guanidine hydrochloride-pretreated IgG1 immobilized through a streptavidin link than between adsorbed native and denatured IgG1, (3) no differences in antigenicity are seen between adsorbed native and guanidine hydrochloride-treated IgG1, and (4) some antibodies distinguish differences in antigenicity depending on the adsorptive surface used. For example, monoclonal antibody M23 readily detects IgG1 adsorbed on polydimethylsiloxane elaster (PEP) but not when adsorbed on Imm 2 (**Fig. 5**).

Table 2
Adsorption-Induced Conformational Change

Protein	Phenomenon	Authors
Albumin	Conformational change after adsorption on glass	Bull, 1956 (*32*)
IgG	Concentration-dependent allosteric conformers after adsorption on polystyrene	Oreskes and Singer, 1961 (*38*)
IgG	Molecule unfolding and changes in antigenicity when adsorbed on polystyrene	Kochwa et al., 1967 (*39*)
IgG	Thermodynamic evidence for conformational change	Nyilas et al., 1974 (*40*)
Monoclonal Ab	Altered specificity after adsorption	Kennel, 1982 (*41*)
Tryptophan synthase	Altered enzymic and antigenic activity after adsorption	Friquet et al., 1984 (*42*)
Lactic dehydrogenase	Conformational alteration after dehydrogenase adsorption on polystyrene	Holland and Katchalski-Katzir, 1986 (*43*)
Monoclonal Ab	Loss of activity after adsorption on polystyrene	Suter and Butler, 1986 (*44*)
IgG, IgA	Loss of antigenicity after adsorption to polystyrene	Dierks et al., 1986 (*12*)
Ferritin	Cluster formation on silica wafers	Nygren, 1988 (*19*)
Antifluorescein	Functional monoclonal antifluorescein adsorbed on polystyrene is clustered	Butler et al., 1992 (*7*)
Antifluorescein	Adsorbed MAbs lose 90% of their activity on polystyrene	Butler et al., 1993 (*15*)
Antitheophylline	MAb adsorbed on polystyrene loses 90% of its activity	Plant et al., 1991 (*37*)
Antiferritin	Adsorbed functional antiferritin is clustered on the surface of polystyrene	Davis et al., 1994 (*8*)
Bovine IgG1	Antigenicity of IgG1 or Gu-HCl denatured IgG1 is similar and much less than IgG1 immobilized through a streptavidin linkage	Butler et al., 1977 (*16*)
Bovine IgG1	Superficial layer of IgG1 adsorbed in multilayers is most antigenic	Butler et al., 1977 (*16*)
Myoglobin	Adsorption of myoglobin effects reactivity of conformation-specific monoclonal antibody	Darst et al., 1988 (*45*)

We have also observed that a MAb cloned against the IgG2^b allotypic variant, when adsorbed on Imm 2, recognizes IgG2^b better when adsorbed than when immobilized via a streptavidin linkage (16) and that differential recognition of IgG2^a and IgG2^b by polyclonal 172 is best if the two allotypic variants are adsorbed (16). The latter experiment suggests that the allotope(s) recognized by certain antibodies become preferentially accessible when IgG2^a is adsorbed. This implies an ACC of IgG2^a. More recently, using monophages cloned from a synthetic antibody phage display library, we observed that most monophages obtained by cloning against adsorbed fibrinogen preferentially recognize the adsorbed form of fibrinogen (**Fig. 9**). Kennel and colleagues, using MAbs prepared against fibrinogen (46), reported similar findings.

Many popular SPIs, such as sandwich ELISAs and ELISpot assays, involve the adsorption of antibody, not antigens, on a hydrophobic surface. When we compared the ability of a panel of adsorbed polyclonal and monoclonal antibodies to fluorescein to recognize fluorescein–protein conjugates, we observed that MAbs especially had lost more than 90% of their binding sites for capturing antigen (CAbeqv) (**Fig. 4**). Plant et al. (37) observed a similar loss of activity. Thus, adsorption affects not merely antigenicity of adsorbed proteins but also the activity of adsorbed antibodies.

These more recent studies (**Figs. 4–6**), together with historic observations (**Table 2**), are consistent with the concept that the adsorption of proteins on hydrophobic polymers like silicone and polystyrene is a denaturation event not dissimilar from treatment of proteins with 6 M guanidine hydrochloride (**Fig. 5**). Thus, investigators using ELISAs and other SPIs need to take this into consideration in the design of their assays and in the interpretation of the data obtained with such SPIs. These observations on the effect of protein adsorption should also be a stimulus for investigators and biotech companies to develop convenient alternatives to simple protein adsorption on hydrophobic surfaces. This is now being recognized (47), and Pierce offers microtiter plates with hydrophilic surfaces.

An observation made in our laboratory that fits very nicely with earlier studies and theories of protein adsorption on medical prostheses in vivo is noteworthy. Protein antigens deliberately adsorbed as multilayers are more antigenically active than those in the surface monolayer (**Fig. 6**). This is consistent with the multilayer passivation theory (34), which predicts that proteins adsorbed in succeeding layers are progressively more native. Such multilayers may exist as clusters that others and ourselves have visualized by scanning electron microscopy (6,7,19). As a principle, moving the reactive molecule away from the solid-phase surface increases the probability that the molecule will remain biologically active.

The biological consequences of adsorption on various blotting membranes appear to have been of less concern to investigators than adsorption on polystyrene. Even if denaturation is on a par with that occurring on polystyrene, much more protein binds to NC, so a greater amount (not proportion) of native protein may survive and denaturation effects are overlooked. Moreover, evidence that hydrophilic as well as hydrophobic forces are involved in adsorption on immunoblotting membranes suggests that conformational alterations on membranes may not be as severe as those that occur on polystyrene.

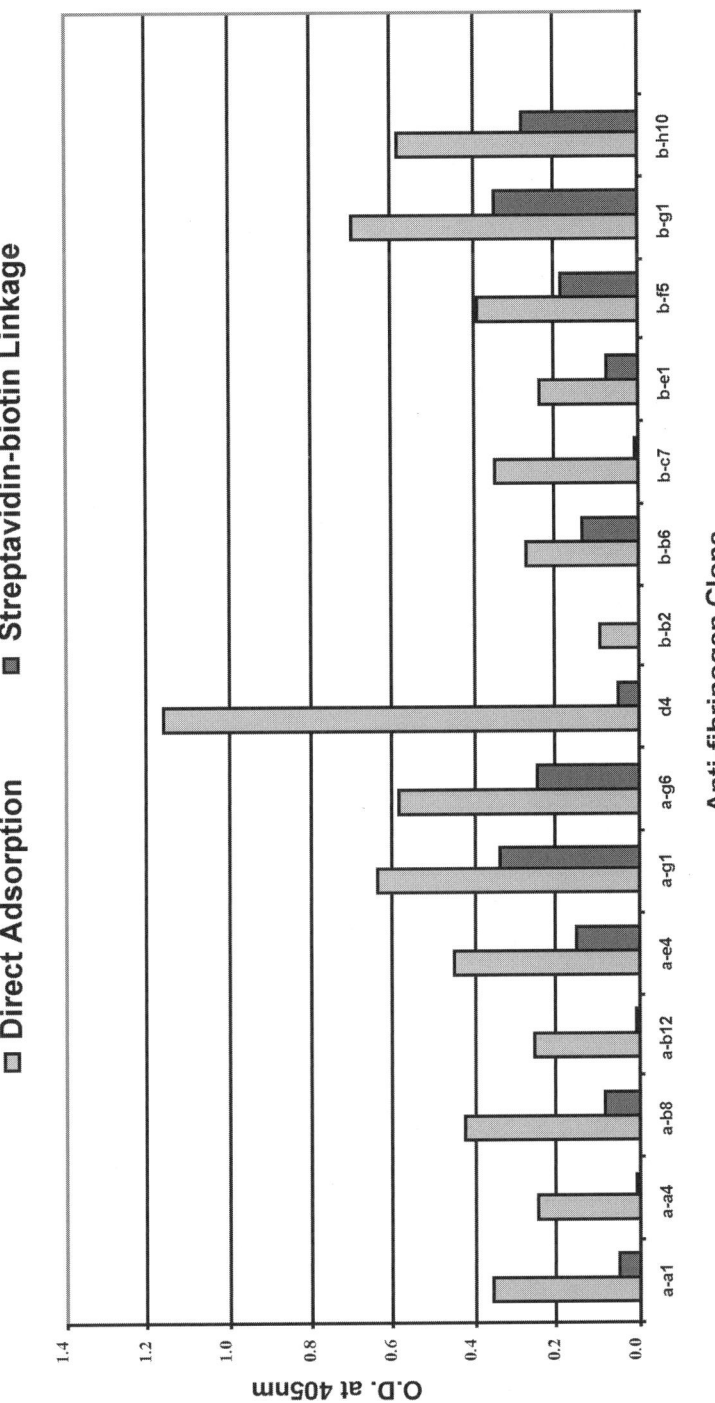

Fig. 9. The specificity of monophage for fibrinogen adsorbed on polystyrene versus for fibrinogen immobilized using a streptavidin–biotin linkage. The phage were cloned from an antibody phage display library on adsorbed fibrinogen (Sun and Butler, unpublished).

If adsorption on hydrophobic surfaces is so destructive for proteins, why is there continued use of this procedure? The answer is simply that enough molecules survive to make the assay work. For example, if 6% of high-affinity capture antibodies adsorbed on a microtiter well survive in a functional state, it is sufficient to provide a sandwich ELISA with a dynamic range of 2–200 ng/mL; this two log range is adequate for most applications and typical of assays reported in the literature.

2.7. Immunochemical Immobilization

Immunochemical immobilization is an alternative method for immobilizing biomolecules that do not lend themselves to direct adsorption. In some cases, investigators have used an adsorbed CAb to immobilize the antigen (*48*; **Fig. 4**). The inherent problem with this approach is that only 10–25% of the initial CAb may itself survive denaturation on polystyrene so that as few as 1 in 10 functional antibody molecules remain to capture the antigen (*15*). This reduces the concentration of antigen much below the total antigen concentration that can be achieved by simple adsorption. Nevertheless, immunochemically immobilized antigen can be 10-fold more active than adsorbed antigen (*49*), and we have seen total retention of activity for CAb immobilized by this method, whereas only 10% were active when they were passively adsorbed (*15*; **Fig. 4**). A major application of immunochemical immobilization is the immobilization of proteins in detergent-containing cell lysates and in samples containing complicating contaminants (*see* **Subheading 2.10.**).

In general, immobilization procedures that can remove the antigen (or antibody) from direct contact with the hydrophobic surface improve their biological activity. This is demonstrated for capture antibodies (**Fig. 4**) and for proteins adsorbed as multilayers that are presumably held only by hydrophilic protein–protein bonds and have no contact with the surface (**Fig. 6**).

2.8. Other Noncovalent, Nonadsorptive Methods

The high affinity of avidin and streptavidin for biotin provides an alternative, nonadsorptive, noncovalent means of immobilizing both antigens and antibodies. Because of the low affinity of streptavidin for polystyrene, it must be (1) covalently bound to the surface (*50*), (2) immobilized by first adsorbing an irrelevant, biotinylated carrier (*44*), or (3) immobilized by biotin that has been covalently attached to functionalized polystyrene (*51*). Each of these methods permits the immobilization of antigens and antibodies that: (1) bind poorly to plastic or NC, and (2) are denatured beyond use by adsorption.

Lectins and the Ig-binding proteins of bacteria, which are readily adsorbed on plastic or other hydrophobic surfaces, can also be utilized as a bridge between the solid phase and the reactant of interest. A prerequisite is that the adsorption process does not destroy or alter their specificity. Concanavalin A adsorbed to microtiter wells is able to immobilize gp120 of HIV, which displays better activity than adsorbed gp120 (*52*). Both protein A of *Streptococcus aureus* and Protein G of *Streptococcus* sp. are capable of stably capturing various IgGs after their adsorption. Recombinant fusion antigen expressing the albumin binding site of Protein G can be captured on adsorbed rat albumin (*53*). Not unexpectedly, background problems are connected with the use

Solid Supports in ELISA and Other Assays

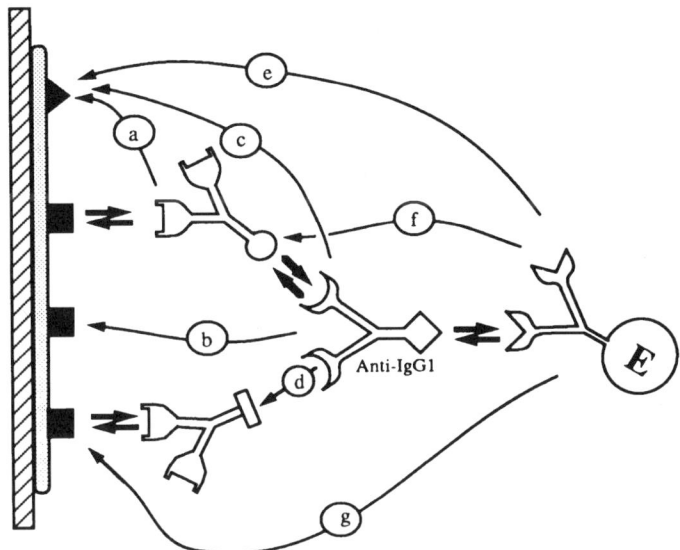

Fig. 10. Short-circuiting produces nonspecific binding (NSB) in ELISA. Antibodies are depicted as bivalent and as Y-shaped molecules. The shape of their Fabs and Fcs are designated to depict their paratope and epitope specificities. The solid, unidirectional arrows with arabic letters depict the various short circuits possible. Double, reciprocal arrows depict the desired or normal reaction pathway. The various short circuits illustrated are as follows: **(a)** primary antibody (analyte) cross-reacts with a second undesirable solid-phase epitope, **(b)** secondary antibody (part of detection system) recognizes the primary antigen on the solid phase, **(c)** secondary antibody recognizes the second undesirable epitope on the solid phase, **(d)** secondary antibody cross reacts with a primary antibody of a different isotype than that for which the assay was designed to measure, **(e)** the tertiary antibody (or reactant that carries the reported enzyme) recognizes an undesirable solid phase epitope; **(f)** the tertiary reactant recognizes and binds the primary antibody (analyte) and thus bypasses the secondary antibody, **(g)** the tertiary reactant binds to the solid phase itself. The latter situation could theoretically occur with any of the reactants. E = enzyme-conjugated antibody.

of Ig binding proteins in ELISAs, especially in assays designed to measure specific antibodies in which the Ig binding protein can also bind the specific antibodies that the assay is designed to measure. Assay "short circuitry" can be a major problem in all immunoassays that involve multiple components like ELISAs (*13*; **Fig. 10**).

2.9. Immobilization of Macromolecules and Peptides by Covalent Bonding and Other Means

Polysaccharides and heavily glycosylated proteins often have low affinity for polystyrene (reviewed in ref. *54*) and often require alternative methods for their immobilization. One alternative is covalent immobilization using a suitable crosslinking agent such as glutaraldehyde, EDAC, or dimethyl suberimidate, to (1) functionalized polystyrenes such as aminostyrene that is commercially available as, e.g., Covalink from

NUNC, (2) polystyrene first treated with surface modifying agents to produce isocyanate groups or amino groups, or (3) poly-L-lysine, phenylalanine-lysine, cotadecylamine, or some irrelevant proteins that are adsorbed beforehand *(55,56;* reviewed in ref. *54).* The same method has been employed for the immobilization of proteins that (1) due to their chemistry cannot bind to a hydrophobic surface, (2) are so small that their adsorptive affinity may be very low, or (3) are conformationally altered by adsorption to the extent that they become a nonfunctional solid-phase ligand *(57–59).* On the other hand, bacterial capsular polysaccharides have been adsorbed successfully on plastic *(60),* and a surprising number of investigators have merely adsorbed peptides passively to polystyrene or synthesized them directly on polystyrene rods *(61).* More efficient immobilization of peptides has been accomplished by γ-irradiation of the polystyrene surface *(62–64)* or photoactivation *(65).* The former is believed to generate carboxyl groups on polystyrene that can form amide bonds with peptide amino groups. Several studies have shown the negative consequences of passive peptide adsorption and have offered alternatives. Sødergard-Anderson et al. *(59)* observed that angiotensin I and II were 5–10-fold more antigenically active when they were covalently attached versus when they were adsorbed. Gregorius et al. *(66)* made similar observations and used tresyl-activated dextran to attach peptides covalently and improve their antigenicity. Others used peptides elongated with lysyl residues *(67)* or improved peptide orientation by deprotecting lysyl cysteine groups *after* covalent attachment to tresyl-activated polystyrene *(68).* Lacroix et al. *(69)* observed that the form of the peptide was important; cyclized peptides appeared to be more active than their linear counterparts.

Another group of biomolecules often studied are those freed from membranes by cell lysis in the presence of detergents. Because detergents are used to block adsorption to hydrophobic surfaces, their presence will inhibit adsorption of the lysate proteins to plastic. Adsorption to NC is possible particularly if detergent concentrations are less than 0.01%. Triton, Tween, and SDS are the most inhibitory detergents; deoxycholate and octylglucoside have a lesser effect *(70).* The fact that detergents are much more inhibitory to adsorption on nylon (Genescreen) than on NC supports the idea that forces other than hydrophobic ones may be involved in adsorption to NC (**Table 1**).

When detergent levels in cell lysates cannot be diluted to permit adsorption, covalent linkages may be required or materials like SM-2 Bio-Beads can be used to reduce the detergent concentration *(71).*

2.10. Cells, Bacteria, and Viruses as Solid-Phase Reactants

Virus particles can be directly adsorbed on polystyrene in the manner of proteins (*see* ref. *54).* Our experiences reveal three potential problems with this approach. First, the virus must be purified free of tissue culture proteins that can competitively inhibit virus adsorption. Second, whole virus may poorly adsorb or the viral epitopes may be altered or destroyed in the process. Third, virus preparations bound on the solid phase must be free of any antigens (normally of media origin) that may have been present in the vaccine preparation, because the sera of immunized animals or vaccinated humans may also contain antibodies to these proteins. The last two problems can be overcome by using an adsorbed, virus-specific capture antibody to immobilize the virus *(54,59)*. Although predictably less virus can be immobilized by this method than by adsorption *(49)*, loss of antigenicity or exposure of internal epitopes due to adsorption-induced denaturation is reduced. NSB is an additional concern with some viruses, e.g., herpes species, that express Ig-binding proteins.

Intact cells and bacteria have also been used as solid-phase ELISA antigens. Their heavily glycosylated membranes can make their stable adsorption difficult so that simultaneously adsorption and crosslinking them to surfaces coated with poly-L-lysine, glutaraldehyde, or the polyaldehyde methyl glyoxal have been used *(72)*. However, stable adsorption of some bacteria can be obtained in phosphate-buffered saline (PBS) *(73,74)*. More serious than in the case of virus, many bacteria have Ig-binding proteins that may increase NSB, thus all but obscuring specific antibody detection.

2.11. "Solution Phase" Solid-Phase Immunoassays

Because immobilization can hinder the availability of epitopes or destroy them altogether, permitting the desired antigen–antibody reaction to occur first in solution, and then capturing and measuring the immune complex, circumvents ACC and steric hindrance. Several variants of this procedure have been described, some with catchy acronyms. The Protein A antibody capture ELISA (PACE) *(75)* involves allowing a biotinylated antigen to bind antibody in solution followed by capture of the complex on solid-phase Protein A and subsequent detection with a streptavidin-enzyme conjugate. A similar system was described for cytochrome c by Schwab and Bosshard *(76)* and for virus by Peterson et al. *(30)*.

3. Practical Rules of Thumb in the Design of Solid-Phase Immunoassays
3.1. Vertrauen ist gut, aber Kontroll ist besser

The diversity of proteins and solid phases that are encountered in ELISA and other SPIs is so large that predicting the optimal conditions for the immobilization of each reactant or its functional activity after immobilization is not possible. This is evident from data presented in **Figs. 5–7**. Similarly, not all antibodies or antigens behave in the same manner when immobilized. This is illustrated by the difference between monoclonal and polyclonal CAbs (**Fig. 4**). The same applies to the issue of *blocking*. Blocking refers to the concept that the use of certain combinations of detergents and proteins can reduce NSB. Whether such a procedure is required or might even be counterproductive, can depend on the surface involved, the biological fluid being tested, or the immunochemical reagents used. Although blind faith often works (*Vertrauen ist gut*), empirical optimization (*Kontroll*) is best when developing or adapting an SPI. A few guidelines and trouble-shooting procedures are provided in this section.

3.2. Immobilization by Adsorption Is the First Choice

Adsorption of proteins on a solid phase is such a simple procedure that it should always be tried first before developing more complicated (and expensive) nonadsorptive immobilization procedures. The following guidelines are useful.

1. Single-use adsorption solutions used at 5 μg/mL are adequate for most proteins that avidly adsorb to polystyrene as a monolayer after 4 h at 37°C or overnight at room temperature (**Fig. 3**). Some investigators (and biotech companies) have been known to reuse the same coating solution. When this is done, higher initial concentrations are required.
2. The role of pH, ionic strength, or even concentration cannot be reliably predicted for proteins (**Fig. 7**), although immunoglobulins reliably adsorb well to polystyrene at pH 7–10 at the concentration indicated above and at physiological ionic strength (**Fig. 3E**). Thus it is recommended that checkerboard matrices be used with variables such as ionic strength and pH that change by dilution in different directions in the matrices. Because biological activity, not the actual amount of adsorbed reactant, is the desired outcome, measurement of antigenic activity (using an antibody at a fixed concentration) to detect an immobilized antigen or the ability of an antibody to capture a labeled antigen (**Fig. 4**) should be the initial test.
3. Failure to detect adequate biological activity in an adsorbed reactant should be followed by tests in which the reactant is first labeled by iodination or a procedure that can provide equivalent sensitivity. When a large quantity of the solid phase is used, the bicinchoninic acid (BCA) assay may even be useful *(37)*. This allows the investigator to determine whether the reactant is actually becoming adsorbed and to what extent. Various investigators have had difficulties with substances like lipopolysaccharide (LPS), but surprisingly few have reported efforts to determine whether the molecules were even being adsorbed! Without such a test, loss of biological activity cannot be distinguished from failed adsorption.
4. Polystyrene surfaces are not created equal. Several companies (NUNC, Dynatech) offer plasma-treated (irradiated) surfaces that generally increase the capacity for protein adsorption. At one time Dynatech offered an entire range of surfaces that could be obtained in small quantities for optimization of adsorption. As is discussed below, certain surfaces are more prone than others to produce NSB. Therefore, before concluding that an assay cannot be done on polystyrene or NSB cannot be blocked, try another source of polystyrene. In an actual situation, a graduate student attempting to duplicate the assay of a colleague away on vacation spent more than 10 frustrating days before discovering that the assay only worked on NUNC Maxisorp wells but not on Immulon 2!
5. Although adsorbing proteins at concentrations beyond the LBR may generate biologically active multilayer clusters, use of these higher adsorption concentrations also encourages greater desorption and therefore lower reproducibility than when a simple monolayer is used. Forming multilayers may be a valuable future technology, but its adoption requires modification of current procedures to reduce reactant loss while maintaining improved biological integrity (*see* **Subheading 2.6.**).

3.3. Blocking and Blocking Reagents

The introduction of a solid phase into reactions involving ligands and their receptors requires that nonspecific interactions, i.e., nonspecific binding (NSB) between the soluble ligand and the solid phase to which the receptor is immobilized, be inhibited or "blocked' so that only the interaction between the receptor and its ligand can occur. Various substances are included in the category of blocking agents, although protein solutions and nonionic detergents are most popular. The choice (or need) for such agents varies with: (1) the solid phase, (2) the empirical experience of the assay designer, and (3) the particular ligand–receptor combination. There are differences of opinion regarding both the need and mechanism of action of blocking agents.

Protein blocking agents, typically solutions of serum albumin, casein, newborn calf serum, or dilute skim milk, are most often added after immobilization of the solid-phase receptor and are believed to "fill in" stretches of the solid phase not occupied by the immobilized receptor. Thus, they reduce NSB by blocking reactive sites on the solid phase. Among the common blockers are those based on casein or skim milk, which appear to be effective on both polystyrene and nitrocellulose surfaces. Polyvinyl alcohol has also been used *(77)*. Their effectiveness is thought to result from low-molecular-weight casein or other small proteins and peptides in skim milk that theoretically "fill in" small areas between the larger immobilized receptor molecules.

Evidence that this is the mechanism is theoretically difficult to explain if indeed a monolayer of receptor already covers the surface (*see* **Subheading 2.4.**). However, such areas of unoccupied surface may indeed still exist if the adsorbed protein that formed the mathematical monolayer was concentrated in "clusters" or "reactive islands," leaving unoccupied plastic between them. If this is the case, the area of reactive, unoccupied surface may vary depending on the configuration of the immunoassay, the manner in which the particular protein became adsorbed, and the choice of solid phase. In immunoassays conducted on nitrocellulose, blocking protein solutions (skim milk) are virtually obligatory to reduce NSB and are typically used during every incubation step after receptor immobilization, whereas on polystyrene, protein blockers, and even nonionic detergents (*see* below) can be omitted from reaction steps that come after receptor immobilization and the initial blocking step. The much larger surface area and membranous matrices of solid phases like nitrocellulose (*see* **Subheading 2.2.**) are probably never covered with anything approaching a monolayer of the receptor so that authentic blocking of unoccupied surfaces is more likely to be the mechanism of action for the blocking agents used on such surfaces.

"Leakiness" of blocking agents refers to their failure to suppress nonspecific interactions as the assay proceeds through subsequent reaction steps. It is believed that such failure does not result from the gradual displacement of the blocking agent but is rather due to the gradual exposure of nonspecific binding sites on the original blocking agent or on the receptor monolayer itself. These observations weaken any concept that the function of protein blocking agents is solely to cover unoccupied reactive sites on the solid phase.

Care must be used in the selection of a blocking agent, if indeed one is necessary, for a particular application. For example, commercial blocking agents like Blotto and Superblock (typically concoctions of skim milk protein, nonionic detergents, and antifoaming agents) either fail to lower NSB or, on silicone, actually increase NSB (J. E. Butler, unpublished data). Thus, there is a need for investigators who work with less ubiquitous surfaces to test empirically the blocking effectiveness of a particular blocking agent on the solid phase being used, rather than extrapolating experiences encountered with other surfaces.

The second category of blocking agents is detergents. These are low-molecular-weight compounds with distinctive symmetry, i.e., hydrophobic at one end and hydrophilic at the other. The combination of their small size and amphipathic nature means they can intercalate protein structure, probably bonding hydrophobically to protein and displaying their hydrophilic ends to the solvent. Detergents can be classified as nonionic, ionic, or zwitterionic. Nonionic forms such as Tween, Triton, and Nonident-40 are those most commonly used in immunoassays. Detergents probably act in various ways, but, in contrast to protein blockers, their action is generally ascribed to *preventing* the hydrophobic adsorption of proteins to the solid phase and decreasing the hydrophobic aggregation of macromolecules rather than "filling in" unoccupied sites on the solid phase. There is, however, little doubt that detergents are capable of dislodging weakly immobilized receptors (**Fig. 2**).

Among the detergents commonly used, Tween-20 appears to be the most popular and is unusual in having the capacity both to prevent nonspecific adsorption to polystyrene and to prevent nonspecific protein–protein interactions during subsequent steps in an immunoassay conducted on polystyrene. Most other nonionic detergents appear to act principally in preventing nonspecific protein–protein interactions. Therefore, many of the investigators who use polystyrene as a solid phase do not use a separate protein blocking step, but merely conduct all steps of their SPI (e.g., ELISA) after the initial immobilization of the receptor, with only Tween-20 present in the washing and reaction buffers at a concentration of 0.05%. By contrast, detergents alone are usually ineffective in preventing NSB on membranous solid phases.

The use of detergents with or without protein blocking agents also has other consequences in the study of receptor–ligand reactions. For example, Qualtiere et al. (*78*) tested the ability of various detergents to interfere with antibody–antigen interactions. Ionic detergents like SDS have inhibitory effects on antibody–antigen reactions. The zwitterionic detergent 3-[(3-cholamidopropyl)dimethylammonio]-1-propanesulfonate (CHAPS) may have the least effect on such interactions but is also less able to block nonspecific interactions, especially hydrophobic adsorption to the solid phase itself. CHAPS is an example of a detergent that should be used with blocking proteins.

It has been suggested by Avrameas and Terynich (*79*) that spontaneous autoreactive antibodies, like those encoded in the preimmune repertoire and observed in sera of nonimmunized individuals, typically: (1) have binding sites that recognize hydrophilic determinants, e.g., CHO, (2) have positively charged residues in their binding sites, and (3) are often high-avidity IgM antibodies. Immunization triggers an adaptive immune response, resulting in antibodies of higher intrinsic affinity that may include antibody binding sites with affinity for hydrophobic epitopes. Perhaps this explains the affinity of IgG from mice infected with lactic dehydrogenase virus (LDV) for polystyrene (*80*) and of MAbs like 4-4-20 that depend on aromatic residues for high-affinity binding (*81*). Hence, unwanted "nonspecific" binding of preimmune repertoire antibodies that interact hydrophilically might be less inhibitable with nonionic detergents than hydrophobic antibodies resulting from "affinity maturation." Thus, "nonspecific" intrinsic or autoreactive antibody may occur and may contribute to NSB.

The "take-home message" on blocking agents is that: (1) the exact mechanism(s) of their actions in each situation remains unknown; (2) there are two major categories, proteins, and detergents; (3) each blocker may behave differently depending on the nature of the ligand receptor interaction and the chemistry of the solid phase; and (4) blockers may interfere with specific as well as background (nonspecific) interactions.

Finally, some antibody binding to synthetic solid phases may not be "nonspecific." Although it is unproved, the IgG in the sera of arterivirus-infected mice *(80)* may be an example. Thus, "blocking" with nonspecific ligands such as detergents and irrelevant proteins may not be possible. This issue of specific antibodies to be solid-phase supports used in SPI appears to be under-researched.

3.4. Consideration in the Choice of Nonadsorptive Immobilization Methods

Evidence that the adsorption of proteins on hydrophobic surfaces leads to ACC (*see* **Subheading 2.6.** and **Table 2**) initially leads to the conclusion that nonadsorptive immobilization is superior. There are caveats in this reasoning, however:

1. Immobilization using an initial monolayer of capture antibody or biotinylated protein means that the second layer will contain fewer molecules than the monolayer; empirically this is less than 1/10 as many. Therefore, one can denature 90% of an initial adsorbed monolayer and still have more reactive molecules than in the second, nonadsorbed layer.
2. Surfaces that have been chemically functionalized to provide $-NH_2$ or COO^- groups for covalent attachment still offer a rather large surface area for adsorption. Hence, unless serious control methods are used, molecules covalently immobilized on such surfaces may represent a minority; hydrophobically adsorbed molecules may comprise the majority.
3. Molecules that are secondarily immobilized by nonadsorptive means must be modified either by biotinylation, by the attachment of a chemical crosslinking agent, or by capture by an antibody that could block or mask a functional group. Therefore, the number of functionally immobilized molecules could be a great deal lower than the total number of immobilized molecules.

In addition to the above considerations, nonadsorptive methods of immobilization are more expensive and more time-consuming to develop. Therefore, selecting nonadsorptive immobilization is best done only after attempts to obtain a biologically active surface by simple adsorption have failed. This general statement is of course predicted from current state-of-the-art solid-phase technology. Should a simple, inexpensive means performing nonadsorptive immobilization become available, the above-quoted statement would require amendment.

3.5. Reagents: Immobilized Antibodies and Antigens

The success of any SPI depends not only on the choice of solid phase but on the reactant to be immobilized and the reagents that make up the detection system. The latter is discussed in **Subheading 3.6.** Several criteria need to be considered:

1. If a reliable SPI can be developed by using a low-molecular-weight determinant, the problem of adsorption-induced conformational change can be avoided. Small haptens like fluorescein, dinitrophenyl, phosphorylcholine, and polyribitol phosphate can merely be conjugated to a carrier macromolecule (ferritin, albumin, fibronectin, and so on) and the conjugate adsorbed. Use of highly purified albumin from the species is a good choice, but contaminating IgG must be removed. Although some of the conjugated epitopes may be buried, those left exposed will be in native conformation because it is the carrier that will be the victim of ACC. The selection of carrier is also important. It is usually wise to avoid using an immunoglobulin as a carrier, because antiglobulins used in subsequent steps of, e.g., an ELISA, may crossreact with the carrier immunoglobulin.

2. The immobilized antigen must be free of contaminants that could interfere with assay reliability. A good example is virus that may contain protein from the culture in which it was grown, e.g., egg proteins, calf serum. If the same virus preparation had been used for immunization, the potential exists that the serum of the immunized individual or animal contains antibodies both to the virus and to the culture proteins. As discussed in **Subheading 2.10.**, this may be an argument for using an immobilized capture antibody, streptavidin, or a bacterial binding protein to immobilize the virus or recombinant proteins of interest *(53)*. An additional problem with contaminants in antigen preparations is that they may be less denatured during adsorption than the antigen being studied. Thus, the intensity of a crossreaction to a contaminant can be out of proportion to the amount of the contaminant(s) present (**Fig. 10**).
3. Whenever possible, choose an antigen preparation that others, or the biotech company supplying the reagent, have tested and shown to be biologically active when adsorbed on the hydrophobic surface selected for use in the SPI.
4. Polyclonal antibodies adsorbed as CAbs typically function better than most MAbs (**Fig. 4**). Although the reason for this is unknown, it is theorized that since polyclonals contain a diverse spectrum of (usually) IgG antibodies, the probability that *all* clonal variants will become functionally inactivated by adsorption is remote. In contrast, MAb are homogeneous, so if one antibody fails to survive adsorption, the probability is high that all will suffer the same fate. When MAbs are needed, it is necessary to select those that remain functional after adsorption.
5. Antibodies for adsorption should be purified or semipurified because any nonantibody IgG or other protein present will simply decrease the number of functional antibodies that comprise the adsorbed monolayer. Because this reduces the functional antibody concentration, it proportionally reduces the useful range of the immunoassay. Caution must be used in the selection of *affinity-purified* antibodies, as certain types of affinity matrices also "leach" their ligands, which then become part of the adsorbed monolayer containing the affinity-purified antibodies so that the threat of crossreactivity, i.e., short circuitry, exists (**Fig. 10**).
6. Antibodies to be immobilized should be purchased from suppliers that have already shown them to be functional when used in this manner in SPI. If this is not the case, the investigator becomes the "test pilot" for every antibody chosen. Many companies specifically indicate that certain antibodies were developed for use in ELISA and other SPIs and may allow the investigator to sample the product or obtain a refund if it does not perform. Buyer beware if this is not the case.

3.6. Reagents: Detection System Antibodies and Conjugates

Successful SPIs rely not only on the integrity and performance of the solid-phase reactant but also on the antibodies and conjugates used to detect the analyte.

1. A major problem in all immunoassays in which a "string" of antibodies is employed is "short circuitry" (*13*; **Fig. 10**). Thus, it is important to choose an enzyme–antibody conjugate that does not recognize the solid-phase antigen, or the primary antibody (**Fig. 10**). Some suppliers indicate that their conjugated antibodies have been absorbed with the sera of various species, and others do not. Investigators should work with suppliers to obtain conjugates that do not produce short circuits. Often if conjugates can be used at very high dilution (1:30,000–1:50,000) their "short-circuit activity" becomes negligible.
2. Conjugate size determines the dynamic range of a solid-phase immunoassay (**Figs. 11 and 12**, and ref. *17*). Thus, glutaraldehyde crosslinking of antibodies and enzymes that

Solid Supports in ELISA and Other Assays 363

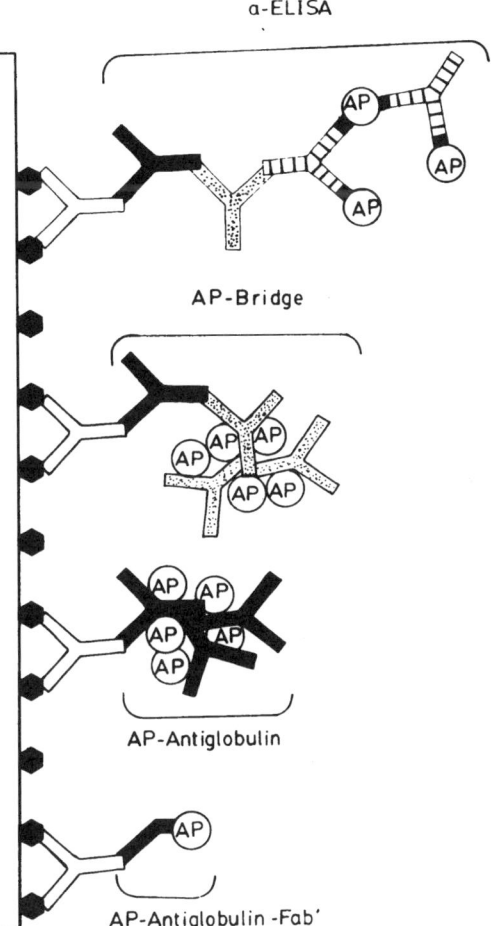

Fig. 11. Comparison of various detection systems used in ELISAs. AP = alkaline phosphatase, hexagon = solid-phase antigen, open antibody = primary antibody; solid antibody = isotype-specific antiglobulin; dotted antibody = tertiary antiglobulin in one case conjugated with AP; hatched antibody = anti-AP. Conjugates depicted in middle two examples are one-step glutaraldehyde conjugates. a-ELISA = a highly amplified ELISA based on use of a soluble enzyme-antibody immune complex. From Koertge and Butler *(17)*.

produce complexes of more than 1,000,000 kDa, causing steric hindrance, typically limits the dynamic range of the assay. Outside this range, the overall reaction changes with the dilution of the primary antibody. Among the common detection systems, the use of horseradish peroxidase (HRP)–antibody complexes gives a broad dynamic range (**Fig. 12**) and works best in ELISA-based assays for measurement of primary antibody affinity *(82)*. Naturally, these conjugates do not have the amplification power of large antibody–enzyme conjugates, so one must decide on whether *sensitivity*, *dynamic range*, or *accurate quantitation of primary antibody binding* is most important.

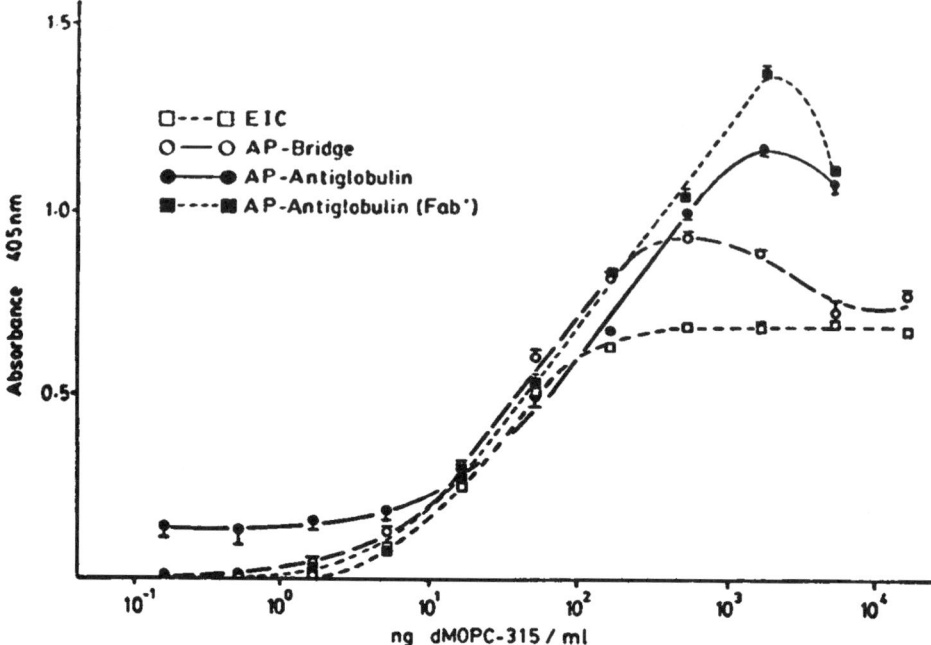

Fig. 12. The influence of the size of the detection system used for antigen-specific ELISA. The legend is on the figure and corresponds to the detection systems depicted in **Fig. 11**. Reaction times were adjusted so that the magnitude of the OD 405nm in the linear regions of the various plots were similar. From Koertge and Butler *(17)*.

3. There are numerous methods to increase assay sensitivity by increasing the signal output or by using fluorescence and chemiluminescence as signal generators. These are discussed in detail elsewhere *(13)*.

3.7. Science and "Old Wives Tales"

A surprisingly large number of the day-to-day practices used in laboratories performing ELISA and other SPIs are more folklore than science. Seldom are these practices verified by new users through empirical testing. Since immunochemistry nowadays is seldom taught to students in either immunology or biochemistry, it is easy to understand how terms like concentration, sensitivity, and affinity have become more vernacular than biochemical. This has led to the following dicta, without an understanding of the basic principles involved:

1. *"Microtiter wells must be blocked after adsorption of protein."* Many ELISA protocols routinely require a blocking step after antigen (or antibody) adsorption. As discussed in **Subheading 3.3.**, this typically involves treatment with an albumin solution, skim milk, or a commercial product (e.g., Blotto, Superblock). The purported theory behind this procedure is that sites on the synthetic surface unoccupied by the adsorbed protein must be "filled in" with the blocking proteins. Because concentrations of protein of approx

5 μg/mL result in the formation of a monolayer, where are the empty sites? Perhaps they exist because the so-called monolayer is a series of "reactive islands" with raw plastic in between. If the adsorbed ligand did not bind these raw sites, why should they bind blocking protein? Perhaps it occurs because blocking protein solutions are used at much higher concentrations, which, through mass law considerations, promotes adsorption to low-affinity plastic sites. Some protocols suggest applying the blocking solution in nonionic detergent. Typically when a plastic surface is incubated with protein in the presence of nonionic detergents, e.g., Tween 20, protein adsorption is prevented. Therefore, if the reaction buffer used in the subsequent SPI contains Tween 20, how can any protein become adsorbed to "empty plastic sites" in any case? Therefore, users of polystyrene microtiter plates should test whether the use of a blocking step improves the "signal to noise ratio" and thus lowers the limit of detection (LOD; *see* **Subheading 3.7., step 2**) before accepting it as standard practice. Remember that for each additional step added to an assay, you need an additional control. The more complicated the system, the more complicated the controls. Finally, all surfaces used in SPI are not created equal (*see* **Subheadings 2.2. and 3.3.**), and blocking has both a theoretical basis and practical value when membranous solid phases are used.

2. "*Radioimmunoassay (RIA) and 'amplified ELISAs' are more sensitive than simple ELISAs.*" The term sensitivity is not a quantitative term in immunoassay, giving it little more value in describing performance in SPI than it has in personal relationships. What needs to be addressed is the LOD, or the smallest amount of analyte that can be detected at a level significantly above background *(13)*. The subject has also been reviewed by Ekins *(83)*. The American Chemical Society (ACS) guidelines define LOD at an analyte concentration at which the mean sample signal should be equal to or greater than the mean blank signal by 3 SDs. Thus, LOD is highly dependent on the mean blank signal, which is often the consequence of NSB. Therefore, the magnitude of signal produced by an enzymatic reaction, fluorescent compound, or radiolabeled reporter does not in itself improve LOD because the ratio of signal to noise typically remains constant. Typically "amplified assays," which use multiple steps in the detection phase, do indeed generate more signal, but each added layer of reagents also has the potential for adding an additional layer of NSB. Three SDs is empirical and depends on the degree of confidence needed for the assay. A high level of confidence would be needed before concluding that a patient is HIV positive.

The original statement cited above originated with attempts to replace classical competitive RIAs with competitive ELISAs. Because NSB was often much higher in first-generation competitive ELISAs, the statement carried some truth. The introduction of "amplified ELISA" improved signal but not always LOD. Currently, many ELISAs use noncompetitive configurations and refined detection systems. Thus, an amplified ELISA may have the same sensitivity as RIAs or a simple ELISA, or be more sensitive, or be less sensitive. It is all a matter of LOD and NSB, *not* the type of detection or reporter system.

3. "*The serum antibody concentration was 20 μg/mL as determined by ELISA.*" Although data on serum antibody concentration (μg/mL) may be found in publications from the best immunology journals, they are often misleading. In any noncovalent bimolecular interaction, a portion of the reactants will remain free. ELISAs measure the amount of serum antibody that is bound to an adsorbed antigen. Since a substantial proportion of the antibody may remain free either because it is of low affinity or because the adsorbed antigen presents altered epitopes, ELISAs measure only bound antibody, *not total antibody*.

Absolute antibody concentrations (mg/mL) are often determined by comparing their ELISA "titers" with those obtained with a known amount of a standard antibody. This is certainly valid if: (1) the unknown antibody has the same affinity for the antigens as the standard so the ratio of bound to free is the same for both, and (2) the proportion of active antibody in the standard is the same as that in the unknown samples. Simply knowing the IgG concentration of a purified antibody standard provides no information on the proportion that is active antibody.

Therefore, unless sound biochemical and immunochemical evidence is presented to refute any of the above criticisms, it is more honest to present "serum antibody concentration" as *"serum antibody activity"*; the latter makes no assumptions about affinity or total versus bound antibody. From a functional point of view, *activity* (a product of both affinity and concentration) is a more *biologically relevant* parameter. However, using an affinity-purified antibody to determine the *approximate range of antibody concentration in mg/mL* should not be discouraged and can be very useful. The most important point is that *investigators who use this procedure should be aware that absolute values cover a range because of differences in affinity.*

Presentation of an antibody response as "specific activity" is a useful alternative *(73,83,84)*. Specific activity is defined in the same manner as for radioisotype, i.e., ELISA activity per microgram of immunoglobulin. This is particularly useful for evaluating antibodies in exocrine body fluids *(83,84)* and can help distinguish between increased ELISA activity resulting from polyclonal activity versus antigen-driven responses *(73)*. If measurements of specific antibody in absolute units is questioned, why shouldn't this apply to absolute units of antigen measured by sandwich ELISA? The answer is that in sandwich ELISA the CAb has a fixed affinity, and in most all cases, the analyte is antigenically homogeneous and therefore identical to the standard.

4. *"All binding of antibodies to the solid phase represents nonspecific protein–polymer interaction."* Efforts by biotech suppliers to simplify the mechanism of SPI, combined with the removal of immunochemical training from most biochemistry and immunology training programs, has resulted in certain misconceptions. At issue is the belief that if antibodies bind to solid-phase supports, it is through "nonspecific" interactions. Since the antibody repertoire is extremely broad, why shouldn't there be antibodies that recognize synthetic polymers like polystyrene or nitrocellulose? Thus, NSB could also represent antibodies that specifically recognize the solid-phase support. This may explain the apparent antibodies to polystyrene in mice infected with the LDV virus *(80)*. The phenomenon received national attention with reports that patients implanted with various silicone devices developed specific antibodies to the silicone elastomer *(85,86)* and, as a result, were believed to develop debilitating immunopathology. Although many of these studies were driven by a legal profession looking for an avenue to financial success *(87)*, one should not ignore the possibility that synthetic polymers can be antigens for the immune system since patients suffering from autoimmunity have antibodies to many native and synthetic substances *(79)* so that such polymer–antibody interactions may not be merely NSB and may not be inhibitable by nonionic detergents (*see* **Subheading 3.4.**, above).

5. *"Antibody affinity cannot be measured by ELISA."* Like total antibody concentration, most "affinity determinations" made using an ELISA are really measures of relative affinity, using standards of known affinity for the antigen. This approach is entirely legitimate and can be exceedingly valuable as long as the procedure used to collect such data are described and are interpreted accordingly. In fact, when monitoring changes in the immune response of individuals or study groups, *relative changes* in affinity are often more useful in demonstrating immunobiological significance; reporting actual equilib-

rium constants may simply raise technical issues as to how such absolute values were obtained and thereby distract from the real significance of the study. However, Friquet and colleagues *(87)* made an effort to describe an assay in which the necessary parameters for the determination of affinity (i.e., bound antibody, free antibody, and their ratio) could be determined in the context of an ELISA-based SPI. This method allows an antigen–antibody reaction to reach equilibrium in free solution, after which the mixture is transferred to antigen-coated microtiter wells where 10% or less of the free antibody is captured. The latter is a necessary requirement to ensure that the original equilibrium is not disturbed and the likelihood for a soluble complex to become crosslinked to the surface is nil. Provided that the total antibody concentration is known, the method allows the free antibody concentration at equilibrium to be determined. Furthermore, the assay is configured so that the total antigen concentration is very much larger than the total antibody, so that free antigen is approximately total antigen. By plotting values for the *fraction of bound antibody* (α)/the free antigen concentration (SC) against α, a Scatchard-like graph results in a negative slope equal to $-K$ (equilibrium constant). The Friquet method represents a case in which SPI can legitimately be used to measure affinity, provided that all the requirements are satisfied. Others have described similar methods *(82)*.

In conclusion, users of ELISA and SPI technology should not lose sight of the fact that solid-phase ligand–receptor interactions can be used as valuable *serological* and highly sensitive *qualitative* tools. They can also be used *quantitatively* in the spirit of biochemical or chemical assays. The important issue is to know the difference, to report data in the proper units, and to rely on empirical science (*"Vertrauen ist gut aber Kontrolle ist besser"*) rather than unproven assumptions or folklorist "old wives tales."

4. Notes

1. Microparticles are less than 1 µm and behave as a colloidal suspension during the assay; "particles" or "beads" are larger and settle out by gravity during nonagitated assays.
2. The volume of the reactive interface, or reaction volume, does not refer to the total volume in which the SPI is conducted, e.g., the bead volume or the volume of a plastic microtiter well, rather, it is a theoretical volume at the interface in which the soluble and immobilized reactants interact on the surface of the solid phase.
3. Subtle differences can affect such things as background, preferential adsorption of certain molecules, and well-to-well variation, but these differences are minor in relation to differences seen among other categories of solid phases (**Table 1**). On the other hand, irradiated plastics like, e.g., NUNC Maxisorp and Immulon 2 (Dynatech), have a greater affinity for macromolecules than do untreated polystyrenes.
4. Plasma discharge or irradiation results in hydroxyl formation, rendering the surface of hydrophobic polymers more hydrophilic.
5. Plasma-treated plastics (*see* above) and materials such as nitrocellulose and especially PVDF display charged groups on their surfaces. Such groups are capable of ionic or weak hydrogen bonding, some of which may even lead to covalent bonds. These are collectively, but perhaps in some case inappropriately, referred to as hydrophilic bonding.

Acknowledgments

The research was supported by a grant from Dow-Corning (Midland, MI) to study the adsorption of proteins on synthetic polymers. The author thanks Ms. Marcia Reeve for preparation of the typescript.

References

1. Franz, B. and Stegemann, M. (1991) The kinetics of solid-phase microtiter immunoassays, in *Immunochemistry of Solid-Phase Immunoassay* (Butler, J. E., ed.), CRC Press, Boca Raton, FL, pp. 277–284.
2. Park, H. (1978) A new plastic receptable for solid-phase immunoassay. *J. Immunol. Methods* **20,** 349–355.
3. Joshi, K. S., Hoffmann, L. G., and Butler, J. E. (1992) The immunochemistry of sandwich ELISAs. V. The capture antibody performance of polyclonal antibody-enriched fractions prepared by various methods. *Mol. Immunol.* **29,** 971–981.
4. Butler, J. E., Lü, E. P., Navarro, P., and Christiansen, B. (1997) Comparative studies on the interaction of proteins with polydimethylsiloxane elastomer. I. Monolayer protein capture capacity (PCC) as a function of proteins pI, buffer pH and buffer ionic strength. *J. Mol. Recognit.* **10,** 36–51.
5. Azimzadeh, A. and van Regenmortel, M. H. V. (1990) Antibody affinity measurements. *J. Mol. Recognit.* **3,** 108–116.
6. Metzger. H. (1992) Transmembrane signalling: the joy of aggregation. *J. Immunol.* **149,** 1477–1487.
7. Butler, J. E., Ni, L., Nessler, R., et al. (1992) The physical and functional behavior of capture antibodies adsorbed on polystyrene. *J. Immunol. Methods* **150,** 77–90.
8. Davies, J., Dawkes, A. C., Haymes, A. G., et al. (1994) A scanning tunnelling microscopy comparison of passive antibody adsorption and biotinylated linkage to streptavidin on microtiter wells. *J. Immunol. Methods* **167,** 263–269.
9. Ehrlich, P. H., Moyle, W. R., and Moustafa, Z. A. (1983) Further characterization of cooperative interactions of monoclonal antibodies. *J. Immunol. Methods* **131,** 1906–1912.
10. Lehtonen, O. P. (1981) Immunoreactivity of solid phase hapten binding plasmacytoma protein (ABPC 24). *Mol. Immunol.* **18,** 323–329.
11. Butler, J. E., Feldbush, T. L., McGivern, P. L., and Steward, N. (1978) The enzyme-linked immunosorbent assay (ELISA): a measurement of antibody concentration or affinity? *Immunochemistry* **15,** 131–136.
12. Dierks, S., Butler, J. E., and Richerson, H. B. (1986) Altered recognition of surface-adsorbed compared to antigen bound antibodies in the ELISA. *Mol. Immunol.* **23,** 403–411.
13. Butler, J. E. (1994) ELISA, in *Immunochemistry* (van Regenmortel, M. H. V. and van Oss, C. J., eds.), Marcel Dekker, New York, pp. 759–803.
14. Cantarero, L. A., Butler, J. E., and Osborne, J. W. (1980) The binding characteristics of various proteins to polystyrene and their significance for solid-phase immunoassays. *Anal. Biochem.* **105,** 375–383.
15. Butler, J. E., Ni, L., Brown, W. R., et al. (1993) The immunochemistry of sandwich ELISAs. VI. Greater than 90% of monoclonal and 75% of polyclonal anti-fluorrescyl capture antibodies (CAbs) are denatured by passive adsorption. *Mol. Immunol.* **30,** 1165–1175.
16. Butler, J. E., Navarro, P., and Lü, E. P. (1997) Comparative studies on the interaction of proteins with a polydimethylsiloxane elastomer. II. The comparative antigenicity of primary and secondary adsorbed IgG1 and IgG2a and their non-adsorbed counterparts. *J. Mol. Recognit.* **10,** 52–62.
17. Koertge, T. E. and Butler, J. E. (1985) The relationship between the binding of primary antibody to solid-phase antigen in microtiter plates and its detection by the ELISA. *J. Immunol. Methods* **83,** 283–299.

18. Michaeli, I., Absolm, D. R., and van Oss, C. J. (1980) Diffusion of adsorbed protein with the plane of adsorption. *J. Colloid Interface Sci.* **77,** 586–587.
19. Nygren, H. (1988) Experimental demonstration of lateral cohesion in a layer of adsorbed protein and layers of gold-antibody complexes bound to surface immunobilized antigen. *J. Immunol. Methods* **114,** 107–114.
20. Nygren, H., Werthen, M., Czerkinsky, C., and Stenberg, M. (1985) Dissociation of antibodies bound to surface-immobilized antigen. *J. Immunol. Methods* **85,** 87–95.
21. Mason, D. W. and Williams, A. F. (1980) The kinetics of antibody binding to membrane antigens in solution and at the cell surface. *Biochem. J.* **187,** 1–20.
22. Stenberg, M. and Nygren, H. (1982) A receptor ligand reaction studied by a novel analytica tool—the isoscope ellipsometer. *Anal. Biochem.* **127,** 183–192.
23. Crothers, D. M. and Metzger, H. (19720 The influence of polyvalency on the binding properties of antibodies. *Immunochemistry* **9,** 341–357.
24. Azimzadeh, A. and Regenmoretl, M. H.V. (1991) Measurement of affinity of viral monoclonal-antibodies by ELISA titration of free antibody in equilibrium mixtures. *J. Immunol. Methods* **141,** 199–208.
25. Berson, S. A. and Yallow, R. S. (1959) Quantitative aspects of the reaction between insulin and insulin binding antibody. *J. Clin. Invest.* **38,** 1996–2016.
26. Andresen, L. O., Klausen, J., Barfod, K., and Sorensen, V. (2002) Detection of antibodies to Actinobacillus pleuropneumoniae serotype 12 in pig serum using a blocking enzyme-linked immunosorbent assay. *Vet. Microbiol.* **89,** 61–67.
27. Gutierrez, J. E., Dolcini, G. L., Arroyo, G. H., Rodriguez Dubra, C., Ferrer, J. F., and Esteban, E. N. (2001) Development and evaluation of a highly sensitive and specific blocking enzyme-linked immunosorbent assay and polymerase chain reaction assay for diagnosis of bovine leukemia virus infection in cattle. *Am. J. Vet. Res.* **62,** 1571–1577.
28. Chenard, G., Giedema, K., Moonen, P., Schrijiver, R. S., and Dekker, A. (2003) A solid phase blocking ELISA for detection of type O foot-and-mouth disease virus antibodies suitable for mass serology. *J. Virol. Methods* **107,** 89–98.
29. Catt, K. and Tregear, G. W. (1967) Solid-phase radioimmunoassay immunoassay in antibody coated tubes. *Science* **158,** 1570–1572.
30. Peterson, J. D., Kim, J. Y., Melvold, R. W., Miller, S. D., and Waltenbaugh, C. (1989) A rapid method for quantitation of anti-viral antibodies. *J. Immunol. Methods* **119,** 83–94.
31. Qian, W., Yao, D., Yu, F., et al. (2000) Immobilization of antibodies on ultraflat polystyrene surface. *Clin. Chem.* **46,** 1456–1463.
32. Bull, H. B. (1956) Adsorption of bovine serum albumin on glass. *Biochem. Biophys. Acta* **19,** 464–471.
33. Karush, F. (1978) The affinity of antibody: range, variability and role of multivalence, in *Immunochemistry, an Advanced Textbook* (Glynn, L. E. and Steward, M. W., eds.), Wiley, Chichester, pp. 233–262.
34. Matusda, T., Takano, H., Hayashi, K., et al. (1984) The blood interface with segmented polyurethanes multi-layer protein passivation mechanism. *Trans. Am. Soc. Artif. Intern. Organs* **30,** 353–358.
35. Stevens, P. W. and Kelso, D. M. Estimation of the protein-binding capacity of microplate wells using sequential ELISAs. *J. Immunol. Methods* **178,** 59–70.
36. van Erp, R., Linders, Y. E., van Sommeren, A. P., and Gribnau, T. C. (1992) Characterization of monoclonal antibodies physically adsorbed onto polystyrene latex particles. *J. Immunol. Methods* **152,** 191–199.

37. Plant, A. L., Locascio-Brown, L., Haller, W., and Durst, R. A. (1991) Immobilization of binding proteins on nonporous supports. Comparison of protein loading, activity and stability. *Appl. Biochem. Biotechnol.* **20,** 83–98.
38. Oreskes, I. and Singer, J. M. (1961) The mechanism of particulate carrier reactions: adsorption of human γ-globulin to polystyrene latex particles. *J. Immunol.* **86,** 338–344.
39. Kochwa, S., Brownell, M., Rosenfield, R. E., and Wasserman, L. R. (1967) Adsorption of proteins by polystyrene particles. I. Molecular unfolding and acquired immunogenicity of IgG. *J. Immunol.* **99,** 981–986.
40. Nyilas, E., Chiu, T.-H., and Herzlinger, G. A. (1974) Thermodynamics of native protein/foreign surface interactions. I. Colorimetry of the human γ-globulin/glass system. *Trans. Am. Soc. Artif. Intern. Organs* **20,** 480–490.
41. Kennel, S. (1982) Binding of monoclonal antibody in fluid phase and bound to a solid support. *J. Immunol. Methods* **55,** 1–12.
42. Friquet, B., Djavadji-Ohaniance, L., and Goldberg, M. E. (1984) Some monoclonal antibodies raised with a native protein bind preferentially to the denatured antigen. *Mol. Immunol.* **21,** 673–677.
43. Holland, Z. and Katchaliski-Katzir, E. (1986) Use of monoclonal antibodies to detect conformational alterations in lactate dehydrogenase isoenzyme 5 on heat denaturation and on adsorption on polystyrene plates. *Mol. Immunol.* **23,** 927–934.
44. Suter, M. and Butler, J. E. (1986) The immunochemistry of sandwich ELISAs. II. A novel system prevents denaturation of capture antibodies. *Immunol. Lett.* **13,** 313–317.
45. Darst, S. A., Rovertson, C. R., and Berzofsky, J. A. (1988) Adsorption of the protein antigen myoglobin affects the binding of conformation-specific monoclonal antibodies. *J. Biophys. Soc.* **53,** 533–539.
46. Kennel, S. J, Chen, J. P, Lankford, P. I., and Foote, L. J. (1981) Monoclonal antibodies from rats immunized with fragment D of human fibrinogen. *Thromb. Res.* **22,** 3090–320.
47. Shmani, V. V., Nikolayeva, T. A., Winokurova, L. G., and Litoshka, A. A. (2001) Oriented antibody immobilization to polystyrene macrocarriers for immunoassay modified with hydrazide derivatives of poly (meth) acrylic acid. *BMC Biotechnol.* **1,** 4.
48. Zeiss, C. R., Pruzansky, J. J., Patterson, R., and Roberts, M. (1973) A solid phase radio-immunoassay for the quantitation of human reagenic antibody against ragweed antigen. *J. Immunol.* **110,** 414–421.
49. Herrmann, J. E., Hendry, R. M., and Collins, M. F. (1979) Factors involved in enzyme-linked immunoassay for viruses and evaluation of the method for identification of enteroviruses. *J. Clin. Microbiol.* **10,** 210–217.
50. Peterman, J. H., Tarcha, P. J., Chu, V. P., and Butler, J. E. (1988) The immunochemistry of sandwich ELISAs. IV. The antigen capture capacity of antibody covalently attached to bromoacetyl polystyrene. *J. Immunol. Methods* **111,** 271–275.
51. Bugari, G., Poiesi, G., Beretta, A., Ghielmi, A., and Albertini. A. (1990) Quantitative immunoenzymatic assay of human lutropin, with use of a bi-specific monoclonal antibody. *Clin. Chem.* **36,** 47–52.
52. Robinson, J. E., Holton, O., Liu, J., McMurdo, H., Murciano, A., and Gohd, R. (1990) A novel enzyme-linked immunosorbent assay (ELISA) for the detection of antibodies to HIV-1 envelope glycoproteins based on immobilization of viral glycoproteins in microtiter wells coated with concanavalin A. *J. Immunol. Methods* **132,** 63–71.
53. Baumann, S., Grob, P., Stuard, F., Pertlik, D., Ackermann, M., and Suter, M. (1998) Indirect immobilization of recombinant proteins to a solid phase using the albumin binding domain of streptococcal protein G and immobilized albumin. *J. Immunol. Methods* **221,** 95–106.

54. Butler, J. E. (1991) The behavior of antigens and antibodies immobilized on a solid-phase, in *Structure of Antigens* (Van Regenmortel, H. M. V., ed.), CRC Press, Boca Raton, FL, pp. 208–258.
55. Spoljar, B. H. and Tomasic, J. (2000) A novel ELISA for determination of polysaccharide specific immunoglobulins. *Vaccine* **19,** 924–930.
56. Takahashi, K., Fukada, M., Kawai, M., and Yokochi, T. (1992) Detection of lipopolysaccharide (LPS) and identification of its serotype by an enzyme-linked immunosorbent assay (ELISA) using poly-L-lysine. *J. Immunol. Methods* **153,** 67–71.
57. Shirahama, H. and Suzawa, T. (1985) Adsorption of bovine serum albumin onto styrene/acrylic acid copolymer latex. *Colloid Polymer Sci.* **263,** 141–146.
58. Lauritzen, E., Masson, M., Rubin, I., and Holm, A. (1990) Dot immunobinding and immunoblotting of picogram and nanogram quantities of small peptides on activated nitrocellulose. *J. Immunol. Methods* **131,** 257–267.
59. Søndergaard-Andersen, J., Lauritzen, E., Lind, K., and Holm, A. (1990) Covalently liked peptides for enzyme-linked immunosorbent assay. *J. Immunol. Methods* **131,** 99–104.
60. Grantstrom, M., Wretlind, B., Markman, B., and Cryz, S. (19898) Enzyme-linked immunosorbent assay to evaluate the immunogenicity of a polyvalent Klebsiella capsular polysaccharide vaccine in humans. *J. Clin. Microbiol.* **26,** 2257–2261.
61. Geysen, H. M., Meloen, R. H., and Barteling, S. J. (1984) Use of peptide synthesis to probe viral antigens for epitopes to a resolution of a single amino acid. *Proc. Natl. Acad. Sci. USA* **81,** 3998–4002.
62. Boudet, F., Theze, J., and Zouali, M. (1991) UV-treated polystyrene microtitre plates for use in an ELISA to measure antibodies against synthetic peptides. **142,** 73–82.
63. Dagenais, P., Desprez, B., Albeit, J., and Escher. E. (1994) Direct covalent attachment of small peptide antigens to enzyme-linked immunosorbent assay plates using radiation and carbodimide activation. *Anal. Biochem.* **222,** 149–155.
64. Hofstetter, O., Hofstetter, H., Then, D., Schurig, V., and Green, B. S. (1997) Direct binding of low molecular weight haptens to ELISA plates. *J. Immunol. Methods* **210,** 89–92.
65. Bora, U., Chugh, L., and Nahar, P. (2002) Covalent immobilization of proteins onto photoactivated polystyrene microtiter plates for enzyme-linked immunosorbent assay procedures. *J. Immunol. Methods* **268,** 171–177.
66. Gregorius, K., Mouritsen, S., and Elsner, H. I. (1995) Hydrocoating: a new method for coupling biomolecules to solid phases. *J. Immunol. Methods* **181,** 65–73.
67. Loomans, E. E., Petersen-van Ettehoven, A., Bloemers, H. P., and Schielen, W. J. (1997) Direct coating of poly (lys) or acetyp-thio-acetyl peptides to polystyrene: the effects in an enzyme-linked immunosorbent assay. *Anal. Biochem.* **248,** 117–129.
68. Gregorius, K. and Theisen, M. (2001) In situ deprotection: a method for covalent immobilization of peptides with well-defined orientation for use in solid phase immunoassays such as enzyme-linked immunosorbent assay. *Anal. Biochem.* **299,** 84–91.
69. LaCroix, M., Dionne, G., Zrein, M., Dwyer, R. J., and Chalifour, R. J. (1991) The use of synthetic peptides as solid-phase antigens, in *Immunochemistry of Solid-Phase Immunoassay* (Butler, J. E., ed.), Boca Raton, FL, pp. 261–268.
70. Palfree, R. G. and Elliott, B. E. (1982) An enzyme linked immunosorbent assay (ELISA) for detergent solubilized Ia glycoproteins using nitrocellulose membrane discs. *J. Immunol. Methods* **52,** 395–408.
71. Drexler, G., Eichinger, A., Wolf, C., and Sieghart, W. (1986) A rapid and simple method for efficient coating of microtiter plates using low amounts of antigen in the presence of detergent. *J. Immunol. Methods* **95,** 117–122.

72. Czerkinsky, C., Rees, A. S., Burgmeier, L. A., and Challacombe, S. J. (1983) The detection and specificity of class specific antibodies to whole bacterial cells using a solid phase radioimmunoassay. *Clin. Exp. Immunol.* **53**, 192–200.
73. Butler, J. E., Weber, P., Sinkora, M., et al. (2002) Antibody repertoire development in fetal and neonatal piglets. VIII. Colonization is required for newborn piglets to make serum antibodies to T-dependent and type 2 T-independent antigens. *J. Immunol.* **169**, 6822–6830.
74. Verschoor, J. A., Meiring, M.J., van Wyngaardt, S., and Weyer, K. (1990) Polystyrene, poly-L-lysine and nylon as adsorptive surfaces for the binding of whole cells of *Mycobacterium tuberculosis* H37 RV to ELISA plates. *J. Immunoassay* **11**, 413–428.
75. Ngai, P. K., Ackermann, F., Wendt, H., Savoca, R., and Bosshard, H. R. (1993) Protein A antibody-capture ELISA (PACE): an ELISA format to avoid denaturation of surface-adsorbed antigens. *J. Immunol. Methods* **158**, 267–276.
76. Schwab, C. and Bosshard, H. R. (1992) Caveats for the use of surface-adsorbed protein antigen to test the specificity of antibodies. *J. Immunol. Methods* **147**, 125–134.
77. Rodda, D. J. and Yamazaki, H. (1994) Poly(vinyl alcohol) as a blocking agent in enzyme immunoassay. *Immunol. Invest.* **23**, 421–428.
78. Qualtiere, L. F., Anderson, A. C., and Meyer, P. (1997) Effects of ionic and non-ionic detergents on antigen-antibody reaction. *J. Immunol.* **119**, 1645–1651.
79. Avrameas, S. and Terynich, K. T. (1993) The natural autoantibody system: between hypotheses and facts. *Mol. Immunol.* **30**, 1133–1142.
80. Cafruny, W. A., Heruth, D.A., Jaqua, M. J., and Plagemann, P. G. W. (1986) Immunoglobulins that bind uncoated ELISA plate surfaces: appearance in mice during infection with lactate dehydrogenase-elevating virus and in human anti-nuclear antibody positive serum. *J. Med. Virol.* **19**, 175–186.
81. Voss, E. W. Jr. (1990) Anti-fluorescein antibodies as structure-function models to examine fundamental immunological and spectroscopic principles. *Comments Mol. Cell Biophys.* **6**, 197–221.
82. Azimzadeh, A., Weiss, E., and van Regenmortel, M. H. (1992) Measurement of affinity of viral monoclonal antibodies using Fab'-peroxidase conjugate. Influence of antibody concentration on apparent affinity. *Mol. Immunol.* **29**, 601–608.
83. Butler, J. E. and Hamilton, R. G. (1991) Quantitation of specific antibodies: methods of expression, standards, solid-phase considerations and specific applications, in *Immunochemistry of Solid-Phase Immunoassays* (Butler, J. E., ed.), CRC Press, Boca Raton, FL, pp. 173–198.
84. Butler, J. E., Spradling, J. E., Peterman, J. H., Joshi, K. S., and Challacombe, S. J. (1990) Humoral immunity in root caries in an elderly population. I. Development of reliable solid-phase immunoassays for routine measurement of antibodies to oral bacteria and quantitation of secretory immunoglobulin. *Oral Microbial Immunol.* **5**, 98–107.
85. Wolf, L. E., Lappe, M., Peterson, R. D., and Ezrailson, E. G. (1993) Human immune response to polydimethylsiloxane (silicone): screening studies in a breast implant populations. *FASEB J.* **7**, 1265–1268.
86. Goldblum, R. M., Pelley, R. P., O'Donnell, A. A., Pyron, D., and Heggers, J. R. (1992) Antibodies to silicone elastomers and reactions to ventriculoperitoneal shunts. *Lancet* **340**, 519–513.
87. Friguet, B., Chaffotte, A, F., Djavadi-Ohaniance, L., and Goldberg, M. E. (1984) Measurements of the true affinity constant of antigen-antibody complexes by enzyme-linked immunosorbent assay. *J. Immunol. Methods* **77**, 305–319.

22

Design and Preparation of Recombinant Antigens as Diagnostic Reagents in Solid-Phase Immunosorbent Assays

Alan Warnes, Anthony R. Fooks, and John R. Stephenson

Abstract

Analysis of the humoral immune response to infectious diseases has played, and will to continue to play, a key role in their diagnosis and immune surveillance. Although rapid genome detection methodologies, such as PCR, are beginning to replace immune assays for disease diagnosis, they are not suitable for all applications, especially the surveillance of the immune status of human populations. Here we review the limitations of current conventional tools for measuring immune responses and outline principles for the design and production of novel diagnostic reagents.

Methods for the production of viral diagnostic antigens by a variety of recombinant systems are described and their relative merits and disadvantages discussed. Protocols for the production of viral diagnostic antigens in eukaryotic, insect and mammalian systems are described using measles nucleocapsid antigen as a model. Indirect ELISA protocols which can differentiate immunoglobulin classes and subclasses are also described. Examples of the use of these analyses in research and surveillance are given.

Key Words: Virus diagnosis; measles; recombinant DNA; ELISA; baculovirus; adenovirus.

1. Introduction

1.1. Introduction of Recombinant Proteins for Use in Immunosorbent Assays

The accurate and sensitive analysis of the humoral immune response to infectious diseases has played, and will to continue to play, a key role in their diagnosis and immune surveillance. Although rapid genome detection methodologies, such as polymerase chain reaction (PCR), are beginning to replace immune assays for the diagnosis of microbial agents present in blood and mucosal secretions, they are not suitable for all applications. Several viruses and bacteria replicate exclusively in the tissues of the host and therefore cannot conveniently be detected in body fluids. Furthermore, measurements of the efficacy of vaccination strategies and the surveillance of the immune status of populations are entirely dependent on immune assays. In recent years laborious and sometimes insensitive bioassays such as hemagglutinin inhibition and complement fixation have been replaced by solid-phase assays that have accelerated the analysis of immune responses and enabled the introduction of automated or semiautomated technologies. The first of these to be widely adopted were radioimmune assays, but concerns over handling radioactive material caused them to be replaced with enzyme-linked immunosorbent assays (ELISA) in all but a few situations. Technologies involving antigens bound to solid matrices and magnetic beads have been introduced more recently, and several groups are assessing the use of immune assays on microchips.

Until recently these assays have depended on the use of crude antigens prepared from whole microorganisms or infected cells, but the introduction of genetic engineering techniques has allowed the controlled and efficient production of microbial proteins. This presents scientists with the opportunity to use a wide range of proteins previously unavailable due to problems relating to the expression, purification, or stability of the native proteins. Consequently, recombinant proteins have been adopted as key antigens in some diagnostic assays, but problems with background interference or crossreactivity from host proteins were frequently encountered *(1)*, because many of the recombinant proteins used in these assays were initially produced from prokaryotic hosts, primarily *Escherichia coli (2,3)*. Consequently, the potential use of recombinant proteins in routine diagnostic assays was not fully realized. Background interference and the associated loss of sensitivity could, however, be overcome by several means, including:

1. Purification of the recombinant protein (which may be facilitated by construction of fusion proteins).
2. Redesigning the assay format, e.g., utilizing antibody capture or competitive assays.
3. Absorbing test sera with bacterial host-cell proteins.

Generally, the delay in adopting recombinant proteins as diagnostic antigens for the detection of antibody responses arose because most users operated in a clinical setting in which the prime concern was the isolation of the pathogenic organism and the determination of its antibiotic sensitivity to permit a rapid suitable treatment. Furthermore, extensive work had to be undertaken to ensure that the correct assay results were obtained compared with those using wild-type antigens. The first successful commercial assay using recombinant proteins was in the detection of syphilis. This was developed as the traditional method of producing antigens was laborious and involved the use of experimental animals *(4)*.

In contrast, virologists were confronted with utilizing antigens produced from prokaryotic systems that had frequently been incorrectly folded or incompletely modified after translation (e.g., by glycosylation), thus causing concern regarding the fidelity of the three-dimensional structure of such proteins when used as diagnostic antigens.

The advent of the baculovirus expression system in the early 1980s enabled eukaryotic proteins to be expressed in a eukaryote system, both at high levels and with a high degree of post-translational modification (albeit with incomplete glycosylation compared with mammalian systems). A number of recombinant proteins produced from recombinant baculovirus-infected insect cells have now been used in indirect ELISA based systems for the detection of antibodies to viral pathogens *(5–8)*. Even so, the acceptance of recombinant proteins in diagnostic assays as commercial products was not automatic, and much validation was required to satisfy the regulatory authorities.

Although the baculovirus expression system made possible the general use of recombinant proteins as diagnostic antigens (especially with regard to nonglycosylated proteins), there was increasing concern about the antigenicity of the glycoproteins, when expressed in this system. At the same time, mammalian expression systems were being developed for the production of authentic proteins (e.g., vaccinia viruses, adenoviruses, avian poxviruses, Chinese hamster ovary cells), with many now being available commercially even though low-level expression was still a problem with certain proteins. Although several academic and commercial groups have been keen to utilize this technology with a view to superseding existing assays, certain sectors have been slow to introduce recombinant proteins into diagnostic assays for legitimate reasons, including:

1. Development and research costs are high when initiating new technologies.
2. Hidden costs exist such as production scale-up for eukaryotic cell cultures, patent protection, and containment facilities for handling recombinant organisms.
3. Antibodies to a single protein or part of a protein may not give an accurate indication of the severity of disease or a complete picture of the immune status of the host.
4. If the wrong antigen is chosen, problems with either specificity or crossreactions may arise.
5. More than one recombinant protein may be required to produce the desired results, thus greatly increasing the costs.
6. It can be difficult to compare results with established biological assays.
7. Approval for the widespread use of recombinant products by regulatory authorities or education of the end-users may be slow.

1.2. Limitations of Current Assays

As the technologies of protein expression have been further studied, so the use of recombinant proteins as diagnostic antigens has become commercially viable, driven by the limitations of existing assays, for example:

1. Some assays, such as complement fixation or hemagglutination inhibition are not very sensitive and may be restricted to the detection of certain antibody subclasses.
2. Plaque reduction neutralization tests (PRNT), in particular, although sensitive, are difficult and laborious to perform.
3. Nearly all biological assays are not suitable for automation, and thus their costs can be relatively high.
4. Current commercial ELISAs using whole organisms as antigens may contain contaminating cellular material, causing background interference or crossreactions resulting in poor-specificity and false-positive results *(1)*.
5. The need to handle large volumes of pathogenic material not only may be hazardous but also adds to production costs by demanding high-level containment facilities and rigorous quality control.

1.3. Designing New Diagnostic Antigens

1.3.1. Genetic Stability

One of the central debates in designing improved diagnostics is over whether a reductionist approach should be taken and only the simplest molecules be used to detect an immune response, or whether nature should be copied and the antigens used should resemble as closely as possible the agents causing disease. The most well-known example of the reductionist approach is the development of peptides or carbohydrates, produced by *de novo* chemical synthesis. This approach has the advantage of reducing backgrounds and undesirable crossreactions. However, most antigenic molecules are complex organic compounds and are exquisitely sensitive to stereochemical conformation. The chemical synthesis of complex organic molecules is frequently difficult and often produces a mixture of stereochemical isomers, which at best can reduce the molar efficacy of a compound and at worst can inhibit the action of the biologically active isomers or cause unwanted crossreactions. Biologically produced antigens, however, nearly always produce correctly folded molecules with the correct stereochemistry, as they closely mimic the natural disease agent. Although the production of these antigens is frequently simple and inexpensive, it is vulnerable to microbial contamination and genetic instability. Genetic plasticity is much more frequent in antigens derived from RNA viruses *(9)* (e.g., poliovirus, rhinoviruses, and the retroviruses) than in vaccines derived from DNA viruses (e.g., vaccinia and adenovirus) as RNA-dependent replicases do not have the error-correction mechanisms associated with DNA-dependent replicases *(10)*. With these considerations in mind, the designer of new antigens may conclude that "Nature knows best" and employ natural products such as viruses to make complex efficient antigenic reagents. However, the advantage of having a biological system that can make complex reagents from simple starting materials can be compromised if the genetic code directing the process is error-prone. Therefore, viruses with a DNA genome would seem to be the agents of choice, and designers should ensure that these vectors either contain their own error-correction mechanisms or can utilize those of the host cell.

1.3.2. Optimal Production of the Desired Antigens

One of the major advantages of using biological systems is that they can synthesize complex molecules with the correct secondary, tertiary, and quaternary structure and the appropriate stereochemistry. To achieve this, however, certain stringent criteria must be met. Simplistically, proteins synthesized by eukaryotic cells can be divided into two categories, those retained in the cytosol and those present on cell membranes or secreted from the cell (reviewed in ref. *11*). Proteins making up the virions of naked viruses and the nucleocapsids of enveloped viruses fall into the first category, and the coat proteins of the enveloped viruses fall into the second. Monocistronic viral messenger RNAs that encode the nucleocapsid or replicase molecules normally contain all the information necessary for the accurate and efficient expression of these proteins. However, vaccine designers who may wish to use single proteins normally encoded by a polycistronic message may have to think carefully about their design and ensure that appropriate promoters, ribosome binding sites, and termination sites are included.

Post-translational modifications such as phosphorylation and acylation can be important for protein stability and should also be taken into account when modifying genetic information. Even so, success may not be achieved, as the folding of these proteins may be dependent on the structure of precursors and the action of sequence-specific proteases, coded either by the virus or the host cell (reviewed in refs. *12* and *13*).

The synthesis of viral membrane proteins and virally encoded secreted proteins follows similar synthetic pathways, but these are more complex than those taken by proteins that remain in the cytosol. In addition to the criteria for cytosolic proteins, care must be taken to ensure that signal sequences and intracellular location sequences are included, as well as glycosylation sites *(14)*. Again, this may be difficult to achieve if proteins encoded by polycistronic viral messengers are to be produced, or if extensive site-specific mutagenesis is required to meet other criteria. As different cell types express a variety of glycosylation enzymes and use different location signals, expression of viral antigens in cell lines and tissues may vary considerably. For example, insect cells infected with genetically engineered baculoviruses have become popular as they can synthesize very high levels of foreign protein. However, their glycosylation pathways are different from those of mammalian cells, and the expression of some viral glycoproteins has met with difficulties *(15)*. As well as ensuring that the desired product resembles the native antigen as closely as possible, other steps can be taken to enhance production and/or purification. The inclusion of powerful constitutive promoters, such as those from the immediate early region of the human and murine cytomegaloviruses, have been used on many occasions to boost dramatically the production of proteins from recombinant viruses *(16)*. Purification can be facilitated to removing membrane-binding sites to enable proteins to be secreted from the cell or by adding motifs to enhance binding to affinity matrices or antibodies.

1.3.3. The Importance of Conserved Epitopes

Genetic variation, leading to evasion of the immune response, is a major weapon in the virus's armory for survival, but it poses a significant problem for the antigen designer. Genetic plasticity is an important feature of RNA genomes, and mutation alone can give rise to the rapid evolution of genetic clades in viruses such as HIV *(17)*, or a plethora of serotypes as seen for rhinoviruses, the cause of the common cold *(18)*. In addition, genetic reassortment by viruses with segmented genomes can lead to dramatic changes in serotype by the introduction of complete genes, as is seen with influenza virus *(19)*. Furthermore, recombination between virus genomes can introduce several new genes in a single event, as has been shown to occur with foot-and-mouth-disease virus and the alphaviruses *(20)*. It is important therefore that antigenically conserved proteins, or protein domains, which are essential for virus function, be identified. Evolution, however, would tend to ensure that these proteins are hidden from the immune response or the key functional domains are protected within the core of these proteins. Thus, surface epitopes on external proteins may initially seem to be suitable targets for vaccine design, as they readily generate antibodies that can be easily measured in most diagnostic laboratories. However, such epitopes will be subject to immune selection and may not be suitable candidates for antigens that could detect immune responses to all virus variants. For the enveloped viruses, the nucleocapsid proteins would be the most obvious choice as they would be protected from the humoral immune response and should therefore be less vulnerable to immune selection. Indeed, it has been shown with many enveloped viruses, e.g., influenza *(21)* and measles *(22)*, that antibodies to the nucleocapsid protein are more crossreactive than those that bind to virion envelope proteins.

The wide variety and sophistication of current genetic engineering techniques now enable the design and efficient production of virtually any antigen required for a given diagnostic purpose. Thus, these "designer antigens" can supersede many established products, and the properties of a new generation of ELISAs utilizing recombinant protein can fulfill the following criteria:

A single antigen should be specific in identifying all pathogenic isolates but capable of distinguishing them from nonpathogenic forms.

1. The assay format should be simple, robust, rapid (<2 h from receipt of sample to analysis of data), sensitive, specific, and not suffer background interference.
2. When reacting with the recombinant protein, sera (or other biological fluids) should not produce false-positive or false-negative results.
3. The assay format should be capable of distinguishing antibody types and subtypes, when necessary.
4. The purification of the recombinant proteins should be minimal.
5. The assay format should be readily adaptable to handling large sample numbers, preferably by robotic analyzers.

The careful construction of the gene or genes to be expressed is important, as in order to maximize the expression and the functionality of the desired protein it is frequently necessary to mimic the genetic organization of the original virus as much as possible. For example, it is nearly impossible to express the flavivirus E protein by itself as its transport through the endoplasmic reticulum (ER) and subsequent posttranslational modification is dependent on the ion-regulating functions of the pre-M protein. Similarly, work on expression of the NS1 protein *(23)* has shown that correct construction of the 5' end of the messenger RNA is essential for expression. The NS1 protein utilizes the C-terminus of the E protein as a signal sequence, and unless the nucleic acid sequence that codes for this signal sequence is included along with the NS1 gene, no expression is observed. Conversely, however, genetic constructs containing only the NS1 gene without any NS2a sequences seem capable of producing high levels of authentic NS1, even though most of the NS2a protein is required for NS1 production in vivo *(24)*. Presumably the role of NS2a in vivo is to ensure correct cleavage of NS1 from downstream amino acid sequences. If the gene is artificially terminated at the end of the coding region for NS1, NS2a is no longer necessary.

The use of recombinant proteins as antigens in diagnostic assays has benefits other than commercial ones; there are significant advantages to be gained by immunosurveillance of the response to vaccination. Information about the antibody subclass response to individual proteins in vaccine preparations can now be assessed using recombinant proteins in modified assays, which is crucial when determining those antigens required to produce a protective, long-lasting immune response. Furthermore, the humoral response to specific antigens of infectious agents may also be determined,

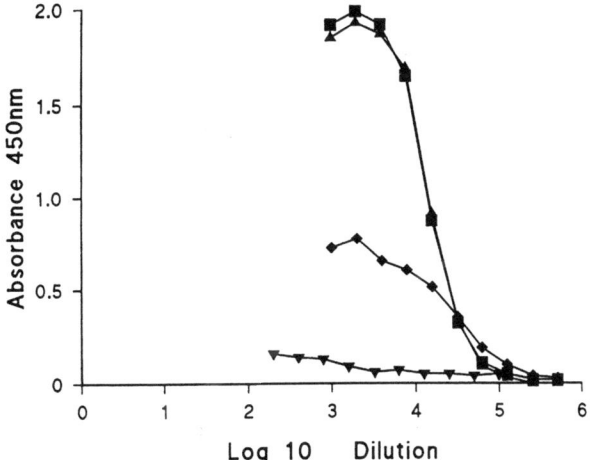

Fig. 1. Human measles ELISA antibody titers using antigens from a variety of sources.

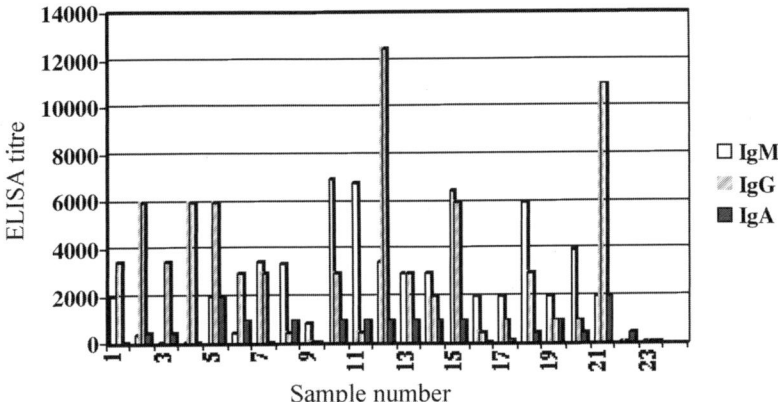

Fig. 2. Example of the use of defined recombinant measles antigen in an ELISA assay to detect IgG, IgM, and IgA. (Reprinted from *Journal of Virological Methods*, Vol. 49, pp. 257–268, 1994, with permission from Elsevier.)

which may not only help in the understanding of the immune system, but also allow the development of treatment or preventive measures. Other advantages may also include the production of recombinant proteins that may be difficult to obtain. For example, numerous viruses and parasites are difficult or hazardous to grow. Thus, recombinant proteins can now be used as diagnostic antigens to confirm infection, including investigations into small, round, structured viruses that may rely entirely on the use of recombinant proteins in ELISA for both detection and taxonomic purposes.

Other assays utilizing recombinant proteins have also been developed including antibody-capture ELISAs and radioimmunoassays. This chapter will, however, focus on the use of recombinant measles virus nucleocapsid protein as a diagnostic antigen in a number of assays ranging from a simple indirect ELISA format to amplified systems involving the determination of antibody subclass responses, to demonstrate the strengths and weaknesses of systems currently available. A comparison of results obtained using the same protein, but expressed in a variety of expression systems, is demonstrated in **Fig. 1**. An example of the utility of using recombinant antigens in assays that simultaneously measure IgG, IgM, and IgA is given in **Fig. 2**.

2. Materials

1. ELISA reader: Flow Laboratories (Oxford, UK) or equivalent.
2. ELISA washer: Skatron (Helisbro, Newmarket, UK) or equivalent.
3. ELISA trays: 96-well Nunc imuno plates (Maxisorp, cat. no. 439454, Invitrogen).
4. Coating buffer: 15 mM Na$_2$CO$_3$, 35 mM NaCO$_3$, 3 mM NaN$_3$, pH 9.8.
5. Phosphate-buffered saline (PBS): 137 mM NaCl, 37 mM KCl, 10 mM NaHPO$_4$, 1.8 mM KH$_2$PO$_4$, pH 7.4.
6. PBS with Tween-20 (PBST): add 0.1% Tween-20 (Sigma, cat. no. P1379) to 1X PBS and mix well.
7. Ready-to-use prestained substrate (Kem En Tek, Denmark).
8. 0.5 M sodium acetate buffer, pH 6.0.
9. STE: 10 mM Tris-HCl, 100 mM NaCl, 5 mM EDTA. pH 7.2. Store at room temperature.
10. Lysis buffer: 1% (v/v) Nonidet P40 in PBS.
11. Polyclonal antibodies: anti-human Ig horseradish peroxidase (HRP) linked F(ab)2.
12. Fragment NA (cat. no. 9330, Amersham International, UK).
13. Monoclonal antibodies (MAbs): anti-human IgG biotin label (cat. no. B3773, Sigma, Poole, UK), anti-human IgG 1 biotin label (cat. no. B6775, Sigma), anti-human IgG2 biotin label (cat. no. B3398, Sigma), anti-human IgG3 biotin label (cat. no. B3523, Sigma), and anti-human IgG4 biotin label (cat. no. B3648, Sigma).

3. Methods

3.1. Construction of Expression Systems Containing Recombinant Genes

A number of expression systems can be used to produce recombinant proteins (**Table 1**); however, protein purification needs to be optimized, the functional activity needs to be established, where the protein will be localized, and the structural integrity determined. A strategy for deciding which system to use is shown in **Fig. 3**.

The cloning vehicle should allow the rapid cloning of the recombinant gene to that of any of the above expression systems with minimal changes to cloning strategy.

3.1.1. Escherichia coli

E. coli is easy to use and is very well understood; although there is no post-translational modification (PTM), there are frequently problems with solubility. Problems associated with solubility can be overcome by the use of fusion proteins and can also aid in purification, e.g., His tagging *(25)*. Screening can be carried out to ascertain high-level expression very easily, and the production costs are the most cost-effective for all the expression systems.

Also, novel systems are now being introduced so that the gene of interest can be cloned into a carrier plasmid that can then be inserted readily into a destination vector of

Table 1
Comparison of Host Expression Systems for the Production of Recombinant Proteins for Use in Diagnostic Assays

Host organism	Ease of use	Defined system	PTM	Background interference	Solubility issues	High[a] expression	Cost
E. coli	Yes	Yes	No	Frequent	Yes	No	Low
Yeast	No	No	Yes[b]	Rare	No	No	Low
Insect	Yes	Yes	Yes[b]	Rare	No	Yes	High
Mammalian	No	Yes	Yes	Rare	No	Variable	High
None (cell-based system)	Yes	No	No	Frequent	No	Yes	High

[a]High level expression can never be guaranteed but frequently is determined by the structure, modification, and synthetic pathway of the protein in each particular system.

[b]Although insect cells carry out glycosylation, there are subtle differences from mammalian cells that can affect the antigenicity of some proteins.

Abbreviation: PTM, post-translational modification.

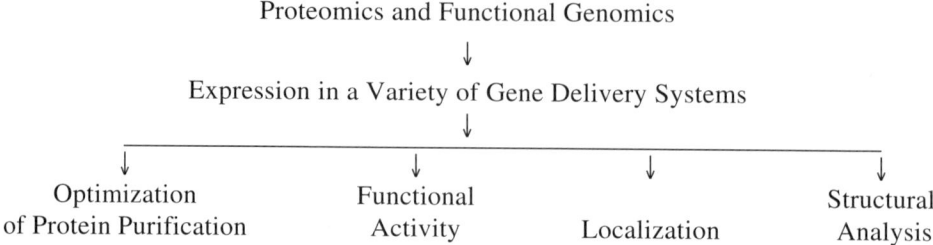

Fig. 3. A strategy for determining a gene delivery system.

choice, e.g., insect, mammalian, or *E. coli* expression system. Different systems can be used in parallel so that screening and authenticity can be monitored at the same time.

3.1.2. Yeast Expression Systems

Yeast systems can produce high levels of expression but are glycosylated differently from that of mammalian expression systems. They are, however, easy to screen for high-level expression. Their use as vehicles to produce diagnostic reagents for the detection of viral diseases is not acknowledged as a usable alternative. However, there has been some success with hepatitis C virus *(26)*. Problems have surrounded the breakage of the cells and also the excessive glycosylation that follows protein translation.

3.1.3. Insect Expression Systems

Insect cells can produce high yields in certain cell lines (e.g., SF9 cells), although, due to the nature of these systems, it is difficult to screen large numbers of constructs. The glycosylation laid down by insect cells can cause differences in antigenicity when one is designing diagnostic assays for the detection of certain diseases. The design of these systems from the molecular biologist's point of view has moved on significantly, and they are now relatively easy to construct, with a high success rate in gene cloning.

These new systems have a tremendous effect on cloning strategies as they can be designed to be moved between different expression systems, giving added versatility and reducing the need for large numbers of restriction enzymes, gel purification, and sequencing *(27,28)*.

3.1.4. Mammalian Expression Systems

Mammalian cells can produce perfect proteins with regards to protein glycosylation and folding, but they generally produce lower yields compared with the other systems. It is difficult to screen large numbers of cell cultures for high-level expression, as expression can vary from cell to cell within a single transformation.

The viral systems available are now well understood, with a number of features being added primarily as safety issues. The major differences between the three systems described below and in **Table 2** are that the adenoviral expression systems do not introduce the recombinant gene into the host chromosome unlike that of the retroviral and lentiviral expression systems. Thus, although expression in the adenovirus sys-

Table 2
Characteristics of Three Viral Expression Systems Commonly Available

	Transient expression		Stable expression			
Virus	Dividing cells	Non-dividing cells	Dividing cells	Neural cells	Growth-arrested cells	Contact-inhibited cells
Adenovirus	+	+	−	−	−	−
Retrovirus	+	−	+	−	−	−
Lentivirus	+	+	+	+	+	+

tems can often produce higher levels of recombinant end product, this is over a significant time period. As the cells divide, the copies will not be carried over to the daughter cells, and eventually expression will be lost.

3.1.4.1. ADENOVIRAL EXPRESSION SYSTEMS

Adenoviral constructs are made containing the recombinant gene, which can be up to 7.5 kb in size due to packaging constraints. To allow expression of the recombinant gene inside the transfected cell and not replication of the adenovirus itself, the E1 and E3 genes from the virus have been deleted, which makes it replication-deficient. The virus still has an active nuclear transport system, but the DNA does not integrate into the host genome. Usually a high-level promoter, such as the cytomegalovirus (CMV) immediate early promoter, drives the gene. Adenoviruses are very useful as they can infect most cell types and can grow in 293 cells, which compensates for the E1 and E3 gene deletions thus allowing growth of stock virus *(29–31)*.

3.1.4.2. LENTIVIRAL EXPRESSION SYSTEMS

Lentiviral constructs are derived from parts of the HIV-1 virus, and as they are inserted into the host chromosome, they require cell division to ensure that good levels of expression are achieved. Presently constructs have used the *gag*, *pol*, and *rev* genes, with the VSV-G envelope: rev and gag are structural proteins and the pol encodes the enzyme RNA reverse transcriptase. Again, the CMV promoter is used in such systems. The average numbers of colony forming units are 10^5–10^7/mL for these constructs. However, when these constructs are used to transfect cell types, the multiplicity of infection (MOI) can be used to control the level of entry into the cells. For example, an MOI of 1 will entail half of the cells becoming transfected with the construct. The effects of cell expression can vary between cells, with some being stable and having relatively high expression, and others exhibiting low levels of expression. These cells can produce stable expression for up to 6 wk. These systems also have an active nuclear transport system, which is essential to ensure that the proteins are exported correctly. A major safety issue is that there is no HIV viral envelope, and only three genes are present from the HIV-1 virus, thus reducing to acceptable levels the use of such constructs, and allowing Food and Drug Administration (FDA) approval as each of the genes used in the expression systems (*gag*, *rev*, and *pol*) are held on separate plasmids *(32)*.

3.1.4.3. RETROVIRUSES

The retroviral systems have been employed to express recombinant proteins primarily for use as vaccines and immunotherapeutics, primarily because, again, the recombinant gene is transferred into the host chromosome and, although high-level expression can be achieved, this is particularly variable from cell to cell. As cells will usually have one copy of the gene, again the levels of expression do not reach those of the adenoviral expression systems. However, high-level expression of p55 has been shown, which is a 10-fold increase in natural cell expression levels *(33)*.

3.1.5. Non-Cell Based Expression Systems: In Vitro Expression Systems

In vitro expression systems can generate soluble material free from proteolytic enzymes, and it can also be used to incorporate unnatural proteins (used in labeling), as it is an open system.

The system uses *E. coli* lysates, as they are low in complexity, no animals are required, and they are easy to manipulate genetically. You can use supercoiled or linear templates, which can yield high levels of protein in 2 h. The same cloning vector used to introduce your gene into different systems can be used here. These new systems have a tremendous effect on cloning strategies as they can be designed to be moved between different expression systems, giving added versatility and reducing the need for large numbers of restriction enzymes, gel purification, and sequencing.

The following procedures can be performed.

1. Protein–protein interactions.
2. Protein–RNA/DNA interactions.
3. Post-translational modification can be introduced with the addition of kinases.
4. Verification that the cloned gene product is active is easy.
5. Large-scale screening can be performed efficiently.
6. Toxic proteins can be generated.
7. Effects on solubility can be determined.
8. Effects of protein chaperones on protein folding can be studied.

Although a number of expression systems are available for the production of recombinant proteins that have potential as diagnostic antigens, we propose to concentrate on three systems that have been evaluated together. These include expression systems suitable for use in *E. coli* for which there are a number of vectors, although those using the well-characterized T7 or tac promoters have produced high levels of expression of intact proteins. More recently, we have also evaluated the baculovirus and adenovirus expression systems, both of which are well documented *(8–12)* and are commercially available. We do not propose to detail the methods for the construction of those systems that are readily available from the literature or from commercial companies (Invitrogen for baculovirus: Microbix, Ontario, Canada by adenovirus). However, we have simplified the major differences among the hosts when considering their use in the production of diagnostic proteins (*see* **Table 1**).

3.2. Preparation of Recombinant Measles Virus Nucleocapsid Protein

The measles virus nucleoprotein has been successfully used as a model in the production of recombinant proteins for use in diagnostic assays *(1,34,35)*. Therefore, we have focused on the use of this protein, which illustrates many of the general issues surrounding the use of recombinant proteins. Although the protocols outlined below are based on simple centrifugation steps to purify the recombinant material, we have found this to be the most successful. The nature of the recombinant protein dictates what types of purification process can be used. Owing to the hydrophilic nature of the nucleocapsid protein, we avoided the use of membrane technologies in the purification process, although this has been used successfully with other proteins.

3.2.1. Antigens Produced in E. coli

1. It is crucial to determine experimentally the optimum time-point for protein expression in each system, i.e., the point at which expression is maximal and degradation is minimal.
2. Centrifuge organisms approx 1.0×10^6 at $3000g$ for 10 min at 4°C, and wash twice in PBS.
3. Resuspend the organisms in lysis buffer for 10 min on ice.
4. Centrifuge lysed cells at $15,000g$ for 10 min at 4°C, remove supernatant, and store as a stock antigen at –20°C (*see* **Note 1**).

3.2.2. Antigens Produced in Insect Cells Infected with Recombinant Baculoviruses

1. Harvest infected cells at an optimum time-point for protein expression determined experimentally for each system. Remove the supernatant from the cell monolayer of 175-cm² flasks and wash the monolayer twice in PBS.
2. Recover the cells (approx 1×10^6) in 2 mL of PBS using glass beads with gentle agitation or by scraping the cells from the surface of a Petri dish using a plastic pipet.
3. Transfer the cell solution to a plastic polypropylene tube. Centrifuge the cells gently at $6500g$ at 4°C for 2 min. Discard the supernatant.
4. Resuspend the cell pellet in 1000 µL of lysis buffer.
5. Lyse the cells by vigorous vortexing for 30 s.
6. Remove the cell debris by low-speed centrifugation at $1000g$ for 2 min at 4°C.
7. Remove the supernatant and store as a stock antigen at –20°C.

3.2.3. Antigens Produced in Insect Cells Infected with Recombinant Adenoviruses

1. Harvest cells (approx 1×10^6) at an optimum time-point for protein expression, by centrifugation at $1000g$ for 10 min at 4°C.
2. Wash the cell pellet once in sterile PBS and pellet as in **step 1**.
3. Resuspend the cell pellet in 1 mL sterile PBS.
4. Freeze/thaw the cell pellet three times by transferring the tube from –70°C to 4°C.
5. Homogenize the cells in a Dounce homogenizer 20 times while keeping them at 4°C.
6. Clarify the supernatant by centrifugation at $10,000g$ for 10 min at 4°C.
7. Remove the supernatant and store at –20°C.
8. Add 1 mL sterile PBS. Mix with the cells and repeat **step 6**.
9. Combine the two supernatants from **steps 7** and **8** and store as stock antigen at –20°C.

3.2.4. Preparation of Purified Recombinant Measles Virus Nucleocapsid Protein from Insect or Mammalian Cells

We have included the preparation of purified recombinant proteins because this product can be used to reduce any background problems encountered and may also be used as antigen in antibody capture assays.

1. Harvest infected cells (approx 1×10^6) at an optimum time-point for protein expression.
2. Remove the supernatant from the cell monolayer and wash twice in PBS.
3. Recover the cells in PBS using glass beads with gentle agitation.
4. Harvest cells by centrifugation at 2000g for 10 min at 4°C.
5. Resuspend the cell pellet in lysis buffer and leave the cells on ice for 15 min.
6. Homogenize the cells as described above.
7. Clarify the supernatant by centrifugation at 10,000g for 10 min at 4°C.
8. Remove the supernatant and store at 4°C.
9. Load the supernatant onto a step gradient of sucrose containing 4 mL 65% (w/w) sucrose and 20 mL 25% (w/w) sucrose in STE butler with the addition of 1% (v/v) Nonidet P40.
10. Centrifuge in a swing-out rotor for 6 h at 25,000g at 4°C.
11. Remove the material at the sucrose interphase.
12. Dialyze the material against STE buffer at 4°C.
13. Add an equal volume of lysis buffer to the dialyzed material and gently mix by inversion.
14. Centrifuge for a second time as in **steps 8–10**.
15. Harvest interface into a 5 mL volume.
16. Remove the final 5 mL and dialyze against STE buffer at 4°C.
17. Load the sample (5 mL) onto a third discontinuous step-sucrose gradient containing 2 mL 65% (w/w) sucrose and 4 mL 25% (w/w) sucrose in STE buffer with the addition of 1% Nonidet P40.
18. Centrifuge in a swing-out rotor for 6 h at 25,000g at 4°C (*see* **Note 2**).
19. Fractionate the gradient into 1-mL fractions.
20. Analyze 10 µL of each fraction using a Coommassie-stained polyacrylamide gel (PAGE). Pool the fractions containing the recombinant protein and dialyze for 16 h against three changes of STE, remove, and store at –20°C as the stock antigen (*see* **Note 3**).

3.3. Optimization of the Recombinant Antigen Required to Coat a Plate

The amount of recombinant antigen coated on the plate is optimized to ensure that sufficient antigen is available for binding and that excess does not cause problems with crossreactivity or problems with background interference. This is usually performed using a checkerboard titration method, as outlined below.

1. Dilute the antigen at ranges from 1:100 to 1:10,000 in coating buffer and pipet 100 µL of each dilution into the wells of each row in the ELISA tray (rows A–G) using column control with PBST. Incubate for 1 h at room temperature.
2. Wash three times in PBST and blot dry.
3. Add 100 mL of a known serum diluted appropriately in PBST to wells in columns 1–11 with PBST in column 12 as a negative control. Incubate for 1 h at room temperature.
4. Wash three times in PBST and blot dry.
5. Dilute the first-stage HRP-conjugated antibody at ranges from 1:100 to 1:10,000 in PBST and pipet 100 µL of each dilution into each of rows 1–11; as a negative control, add 100 µL PBST to column 12 (*see* **Note 4**).

6. Add 100 µL substrate to each of the wells. Incubate for 30 min.
7. Stop the reacting with 25 µL of 1 N H_2SO_4 and read at 450 nm.
8. The optimum amount of both antigen and primary antibody can then be determined at a combination that gives the highest antibody index value and also a low background reading for the negative control. This will produce a good signal to background ratio, while conserving reagents.

3.4. Indirect ELISA for the Detection of IgG in Serum

The indirect ELISA is a basic assay with the recombinant antigen coated onto the plate for the detection both of IgG and IgA. We have established that this type of ELISA can determine both negative and strong positive sera, although problems with sensitivity can occur when evaluating low-titer sera. However, such problems are significantly fewer than when conventional antigens are used.

1. Coat the ELISA plate with 100 µL per well with the appropriate dilutions of antigen (determined in **Subheading 3.3.**). in columns 1, 2, 4, 6, 8, and 10 and with 100 µL of control antigen in columns 3, 5, 7, 9, and 11 in ELISA-coating buffer. Incubate for 1 h at room temperature.
2. Prepare dilutions of serum as required in PBST (usually 1:100 or 1:1500) and make doubling dilutions in PBST in columns down a dilution plate using two columns for each sample (positive and negative). A standard serum should always be used to determine error.
3. Wash the reaction plate three times with PBST and blot dry on paper towels.
4. Transfer 80 µL of the diluted antiserum to the reaction plate in the corresponding wells and add 80 µL PBST to columns 1 and 12, which are the negative controls. Incubate for 1 h at room temperature.
5. Wash the reaction plate three times with PBST and blot dry on paper towels.
6. Prepare an appropriate dilution of polyclonal or monoclonal antihuman IgG labeled with HRP as described in **Subheading 3.3.** in PBST and add 100 µL to each of the wells of the reaction plate.
7. Incubate for 1 h at room temperature.
8. Wash the plate three times with PBST and blot dry on paper towels.
9. Add 100 µL of substrate. Incubate for 30 min.
10. Stop the reaction by adding 25 µL of 1 N H_2SO_4.
11. Read the plate at an absorbance of 450 nm.

3.5. Indirect ELISA for the Detection of IgG in Serum

The indirect ELISA utilizing an amplified system has the advantage of increased sensitivity compared with the basic system. If a primary monoclonal antibody is used instead of a polyclonal antibody, then specificity is also increased, which can overcome most problems encountered with background interference.

1. Coat the ELISA plate with 100 µL per well with the appropriate dilution of antigen (determined in **Subheading 3.3.**) in columns 1, 2, 4, 6, 8, and 10 and with 100 µL of control antigen in columns 3, 5, 7, 9, and 11 in ELISA-coating buffer. Incubate for 1 h at room temperature.
2. Prepare dilutions of serum as required in PBST (usually 1:100 or 1:500) and make double dilutions in PBST in columns down a dilution plate using two columns for each sample (positive and negative). A standard serum should always be used to determine error.

3. Wash the reaction plate three times with PBST and blot dry on paper towels.
4. Transfer 80 µL of the diluted antiserum to the reaction plate in the corresponding wells and add 80 µL PBST to columns 1 and 12, which are the negative controls. Incubate for 1 h at room temperature.
5. Wash the reaction plate three times with PBST and blot dry on paper towels.
6. Prepare a 1:1000 dilution of monoclonal antihuman IgG labeled with biotin in PBST and add 100 µL to each of the wells of the reaction plate. Incubate for 1 h at room temperature.
7. Wash the reaction plate three times with PBST and blot dry on paper towels.
8. Prepare a 1:1000 dilution of streptavidin-HRP and add 100 µL to each well. Incubate for 1 h at room temperature.
9. Wash the reaction plate three times with PBST and blot dry on paper towels.
10. Add 100 µL of substrate. Incubate for 30 min at room temperature.
11. Stop the reaction by adding 25 µL of 1 N H_2SO_4.
12. Read the plate at an absorbance of 450 nm.

3.6. Indirect ELISA for the Detection of IgG

The ELISA systems outlined in **Subheading 3.5.** can be used as a highly sophisticated research tool, to evaluate the precise humoral response to both vaccination and infectious disease, thus helping us to understand further the immune response not only to whole pathogens but also to individual antigens encoded by them. In addition, MAbs specific for the IgG subclasses can be used separately instead of a labeled MAb specific for the total IgG, as outlined below:

1. Coat the ELISA plate with 100 µL per well with the appropriate dilution of antigen (determined in **Subheading 3.3.**) in columns 1, 2, 4, 6, 8, and 10 and with 100 µL of control antigen in columns 3. 5, 7, 9, and 11 in ELISA-coating buffer. Incubate for 1 h at room temperature.
2. Prepare dilutions of serum as required in PBST (usually 1:100 or 1:500) and make double dilutions in PBST in columns down a dilution plate using two columns for each sample (a positive and negative). A standard serum should always be used to determine errors.
3. Wash the reaction plate three times with PBST and blot dry on clean paper towels.
4. Transfer 80 µL of the diluted antiserum to the reaction plate in the corresponding wells and add 80 µL PBST to columns 1 and 12, which are the negative controls. Incubate for 1 h at room temperature.
5. Wash the reaction plate three times with PBST and blot dry on paper towels.
6. Prepare a 1:1000 dilution of biotin-labeled monoclonals anti-human IgG1, IgG2, IgG3, IgG4, and IgG in PBST and add 100 µL of an individual monoclonal to the appropriate well of the reaction plate. Incubate for 1 h at room temperature.
7. Wash the reaction plate three times with PBST and blot dry on paper towels.
8. Prepare a 1:1000 dilution of streptavidin-HRP and add 100 µL to each well. Incubate for 1 h at room temperature.
9. Wash the reaction plate three times with PBST and blot dry on clean paper towels.
10. Add 100 µL of substrate. Incubate for 30 min at room temperature.
11. Stop the reaction by adding 25 µL of 1 N H_2SO_4.
12. Read the plate at an absorbance 450 nm.

4. Notes

1. It is important to note that waste materials should be disposed of using the worker's country's guidelines on the handling of dangerous pathogens.
2. The ultracentrifuge rotors used are a TST 41.14 and an AH629 swing-out rotor (Sorval). The ultracentrifuge is a OTD-75B. The equipment was purchased from Sorval Instruments, DuPont.
3. The recombinant protein purified in **step 18** is more than 90% pure when analyzed by PAGE and produces a single protein band of the correct molecular weight after Coommassie blue staining.
4. It is important to include the relevant negative controls to ascertain the background levels produced in each particular system. In all cases we designed the expression systems to produce β-galactosidase, which could be used as appropriate negative control.
5. Room temperature is defined as 20–25°C throughout.

References

1. Warnes, A., Fooks, A. R., and Stephenson, J. R. (1994) Production of measles nucleoprotein in different expression systems and its use as a diagnostic reagent. *J. Virol. Methods* **49,** 257–268.
2. Warnes. A., Fooks, A. R., Wilkinson, G. W. G., Dowsett A. B., and Stephenson, J. R. (1995) Expression of measles virus nucleoprotein in *Escherichia coli* and assembly of nucleocapsid-like complexes. *Gene* **160,** 173–178.
3. Johnson, N., Mansfield, K., and Fooks, A. R. (2002) Canine vaccine recipients recognize an immunodominant region of the rabies glycoprotein. *J. Gen. Virol.* **83,** 2635–2661.
4. Young, H., Moyes, A., Seagar, L., and McMillan, A. (1998). Novel recombinant-antigen enzyme immunoassay for serological diagnosis of syphilis. *J. Clin. Microbiol.* **36,** 913–917.
5. Barber, G. N., Clegg, J. C. S., and Lloyd, G. (1990) Expression of the Lassa virus nucleocapsid protein in insect cells infected with a recombinant baculovirus: application to diagnostic assays for Lassa virus infection. *J. Gen. Virol.* **71,** 19–28.
6. Lukashevich, I. S., Clegg, J. C. S., and Sidibe, K. (1993). Lassa virus activity in Guinea: distribution of human antiviral antibody defined using enzyme-linked immunosorbent assay with recombinant antigen. *J. Med. Virol.* **40,** 210–217.
7. Libeau, G., Prehaud, C., Lancelot, R., et al. (1995) Development of a competitive ELISA for detecting antibodies to the peste dos petits ruminants virus using a recombinant nucleoprotein. *Res. Vet. Sci.* **58,** 50–55.
8. Groen, J., van den Hoogen, B. G., Burghoorn-Maas, C. P., et al. (2003). Serological reactivity of baculovirus-expressed Ebola virus VP35 and nucleoproteins. *Microbes Infect.* **5,** 379–385.
9. Temin, H. M. (1993) Retrovirus variation and reverse transcription: abnormal strand transfers result in retrovirus genetic variation. *Proc. Natl. Acad. Sci. USA* **90,** 6900–6903.
10. Gmyl, A. P., Pilipenko, E. V., Maslova, S. V., Belov, G. A., and Agol, V. I. (1993) Functional and genetic plasticities of the poliovirus genome: quasi-infectious RNAs modified in the 5'-untranslated region yield a variety of pseudorevertants. *J. Virol.* **67,** 6309–6316.
11. Mellman, I. and Warren, G. (2000) The road taken: past and future foundations of membrane traffic. *Cell* **100,** 99–112.
12. Ansardi, D. C., Porter, D. C., Anderson, M. J., and Morrow, C. D. (1996) Poliovirus assembly and encapsidation of genomic RNA. *Adv. Virus Res.* **46,** 1–68.

13. Yamshchikov, V. F. and Compans, R. W. (1995) Formation of the flavivirus envelope: role of the viral NS2B-NS3 protease. *J. Virol.* **69,** 1995–2003.
14. Stephenson, J. R. (2001) Genetically modified viruses; vaccines by design. *Curr. Pharmaceuti. Biotechnol.* **2,** 47–76.
15. Takehara, K., Ireland, D., and Bishop, D. H. L. (1988) Co-expression of the hepatitis B surface and core antigens using baculovirus multiple expression vectors. *J. Gen. Virol.* **69,** 2763–2777.
16. Wilkinson, G. W. G. and Akrigg, A. (1992) Constitutive and enhanced expression from the CMV major IE promoter in a defective adenovirus vector. *Nucleic Acids Res.* **20,** 2233–2239.
17. Crandall, K. A., Vasco, D. A., Posada, D., and Imamichi, H. (1999) Advances in understanding the evolution of HIV. *AIDS* **13(suppl A),** S39–47.
18. Horsnell, C., Gama, R. E., Hughes, P. J., and Stanway, G. (1995) Molecular relationships between 21 human rhinovirus serotypes. *J. Gen. Virol.* **76,** 2549–2555.
19. Zambon, M. C. (1999) Epidemiology and pathogenesis of influenza J. *Antimicrob. Chemother.* **44(suppl B),** 3–9.
20. Weaver, S. C., Hagenbaugh, A., Bellew, L. A., et al. (1993) A comparison of the nucleotide sequences of eastern and western equine encephalomyelitis viruses with those of other alphaviruses and related RNA viruses. *Virology* **197,** 375–390.
21. Zweerink, H. J., Askonas, B. A., Millican, D., Courtneidge, S. A., and Skehel, J. J. (1977) Cytotoxic T cells to type A influenza virus; viral hemagglutinin induces A-strain specificity while infected cells confer cross-reactive cytotoxicity. *Eur. J. Immun.* **7,** 630–635.
22. Gershon, A. and Krugman, S. (1979) Measles, in *Diagnostic Procedures for Viral and Rickettsial Infections* (Schmidt, N. J. and Emmons, R. W., eds.), American Public Health Association, Washington, DC.
23. Jacobs, S. C., Stephenson, L. R., and Wilkinson, G. W. G. (1992) High-level expression of the tick-borne encephalitis virus NSI protein by using an adenovirus-based vector: protection elicited in a murine model. *J. Virol.* **66,** 2086–2095.
24. Falgout, B., Channock, R., and Lau, C. J. (1989) Proper processing of dengue virus nonstructural glycoprotein NS1 requires the N-terminal hydrophobic signal sequence and the downstream nonstructural protein NS2a. *J. Virol.* **63,** 1852–1860.
25. Das, T. and Banerjee, A. K. (1993) Expression of the vesicular stomatitis nucleoprotein gene in *Escherichia coli* analysis of its biological activity *in vitro*. *Virology* **193,** 340–347.
26. Chien, D. Y., Choo, Q. L., Tabizi, A., et al. (1992). Diagnosis of hepatitis C virus (HCV) infection using an immunodominant chimeric polyprotein to capture circulating antibodies: reevaluation of the role of HCV in liver disease. *Proc. Natl. Acad. Sci. USA* **89,** 10,011–10,015.
27. Fooks, A. R., Stephenson, J. R., Warnes, A., Rima, B. K., Dowsett, B. A., and Wilkinson, G. W. G. (1993). Measles virus nucleocapsid protein expressed in insect cells assembles into nucleocapsid-like structures. *J. Gen. Virol.* **14,** 1439–1444.
28. Kost, T. A. and Condreay, J. P. (2002) Recombinant baculoviruses as mammalian cell gene-delivery vectors. *Trends. Biotechnol.* **20,** 173–180.
29. Graham, F. L. and Prevec, L. (1991) Manipulation of adenovirus vectors, in *Gene Transfer and Expression Protocols* (Murray, E. J., ed.), Humana, Totowa, NJ.
30. Fooks, A. R., Schadeck, E., Liebert, U. G., et al. (1995) High-level expression of the measles virus nucleocapsid protein by using a replication-deficient adenovirus vector: induction of an Ml'ltEr-l-restricted CTL response and protection in a murine model. *Virology* **210,** 456–465.

31. Warnes, A. and Fooks, A. R. (1996) Live viral vectors: construction of a replication-deficient recombinant adenovirus, in *Vaccine Protocols* (Robinson, A., Farrah, G., and Wiblin, C., eds.), Humana, Totowa, NJ.
32. Jolly, D. J. and Warner, J. F. (1990). Retroviral vectors as vaccines and immunotherapeutics. *Semin. Immunol.* **2,** 329–339.
33. Kato, H., Velu, T., and Cheng, S. Y. (1989). High level expression of p55, a thyroid hormone binding protein which is homologous to protein disulfide isomerase in a retroviral vector. *Biochem. Biophys. Res. Commun.* **164,** 238–244.
34. Racher, A. J. S., Folks, A. R., and Griffins, L. B. (1995) Culture of 293 cells in different culture systems: cell growth and recombinant adenovirus production. *Biotechnol. Techn.* **9,** 169–174.
35. Hutchinson, I., Fooks, A. R., Smith, R., and Stacey, G. N. (1995) Selection of host cell lines for scale-up growth of conditions of a recombinant baculovirus vector expressing measles virus nucleocapsid protein. *Biotechnol. Techn.* **9,** 907–912.

23

Basic Problems of Serological Laboratory Diagnosis

Walter Fierz

Abstract

Serological laboratory diagnosis of infectious diseases is inflicted with several kinds of basic problems. One difficulty relates to the fact that the serological diagnosis of infectious diseases is double indirect: The first indirect aim in diagnosing an infectious disease is to identify the microbial agent that caused the disease. The second indirect aim is to identify this infectious agent by measuring the patient's immune response to the potential agent. Thus, the serological test is neither measuring directly disease nor the cause of the disease, but the patient's immune system. The latter poses another type of problem, because each person's immune system is unique. The immune response to an infectious agent is usually of polyclonal nature, and the exact physicochemical properties of antibodies are unique for each clone of antibody. The clonal makeup and composition and, therefore, the way an individual's immune system sees an infectious agent, depends not only on the genetic background of the person but also on the individual experience from former encounters with various infectious agents. In consequence, the reaction of a patient's serum in an analytical system is not precisely predictable. Also, the antigenic makeup of an infectious agent is not always foreseeable. Antigenic variations leading to different serotypes is a quite common phenomenon. Altogether, these biological problems lead to complexities in selecting the appropriate tests and strategies for testing, in interpreting the results, and in standardizing serological test systems. For that reason, a close collaboration of the laboratory with the clinic is mandatory to avoid erroneous conclusions from serological test results, which might lead to wrong decisions in patient care.

Key Words: Serology; test performance; antigen; antibody; standardization.

1. Introduction

Serological testing and the use of immunochemical techniques are well established in the clinical laboratory, and their uses are still being developed and refined. There are two basic types of immunoserological tests: those that test for the presence of antigen(s) and those that measure antibody response to such antigens. Serological tests are applied for the diagnosis of infectious diseases, autoimmune disorders, allergies, and malignancies. Despite their widespread use, immunoassays are inflicted with a number of significant problems that lead to a substantial variation in the reliability and accuracy of currently available tests. A particular problem ensues from the increasing use of bedside diagnostic kits. In the recent process of establishing procedures and criteria for quality assurance in the clinical laboratory, an increased awareness of these problems has arisen among the clinical and reference laboratories as well as among the manufacturers of commercial test kits. The recognition of this situation prompted the U.S. National Committee for Clinical Laboratory Standards (NCCLS) to create specific guidelines that address "the generic problems of preparation and characterization of antigens and antibodies, testing using these reagents, and understanding the results." These guidelines are made public in the NCCLS document I/LA18-A2 *Specifications for Immunological Testing for Infectious Diseases; Approved Guideline-Second Edition*, issued in September 2001 *(1)*. References to this document are made in this chapter where appropriate; in particular, attempts are made to adhere to the definitions of terms and procedures as they are laid down in the NCCLS consensus document.

2. Relationship Between Interpretation of Test Results and Clinical Questions Asked

For all but the most trivial laboratory tests and, in particular, for serological analyses, it is of utmost importance for both the clinician and the laboratory staff to realize that a particular test result can have very different meanings depending on the question asked by the clinician and on the clinical situation. One has to distinguish among tests suitable for answering the following clinical problems:

1. Screening for disease.
2. Confirmation of disease.
3. Exclusion of disease.
4. Activity of disease.
5. Control of therapy.
6. Evaluation of prognosis.
7. Assessment of epidemiology.

In **Table 1** these categories are exemplified for the case of HIV infection.

It might well be that the same method can fulfill more than one of the above functions depending on:

1. The question and clinical situation.
2. The material tested [serum, cerebrospinal fluid (CSF), urine, saliva].
3. The parameter settings of the test (e.g., cutoff).

For example, a test for detection of antibodies to herpes simplex virus (HSV) in the serum might be not more than a screening test for carriers of the HSV. The same test applied in a quantitative way to measure antibody levels in the CSF compared with the serum will provide the clinician with direct information about intrathecal antibody production in the case of acute HSV-encephalitis.

Table 1
Tests to Answer Different Clinical Questions in the Case of HIV Infection

Clinical question	Test
Screening, exclusion, epidemiology	Anti-HIV antibody and antigen EIA
Confirmation	Immunoblot (Western blot) antigen neutralization EIA HIV DNA/RNA
Activity, control of therapy	CD4 cell count, quantitative HIV RNA
Prognosis	CD4 cell count, quantitative HIV RNA

At the same time, the results of such tests have to be interpreted in the context of the particular clinical, and perhaps also epidemiological, situation. The following patient-related factors are to be considered:

1. Age, sex, and constitution.
2. Life style and habits (e.g., drug consumption).
3. Actual life circumstances (e.g., pregnancy, traveling).
4. Other concurrent diseases.
5. Race and genetic predisposition.

In a newborn, for example, specific IgG antibodies detected in the serum originate with a high probability from the mother and rarely result from an immune response of the baby.

In addition, factors related to the patient's surroundings might also be of importance:

1. Epidemiological situation.
2. Environmental influence.
3. Professional exposure.
4. Sexual partner(s).

A positive test for tick-borne encephalitis (TBE) that indicates an infection with TBE virus in the European population might indicate an infection with Dengue virus if the patient is returning from a tropical area. A positive *Treponema pallidum* hemagglutination (TPHA) test in a patient coming from a tropical area might be associated with yaws, whereas it usually is an indicator for syphilis.

Not all these factors can be dealt with in this context, but some of the basic problems will be discussed in more detail. First, three main classes of tests have to be distinguished, as outlined below.

2.1. Screening Tests

Screening tests are used in two situations:

1. *Exploratory tests.* In the case of an unknown disease, possible causes are screened in order to narrow the range of further diagnostic procedures. The spectrum of tests to be used depends greatly on the clinical and epidemiological situation. Often, a defined combination of tests (screening blocks) is used for a particular circumstance (e.g., fever of unknown origin). To choose the right range of tests, it is important not to miss the actual

cause of the disorder. However, the problem usually is to find the right balance between costs and efficacy.

2. *Security tests.* In the situation in which particular causes of disease have to be excluded, screening tests are applied in order not to miss a possible treatment or preventive measure. Typical examples are screening tests in blood transfusion centers. The range of tests to be used is usually well defined, and the residual risks are well known and balanced against costs.

In any case, screening tests should ideally produce no false-negative values, i.e., they should be as sensitive as possible. They should also be rapid and cheap. In order to capture as many cases as possible, the specificity of serological screening tests should be broad, should encompass various serotypes and epitopes of antigens, and should usually detect all classes of immunoglobulins. The inherent problem ensuing from these characteristics of screening tests is, however, the potentially high rate of false-positive results, which is a particularly serious challenge in bedside diagnostics.

The clinically relevant parameter for the performance of a screening test is the negative predictive value (NPV) (*see* **Subheading 3.**). As will be discussed later, the NPV is dependent not only on the sensitivity and specificity of the test but also on the clinical prevalence of the disease. The higher the prevalence, the lower the NPV of the test.

2.2. Confirmatory Tests

Because screening tests inherently lack high specificity, it is often mandatory to confirm a positive screening result with a more specific test. Such confirmatory tests ideally should not produce any false-positive results, i.e., they should be as specific as possible. To reach this goal, one usually accepts that these tests are more time-consuming and more costly. For the purpose of specificity, it is often advantageous to detect antibodies to single epitopes and to identify the Ig class involved. To maintain sensitivity, many different antigens might have to be tested. One of the suitable techniques is the immunoblot (Western blot), in which various antigenic proteins are separated by gel electrophoresis and used to detect corresponding antibodies in a single assay. Natural antigens are increasingly replaced by recombinant proteins for this purpose.

Conformation of a positive antigen test is usually done by neutralization assays with well-defined (monoclonal) antibodies.

The clinically relevant parameter for the performance of a confirmatory test is the positive predictive value (PPV) (*see* **Subheading 3.**). Again, the PPV, like the NPV, is dependent on the clinical prevalence of the disease. The lower the prevalence, the lower the PPV of the test.

2.3. Tests for Evaluation of Disease Activity

Once the specific diagnosis of a disease has been reached by clinical and laboratory measures, a further application for laboratory tests will be the evaluation of disease activity in order to follow the course of the disorder, judge the prognosis, and monitor the effects of a therapeutic intervention. Such tests do not necessarily need to be very specific, but they preferably will provide a quantitative result that allows assessment

Problems of Laboratory Diagnosis

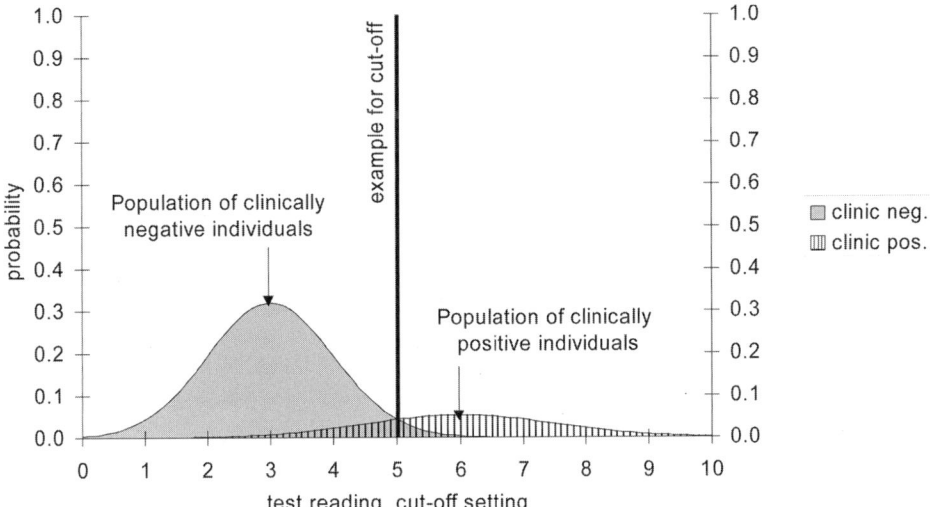

Fig. 1. Test reading for clinically negative and positive populations. Theoretical example of normally distributed test readings from two populations of clinically negative (80%) and positive (20%) individuals. The cutoff is arbitrarily set at a test reading of 5 (arbitrary units).

of disease activity. For many purposes, nonspecific inflammatory parameters will be used. The distinction of antigen-specific Ig classes, i.e., IgM, IgG, and IgA, will help to estimate the time-course of the disease. Newer techniques try to exploit a particularity of the immune system, called affinity maturation, for that aim (*see* **Subheading 5.2.**): the measurement of antibody avidities allows one to follow the development of the immune response after an acute encounter with a new antigen.

The clinically relevant parameter for the performance of an activity test will be the quantitative correlation of the test results with the clinical activity of the disease. For that purpose, the quantitative reproducibility of the test result, preferably expressed in international units per milliliters will be of paramount importance (*see* **Subheading 6.**).

3. Measures of Test Performance

The performance of any given laboratory test can be expressed in terms of its sensitivity, specificity, positive and negative predictive values, and likelihood ratios. These parameters are not independent of each other and are based on the characteristics of the test as well as on the clinical prevalence of the entity (e.g., disease or immune status) to be tested. In the following, the mathematical interrelations among these parameters will be illustrated with the help of a fictional example based on two populations of negative and positive individuals with normally distributed test results (**Fig. 1**).

3.1. Cutoff Between Negative and Positive Results

For every test a decision criterion (cutoff) has to be defined that separates the negative from the positive test readings (**Fig. 1**). Since the test readings for the clinically

**Table 2
Definition of True/False Positive/Negative Results**

		clinical classification +	−	TP true positive FP false positive
test result	+	TP	FP	FN false negative
	−	FN	TN	TN true negative

negative and the clinically positive populations almost always overlap to a certain extent, the setting of the cutoff is arbitrary and will define, together with the biochemical test characteristics, the sensitivity and specificity of the test.

3.2. (False) Positive and (False) Negative Results: Sensitivity and Specificity

True/false-positive and true/false-negative results are defined in **Table 2**. The sensitivity (Se) indicates the probability with which a test identifies clinical positive cases. It is defined by the fraction of true test positives over all clinical positives:

$$Se = \frac{TP}{TP + FN} \tag{1}$$

The specificity (Sp) indicates the probability with which a test identifies clinical negative cases. It is defined by the fraction of true test negatives over all clinical negatives:

$$Sp = \frac{TN}{FP + TN} \tag{2}$$

Specificity is perhaps more often thought of as the complement to nonspecificity, which is the fraction of false positives over all clinical negatives:

$$Sp = 1 - \frac{FP}{FP + TN} \tag{3}$$

The clinical prevalence ($P_{Disease}$) is given by

$$P_D = \frac{TP + FN}{TP + FP + FN + TN} \tag{4}$$

and is set in this example to 0.2.

Depending on the cutoff value used to separate the positive from the negative test readings, the fraction of (false) positive and (false) negative results and with it the sensitivity and specificity vary. Therefore, sensitivity and specificity of a test are dependent not only on the biochemical test characteristics but also on the cutoff used to define positive and negative test readings. In **Fig. 2** the sensitivity and specificity are expressed as a function of the cutoff setting.

Problems of Laboratory Diagnosis 399

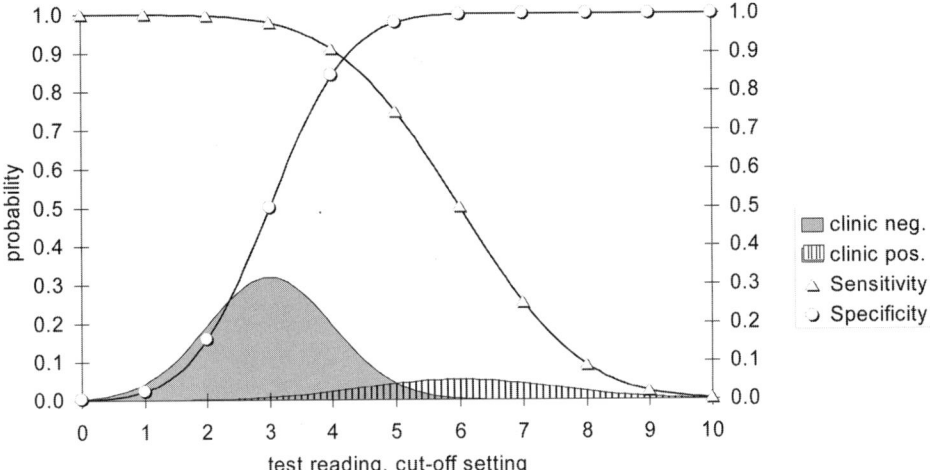

Fig. 2. Relation between cutoff and sensitivity/specificity. The sensitivity and specificity of testing the populations of **Fig. 1** are expressed as a function of the cutoff setting. A higher cutoff setting results in higher specificity and lower sensitivity and vice versa.

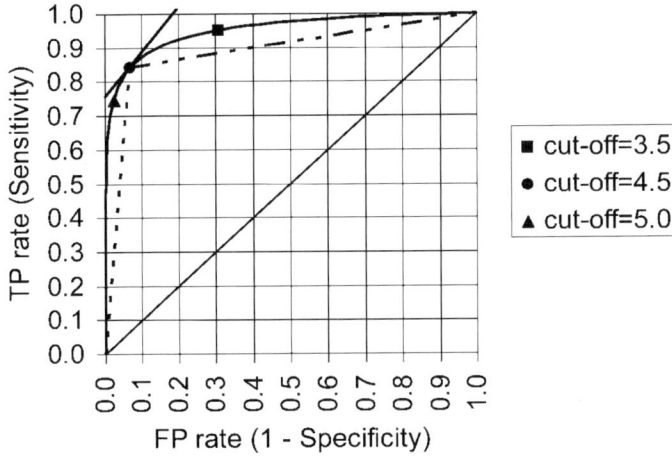

Fig. 3. Receiver operating characteristics (ROC). Plot of true-positive rate (sensitivity) and false-positive rate (nonspecificity) for all cutoff settings.

3.3. Receiver Operating Characteristics

A common way to depict the dependence of sensitivity and specificity on the cutoff is called receiver operating characteristics (ROC), a term coming from the field of radio transmission. As demonstrated in **Fig. 3**, the true-positive rate (sensitivity) is plotted against the false-positive rate (nonspecificity) across all cutoff settings.

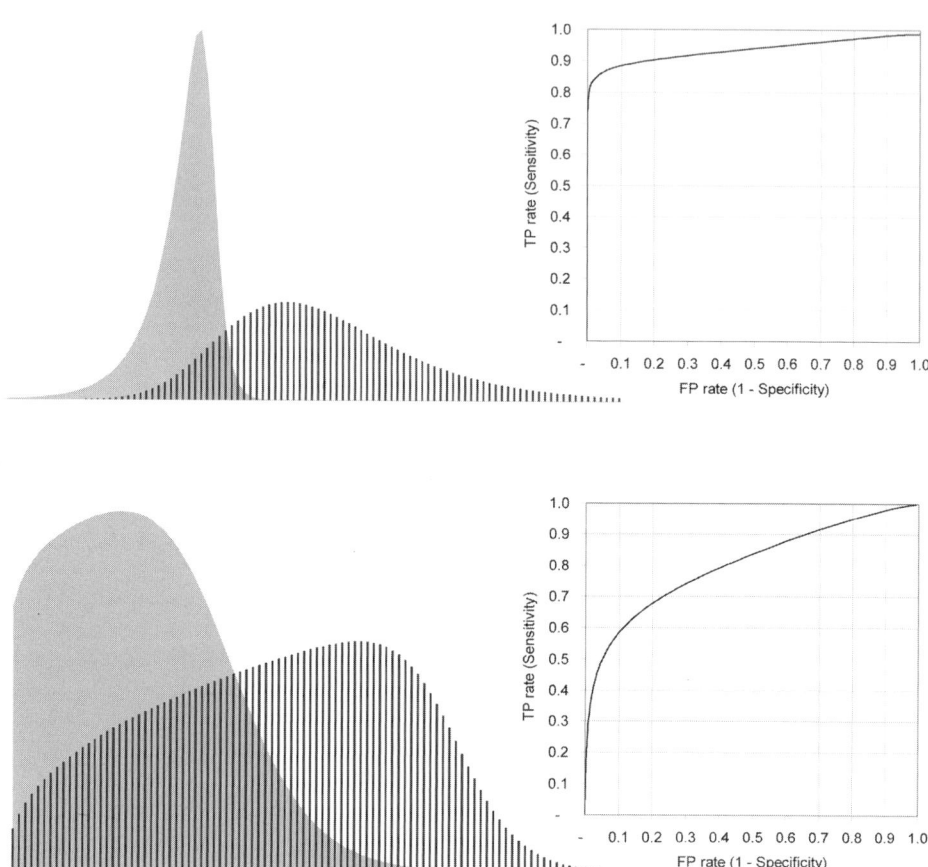

Fig. 4. Examples of *S*-distribution-based ROC curves.

As examples, *TP* and *FP* rates are indicated for three different cutoffs. This gives an overall illustration of test performance. The area under the curve can be used as a summary measure for diagnostic test assessment (provided that the ROC curves of different tests do not cross each other) *(2)*.

When constructing ROC curves from sample data of a continuous diagnostic test, one might want to model the data with a particular distribution, so that a continuous rather than a stepwise ROC curve can be calculated. However, a normal distribution does not always fit the data as in our theoretical model, the distribution underlying the sample data might not be known, and the data might be distributed in a highly asymmetrical way. Recently, a new parametric method based on *S*-distributions for computing ROC curves was published *(3)*. Methods to estimate *S*-distributions from sample data and to calculate confidence bands for the ROC curve and the area under the ROC curve are given. **Figure 4** shows two examples of fictitious *S*-distributions of clinically negative and positive populations and the corresponding ROC curves.

3.4. Positive and Negative Predictive Values, Efficiency

For the clinician, the positive and negative predictive values are the most important characteristics of a test. The clinician wants to know how many of the positive or negative test results are clinically true. It is important to realize, however, that these values depend not only on the laboratory test but also to a large degree on the clinical prevalence of the entity to be tested.

The PPV indicates the probability of disease in the case of a positive test result. It is defined by the fraction of true test positives over all test positives:

$$PPV = \frac{TP}{TP + FP} \quad (5)$$

The negative predictive value (NPV) indicates the probability of absence of disease in case of a negative test result. It is defined by the fraction of true test negatives over all test negatives:

$$NPV = \frac{TN}{TN + FN} \quad (6)$$

An overall measure of the test performance is given by its efficiency (E), which is defined by the fraction of true test readings over all test results:

$$E = \frac{TP + TN}{TP + FP + TN + FN} \quad (7)$$

All three parameters can be expressed as a function of sensitivity, specificity, and clinical prevalence (P_D):

$$PPV = \frac{Se \cdot P_D}{Se \cdot P_D + (1 - Sp) \cdot (1 - P_D)} \quad (8)$$

$$NPV = \frac{Sp \cdot (1 - P_D)}{Sp \cdot (1 - P_D) + (1 - Se) \cdot P_D} \quad (9)$$

$$E = \frac{Se \cdot P_D}{Sp \cdot (1 - P_D)} \quad (10)$$

For our example, **Fig. 5** illustrates the strong dependence of the positive predictive value on the clinical prevalence, which is rarely known exactly but can be estimated. The PPV is also highly susceptible to changes in specificity, but to a much lesser extent to changes in sensitivity (**Fig. 5**).

To give an example of the dependence of the PPV on the prevalence, the case of HIV testing can be considered. We might assume a sensitivity of 100% and a specificity of 99% for an HIV test. In a population of drug abusers, the prevalence of HIV infection might be approx 10%, which gives a PPV of approx 90%. In contrast, in a

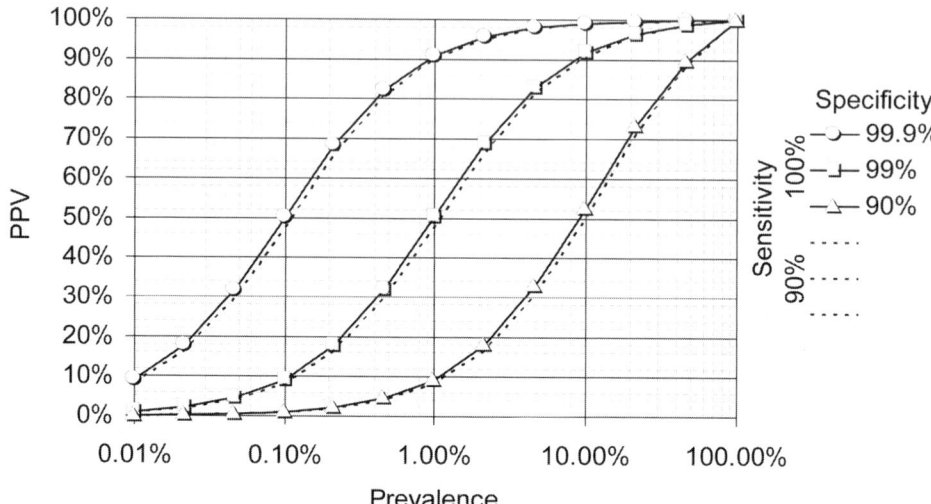

Fig. 5. Relation between prevalence and PPV for different specificities and sensitivities. PPVs are expressed as a function of the clinical prevalence for three different levels of specificity (90, 99, and 99.9%) and two levels of sensitivity (90 and 100%).

population of blood donors in which the prevalence is lower than 0.01%, the PPV will be below 1%, i.e., 99 of 100 positive test results will be false positive.

3.5. Likelihood Ratios

A different way to look at the efficacy of a laboratory test is to ask how much the knowledge of a test result helps the clinician to judge the odds for disease in the patient tested. The pretest odds are given by the prevalence of the disease in the population to which the patient belongs. The post-test odds are given by the predictive values of the tests. The efficacy of a test can then be expressed by the ratio of the post-test odds over the pretest odds, which is equal to the likelihood ratio (LR). In recent years, likelihood ratios have been recommended as a measure of test performance, as an alternative to the concept of sensitivity and specificity *(4)*.

It is important that likelihood ratios be (like sensitivity and specificity) prevalence-independent and can be calculated with data from test evaluations (but *see* **Subheading 3.6.**). LRs are defined as the ratio of two probabilities (likelihoods):

$$LR = \frac{\text{Likelihood of test outcome given clinically positive cases}}{\text{Likelihood of test outcome given clinically negative cases}} \quad (11)$$

LR for positive test results: $LR(+) = \dfrac{TP \text{ rate}}{FP \text{ rate}} = \dfrac{Se}{1-Sp}$ \quad (12)

LR for negative test results: $LR(-) = \dfrac{FN \text{ rate}}{TN \text{ rate}} = \dfrac{1-Se}{Sp}$ \quad (13)

Problems of Laboratory Diagnosis

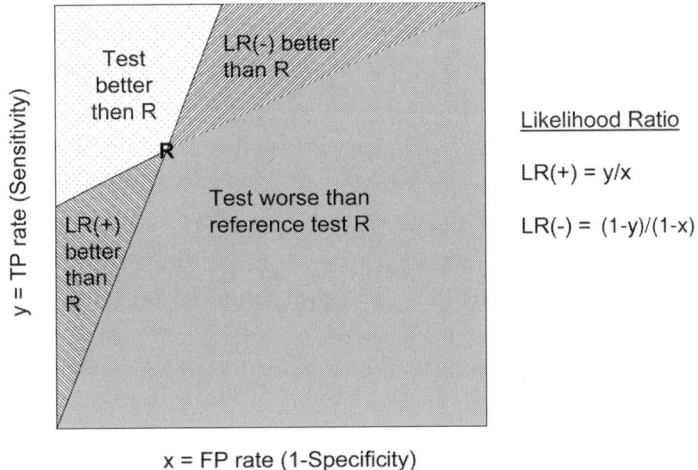

Fig. 6. Comparing diagnostic tests by graphic display of likelihood ratios.

From the above formulas, it can easily be seen that for a given cutoff in the ROC curve the LR(+) equals the slope of the line connecting the origin and the point of the ROC curve that corresponds to the cutoff, and the LR(–) is the slope of the line between that point and the upper right corner of the ROC (*see* **Fig. 3**). In this way, likelihood ratios can be used to compare the performance of different tests without knowing the whole ROC curve. **Figure 6** shows how by plotting the *TP* ratio against the *FP* ratio of a given reference test (*R*), areas are defined where the LR(+) or the LR(–) of a new test or both are better or worse than for the reference test *(5)*.

Likelihood ratios obviously contain the same information as sensitivity and specificity, but they are more convenient for calculating certain relationships and to evaluate sequential testing. According to Bayes' theorem (*see* **Appendix 1**), if a disease's prevalence (P_D) is known, likelihood ratios permit direct calculation of post-test probabilities (predictive values):

$$\text{Post-test odds} = \text{LR} \times \text{pre-test odds} \qquad (14)$$

with

$$\text{Pre-test odds} = \frac{P_D}{1 - P_D} \qquad (15)$$

and

$$\text{Post-test odds} = \frac{PPV}{1 - PPV} = \frac{TP}{FP} \text{ in the case of LR(+)} \qquad (16)$$

or

$$\text{Post-test odds} = \frac{1 - NPV}{NPV} = \frac{FN}{TN} \text{ in the case of LR(–)} \qquad (17)$$

For serial testing, the final predictive value can easily be calculated by multiplying the post-test odds of the previous test with the likelihood ratio of the following test.

LRs can also be calculated for quantitative test results. This LR(q) equals the slope of the tangent to the ROC curve at the point where the cutoff is set to the quantitative result *(6)*. The important point is that the likelihood ratio of a quantitative result with a value just above the cutoff may be much smaller then the LR(+) of a positive test result in general, as can easily be seen from **Fig. 3**. If in our example the cutoff is set to a low value, like 3.5, to reach a high sensitivity of 95% (on the cost of a low specificity of 70%), a test result between 3.5 and 4.4, although positive, would even yield a LR(q) lower than 1, therefore reducing the probability of disease. Consequently, one should avoid setting the cutoff of a test lower than the point at which the slope of the tangent becomes smaller than 1.

3.6. Causal Modeling to Estimate Sensitivity and Specificity

The following ideas are based on a paper by Choi *(7)*.

3.6.1. Diagnostic Model

In the above considerations about parameters for test performance, the underlying assumption was that there is a direct causal relation between clinical disease and test result. However, to be precise, this assumption only holds for diagnostic tests in which the disease itself causes some change of a biomarker that is measured with the test. In this model, disease prevalence (P_D) is an independent variable, whereas test prevalence (P_T) is a dependent variable. Sensitivity, specificity, and LRs are constant, whereas predictive values are dependent on the disease prevalence.

3.6.2. Correlational Model

However, for most serological tests, the diagnostic model is an oversimplification. The augmentation of a specific antibody concentration is hardly caused by the clinical disease, but rather is the effect of an infection by some micro-organism that is the cause of the disease. Clinical disease and test result correlate because of an underlying causal factor. In this correlational model (**Fig. 7**), both disease prevalence and test prevalence are dependent variables, and only the prevalence of the underlying cause is independent. Choi *(7)* has shown that in this—more realistic—model, sensitivity, specificity, and likelihood ratios are no longer constant but vary with the prevalence of the underlying causal factor and with the degree of correlation between causal factor and disease on one hand, and between causal factor and test result on the other hand:

$$Se = \frac{Se_D \cdot Se_T \cdot P + (1 - Sp_D) \cdot (1 - Sp_T) \cdot (1 - P)}{Se_D \cdot P + (1 - Sp_D) \cdot (1 - P)} \quad (18)$$

$$Sp = \frac{(1 - Se_D) \cdot (1 - Se_T) \cdot P + Sp_T \cdot Sp_D \cdot (1 - P)}{(1 - Se_D) \cdot P + Sp_D \cdot (1 - P)} \quad (19)$$

Problems of Laboratory Diagnosis

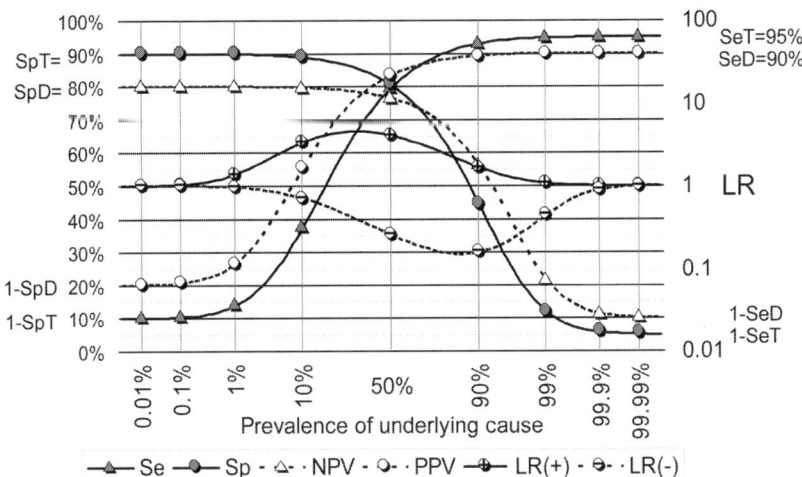

Fig. 7. Correlational model: relation between prevalence of underlying cause and test parameters. The parameters of a test to detect clinical disease is expressed as a function of the prevalence of the common underlying cause. The intercepts of the curves at prevalence = 100% indicate the sensitivity of the test-cause (Se_T) and disease-cause (Se_D) relation. The intercepts of the curves at prevalence = 0% indicate the specificity of the test-cause (Sp_T) and disease-cause (Sp_D) relation.

with

P = prevalence of underlying cause
Se_D = sensitivity of the disease relation to the underlying cause
Se_T = sensitivity of the test relation to the underlying cause
Sp_D = specificity of the disease relation to the underlying cause
Sp_T = specificity of the test relation to the underlying cause

To give an example, Lyme arthritis, caused by an infection with *Borrelia burgdorferi*, correlates with the presence of IgG specific for *B. burgdorferi*. Let us consider two hypothetical (extreme) situations:

1. Assuming *the underlying cause is present*, i.e., infection with *B. burgdorferi* is true ($P = 1$; see right side of **Fig. 7**), the sensitivity to detect disease equals the sensitivity to detect infection (formula 18 collapses to $Se = Se_T$). This is evident because, under this assumption, infection is present in all cases and, consequently, also in those with arthritis. Therefore, the proportion of detected cases among diseased cases equals the proportion of total detected cases (Se_T). However, not all infected cases have arthritis, and the PPV equals the sensitivity of the disease–cause relation (PPV = Se_D), which is the probability of having arthritis with *B. burgdorferi* infection.

   All test-negative cases are an expression of the nonsensitivity of the test (formula 19 collapses to $Sp = 1 - Se_T$). The proportion of test-negative cases among those without disease (Sp) equals the proportion of total (false) negative cases ($1 - Se_T$). Apparently true-negative cases (test-negative and disease-absent) are merely accidental (not causally related). The proportion of these "true" negative cases among all test-negative cases (NPV) equals the proportion of cases without disease despite of infection (NPV = $1 - Se_D$).

The LRs collapse to 1 because, according to the above relations:

$$LR(+) = \frac{Se}{1-Sp} = \frac{Se_T}{Se_T} = 1$$

and

$$LR(-) = \frac{1-Se}{Sp} = \frac{1-Se_T}{1-Se_T} = 1$$

(*See* also **Fig. 7**). This is in agreement with expectation, because in the presence of infection the likelihood of a test outcome is the same for diseased and not diseased cases.

2. Assuming *the underlying cause is absent*, i.e., there is no infection with B. burgdorferi ($P = 0$; see left side of **Fig. 7**), the specificity to detect disease equals the specificity to detect infection (formula 19 collapses to $Sp = Sp_T$). This is evident because, under this assumption, infection is absent in all cases and, consequently, also in those without arthritis. Therefore, the proportion of test negative cases among those without disease (Sp) equals the proportion of total test-negative cases (Sp_T). However, not all cases lack arthritis, and the NPV equals the specificity of the disease–cause relation (NPV = Sp_D), which is the probability that arthritis is not caused by something other than infection with *B. burgdorferi*.

All test-positive cases are an expression of the nonspecificity of the test (formula 18 collapses to $Se = 1 - Sp_T$). The proportion of test-positive cases among the diseased (Se) equals the proportion of total (false) positive cases ($1 - Sp_T$). Apparently true-positive cases (test-positive and arthritis-present) are merely accidental (not causally related). The proportion of these "true" positive cases among all test-positive cases (PPV) equals the proportion of cases with disease despite absence of infection (PPV = $1 - Sp_D$).

The LRs collapse to 1 because, according to the above relations:

$$LR(+) = \frac{Se}{1-Sp} = \frac{1-Sp_T}{1-Sp_T} = 1$$

and

$$LR(-) = \frac{1-Se}{Sp} = \frac{Sp_T}{Sp_T} = 1$$

(*See* also **Fig. 7**). This is in agreement with expectation, since in the absence of infection, the likelihood of a test outcome is the same for diseased and not diseased cases.

In real life, in which infection status is not known, we have a superposition of situations 1 and 2 (middle of **Fig. 7**). In the case of Lyme arthritis, this superposition will rarely be resolved, because direct evidence of infection with *B. burgdorferi* is difficult to obtain. Therefore, sensitivity and specificity to detect Lyme arthritis depend on the prevalence of infection with *B. burgdorferi*. High prevalence of the cause leads to high sensitivity and low specificity to detect disease (**Fig. 7**). Many cases without arthritis will have positive test results. Low prevalence of the cause leads to low sensitivity and high specificity to detect disease (**Fig. 7**). Many cases with arthritis will have negative test results. For low prevalence of infection, PPV and NPV are strongly influenced by the specificity

Problems of Laboratory Diagnosis

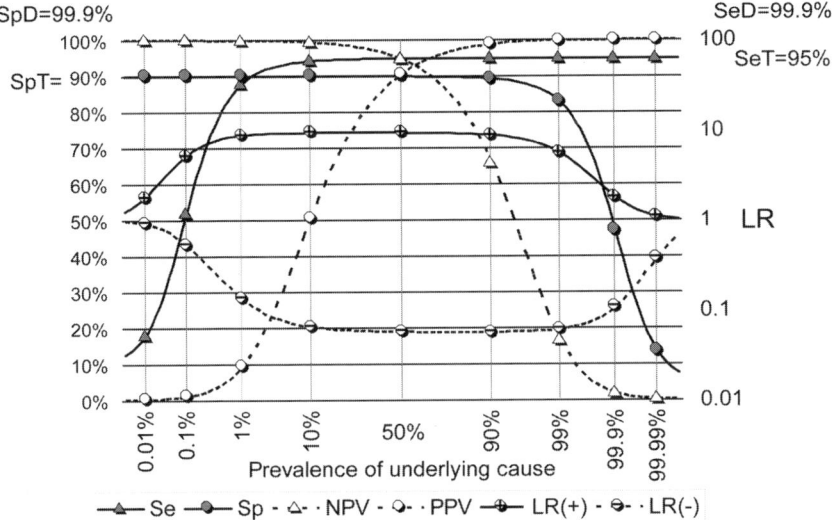

Fig. 8. Correlational model: approaching the diagnostic model. The parameters of a test to detect clinical disease are expressed as a function of the prevalence of the common underlying cause. Sensitivity (Se_D) and specificity (Sp_D) of the disease-cause relation are put at 99.9%.

(Sp_D) of the arthritis to be caused by infection with *B. burgdorferi*. For high prevalence of infection, PPV and NPV are strongly influenced by the probability (Se_D) of arthritis after infection with *B. burgdorferi*.

With sensitivity and specificity being dependent on the prevalence of the underlying cause, LRs are not constant either. As discussed above, they tend toward 1 on both ends of the prevalence scale (**Fig. 7**), and they reach a maximum at lower prevalence ranges (for positive test results) or a minimum at higher prevalence ranges (for negative test results).

Some special cases can be considered:

1. If infection would always lead to disease (as is the case, e.g., with yellow fever; $Se_D = 1$), the specificity to detect disease equals the specificity to detect infection (formula 19 collapses to $Sp = Sp_T$), independent of the prevalence of the cause. LRs remain different from 1 up to very high prevalence.
2. If there is no other cause for disease (as is the case, e.g., with a typical erythema migrans; $Sp_D = 1$), the sensitivity to detect disease equals the sensitivity to detect infection (formula 18 collapses to $Se = Se_T$), independent of the prevalence of the cause. LRs stay different from 1 down to very low prevalence.
3. If both Se_D and Sp_D would be equal to 1, i.e., all clinically diagnosed Lyme arthritis cases would be surely caused by *B. burgdorferi* and all infections by *B. burgdorferi* would cause Lyme arthritis ($P = P_D$), the *diagnostic model* applies. Sensitivity, specificity, and LRs are constant (independent of prevalence of cause/disease), but the predictive values depend on the prevalence of the cause/disease. *However, even small deviations of* Se_D *and* Sp_D *from 1 make the classical diagnostic model inappropriate.* As shown in **Fig. 8**, with $Se_D = 99.9\%$ and $Sp_D = 99.9\%$, sensitivity and specificity start to deviate substan-

tially from the classical model when prevalence is lower than 10% or higher than 90%, respectively.
4. If we are dealing with a highly cause-specific test (as, e.g., with bacterial culture; $Sp_T = 1$), the PPV equals the sensitivity of the disease–cause relation (PPV = Se_D), independent of the prevalence of the cause.
5. If we are dealing with a highly cause-sensitive test (as, e.g., with an HIV screening test; $Se_T = 1$), the NPV equals the specificity of the disease–cause relation (NPV = Sp_D), independent of the prevalence of the cause.
6. If both Se_T and Sp_T would be equal to 1, i.e., all test-positive cases would represent infection and all infections would produce a positive test ($P = P_T$), the *prognostic model* applies (see below), and the predictive values are constant (independent on prevalence of cause, i.e., biomarker-tested), but sensitivity and specificity depend on prevalence of cause (*see* **Subheading 3.6.2.**).

One might rightly argue that, in the case of Lyme arthritis, the test is not used to detect disease but rather to detect infection. However, when sensitivities and specificities of serological tests for *B. burgdorferi* are determined, the case definition for Lyme arthritis is usually based purely on clinical grounds, and direct evidence for infection is rarely available. So the prevalence of infection will influence the outcome of such establishments of the test characteristics, according to the correlational model. Such conditions often prevail also with other infectious diseases, for which serology is the primary diagnostic tool because direct detection of the infectious agent is difficult to obtain.

3.6.3. Prognostic Model

If a biomarker is known to cause a clinical disease directly, a test that measures such a biomarker can predict the later occurrence of the disease. [Contrary to Choi (7), here this assumption is called "prognostic" instead of "predictive" to avoid confusion with the term "predictive value."] In this model disease prevalence is a dependent variable, whereas test prevalence is an independent variable. In contrast to the diagnostic model, predictive values are constant (PPV = Se_D, NPV = Sp_D), whereas sensitivity, specificity, and likelihood ratios are dependent on the prevalence of the biomarker tested:

$$Se = \frac{PPV \cdot P_T}{PPV \cdot P_T + (1-NPV) \cdot (1-P_T)}$$

$$Sp = \frac{P_T + NPV \cdot (1-P_T)}{(1-PPV) \cdot P_T + NPV \cdot (1-P_T)}$$

with P_T = prevalence of biomarker tested.

The equations are analogous to those describing the positive and negative predictive values in the diagnostic model, with the following exchange of parameters:

$$Se \leftrightarrow PPV, SP \leftrightarrow NPV, \text{ and } P_D \leftrightarrow P_T$$

To give an example, in HIV-infected patients, the risk of developing AIDS is directly related to the level of helper (CD4)lymphocytes and to the viral load as biomarkers. In a population with a high prevalence of low CD4 cell numbers or high viral load, i.e., in *untreated HIV patients*, the prognostic sensitivity (Se) of these tests is high, i.e., most patients who develop AIDS have low CD4 cell numbers or a high viral load. However, the sensitivity strongly depends on the NPV, i.e., on the specificity (Sp_D) to develop AIDS under this condition. When many patients without high viral load are still developing AIDS, the test is obviously less sensitive. Conversely, the prognostic specificity (Sp) is low, i.e., many patients not developing AIDS have a high viral load. In addition, the specificity depends on the PPV, i.e., on the sensitivity (Se_D) to develop AIDS under this condition. When many patients with a high viral load are not developing AIDS, the test is even less specific.

On the other hand, in *highly active antiretroviral therapy (HAART)-treated HIV patients* in whom the prevalence of low CD4 cell numbers or high viral load is low, the prognostic sensitivity (Se) of these tests is low, i.e., many patients will develop AIDS without having low CD4 cell numbers or a high viral load. In addition, the sensitivity depends on the NPV, i.e., on the specificity (Sp_D) to develop AIDS under this condition. When many patients without high viral load are still developing AIDS, the test is even less sensitive. Conversely, the prognostic specificity (Sp) is high, i.e., most patients not developing AIDS do not have a high viral load. However, the specificity strongly depends on the PPV, i.e., on the sensitivity (Se_D) to develop AIDS under this condition. When many patients with a high viral load are not developing AIDS, the test is obviously less specific.

3.7. Practical Establishment of Parameters for Test Performance
3.7.1. Classification of Disease

A further ambiguity in establishing parameters for the validity of a test ensues from the binary classification of disease when the underlying trait for classification is not dichotomous but continuous (*8*). In this case, sensitivity (Se_D) and specificity (Sp_D) of the disease–cause relation is dependent on the cutoff within the continuous trait (similar to a test cutoff) and on the classification errors for cases near the cutoff.

3.7.2. Which Population to Study?

From all the above considerations about parameters of test performance, it can be concluded that the validity of a test can typically only be established for a defined population. If we had to evaluate an IgG antibody test for rubella, could we expect results on a population with a past rubella infection comparable to those on a population of vaccinated individuals? Should we select our negative population from blood donors, or should we rather study hospital patients or perhaps drug addicts? The test sensitivity and specificity established with a panel of healthy blood donor sera versus a panel of sera from clinically well-established disorders will not be equal to the parameters established on a so-called challenge panel (*1*) including sera from patients with autoimmune disease, Epstein–Barr virus (EBV) infection, rheumatoid factors, and others. This question of the right study population cannot be answered in a general fashion, because the requirements might be different for each test and each application of a test. However, it is very important that the user of a laboratory test have some knowledge about the selection of the population used to characterize the test. Ideally, the test should have been evaluated on individuals comparable to the actual patient *(9)*. Because this perfect goal will never be reached, the user should have at least an idea how far away he or she is from this aim.

3.8. Crossreactivities and Interferences

Most of the problems related to the validity of a test are of a clinical nature and are not only based on test characteristics. True crossreactivities, e.g., cannot always be eliminated by improving the quality of the test reagents, since they are based on *shared antigenic determinants* between different entities. This is in contrast to reactions toward impurities of the antigen preparation, which are not crossreactivities. However, the presence of shared antigenic determinants in a particular patient is not always predictable and depends on concomitant diseases and environmental influences.

Another generic problem is the commonly assumed association between the presence of antibodies to a certain infectious agent and the presence of the agent in the organism. This assumption is not always valid. B-lymphocytes can also be stimulated by other means than the specific antigen, as it is the case in infections with EBV or *B. burgdorferi*, or in certain autoimmune diseases. Such a *polyclonal B-cell activation* can lead to false-positive test results, even with a perfectly designed test. IgG antibodies to a particular infectious agent might still be present years after the resolution of an infection, a so-called *sero-scar*, or they might be the result of *vaccination*. In newborns, IgG antibodies might well be produced by the mother and transferred to the fetus via the placental barrier and not be a sign of infection of the baby.

Conversely, the presence of antibodies to a certain agent might be concealed by *sequestration* of the antibodies within the organism, e.g., as *immune complexes*. The latter problem can partly be solved by dissociating potential immune complexes before the test is performed. However, this technique has to be evaluated for each particular test.

Rheumatoid factors or substances other than antibodies (such as complement, hemoglobin, bilirubin, or lipids) might interfere with some assays. The effects of these commonly occurring interfering substances are usually well evaluated, and certain measures are taken to minimize their effect. However, the influences of more rarely occurring agents like certain drugs are less often studied.

4. Antigen

Antigens are used as immunogens either to produce antibodies during vaccinations or to produce antibodies as analytical tools (usually in non-human species) for the detection of similar antigens in patients' specimens. Antigens are also used as reagents to detect the presence of antibodies in such specimens. In all cases, the selection of antigens affects the performance of the immunological tests. The criteria and methods for preparation, purification, and testing of these antigens are crucial *(1)*.

4.1. Antigen Definition

Antigens have originally often been defined as *functional entities* without knowledge of their biochemical nature. This is true for many viral antigens, such as the

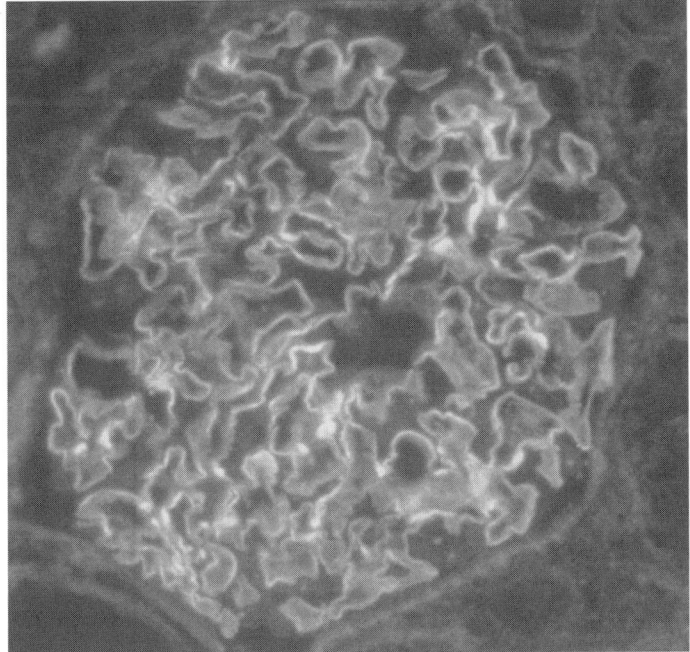

Fig. 9. Immunofluorescent detection of antibodies to the kidney glomerular basal membrane. The morphological information gained from characteristic structures in the microarchitecture of an immunofluorescent kidney specimen helps to identify the antigen (glomerular basal membrane) that is recognized by the patient's antibodies.

"Australia" antigen (hepatitis B virus), or for the histocompatibility antigens and for many others. Tests for antibodies to functionally defined antigens are consequently also of a functional nature if the patient's serum is tested with preparations that are functionally defined as antigenic positive without knowing the antigenic molecules involved. The critical point in establishing detection systems for antibodies to such functional antigens is to use the right control preparation that does not contain the antigen but is otherwise identical to the preparation containing antigen. If we look, e.g., at an in vitro viral culture system to define the antigen, we will compare antibodies that bind to cells or extracts of cells that have been infected with the virus versus noninfected cells. In such a test system, however, we cannot be sure whether the differential antibody binding is actually due to viral antigens or whether cellular antigens not expressed in the noninfected control cells (e.g., activation antigens, interferons, and others) are detected. To confirm the specificity of the detected antibodies, a competitive assay might be used, in which the binding of critical (human) antibody is inhibited by a competing (mouse) monoclonal antibody to the functional antigen, if available. However, it will not always be possible to solve this problem definitively. In histocompatibility testing, a similar difficulty is brought up by the tight linkage of

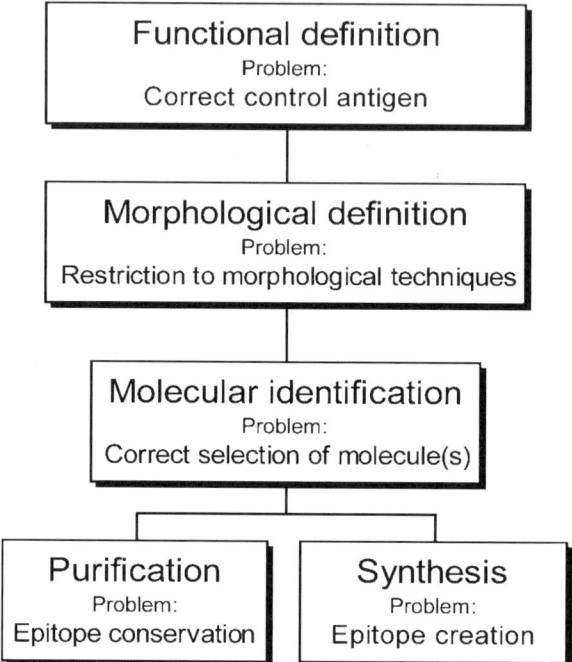

Fig. 10. Antigen: problems for antibody detection systems. For explanation, see text.

alleles of different loci, which might confound the identification of the antigen by functional means.

In addition to the functional definition of antigens, *morphological criteria* might be used to characterize an antigen. In cell culture systems it might be advantageous to use immunofluorescence microscopy as an antibody detection system in which the morphological information about the antigen distribution within the cell sometimes helps to distinguish antibodies binding to characteristic viral structures from antibodies binding to cellular antigens. The morphological information is also used to identify autoantibodies to certain glomerular antigens such as the basal membrane (**Fig. 9**). In enzyme immunoassay (EIA) tests this morphological information cannot be exploited.

To circumvent the problems of functional definition of antigen, one can follow two possible strategies (**Fig. 10**). One way would be to purify the natural antigen chemically. Alternatively, one might choose to use synthetic antigens. However, both tactics invoke their own quandaries.

4.2. Molecular Identification and Purification or Synthesis of Antigens

In either the molecular purification or synthesis of antigens, the crucial question will be which molecules are representative for the functional or morphological antigenic entity. Because a single molecule is rarely involved alone, it might be advantageous to purify a whole spectrum of possible antigenic molecules. One very convenient and widely used technique for this purpose is the *Western blot* analysis, which involves the electrophoretic separation of proteins and their consequent usage as antigens in an immunoblot procedure. Selecting the right protein bands for the diagnostic detection of a certain entity is, however, not always a simple procedure. In HIV antibody testing, various rules have been in use, which define the spectrum of protein bands that have to be recognized to establish the diagnosis of HIV infection. In the case of Lyme disease diagnostics, the debate continues about the protein bands required to characterize a specific antibody response to *B. burgdorferi*.

Whereas testing with multiple antigenic molecules is frequently used in confirmatory antibody tests, this procedure is usually too laborious and expensive for screening tests. For this purpose EIA tests are commonly used, which ideally contain more than one antigenic molecule. One way to make up such antigenic mixtures is to use whole lysates or extracts of the naturally occurring entity. In these preparations, the relative concentrations of various antigenic molecules are usually preserved, which can be advantageous or disadvantageous. One danger is that antibodies against weakly represented molecules might remain undetected. Another pitfall might be that nonspecific antigenic molecules that lead to crossreactivities are copurified in this procedure. This is the case, e.g., in preparations of *B. burgdorferi* in which flagellin as a major antigenic component is responsible for many crossreactivities of the test. Another way to prepare antigens for screening tests is to tailor-make mixtures of defined molecules that have either been purified or synthesized. Such concoctions have the advantage of being adjustable to the needs of the test, but their manufacturing will usually be more expensive.

4.3. Characteristics of Natural or Synthetic Antigens: The Qualification of Epitopes

Any purification of molecules has the inherent danger of denaturing them to a certain extent, and, consequently, of destroying certain epitopes. On the other hand, the synthetic approach carries the uncertainty of whether the crucial antigenic epitopes will be present in the product. Thus, the nature of the epitope is crucial for predicting its behavior, after purification or synthesis, and, consequently, for selecting between the two strategies.

However, in most situations the physicochemical nature of the epitope will not be known. Protein epitopes might involve linear stretches of amino acids or secondary and tertiary structures, so-called conformational epitopes, or even quaternary structures, which are only expressed in protein molecule assemblies such as viral capsids or envelopes *(10)*. However, small peptides in solution can also exhibit conformational structures that influence the binding of antibodies *(11,12)*. *Post-translational modifications* might affect the nature of a protein epitope. An example from the field of autoimmunity is filaggrin, an antigen recognized by antibodies of patients with rheumatoid arthritis. The crucial epitopes in this molecule are citrullin-containing peptides. Citrullin is the result of a post-translational modification of arginin residues. The corollary of this is that recombinant filaggrin cannot be used for diagnostic tests, because it does not contain citrullin, contrary to the natural protein *(13)*.

So-called *hidden or cryptic epitopes* might only be accessible or created by certain physiological or artificial conformational changes of the protein. This is the case, e.g., with the conformational epitopes of the glomerular basement membrane (**Fig. 9**) that are targeted by the autoantibodies in Goodpasture's disease. The two major epitopes are sequestered within the quaternary organization of collagen IV *(14)*. Other epitopes might be of a nucleic acid, carbohydrate, or lipid nature or might contain metal ions. It is also conceivable that parts of the antigenic molecule not directly involved in recognition by the antibody have an indirect influence by modifying the epitope via *allosteric effects*.

To further the complexity, as a consequence of the polyclonal nature of an antibody response, several epitopes might be involved in an antigen–antibody interaction, and because of the genetic diversity of the patients and their individual immunological history, not all individual antibody responses to a given antigen will involve the same epitopes. The way out of this dilemma can only be pragmatic by trying to correlate the serological results produced using various antigen preparations with the clinical condition and by selecting the antigenic preparation with the best capacity to answer the clinical question.

In infectious diseases it is often important to know whether immunoassays detect antibodies to neutralizing epitopes, because the clinician and the patient want to have an indication of whether the measured antibodies are protective. Two different antibodies recognizing different epitopes on an antigen can have completely different efficacies for protection from an infectious organism *(15)*. In autoimmune disorders it is of interest whether the interaction of antibodies with an epitope in vivo has pathogenic consequences *(16)* or whether the amount of antibodies to that antigen correlates with disease activity (*see* **Subheading 5.5.**).

4.4. Serotypes and Mutations

Many microorganisms express antigens that display a considerable degree of variation, leading to various serotypes. In the case of *B. burgdorferi*, e.g., the specific outer surface proteins are coded for by plasmids and show a substantial degree of heterogeneity. Conversely, they might not be expressed at all because of mutations or loss of the particular plasmid *(17)*. If no common antigen is known between such serotypes, it is important that the same serotypes be present in the test system, which are prevalent in the population to be tested. Because the extent of serotypic variation is not always known, this problem is a potential cause of false-negative results. Other events that lead to the same problem are spontaneous mutations that particularly accumulate under the selective pressure of an immune response, leading to so-called *escape variants* of a microorganism.

4.5. Purity of Antigens

An important issue in immunochemical procedures is the purity of the antigen. However, it is not primarily the quantity of impurities that leads to bad performance of a test, but rather the quality of possible contaminants. Specific minor components might be recognized more often by the patients' immune system than certain gross contaminants that rarely evoke an antibody response. However, the extent to which a

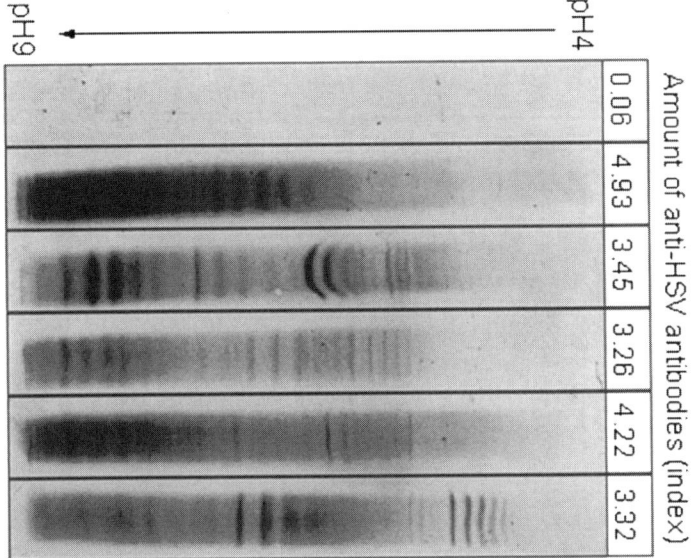

Fig. 11. Different individual immune responses to herpes simplex virus (HSV). Six patient sera with different levels of anti-HSV antibodies (indicated as test index) were separated by isoelectric focusing (IEF) on a pH gradient and blotted on a nylon membrane covered with HSV antigen. Bands of antibodies that recognize HSV are identified by immunostaining for human IgG. Individual patient's immune responses produce different oligoclonal IgG to the same viral antigen.

specific impurity is recognized by the antibodies of a particular patient is not known. Therefore, it is always of advantage to maximize the purity of an antigen preparation, as long as the quality of the epitopes is not compromised. This problem has led in recent years to the prolific use of recombinant antigens or synthetic peptides as antigens. Unfortunately, the label recombinant is often taken as a substitute for purity, which can be entirely misleading because the expression system itself produces abundant contaminants. The advantage of using recombinant antigens is the better definition of the system, i.e., contaminants are better known, and procedures to eliminate or counteract them are easier to standardize. In any case, the last resort for tackling impurities is to recommend absorption of the patient's serum with the particular impurity.

5. Antibodies
5.1. The Polyclonal Nature of Antibody Responses

The complex interaction of antibodies and antigens is not fully understood. The colloquial wording "antibodies are *seeing* antigens" symbolizes a holistic aspect of this interaction. In fact, similar to the visual perception of a picture, various aspects of an antigen are recognized simultaneously. In contrast to the optical recognition of an object, the whole "picture" of an antigen or the epitopes recognized by an individual immune system remain enigmatic. Since antigens are processed in many ways before they are presented to the immune system, the "picture" seen by the immune system differs considerably from our perception of the whole, unprocessed agent. To make it more complex, the selection of the epitopes that the individual immune system will recognize is biased by its previous experience, i.e., antibodies mounted by one individual might "focus" on certain epitopes that another person's antibodies will hardly "see." Finally, the immune system's "picture" of an antigen is evolving over time, i.e., the perception of the antigen by the mature immune response may differ notably from its first "impression." The challenging task of serology is to determine whether the antibodies have "recognized" the agent and to measure the immune system's "impression" of the agent. An example of the diversity of different individual immune responses to a specific antigen is given in **Fig. 11**.

The selection of the relevant "clues" in a picture, to which antibodies from different individuals will similarly and specifically react, is the main problem in the construction of a serological test. The next task is to standardize these "clues" so that the "impressions" they evoke in terms of antibody response can be quantitatively compared (*see* **Subheading 6.**).

5.2. Avidity and Affinities of Antibodies

The binding force of an antibody to an antigen is called avidity. It can be defined as "the net affinity of all binding sites of all antibodies in the antiserum, under specified physicochemical reaction conditions" *(1)*. As implied in this definition, the avidity takes into account the polyclonal nature of an antibody response and the involvement of more than one epitope in antigen recognition. "A measure of the attraction, or force of association, between a single antigenic site and a single antibody to that site" is the affinity *(1)*. The affinity is expressed as the equilibrium constant for the antibody/epitope interaction. Thus, the avidity can be expressed as an average or mean of affinity constants, but other, arbitrary measures are more usual, since the individual affinity constants are rarely known.

During the first encounter of an organism with an antigen, the avidities of the antibodies usually increase over time. This phenomenon, which is termed affinity maturation, extends over a period of weeks to months. This view, which is well accepted, is mainly based on experiments with small haptens used as antigens, and it has its explanatory foundations in the molecular events of somatic mutation and selection during B-cell responses. However, in recent experiments in which antibodies to vesicular stomatitis virus (VSV) in mice have been measured, high-affinity neutralizing antibodies have already been observed at 6 d after infection (the earliest time point examined) *(18)*. This and other reports indicate that the exact timing of affinity maturation can be highly variable and has to be evaluated for each particular test system *(19)*.

These questions are relevant to potential clinical applications for the measurement of affinity maturation. Ideally, one would like to utilize antibody avidity to estimate the time-point of infection in cases in which IgM measurements do not give sufficient information. This applies, e.g., to toxoplasma infections during pregnancies. IgM antibodies against toxoplasma can remain for years and therefore do not allow the clinician to determine whether a positive IgM is caused by infection before or during pregnancy. However, it remains to be seen whether the natural variation in affinity maturation is not too high for the application of avidity measurements to be useful as diagnostic tools.

5.3. Measurement of Binding Activity: Quantification of Antibodies

In contrast to other serum proteins whose concentration is expressed in physical units (g/L), the quantity of antibodies is usually related to the binding activity toward a specific antigen, which is indicated in units per liter (U/L). This binding activity is not a physicochemically defined unit, but rather a characteristic of the particular test system. Therefore, an antibody's "concentration," expressed in arbitrary units per liter, cannot be compared with a quantity measured with a different method. To alleviate this problem, one tries to establish international units (IU/L), which should render the results of different test systems comparable. The main reasons for this inherent dilemma of "units" are evident from the properties of antibodies as mentioned above, which can be summarized by the fact that antibodies are not homogeneous but a mixture of molecules with different specificities, affinities, and other chemical properties (**Fig. 11**).

5.4. Nonlinear Dilution Effects and Prozone Phenomena

A generic problem in serological testing, particularly in quantifying antibodies, is the possibility of nonlinear effects when antibody-containing sera are diluted. The causes of such effects are numerous, but again the heterogeneity of antibody responses is mainly involved. Immunofluorescence analysis is often very illustrative for such effects. In testing for antinuclear antibodies, e.g., a particular prominent fluorescence pattern might dilute out at a low titer, and a different pattern might appear and remain in much higher dilutions. Of course, such phenomena also occur with other techniques, like EIA, but here the problem is not recognized as easily. A more serious problem can turn up when an antibody binding to an antigen induces a conformational change in the antigen *(20)*, and thus, via an allosteric effect, a different epitope on the same antigen gets exposed or hidden, which consequently affects the binding of other antibodies. Such interactions contribute to nonlinear dilution effects. Nonlinear effects that are produced by suboptimal antigen-antibody reaction when the antibody is in overwhelming excess are termed prozone phenomena.

5.5. Biological Activities of Antibodies

For the clinician the quality of antibodies, i.e., their biological activity, is as important as their quantitative levels. Depending on the clinical situation, it is of interest to know whether the antibodies measured are protective, pathogenic, or regulatory. However, this information is rarely available with currently available diagnostic tests. Functional tests that measure biological activities, such as neutralization, blocking, or induction of functions, are usually very elaborate and expensive. To circumvent this problem, immunologists try to evaluate certain qualities of antibodies that correlate with biological activity. In addition to epitope specificity, which has been discussed in

Subheading 4.3., complement binding activity and the Ig isotype (M, G, A, and E), but also more subtle characteristics like the IgG subclass *(21)* and the glycosylation pattern *(22,23)*, can correlate with functional activity. These latter properties are, however, not yet exploited in commercial diagnostic tests. The elucidation of biological structure–function correlations in antibodies is currently a topic of intense research in the field of structural immunology *(24)*. One example is the study of intramolecular allosteric conformational changes in antibodies on complexation with antigen *(25)*.

6. Standardization and Quality Control

From what has been discussed above, it is obvious that standardization of serological tests is a very difficult task. One should obviously not expect a high degree of accuracy from standardizing measurements of poorly defined, nonhomogeneous mixtures of polyclonal antibodies. Clearly documenting the method and the standard reagents used to test and compare a particular serum is the best we can do. In this way, two different sera from the same patient can reasonably be evaluated by comparing the standard units obtained with the identical procedure, with the caveat of possible nonlinear dilution effects and affinity maturation over time. It is, on the other hand, dangerous to assume that the same serum tested with a different method against the same standard would yield the same result. Also, results of testing sera from different patients with the same method against the same standard cannot strictly be compared. What should definitively not be done is to compare the result of a test performed in laboratory L with method M that has been gauged with the manufacturer's standard A, with the result of a test performed in laboratory K with method N that has been gauged with the manufacturer's standard B, even when both results are expressed in International Units (**Fig. 12**).

6.1. Comparability of Different Test Principles

The difficulty in standardizing tests based on different techniques can perhaps best be illustrated by giving an example: In an attempt to standardize measurements for IgG against parvovirus B19, three different EIA tests were applied to several patients' sera (**Figs. 13–15**). Test 1 used recombinant capsid protein VP1 expressed in *E. coli* as antigen. The results are expressed in arbitrary units/mL. Tests 2 and 3 were both based on recombinant capsid protein VP2 expressed in a eukaryotic system but produced by different manufacturers. The results are expressed as indices. Since at the time no international units, which could be used as a common scale, were yet defined for anti-B19 IgG, the three tests are compared on an arbitrary scale, as shown in **Figs. 13–15**.

The results of a time-course of three different patients (A, B, and C) clearly demonstrate that the relative magnitude of IgG against B19 over a certain time period measured with the three different tests is very different in each patient. It seems that each patient has his or her own individual pattern of reactivity among the three different tests. Even tests 2 and 3, which are based on the same type of recombinant antigen, show a completely different behavior in patients A, B, and C. Furthermore, as shown in **Figs. 14** and **15**, even within one patient the pattern of relative magnitudes of IgG against B19 measured by the three different tests shifts over time.

Problems of Laboratory Diagnosis

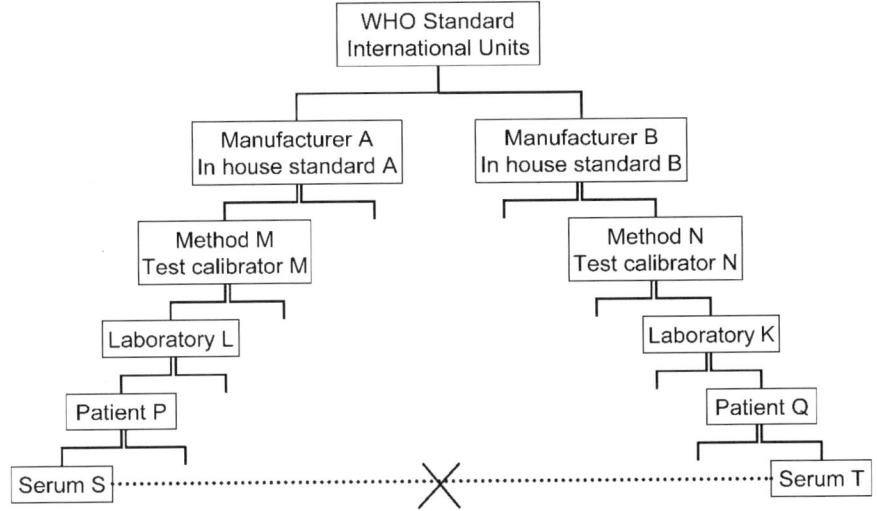

Fig. 12. Hierarchy of standards. For explanation, see text.

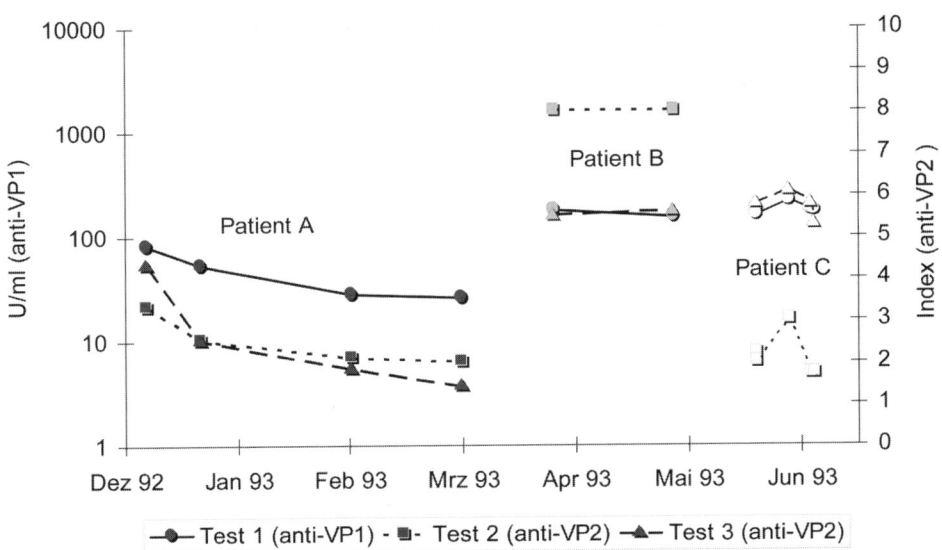

Fig. 13. Comparison of three different tests for IgG against parvovirus B19 in patients A, B, and C. For explanation, see text.

In the time-courses of IgG against B19 from patients D and E, tests 1 and 3 show lanes that are fairly parallel to each other, whereas test 2 shows a different course of IgG, particularly in patient E. Similarly, the relative time-courses as depicted in **Fig. 15** seem not to be in parallel among the three tests in patients F and G.

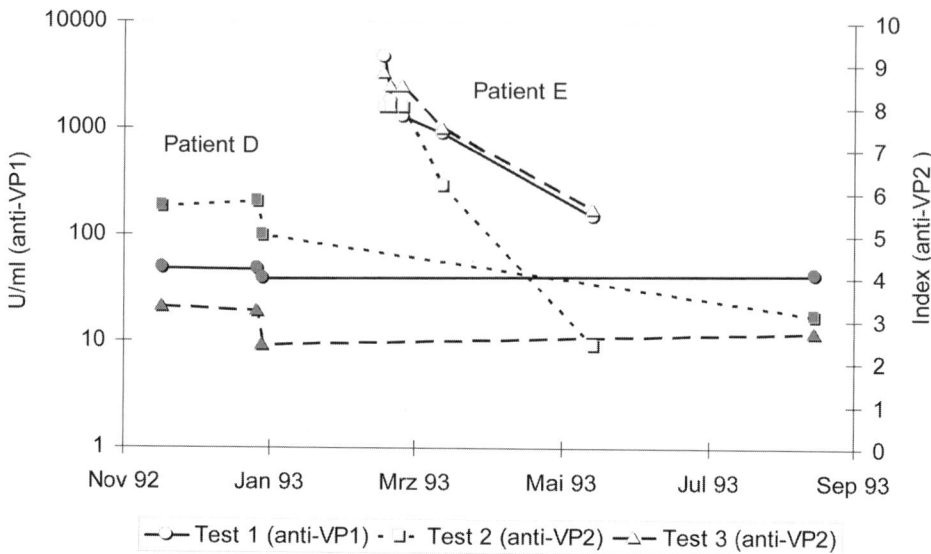

Fig. 14. Comparison of three different tests for IgG against parvovirus B19 in patients D and E. For explanation, see text.

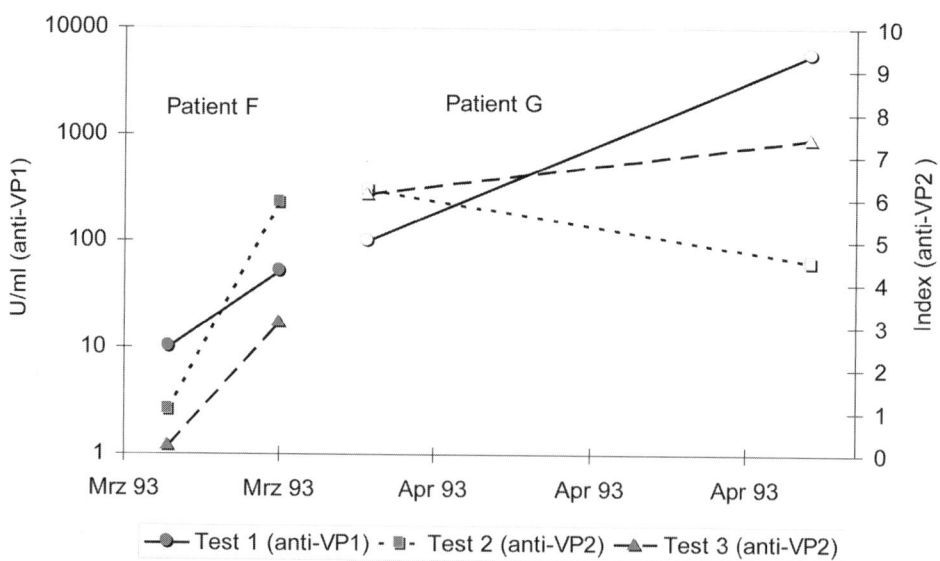

Fig. 15. Comparison of three different tests for IgG against parvovirus B19 in patients F and G. For explanation, see text.

This discrepant behavior of the three tests is difficult to explain. Such factors as different responsiveness of the tests toward antibodies with different avidities and epitope specificities certainly are to be considered. It has recently been described that IgG responses to native or denatured VP2 differ drastically years after infection with parvovirus B19, whereas they are of similar magnitude in the first few month after onset of symptoms *(26,27)*. Whatever the reason, the observation clearly demonstrates the limitations of any attempts to standardize serological tests.

6.2. Comparability Among Different Laboratories: External Quality Control

With an increasing demand for mandatory quality control in the clinical laboratory, the experience with interlaboratory comparisons of diagnostic tests has increased substantially. Proficiency panels are routinely performed in many laboratories as a form of external quality assessment. The emphasis in such endeavors should, however, be less on pinpointing laboratories whose performance is not satisfactory but rather on identifying methodological factors that critically influence test performance and cause systematic imprecisions and inaccuracies. In this sense it is of great value when not only clinical laboratories, but also manufacturers take part in such studies *(1)*.

6.3. Comparability Over Time: Internal Quality Control

In many applications of serological laboratory tests, the desired result can only be reached by comparing at least two sera taken at different time points. This approach is, however, only valid when, from a technical point of view, the results can truly be compared over time. In principle, two methods are possible to perform such comparisons: (1) the first serum is stored and tested later together with the second serum in the same assay; this procedure avoids interassay variation but carries the risk of storage artifacts; and (2) the first serum is tested immediately and the result is later compared with the result of the second serum. There, of course, the interassay variability comes into play and comparability crucially depends on the reliability of the test calibrators, particularly when the two assays are done with different lots. The choice between the two procedures depends on the assay system used and the possibilities for sample storage. In both cases, however, changes in the properties of the antibodies themselves might make a comparison doubtful. The uncertainties are given by possible changes in avidities or epitope specificities of the antibodies, which might confound the results of a particular assay system.

Comparability over time is also a major issue of internal quality control. As in clinical chemistry, internal quality control samples should also be used regularly in diagnostic serology *(1,28,29)*, and the results should be analyzed not only to reject erroneous results *(30)* but also to identify the sources of irregularities. Furthermore, repeated testing of internal control samples will yield the actual coefficient of variation of a test. This is an important factor to determine the significance of quantitative changes of test results from clinical samples over time. Such data are fundamental for the interpretation of test results, but they are often not available or neglected.

6.4. Requirements for International Standards of Antibody Content

Considering the problems discussed above in interassay comparability, the attempt to establish international standards for antibody contents, in the form of reference sera, has to be met with a certain skepticism. Nevertheless, such reference sera, used with the necessary caution, are better than no standards at all, but certain minimal requirements should be observed when they are used: the population from which the reference serum pool was collected should be indicated (blood donors, vaccinated persons, recovering patients). The nature of the antigen used to define the international units should be described as precisely as possible. An indication about the avidity of the antibody standard would be helpful. Data about correlation of the international units with certain clinical states (disease or vaccination) would be desirable. These requirements, of course, should be fulfilled even more when secondary, national, or manufacturer's standards are derived. With such precautions, confusions about comparability of different tests *(31)* could be avoided.

6.5. Automation and the Human Factor

Apart from all the problems of manufacturing and evaluating standardized test systems, a further irregularity comes into play when the test kits are actually applied in the clinical laboratory. Like all human beings, laboratory technicians do make mistakes. Therefore, the overall performance of a test might in fact improve more when the handling of the test is optimized rather than by perfecting the biochemical test parameters. This, apart from other factors, led to increased automation in the clinical laboratory. It has, however, been forgotten that the human factor also has positive aspects: the controlling eyes, ears, and noses of experienced technicians are unsurpassed monitors for detecting irregularities during test performance, and human common sense, with its outstanding ability to check plausibilities, is not easily replaced by machines. It certainly would be prudent to take advantage of both automation and the positive sides of the human factor. To this end, endeavors to improve the interface between robots and humans are urgently needed to reach collaboration rather than animosity between the two.

7. Interpretation

7.1. Unité de Doctrine: Standardization of Interpretation

As already discussed in **Subheading 2.** the interpretation of serological laboratory results is an interactive process between the clinician and the clinical laboratory *(1)*. Attempts should be made to base any interpretation on hard data, i.e., predictive values based on the prevalence data of the population tested should be known or at least estimated. Borderline results, i.e., values around the cutoff point, demand especially careful evaluation. Firm criteria have to be established for further testing when borderline results are produced: repeat assays of the same sample with the same or a different method, assay of a second sample, or assay of a sample obtained after several weeks or months.

Although it will be difficult to standardize completely interpretation of serological laboratory results, it is important that one at least documents the way an interpretation

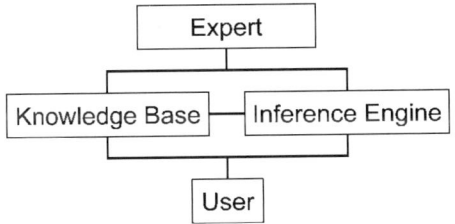

Fig. 16. Expert systems. For explanation, see text.

is reached. This is generally a three-step procedure. First, the manufacturer of a test should provide interpretive guidelines based on published studies. Second, these guidelines should be adapted and expanded by the clinical laboratory in relation to the population being tested. Third, the clinician should adjust the interpretation in the light of clinical information.

7.2. Artificial Intelligence for Decision Support: Expert Systems and Neural Networks

The demand for data-based interpretation of clinical laboratory results can realistically only be fulfilled on a larger scale when computer-supported medical decision making (CMD) systems are employed *(32–34)*. Two major types of such artificial intelligence strategies have been applied in laboratory medicine: expert systems *(35)* and neural networks *(36)*. These two approaches are based on fundamentally different conceptions.

Expert systems are rule-based and can be defined as computer programs that give advice in a well-defined area of expertise and can explain their reasoning. Its major elements are the knowledge base and the inference engine (**Fig. 16**). Both components have to be explicitly constructed by experts. Adaptations to data of the local laboratory are not easily performed, except when an adaptive element is specially provided by the program.

Artificial neural networks (**Fig. 17**), on the other hand, are computer programs that learn by experience and can extract generalized information but that cannot explain their reasoning. Such algorithms do not need the knowledge of the expert. In contrast, it is impossible to integrate directly the knowledge of an expert into the system. However, neural networks are quite suitable to accumulate knowledge from data of the local laboratory and can adapt to changes in the population tested. It is difficult, however, if not impossible to examine the acquired knowledge representation of neural networks. Nevertheless, their performance can be quite spectacular. They are particularly suitable for handling fuzzy data, nonlinear relations, and pattern recognition.

The diverging but complementary properties of expert systems and neural networks make it foreseeable that a combined application of both types of artificial intelligence will take the best advantage of computer assistance in medical decision making. The field of clinical laboratory diagnosis, because of its already huge computerized

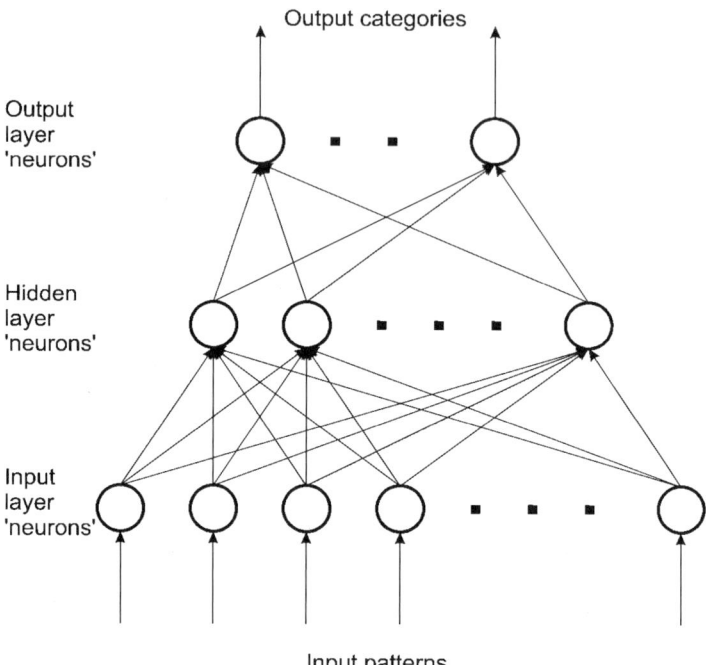

Fig. 17. Neural networks. The principal architecture of a back-propagation neural network program consists of an input layer, an output layer, and at least one hidden layer of "neurons." The processing elements ("neurons") of each layer are fully connected with the elements of the succeeding layer by "synaptic" junctions of variable strengths (weights). For training of the network, various input patterns (laboratory test results) are propagated through the network, and the resulting output pattern (disease categories) are compared with reality. Differences to the desired output are used to adjust the synaptic strengths of individual connections by a process called back-propagation of error. Once the network has minimized the output error, the network can be used to classify new examples of test patterns that have not occurred in the training set.

database, is probably well qualified to take a leading role in the imminent establishment of artificial intelligence in medicine.

7.3. Infoglut and Adaptive Information Systems

The last step in finding the right clinical decision based on diagnostic test data is the connection of individual patient data with the global medical knowledge. This part, which is the main task of the clinician, gets increasingly complex the more diagnostic data are available and the more medical knowledge in the form of consensus-based guidelines and evidenced-based medical information is globally accessible. The challenge of helping the clinician with computer systems is to adjust the search for information to the needs of the patient. Such adaptive provision of information gets even more important when the pharmaceutical industry is providing medicaments that are adjusted to individual types of patients. The aim is to find the right information for the right patient to treat him or her with the right drug *(37)*.

Appendix 1

The classical form of Bayes' theorem is:

$$P(H|D) = P(D|H) \times P(H)/P(D)$$

where the probability (P) of the hypothesis (H), given the data (D), is equal to the probability of the data, given that the hypothesis is correct, multiplied by the probability of the hypothesis before obtaining the data divided by the average probability of the data *(38)*.

By dividing both sides of the equations with their complement to 1, we get

$$\frac{P(H|D)}{1-P(H|D)} = \frac{P(D|H) \times P(H)/P(D)}{1-P(D|H) \times P(H)/P(D)} = \frac{P(D|H) \times P(H)}{P(D)-P(D|H) \times P(H)}$$

Since generally $P(A \cap B) = P(A|B) \times P(B)$, we get

$$\frac{P(H|D)}{1-P(H|D)} = \frac{P(D|H) \times P(H)}{P(D)-P(D \cap H)} = \frac{P(D|H) \times P(H)}{P(D \cap \overline{H})} = \frac{P(D|H) \times P(H)}{P(D|\overline{H}) \times [1-P(H)]}$$

Since

$$\frac{P(H)}{1-P(H)}$$

are the pre-test odds and

$$\frac{P(H|D)}{1-P(H|D)}$$

are the post-test odds, Bayes' equation finally becomes

$$\text{Post-test odds} = \frac{P(D|H)}{P(D|\overline{H})} \times \text{pretest odds}$$

$$\frac{P(D|H)}{P(D|\overline{H})}$$

by definition is the likelihood ratio (LR); therefore

$$\text{Post-test odds} = \text{LR} \times \text{pretest odds}$$

References

1. NCCLS (2001) *Specifications for Immunological Testing for Infectious Diseases; Approved Guideline-Second Edition*, in *NCCLS Document I/LA18-A2* (NCCLS, ed.), NCCLS, Wayne, PA, pp. 1–46.
2. Tosteson, A. N. A., Weinstein, M. C., Wittenberg, J., and Begg, C. B. (1994) ROC curve regression analysis: the use of ordinal regression models for diagnostic test assessment. *Environ. Health Perspect.* **102,** 73–78.

3. Sorribas, A., March, J., and Trujillano, J. (2002) A new parametric method based on S-distributions for computing receiver operating characteristic curves for continuous diagnostic tests. *Stat. Med* **21,** 1213–1235.
4. Dujardin, B., Vandenende, J., Vangompel, A., Unger, J. P., and Vanderstuyft, P. (1994) Likelihood ratios—a real improvement for clinical decision making. *Eur. J. Epidemiol.* **10,** 29–36.
5. Biggerstaff, B. J. (2000) Comparing diagnostic tests: a simple graphic using likelihood ratios. *Stat. Med* **19,** 649–663.
6. Choi, B. C. (1998) Slopes of a receiver operating characteristic curve and likelihood ratios for a diagnostic test. *Am. J. Epidemiol.* **148,** 1127–1132.
7. Choi, B. C. (1997) Causal modeling to estimate sensitivity and specificity of a test when prevalence changes. *Epidemiology* **8,** 80–86.
8. Brenner, H. and Gefeller, O. (1997) Variation of sensitivity, specificity, likelihood ratios and predictive values with disease prevalence. *Stat. Med* **16,** 981–991.
9. Vanderschouw, Y. T., Vandijk, R., and Verbeek, A. L. M. (1995) Problems in selecting the adequate patient population from existing data files for assessment studies of new diagnostic tests. *J. Clin. Epidemiol.* **48,** 417–422.
10. Broder, C. C., Earl, P. L., Long, D., Abedon, S. T., Moss, B., and Doms, R. W. (1994) Antigenic implications of human immunodeficiency virus type 1 envelope quaternary structure: oligomer-specific and -sensitive monoclonal antibodies. *Proc. Natl. Acad. Sci. USA* **91,** 11,699–11,703.
11. Dyson, H. J. and Wright, P. E. (1995) Antigenic peptides. *FASEB J.* **9,** 37–42.
12. Gurunath, R., Beena, T. K., Adiga, P. R., and Balaram, P. (1995) Enhancing peptide antigenicity by helix stabilization. *FEBS Lett.* **361,** 176–178.
13. Union, A., Meheus, L., Humbel, R. L., et al. (2002) Identification of citrullinated rheumatoid arthritis-specific epitopes in natural filaggrin relevant for antifilaggrin autoantibody detection by line immunoassay. *Arthritis Rheum.* **46,** 1185–1195.
14. Borza, D. B., Bondar, O., Todd, P., et al. (2002) Quaternary Organization of the Goodpasture autoantigen, the alpha 3(IV) collagen chain. Sequestration of two cryptic autoepitopes by intraprotomer interactions with the alpha 4 and alpha 5 NC1 domains. *J. Biol. Chem.* **277,** 40,075–40,083.
15. Mukherjee, J., Nussbaum, G., Scharff, M. D., and Casadevall, A. (1995) Protective and nonprotective monoclonal antibodies to *Cryptococcus neoformans* originating from one B cell. *J. Exp. Med.* **181,** 405–409.
16. Prentice, L., Kiso, Y., Fukuma, N., et al. (1995) Monoclonal thyroglobulin autoantibodies: Variable region analysis and epitope recognition. *J. Clin. Endocrinol. Metab.* **80,** 977–986.
17. Lovrich, S. D., Callister, S. M., Lim, L. C. L., Duchateau, B. K., and Schell, R. F. (1994) Seroprotective groups of Lyme borreliosis spirochetes from North America and Europe. *J. Infect. Dis.* **170,** 115–121.
18. Roost, H. P., Bachmann, M. F., Haag, A., et al. (1995) Early high-affinity neutralizing anti-viral IgG responses without further overall improvements of affinity. *Proc. Natl. Acad. Sci. USA* **92,** 1257–1261.
19. Foote, J. and Eisen, H. N. (1995) Kinetic and affinity limits on antibodies produced during immune responses. *Proc. Natl. Acad. Sci. USA* **92,** 1254–1256.
20. Braden, B. C. and Poljak, R. J. (1995) Structural features of the reactions: between antibodies and protein antigens. *FASEB J.* **9,** 9–16.
21. Yuan, R. R., Casadevall, A., Spira, G., and Scharff, M. D. (1995) Isotype switching from IgG3 to IgG1 converts a nonprotective murine antibody to *Cryptococcus neoformans* into a protective antibody. *J. Immunol.* **154,** 1810–1816.

22. Rademacher, T. W., Williams, P., and Dwek, R. A. (1994) Agalactosyl glycoforms of IgG autoantibodies are pathogenic. Proc. Natl. Acad. Sci. USA 91, 6123–6127.
23. Lund, J., Takahashi, N., Pound, J. D., Goodall, M., Nakagawa, H., and Jefferis, R. (1995) Oligosaccharide protein interactions in IgG can modulate recognition by Fc gamma receptors. *FASEB J.* **9,** 115–119.
24. Amzel, L. M. and Gaffney, B. J. (1995) Structural immunology: problems in molecular recognition. *FASEB J.* **9,** 7–8.
25. Guddat, L. W., Shan, L., Fan, Z. C., et al. (1995) Intramolecular signaling upon complexation. *FASEB J.* **9,** 101–106.
26. Soderlund, M., Brown, C. S., Spaan, W. J., Hedman, L., and Hedman, K. (1995) Epitope type-specific IgG responses to capsid proteins VP1 and VP2 of human parvovirus B19. *J. Infect. Dis.* **172,** 1431–1436.
27. Manaresi, E., Gallinella, G., Zerbini, M., Venturoli, S., Gentilomi, G., and Musiani, M. (1999) IgG immune response to B19 parvovirus VP1 and VP2 linear epitopes by immunoblot assay. *J. Med. Virol.* **57,** 174–178.
28. Gray, J. J., Wreghitt, T. G., Mckee, T. A., et al. (1995) Internal quality assurance in a clinical virology laboratory. 1. Internal quality assessment. *J. Clin. Pathol.* **48,** 168–173.
29. Gray, J. J., Wreghitt, T. G., Mckee, T. A., et al. (1995) Internal quality assurance in a clinical virology laboratory. 2. Internal quality control. *J. Clin. Pathol.* **48,** 198–202.
30. Mugan, K., Carlson, I. H., and Westgard, J. O. (1994) Planning QC procedures for immunoassays. *J. Clin. Immunoassay* **17,** 216–222.
31. Dimech, W., Bettoli, A., Eckert, D., et al. (1992) Multicenter evaluation of five commercial rubella virus immunoglobulin G kits which report in international units per milliliter. *J. Clin. Microbiol.* **30,** 633–641.
32. Place, J. F., Truchaud, A., Ozawa, K., Pardue, H., and Schnipelsky, P. (1994) Use of artificial intelligence in analytical systems for the clinical laboratory. *Clin. Chim. Acta* **231,** S5–S34.
33. Place, J. F., Truchaud, A., Ozawa, K., Pardue, H., and Schnipelsky, P. (1994) Use of artificial intelligence in analytical systems for the clinical laboratory. *Ann. Biol. Clin. Paris* **52,** 729–743.
34. Winkel, P. (1994) Artificial intelligence within the chemical laboratory. *Ann. Biol. Clin. Paris* **52,** 277–282.
35. Winkel, P. (1994) Multivariate analysis and expert systems. *Scand. J. Clin. Lab. Invest.* **54,** 12–24.
36. Sharpe, P. K. and Caleb, P. (1994) Artificial neural networks within medical decision support systems. *Scand. J. Clin. Lab. Invest.* **54,** 3–11.
37. Fierz, W. (2002) Information management driven by diagnostic patient data: right information for the right patient. *Expert. Rev. Mol. Diagn.* **2,** 355–360.
38. Malakoff, D. (1999) Bayes offers a 'new' way to make sense of numbers [news] [see comments]. *Science* **286,** 1460–1464.

24

Molecular Diagnostics Resources on the Internet

Larry Winger

Abstract

Busy diagnosticians need to know what is useful, and what is dross, when dealing with the internet. From the comprehensive array of resources that characterizes the offerings available via the world wide web and email correspondence, in particular, this chapter seeks to identify the most useful tools for the diagnostics laboratory. With rapid communications and fast internet consultations only a few keystrokes away, there really is no point in wasting time on fruitless searches when professionals are so accessible. But accessibility carries the weight of responsibility as well, and communications must be engaged with a fair modicum of civility and common courtesy. Responsibility is a crucially important component of public or semiprivate communication in terms of your own identity, and that of the organization that you represent. Recognition of the relative vulnerability of individual machines to the worldwide disseminated computer viruses, worms, or trojan horses currently abounding, for example, is perhaps the most important step in your approach to security issues, but this recognition must go hand in hand with institutional steps to protect the organization of which you are a part. Organizational tools that serve the diagnostician well in the laboratory can also be mobilized in the aid of communications through the net, and always that harbinger of understanding, common sense, should prevail in one's dealings both with machines, and the people who are communicating either directly or indirectly through them.

Key Words: World wide web; email; listserv; library; search; etiquette.

1. Introduction

On the universally adopted Internet search engine Google, key in these words: "molecular"; "diagnostics"; "resources." Now key in this phrase: "molecular diagnostics resources." In the first search, more than 40,000 results might be obtained, whereas in the second, only 10, almost all of which pertain to the first edition of this book, are likely to appear. It is specifically within this window of Internet availability that this chapter is concerned.

From: *Methods in Molecular Medicine, vol. 94: Molecular Diagnosis of Infectious Diseases, 2/e*
Edited by: J. Decker and U. Reischl © Humana Press Inc., Totowa, NJ

How can the busy researcher or clinician in the diagnostics sphere hope to sift the wheat from the chaff, or find the golden needle of utility in the 40,000 strands of the haystack? Let's face it, everybody and everything is on the net these days and can be found with a few simple keystrokes—but in many cases you have to know what you're looking for.

It is important that a strong sense of purposeful organization and objective-oriented search must underpin any understanding of online resources for molecular diagnostics. However, in addition to clever organizing, a credit card can be a useful and sometimes requisite tool to open certain databases, which are important resources in the research diagnostician's armamentarium. Furthermore, it can be useful to realize when a particular search is unlikely to pay dividends, even if the light appears better!

2. Molecular Diagnostics Information: From the Internet to You and Back Again

The sections of this chapter develop from the personal information acquisition and information processing paraphernalia, through the convenient sources of larger institutional information provision and on to private/public networks and commercial information sources. In this chapter, immunologically based diagnostic resources are specifically considered; although there is some natural overlap between nucleic acid-based diagnostic approaches and those that are antibody-based, researchers involved in molecular biology diagnosis will have their own community of resources.

It may be useful to consider the organization of this chapter in terms of the subject headings, as follows:

- 2.1. Personal Hardware and Software for the Portable and Acquisitive Information Age.
- 2.2. Your Institutional Library and Periodical Index Databases.
- 2.3. Internet Compendia of Resources.
- 2.4. Search Engine.
- 2.5. Internet Mailing Lists.
- 2.5.1. Setting up a Mailing List.
- 2.5.2. Pursuing Individual Leads via Mailing List Contacts.
- 2.6. Online Electronic Journals.
- 2.7. Distributor Relay of Information.
- 2.8. Online Database Access for Free.
- 2.9. Online Database Access for Sale.
- 2.10. Electronic Classrooms.
- 2.11. Telemedicine.
- 2.12. Time-Wasters (Newsgroups, Chatrooms, Email Spam) or Nonstarters (Electronic Conferences, Virtual Reality Spaces).
- 2.13. Your Web Presence or Nonpresence.

2.1. Personal Hardware and Software for the Portable and Acquisitive Information Age

Access to the Internet by an institutional or commercial Internet Service Provider (ISP) is a given in today's high-tech world. It is usually a good idea, however, to recognize what constitutes appropriate internet usage in specific environments.

The workplace is not the best place to trace your family tree, nor is it always felicitous to receive and deal with laboratory-related email in the home office. Busy professionals can find it difficult to differentiate between work and a vocational activity, and this differentiation becomes harder when the same Internet tools are a constant feature of both practices. Usually, professional practice dictates that work-related correspondence be performed on workplace-dedicated machines; few workplace machines (yet) are of the laptop or portable variety. However, there are times when you really must access workplace Email from home and vice versa.

Any diagnostics scientist the world over will want to develop home and work-based computer support that covers the best working practice for them. This may mean absolute differentiation between work and home tasks, but perhaps more commonly the acquisition of digital information will occur in some sort of conjunction between the two.

The emergence of the Hotmail hub has been driven because people hesitate to manipulate the settings of their Internet Configuration (whether at home or work), and it's true that you can always access your Hotmail account through any browser from any online computer in the world. Intriguingly, however, you can do exactly the same thing with your own email directed to your regular address(es), if you know your "pop" settings. With these settings, from any online computer you can easily access email sent to a variety of addresses wherever you have your ISP accounts. You could ask your computer support people how this might be facilitated in your particular case, but it's nice to know that it can be done.

There are a variety of electronic means whereby you can achieve even more direct online access, for example, to a work-related machine from your home office. Timbuktu software can give you control of one machine from another, and there are other similar software products on the market. This sort of approach to file transfer has inherent security dangers, of course, but it can be incredibly handy when a crucial email or report is sitting resplendent in its directory in the workplace machine and you are an hour's drive away struggling with a report in the home office.

Alternatives to electronic transfer and remote control abound, of course. People may download important files on either the old 3.5-cm floppy disks, or, given the larger file sizes attendant with PowerPoint presentations, more likely onto CDs, with less worry about magnetic wiping or mechanical failure. Of course, the Zip drives have co-evolved too, but the handy digital storage devices that are now key-ring size, or even the minuscule 20-gigabyte iPod can be used for portability of important files and applications.

Whatever your storage, archival, or data transfer requirements, it is a given that the computer systems you surround yourself with will grow to match. It becomes almost a moot question, in the end, whether your requirements are evolving under pressure from the computer developers, or whether busy professionals like yourself are driving that development. However, the goal appears to be absolute portability and universal access of any information you require at any geographical point you may be.

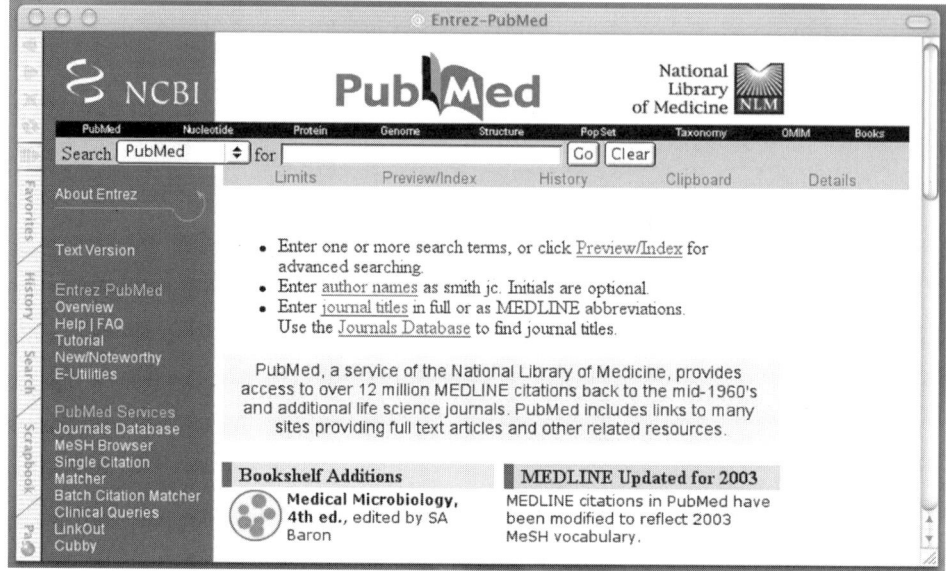

Fig. 1. PubMed.

2.2. Your Institutional Library and Periodical Index Databases

Typically, institutions will maintain links with the main scientific databases that catalog information about publications in all of science and medicine. That's what good libraries are for!

Everyone knows about PubMed (http://www.ncbi.nlm.nih.gov/entrez/query.fcgi), the universally available online index (Medline) of the world's biomedical research literature that is maintained by the American National Library of Medicine (**Fig. 1**). It's easy to find through any search engine and is very handy for the world's searchers in the field of biomedical research. This is probably the number one database used by the general public, who are often desperate to identify current research on their own particular medical condition.

Not everyone, however, can simply access the World of Science (http://wos.mimas.ac.uk/), but institutions typically will have purchased access for their staff and/or students based on individual user identification and password (**Fig. 2**). The Science Citation database there is invaluable for acquiring information not listed in Medline (PubMed). At the time of writing this chapter, a free trial period of the World of Science database was being offered.

Between these two indexing resources, the world's scientific literature should be comprehensively covered. There is, of course, the natural time delay between publication and the actual cataloging of the information/abstract, but this lag period gets shorter with every passing month.

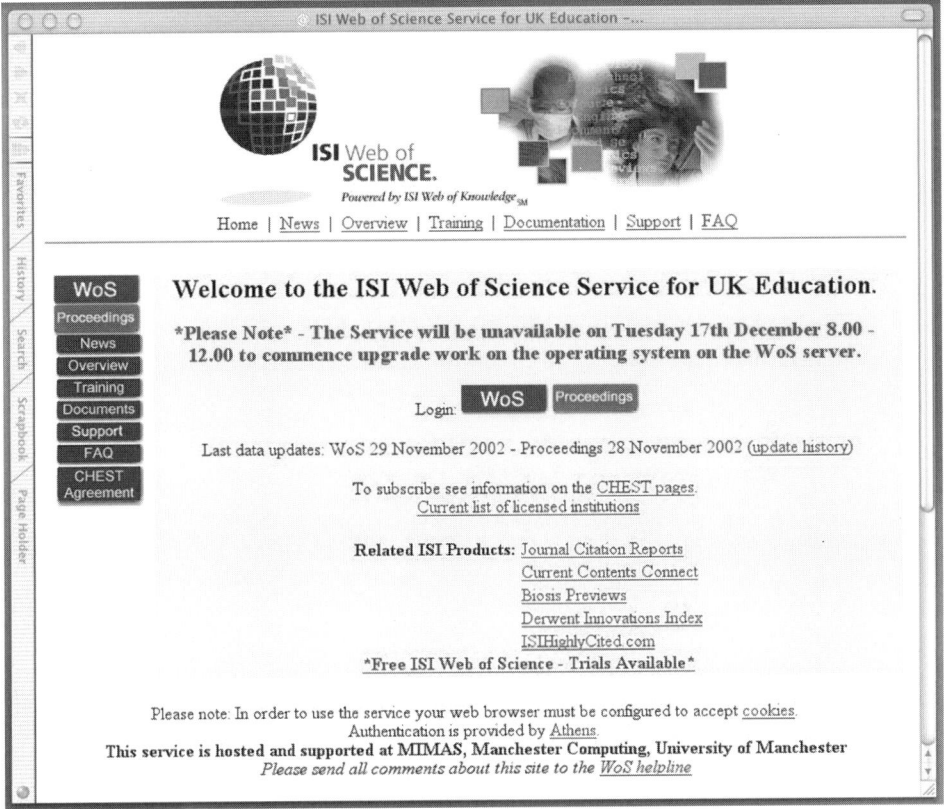

Fig. 2. World of Science.

2.3. Internet Compendia of Resources

A quick search on Google, restricted by the search term "molecular diagnostics" and resources reveals two very useful compilations of resources for the molecular diagnostician.

BioSpace (http://www.biospace.com) (**Fig. 3**) and BioExchange (http://directory.bioexchange.com) (**Fig. 4**) are exhaustive compilations of biotechnology companies and resources for biotechnologists throughout the world.

Similarly, BioMedNet, with over a million members worldwide, is a useful compilation site (http://www.bmn.com) bursting with resources for anyone involved in the biological or medical fields.

2.4. Search Engine for the World Wide Web

Google (http://www.google.com) is, as of this writing, the world's pre-eminent search engine of world wide web-based information (**Fig. 5**). When searching is as good as this, who needs multiple kinds? It's quick, it's clean, and advertising is

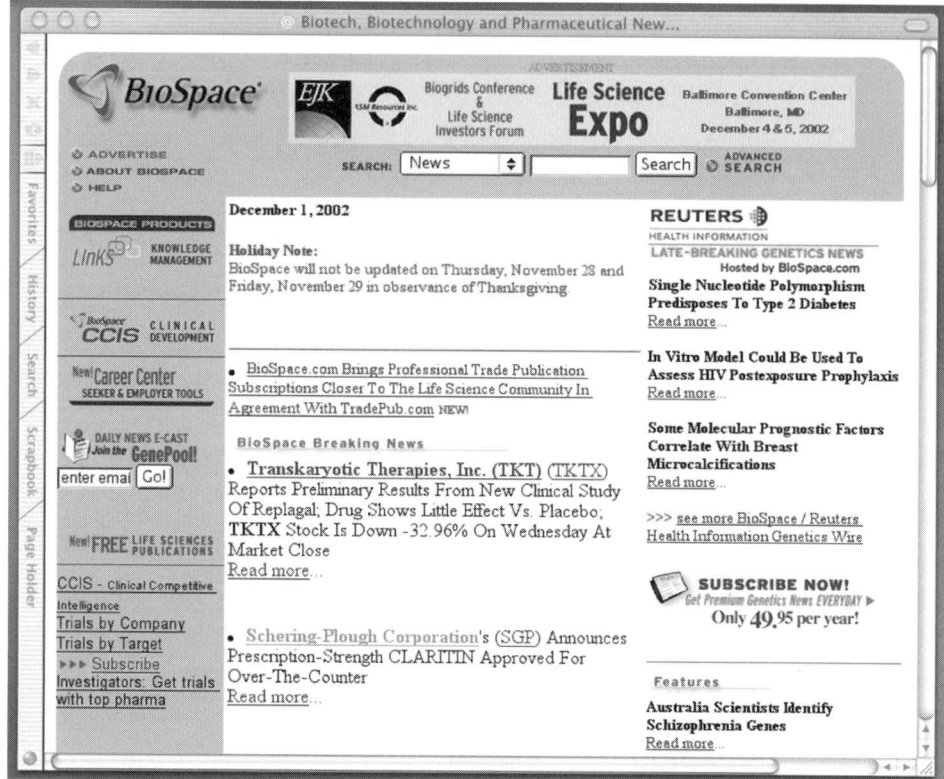

Fig. 3. BioSpace.

restricted to minimum links on a sidebar when you receive your search items; sometimes these links can even be helpful, as they're triggered by the keywords you're using for the search.

2.5. Mailing Lists

Alone at your computer keyboard, you are just one person with isolated knowledge gleaned from your studies. However, in a matter of only a few keystrokes, you can be connected to a world-wide community of individuals similarly concerned with diagnosis using molecular tools.

The very best internet-based mailing list software is that provided by L-Soft, which virtually invented the term LISTSERV. The reknowned CataList site (http://www.lsoft.com/lists/listref.html) gives all 72,000 mailing lists using this software (**Fig. 6**).

There are, as of this writing, some 72,000 mailing lists around the world that use LISTSERV software. You can search by keyword for a mailing list that may be appropriate for your special field of interest.

Of particular interest for the purposes of this volume, of course, are two long-lasting mailing lists that continue to serve the molecular diagnostics community.

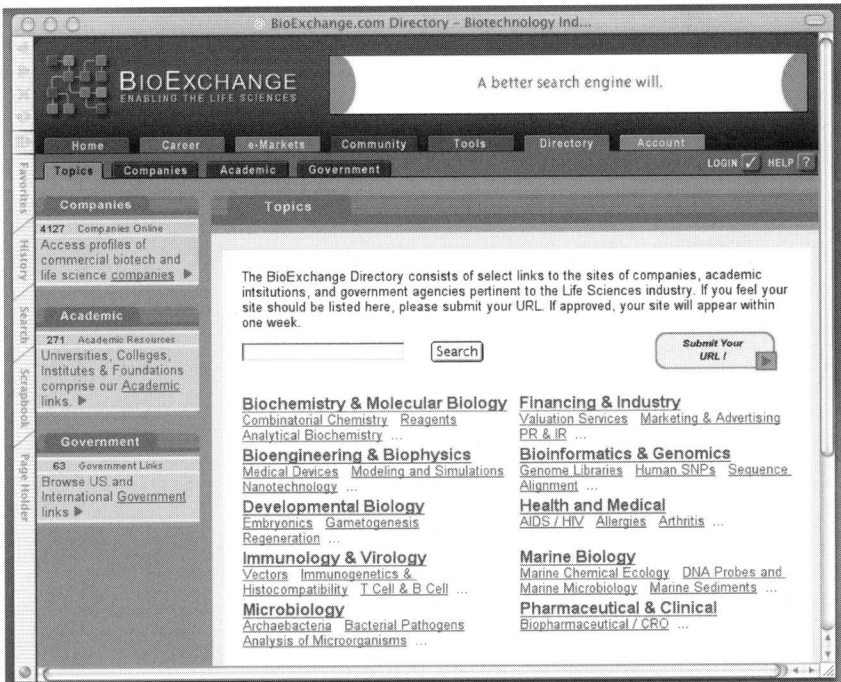

Fig. 4. BioExchange.

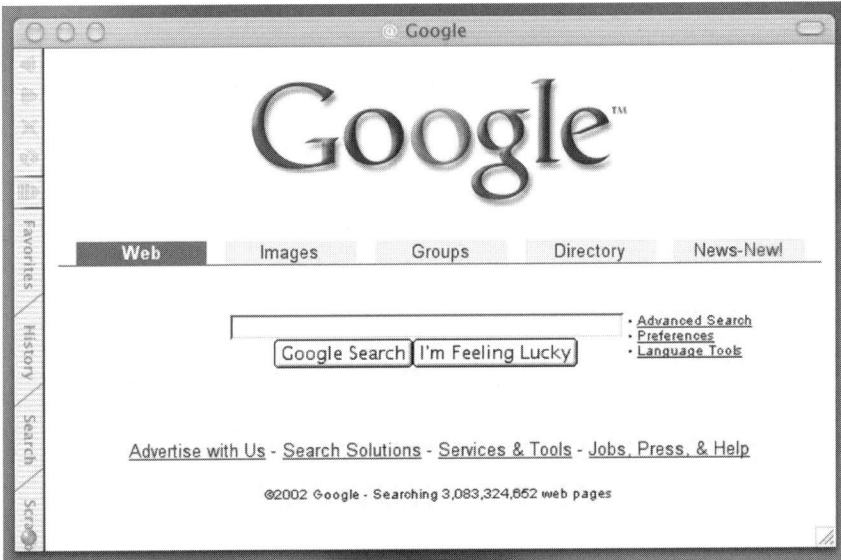

Fig. 5. Google.

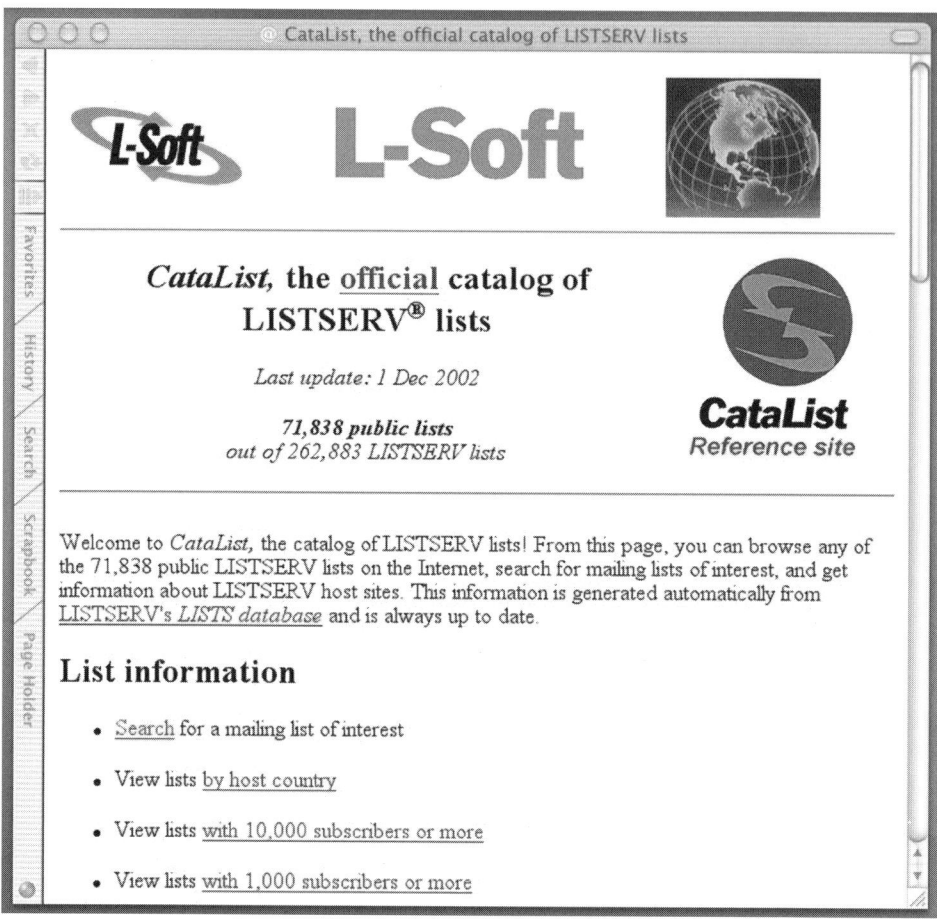

Fig. 6. CataList.

1. MEDLAB-L@listserv.buffalo.edu
   Now encompassing over 2400 subscribers, this mailing list links people working in medical laboratories all over the world (whether they call themselves clinical scientists, or laboratorians, or technicians). You can easily subscribe by sending an email to:
   listserv@listserv.buffalo.edu
   with the following in the body of the email:
   Subscribe MEDLAB-L
   After you have been subscribed by the LISTSERV, your first email posting (which will go out to each of the 2400 subscribers) is vetted, to ensure you are who you represent yourself to be and that you understand basic rules of email list etiquette, but thereafter your posts are passed without moderation.
2. Another mailing list concerned with the molecular genetics of human subjects is the HUM-MOLGEN list:

Molecular Diagnostics Resources on the Internet 437

HUM-MOLGEN@NIC.SURFNET.NL

Subscription, because all LISTSERV-based mailing lists operate similarly, is achieved by sending an email to:

listserv@nic.surfnet.nl

with the following words in the body of the email:

subscribe hum-molgen

Of the 6431 subscribers (as of this writing) to Hum-Molgen, there are sure to be specialists who will appreciate the problem or idiosyncratic nature of the molecular biology diagnostics issue with which you are concerned.

A few internet etiquette issues are important to emphasize when considering mailing lists.

1. Most important of all is to remember that any message sent to the "list" is public: each of the thousands of members will be receiving your correspondence. Is your message important enough for all (or a significant proportion) to read and consider?
2. Second in importance is the matter of archival storage. Most mailing lists maintain an archive of correspondence, which is exhaustive, often covering the entire lifetime of the list. It's surprising how often information from this archive can be found that will answer a specific query you might have.
3. The third most important issue on any mailing list is that you understand the difference between the LISTSERV and the "list." The LISTSERV is the machine-based mechanism whereby subscription options are handled; this is where you work to deal with your subscriptions. The "list" is the set of correspondents who have subscribed. Nobody wants to be informed about your requests to unsubscribe!
4. No matter how important any particular chain-mail message may appear to be (and of the many circulated around the world over the past decade, importance has varied from the plight of Afghanistan's women to famine in Africa, to innumerable computer-virus alerts), it's almost a certainty that its circulation in the mailing list will not be appreciated and will probably result in your removal from the membership by the list coordinator or owner.
5. Harvesting and redistributing of email addresses of individuals subscribing to the list is a definite no-no. Don't be tempted; don't do it for any reason!
6. Ensure that you understand to whom you are sending any reply; some mailing lists are set up so that replies go back only to the person originally posting the first note, whereas others ensure that replies go to everyone on the list.
7. Professionalism is always appreciated by everyone on any list, and this approach will always stand you in good stead when corresponding with other professionals.

2.5.1. Setting Up an Internet-Based Mailing List

You may wish to take advantage of the numerous opportunities to create your own mailing list dedicated to the professionals with whom you are in contact. Your community might, for example, be a geographical unit (like the Bluegrass Medical Librarians) and a restricted list might be convenient to you. In the UK, Mailbase (http://www.mailbase.ac.uk) provides a simple, convenient, and free mailing list setup service for academics (principally those in the UK) who wish to communicate around a particular topic. There is also a huge variety of free and/or nominal fee-based mailing list communities or servers (http://directory.google.com/Top/Computers/Internet/Mailing_Lists/Hosting_Companies/) that you may browse through to find the particu-

2.5.2. Individual Email Contact

One of the benefits of membership in professional mailing lists is, frankly, email access to specific expertise that is demonstrably active and online in the field. If someone posts a note on a particular topic in MEDLAB-L, for example, they tend to be writing authoritatively from their own laboratory. Just listening in to the regular correspondence that goes on in the list identifies people of knowledge/specialist expertise. Don't be tempted to abuse this special privilege, but be aware that people who post items, or respond to others, are a good potential source of further information and may be more likely to respond to professionally identified correspondents than any one specialist identified at random.

As good as email correspondence is, however, it carries a significant amount of misinterpretation risk. Even in the less emotive realm of diagnostics, abrupt mannerisms and enquiries, which can seem less than polite to the recipient can be cause for serious misinterpretation of good will and/or intent.

The sheer volume of email that most people receive can also mitigate against replies. Email work is sort of a self-fulfilling odyssey (I write, therefore I'm working). Be aware of what actual work and product you are generating at your keyboard!

Beware the chain-mail temptation (*see* number 4 in **Subheading 2.5.** above in the context of mailing lists) in your individual correspondence, as well. Blanket mailing of everyone in your email address book (with a petition against a particular war, or a petition supporting striking UK firemen, or a petition decrying South African President Mbeki's stance on AIDS therapeutics, whatever it is) may elicit a range of effects ranging from sharp and savage ripostes to sullen silence thereafter from your valued correspondents.

Indeed, the sort of chain-mail correspondence that occurs today is often filtered out even before it reaches its target computer user, as unwanted "spam." The term *spam* comes from the Monty Python sketch in which you can order anything you want from the café, providing you're willing to have it with spam (a kind of processed meat). Spam is the biggest *bête noir* of the internet today, and these filters will only get more sophisticated, so that end-users will find it easier and easier to ignore. No matter how altruistic your intentions, please do not add to the spam burden!

2.6. Online Electronic Journals

The University of Newcastle has a comprehensive web page detailing available electronic journals and other similar published resources on the world wide web (http://www.ncl.ac.uk/library/guides/jnlsig.html). For massive list of lists of electronic journals, you would be well advised to check out these universal electronic journal listings:

> http://gort.ucsd.edu/ejourn/jdir.html
> http://www.e-journals.org/

Many journals publish papers in a standard format known as "pdf," which helps to ensure that the image received is the same as the image published. To see these papers,

Molecular Diagnostics Resources on the Internet 439

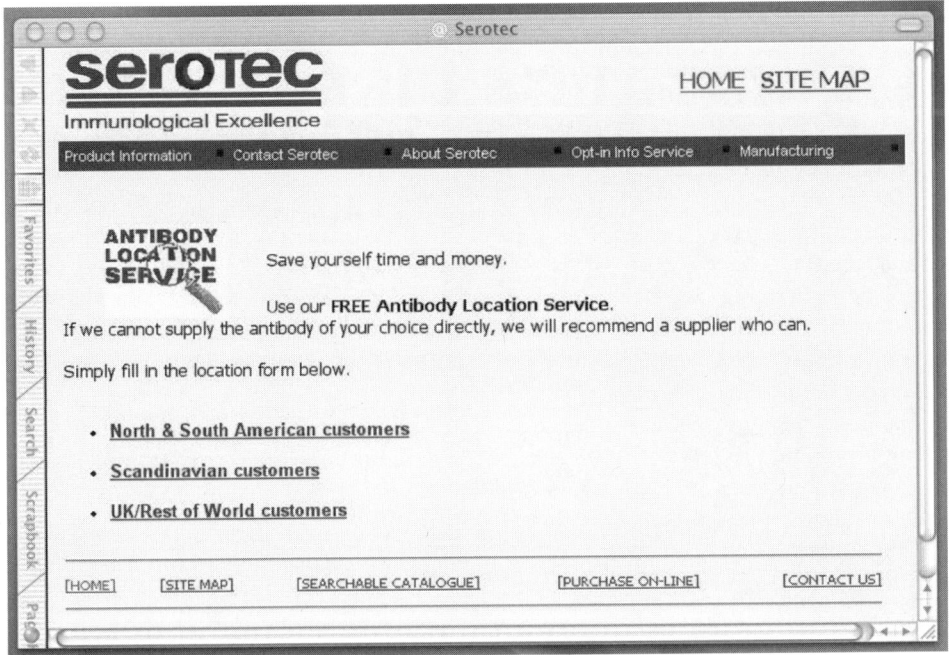

Fig. 7. Serotec.

you will need the familiar Adobe Acrobat reader, which is always available free in its most current update at http://www.acrobat.com.

2.7. Distributor Relay of Information

The immunologically based reagent company Serotec offers an Antibody Location Service that will search for items in the antibody product area (http://www.serotec.co.uk). (**Fig. 7**).

It seems that this service is a useful attraction to customers who may find their antibody of choice as distributed or produced by Serotec itself. Typically, only one or two suppliers of any particular antibody specificity are returned by the friendly Serotec service, but it makes good business sense to build up this sort of goodwill.

Similarly, for enquiries from Germany, Acris Online will do free searches for you, if you ask them nicely (http://www.acris-online.de/english/als001_en.htm), or Cedarlane Laboratories (http://www.cedarlanelabs.com) will open their database of antibody producers for North American enquiries.

2.8. Online Database Access for Free

Promoting itself as the world's antibody gateway, AbCam (http://www.abcam.com) relies on advertisers for its revenue—as of this writing, AbCam gets approx 10,000 hits a week, of which 24% used the gateway for direct ordering of diagnostics-related immunological products. AbCam also provides a single toolbar for searching Google,

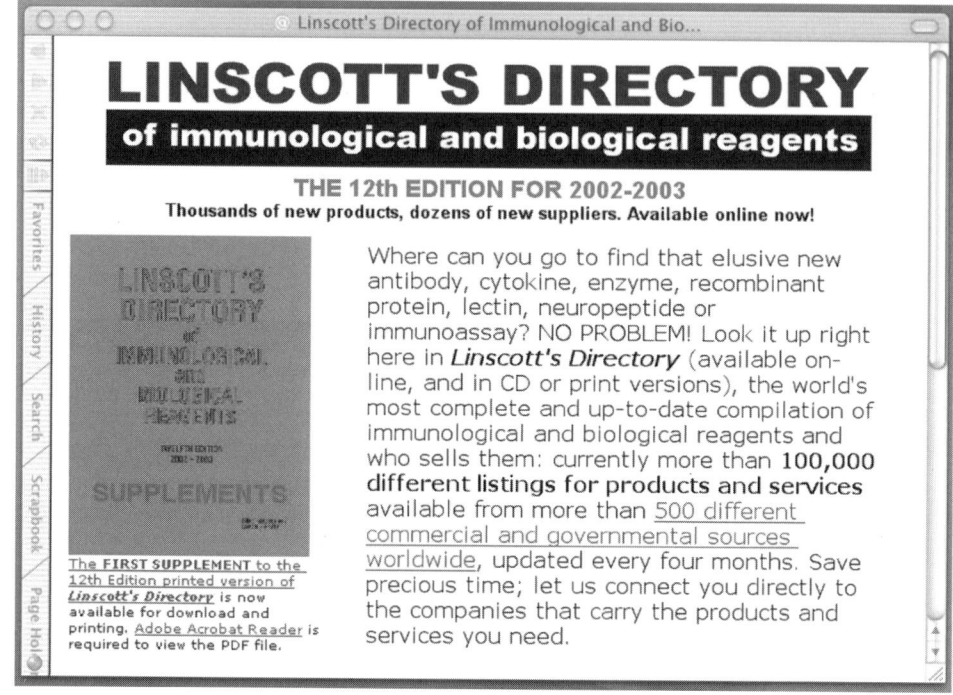

Fig. 8. Linscott's Directory.

Abcam, GenBank, OMIM, SwissProt, and PubMed databases, which can be quite convenient.

2.9. Online Database Access for Sale

Two databases charging annual fees for access are particularly important for molecular diagnostics researchers and clinicians. The familiar Linscott's Directory (http://www.linscottsdirectory.com or .co.uk), now in its 12th edition, is available online for more than 100,000 different listings of antibodies or natural or recombinant proteins from over 500 commercial sources worldwide (**Fig. 8**). Currently single-user subscriptions cost U.S. $90.

Even more antibodies (165,000) are claimed by the MSRS Antibodies/Probes database (http://www.antibodies-probes.com), and annual subscriptions are around the U.S. $78. level (**Fig. 9**).

2.10. Electronic Classrooms

Most universities will have specific software (e.g., electronic black/white board) that can permit uploading of images into an electronic space shared between teacher/facilitator and students. Such boards also usually have drawing tools and a chat panel by which a real learning environment can be facilitated (**Fig. 10**).

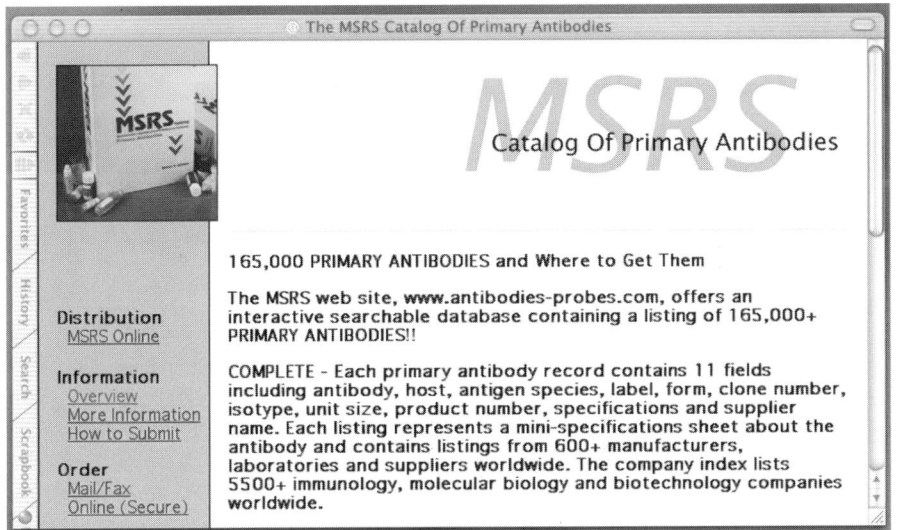

Fig. 9. MSRS database.

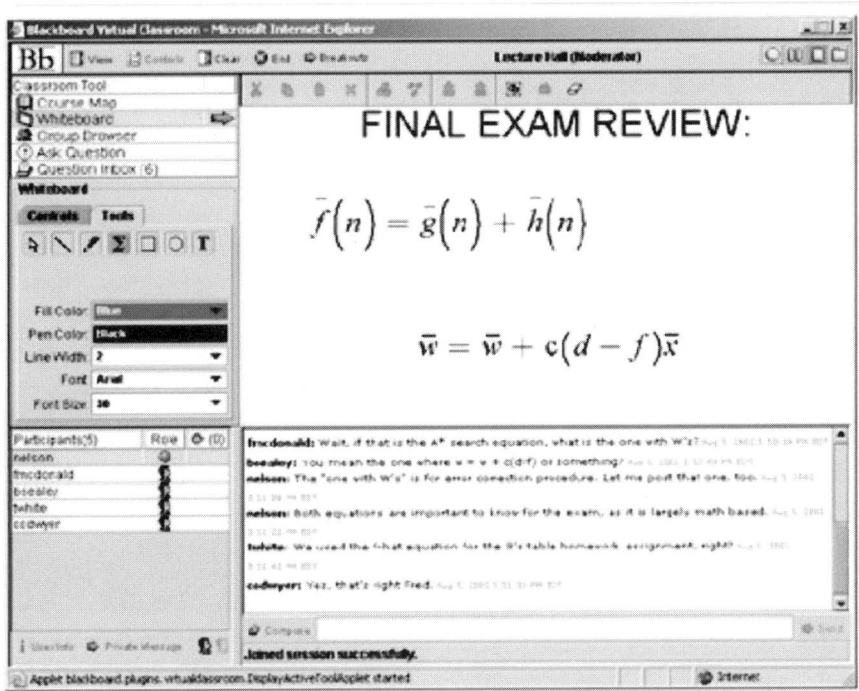

Fig. 10. Example of an electronic classroom board.

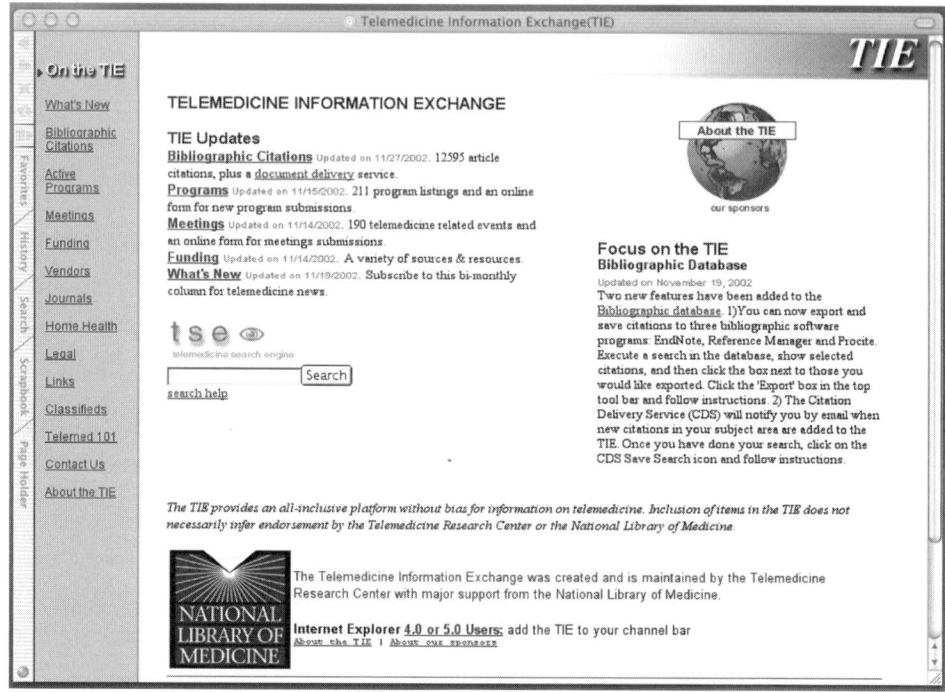

Fig. 11. Telemedicine Information Exchange.

One platform particularly favored by universities (including the University of Newcastle) is the offering by http://www.blackboard.com. These electronic classrooms could be a useful advance for central diagnostic laboratories that have satellite stations.

2.11. Telemedicine Connections

Telemedicine can be roughly defined as an electronic-based communication between two remote locations for clinical diagnostic purposes. In this sense, telemedicine can be an important tool in the clinical diagnostic sphere and will certainly increase in use as centralized laboratories need to communicate with far-distant outposts.

Telemedicine Information Exchange, a service of America's National Library of Medicine, provides exhaustive information on all events relating to telemedicine, worldwide (http://tie.telemed.org) (**Fig. 11**).

2.12. Time Wasters

Some Internet areas are not worthwhile or are counterproductive for the clinical or research diagnostics specialist (newsgroups, chatrooms, email spam) or are of little usefulness (electronic conferences and virtual reality rooms).

What a disappointment the UseNet newsgroups have turned out to be! Five years ago, these internet bulletin boards held the promise of fast and immediate access to the expertise of a world community dedicated to a particular topic. Today most are moribund repositories of tired old rehashed arguments or endless lists of favorites, or the site of intensely vituperative flame wars among angst-ridden or blithely guilt-free individuals. Not a place for dedicated diagnostic professionals. The harvesting of email information from these sites by spammers (mass email advertisers) was one of the death knells of this potentially useful medium, and this practice is more than sufficient inducement to keep careful professionals away, particularly if, as itemized below, you are in a politically turbulent area of work.

If you are a fast typist, chatrooms have an immediacy that can be seductive, but who really wants to communicate in real time through their fingers? These areas are ideal for lonely hearts or teenage romance, but it's difficult to see how they can be useful apart from the very restricted practice of electronic classrooms (as above). Pick up the phone and convey much more information directly from mouth to ear! Or, if convenience between correspondents is the goal (and it is), an email can be as immediate as a chatroom but with less capacity than the phone to intrude on other more pressing involvements.

Electronic conferences and other virtual reality spaces are ideas whose time has not yet come in the molecular diagnostic resource sphere. Although pure chemists seem to enjoy the utility of electronic conferencing, by which displays of virtual molecules can be conveniently disseminated, similar activities do not seem to be organized with the clinical diagnosis field in mind. Perhaps this is because Clinical Chemistry and Clinical Diagnosis conferences tend to be financed by the major clinical chemistry/immunoassay platforms, so that the actual, physical interaction of conference participants is, in effect, subsidized. Certainly electronic conferencing will need to improve by several orders of magnitude before the amount of information exchange can begin to approximate that available on a real conference floor.

2.13. Staying Visible or Nonvisible; Ensuring and/or Safeguarding Your Own Presence, or Nonpresence, on the Net

If you are in academia, your institution will have ensured that your presence in the community is noted, your research interests itemized, and your publication list presented for all to see. Alternatively, you will have been pushing for your own specific website demarcating the research interests of your laboratory. Similarly, if you happen to be working in the commercial sphere, your company will undoubtedly have a web presence touting for custom, advertising your particular diagnostic wares, skill base, and technology/machinery.

In these circumstances, you will probably wish to have a high profile, and you can help to ensure high visibility by informing the search engines of your web pages, by spreading the address of your website around so that interested browsers are constantly reminded of your site. (Email signature files are a simple and highly effective way to identify yourself and your site.)

Alternatively, you may be part of a commercial team or a clinical research institution that wishes to remain beneath the threshold of visibility. If you harvest serum from immunized animals, for example, or are involved in the daily generation of monoclonal antibodies, you will know of the potential ramifications of these practices relative to the activities of today's animal-rights activists. In such circumstances, you may wish to be protected from high visibility in the world-wide domain of the net.

Your computer systems' administrator will undoubtedly set protocols of security as a matter of course, if you are in such a situation. Common sense adherence to the demands of your company or institution (whether they are as painless as a signature file that disclaims any institutional authority for your message, or as fundamental as a multilayered firewall-protected system through which your browser must traverse before any downloaded files reach your computer) will help to ensure that these security demands are met.

3. Conclusions

The Internet is a fundamental component of life for busy scientists and clinical practitioners in the diagnostic sphere. The information resources available can prevent the loss of countless hours in fruitless attempts to "reinvent the wheel." Alternatively, specific information can identify an important tool that could mean the difference between success and failure in the performance of a particular assay. Wise and careful use of this global information conduit that connects diagnostic researchers world-wide must be a matter of daily bread-and-butter activity by any conscientious participant in this exciting field.

VIII

RECOMBINANT RECEPTOR MOLECULES

25

Cloning Single-Chain Antibody Fragments (scFv) from Hybridoma Cells

Lars Toleikis, Olaf Broders, and Stefan Dübel

Abstract

Despite the availability of antibody libraries for the selection of receptor molecules, the large number of established and well-characterized hybridoma lines still represent a useful source for recombinant antibody genes. This protocol describes the PCR amplification, cloning, and a small-scale expression test for the generation of scFv fragments from hybridoma cell lines. Particular emphasis was placed on frequently observed problems and pitfalls of this method.

Key Words: Antibody engineering; proteomics; monoclonal antibodies; phage display.

1. Introduction

The variable region (Fv) portion of an antibody is comprised of the antibody V_H and V_L domains and is the smallest antibody fragment containing a complete antigen binding site. To stabilize the fragment kinetically by generating a very high local concentration for the association of the recombinant V_H and V_L domains, they are usually linked in single-chain Fv (scFv) constructs with a short peptide that bridges the approx 3.5 nm between the carboxy-terminus of one domain and the amino-terminus of the other (1–3). The cloning of these variable domain genes has been well established as a common method for the "immortalization" of valuable mouse or rat hybridoma clones. Furthermore, a large number of protein fusions to the antigen binding variable (Fv) portion of an antibody in *E. coli* has been constructed to add a variety of heterologous functions. The genetic information for the variable heavy and light chain domains (V_H and V_L) is generally amplified from hybridoma cells using polymerase chain reaction (PCR) with immunoglobulin-specific primers. A variety of primer sets for the amplification of mouse variable domains have been developed (4–6). However, several different sequences for each domain may be found in the PCR products amplified from the cDNA of a single hybridoma "clone." The fact that one cell line might express more than one heavy and light chain was also observed by other authors *(7–10)*. Up to nine different V_L and five different V_H sequences with a homology of about 95% have been isolated from a single hybridoma culture of hybridoma cell line PA-1 (S. Deyev, personal communication). Various reasons may be responsible for these heterogenous results. Additional antibody variable domain genes may derive from antibody mRNA transcribed from the genes of the parental cells, e.g., the myeloma fusion partner used for generation of the hybridoma line. These chains, found in older fusion lines, are the easiest to identify by comparison with the known sequences of the original fusion partners.

Allelic dysregulation may lead to the expression of all V regions available, and pseudogenes with an internal stop codon were found as the most abundant PCR product in some cases. Even tandem-like assemblies of parts of two different V region sequences have been observed (J. Görnemann, personal communication). It has to be kept in mind that hybridoma cells are very different from the well-regulated antibody-producing cells in our body. In addition to their doubled ploidy and partial cancer cell origin, they are completely released from the strict regulation of the immune system, allowing for an abundance of gene expression deviations. Mutations may be introduced into the antibody genes during prolonged cell culture, resulting in a heterogenous population containing a set of highly homologous antibody genes. Therefore, it is very important to prepare the cDNA from a freshly subcloned hybridoma clone tested for productivity. Furthermore, mutations may be introduced in the PCR amplification step. However, the use of polymerase with proofreading activity should suppress these mutations. Further sequence variation is frequently introduced by the use of degenerated mixtures of oligonucleotide primers *(6)*. This cannot be avoided completely since it is not possible to use primer sets matching 100% with all possible sequences, simply because of the observed variation in the framework 1 region and the fact that not all somatic antibody sequences can be known. Furthermore, owing to the necessity of amplifying unknown sequences, the hybridization conditions during PCR of antibody sequences from cDNA have to be adjusted to allow mismatches. Under these conditions, even amplification of antibody DNA belonging to several different subgroups with the same primer was observed *(10)*. Particular attention has to be paid to the position 6 of V_H, where a mutational toggling between Gln and Glu has been shown to influence the stability of the product dramatically *(11,12)*.

As a result, the common strategy should be to analyze quite a number of randomly picked V_H and V_L clones after the PCR, in a first screen for heterogenicity. A quick method is to compare the restriction patterns after digestion with *Bst*NI *(13)*. This enzyme has been shown to cleave different sites in V regions frequently. It is not completely conclusive, because it may give a hint that V-region genes of different genetic origin are present, but it will rarely detect point mutations. It is more advisable to sequence the DNA of at least 10–20 clones of each heavy and light chain PCR reaction product. If different sequences are found, their various combinations should be assembled in an expression vector and tested for antigen binding separately.

A protein sequencing step (preferentially of tryptic peptides) may be employed to identify the correct sequence. Peptide sequences containing complementarily-determining regions (CDRs) or their fragments may serve as a template for designing specific oligonucleotide primers for a PCR assembly cloning of the correct sequence in case no product or only incorrect products are amplified by the available sets of primers.

In case all directly cloned combinations of V_H and V_L fail to bind, the construction of a small phage library from each hybridoma by cloning the PCR products into a surface expression phagemid (e.g., pSEX; *see* refs. *14* and *15*) is recommended. This hybridoma library may then be screened for binding activity by panning on antigen. Because in this approach all the possible combinations of all sequences derived from the hybridoma cDNA are directly screened for antigen binding *(16)*, wrong sequences, wrong V_H/V_L pairs, and ligation artifacts are excluded.

However, even if the correct sequences have been cloned, an scFv may fail to bind in an assay optimized for the original monoclonal for several reasons. First, expression levels of individual scFvs derived from hybridomas can vary dramatically depending on the individual V-region sequences, mostly as a result of their folding efficiency, with the consequence of impractically low yields in some cases *(17)*. The periplasmic preparation used for the specificity test should therefore always be analyzed for its content of scFvs, most conveniently by immunoblot using an anti-tag reagent. Second, even though many scFvs are quite rigid molecules, others may be unstable in solution, their activity decaying within a few hours. In contrast to scFvs derived from phage display libraries (which during the panning rounds in addition to specificity are coselected for efficient functional expression in *E. coli*), hybridoma-derived scFv fragments have not undergone this selection. The use of freshly prepared periplasmic fractions can help to prove specificity of such a sequence prior to a genetic stabilization by subcloning the V regions into vectors allowing expression as Fab or IgG. Finally, scFvs designed in the usual regime (with a 15–18 amino acid linker connecting the two regions) may fail to bind in comparison with their monoclonal parent owing to the loss of the avidity effect allowed by densely coated antigen. With 2 (in case of IgG) or 10 (in case of IgM) binding arms, the apparent affinity of multivalent molecules in enzyme-linked immunosorbent assay (ELISA) or on an Immunoblot may be increased by orders of magnitude compared with a single binding domain. The best way to check this would be to use Fab fragments of the parental monoclonal antibody (MAb) for positive control. However, these are not always easily prepared, but the vice versa approach can be tried by dimerizing/multimerizing the scFv. A straightforward approach to achieve this is reduction of the linker length between the V_H and V_L, thus forcing the scFv to assemble in the diabody/triabody/tetrabody format. Alternatively, one of various vectors encoding multimerization domains fused downstream to the scFv gene can be used (for review, *see* ref. *18*).

Finally, a number of scFv with post-translational modifications in the CDR regions have been described. Unpaired cysteines in a CDR may affect the activity, in particular upon long-term storage, as they are prone to oxidation *(19)*. Glycosylation is frequently found in various CDRs, but it does not necessarily affect binding *(20)*. In some cases, the glycosylation decreases the affinity and the antibody benefits from a mutation removing the glycosylation site *(21,22)*. However, when the sugars contribute to the binding, the scFvs may not be functionally produced in *E. coli*. Mammalian cell lines [e.g., Chinese hamster ovary (CHO)] are the expression host of choice in these cases. The contribution of sugars to the binding may be analyzed by glycosidase digestion of the maternal hybridoma antibody.

A large variety of expression vectors for scFv fragments have been developed. They all allow a secretion of the scFv fragment into the periplasmic space of *E. coli*, where the biochemical milieu promotes correct folding and formation of the intrachain disulfide bonds. This is achieved by employing an amino-terminal bacterial leader sequence *(23)*, which is removed during secretion. Most vectors introduce additional motifs suitable for identification and/or purification of the scFv fragments. Most common is the His tag, allowing both IMAC purification and detection on Immunoblots, in ELISA, and so on. An example is the vector pOPE101 *(24)*. Some vectors encode an unpaired Cys residue in the tag region of the expressed polypeptide. These vectors should only be employed if the cysteine residue is required afterward for chemical conjugation because unpaired Cys residues have a negative influence on the yield *(24)*. A collection of complementing protocols on various recombinant antibody selection, expression, and analysis systems as well as alternative hybridoma cloning methods can be found in ref. *25*.

2. Materials
2.1. Isolation of Antibody DNA
1. RNeasy Mini Kit for total RNA minipreps (Qiagen, Hilden, Germany).
2. QIAshredder Homogenizer Kit (Qiagen).
3. Oligotex mRNA Mini Kit for isolation of mRNA from total RNA (Qiagen).
4. SuperScript™ First-Strand Synthesis System for RT-PCR (Invitrogen, Karlsruhe, Germany).
5. DNA polymerase with proofreading function, e.g., ProofStart DNA Polymerase (Qiagen) or Expand High Fidelity PCR System (Roche, Mannheim, Germany).
6. dNTP Mix (MBI Fermentas, St. Leon-Rot, Germany).
7. Thermal cycler.
8. Agarose gel electrophoresis equipment.
9. QIAquick PCR Purification Kit and QIAquick Gel Extraction Kit (Qiagen).

2.2. Cloning into an Expression Vector
1. Restriction endonucleases *Hin*dIII, *Mlu*I, *Nco*I, and *Not*I with appropriate buffers (New England Biolabs, Frankfurt, Germany).
2. Alkaline phosphatase (CIP) for DNA dephosphorylation (Roche).
3. pOPE101 expression vector *(24)*.
4. Agarose gel electrophoresis equipment.
5. QIAquick PCR Purification Kit and QIAquick Gel Extraction Kit (Qiagen).
6. T4 DNA ligase and buffer (Invitrogen).
7. Electroporation equipment.
8. *n*-Butanol (1- or 2-butanol will do).
9. *E. coli* XL1-Blue electroporation-competent cells (Stratagene, Amsterdam, The Netherlands).
10. Luria–Bertani (LB) medium containing 100 mM glucose and 100 µg/mL ampicillin.
11. LB agar plates containing 100 mM glucose and 100 µg/mL ampicillin.

12. QIAprep Spin Miniprep Kit (Qiagen).
13. Glycerol.

2.3. Small-Scale Expression Test

1. LB medium containing 100 mM glucose and 100 µg/mL ampicillin.
2. 100 mM solution of isopropyl-β-D-thiogalactopyranoside (IPTG). Store in aliquots at −20°C in the dark.
3. Sodium dodecyl sulfate-polyacrylamide gel electrophoresis (SDS-PAGE) and immunoblot equipment.
4. Anti-c-myc monoclonal antibody Myc1-9E10 (Invitrogen) and/or anti-penta-His monoclonal antibody (Qiagen).
5. Horseradish peroxidase (HRP)-conjugated goat-anti-mouse IgG antibody (Invitrogen).
6. Tetramethylbenzidine (TMB)-stabilized substrate for HRP (Progen, Heidelberg, Germany).

3. Methods

Methods that are not specifically described step by step can be done according to the manufacturers' protocols (Kits) or according to Sambrook et al. *(26)*.

3.1. Isolation of Antibody DNA

1. Collect up to 5×10^6 hybridoma cells by centrifugation.
2. Isolate the total RNA from the hybridoma cells using a kit (*see* **Note 1**).
3. Isolate the mRNA from the total RNA using a kit (*see* **Note 2**).
4. Prepare the first-strand cDNA from the mRNA using a kit (*see* **Note 3**). A second-strand synthesis is not necessary.

First PCR for the amplification of antibody DNA:

5. For each reaction combine the primer specific for the constant domain and a primer specific for the variable domain (**Table 1**). Perform each reaction in a total volume of 50 µL. Each 50-µL reaction contains 25 pmol of each primer, dNTP mix (10 mM of each), and polymerase buffer as described by the supplier. Use 1 µL of the prepared cDNA for each 50-µL reaction. Add 1 µL (2.5 U) of ProofStart or High Fidelity DNA Polymerase per 50 µL reaction (*see* **Note 4**).
6. Place the PCR tubes in a thermal cycler (*see* **Note 5**). Denature at 95°C for 5 min (*see* **Note 6**).
7. Perform 30 cycles with the following cycling program: 45 s of denaturation at 94°C, 45 s of annealing at the appropriate primer hybridization temperature (*see* **Note 7**), and 90 s of extension at 72°C. After 30 cycles, perform a final extension step for 5 min at 72°C.
8. Take out 1/10 volume of each PCR reaction for analytical gel electrophoresis on a 2% agarose gel.
9. Purify the PCR products using a kit. Elute DNA in H_2O (*see* **Note 8**).

Second PCR for reamplification and introduction of appropriate restriction sites into the antibody DNA:

10. Each 50-µL reaction contains 25 pmol of each primer, dNTP mix (10 mM each), and polymerase buffer as described by the supplier. Use up to 50 ng of purified PCR product from the first PCR for each 50-µL reaction. Add 1 µL (2.5 U) of ProofStart or High Fidelity DNA Polymerase per 50 µL reaction (*see* **Note 4**). Use the primer pairs (with the introduced restriction sites; *see* **Table 2**) that gave the desired PCR product in the first PCR.

Table 1
Oligonucleotides for the Amplification of Mouse Immunoglobulin Variable Region DNA

Heavy chain
  Constant domain
    MHC.F   5'-GGCCAGTGGATAGTCAGATGGGGGTGTCGTTTTGGC-3'
  Variable domain
    MHV.B1   5'-GATGTGAAGCTTCAGGAGTC-3'
    MHV.B2   5'-CAGGTGCAGCTGAAGGAGTC-3'
    MHV.B3   5'-CAGGTGCAGCTGAAGCAGTC-3'
    MHV.B4   5'-CAGGTTACTCTGAAAGAGTC-3'
    MHV.B5   5'-GAGGTCCAGCTGCAACAATCT-3'
    MHV.B6   5'-GAGGTCCAGCTGCAGCAGTC-3'
    MHV.B7   5'-CAGGTCCAACTGCAGCAGCCT-3'
    MHV.B8   5'-GAGGTGAAGCTGGTGGAGTC-3'
    MHV.B9   5'-GAGGTGAAGCTGGTGGAATC-3'
    MHV.B10   5'-GATGTGAACTTGGAAGTGTC-3'
    MHV.B12   5'-GAGGTGCAGCTGGAGGAGTC-3'
Light chain
  Kappa chain
  constant domain
    MKC.F   5'-GGATACAGTTGGTGCAGCATC-3'
  Kappa chain
  variable domain
    MKV.B1   5'-GATGTTTTGATGACCCAAACT-3'
    MKV.B2   5'-GATATTGTGATGACGCAGGCT-3'
    MKV.B3   5'-GATATTGTGATAACCCAG-3'
    MKV.B4   5'-GACATTGTGCTGACCCAATCT-3'
    MKV.B5   5'-GACATTGTGATGACCCAGTCT-3'
    MKV.B6   5'-GATATTGTGCTAACTCAGTCT-3'
    MKV.B7   5'-GATATCCAGATGACACAGACT-3'
    MKV.B8   5'-GACATCCAGCTGACTCAGTCT-3'
    MKV.B9   5'-CAAATTGTTCTCACCCAGTCT-3'
    MKV.B10   5'-GACATTCTGATGACCCAGTCT-3'
  Lambda chain
  constant domain
    MLC.F   5'-GGTGAGTGTGGGAGTGGACTTGGGCTG-3'
  Lambda chain
  variable domain
    MLV.B   5'-CAGGCTGTTGTGACTCAGGAA-3'

Data from ref. 27.

Table 2
Oligonucleotides for the Reamplification of Mouse Immunoglobulin Variable Region DNA and Introduction of Appropriate Restriction Sites[a]

Heavy chain
  Constant domain
    MHC.F.Hind   5'-GGCCAGTGGATA<u>AAGCTT</u>TGGGGGTGTCGTTTTGGC-3'
  Variable domain
    MHV.B1.Nco   5'-GAATAGG<u>CCATGG</u>CGGATGTGAAGCTGCAGGAGTC-3'
    MHV.B2.Nco   5'-GAATAGG<u>CCATGG</u>CGCAGGTGCAGCTGAAGGAGTC-3'
    MHV.B3.Nco   5'-GAATAGG<u>CCATGG</u>CGCAGGTGCAGCTGAAGCAGTC-3'
    MHV.B4.Nco   5'-GAATAGG<u>CCATGG</u>CGCAGGTTACTCTGAAAGAGTC-3'
    MHV.B5.Nco   5'-GAATAGG<u>CCATGG</u>CGGAGGTCCAGCTGCAACAATCT-3'
    MHV.B6.Nco   5'-GAATAGG<u>CCATGG</u>CGGAGGTCCAGCTGCAGCAGTC-3'
    MHV.B7.Nco   5'-GAATAGA<u>CCATGG</u>CGCAGGTCCAACTGCAGCAGCCT-3'
    MHV.B8.Nco   5'-GAATAGG<u>CCATGG</u>CGGAGGTGAAGCTGGTGGAGTC-3'
    MHV.B9.Nco   5'-GAATAGG<u>CCATGG</u>CGGAGGTGAAGCTGGTGGAATC-3'
    MHV.B10.Nco   5'-GAATAGG<u>CCATGG</u>CGGATGTGAACTTGGAAGTGTC-3'
    MHV.B12.Nco   5'-GAATAGG<u>CCATGG</u>CGGAGGTGCAGCTGGAGGAGTC-3'

Light chain
  Kappa chain
  constant domain
    MKC.F.Not   5'-TGACAAGCTT<u>GCGGCCGC</u>GGATACAGTTGGTGCAGCATC-3'
  Kappa chain
  variable domain
    MKV.B1.Mlu   5'-TACAGGATCC<u>ACGCGT</u>AGATGTTTTGATGACCCAAACT-3'
    MKV.B2.Mlu   5'-TACAGGATCC<u>ACGCGT</u>AGATATTGTGATGACGCAGGCT-3'
    MKV.B3.Mlu   5'-TACAGGATCC<u>ACGCGT</u>AGATATTGTGATAACCCAG-3'
    MKV.B4.Mlu   5'-TACAGGATCC<u>ACGCGT</u>AGACATTGTGCTGACCCAATCT-3'
    MKV.B5.Mlu   5'-TACAGGATCC<u>ACGCGT</u>AGACATTGTGATGACCCAGTCT-3'
    MKV.B6.Mlu   5'-TACAGGATCC<u>ACGCGT</u>AGATATTGTGCTAACTCAGTCT-3'
    MKV.B7.Mlu   5'-TACAGGATCC<u>ACGCGT</u>AGATATCCAGATGACACAGACT-3'
    MKV.B8 Mlu   5'-TACAGGATCC<u>ACGCGT</u>AGACATCCAGCTGACTCAGTCT-3'
    MKV.B9.Mlu   5'-TACAGGATCC<u>ACGCGT</u>ACAAATTGTTCTCACCCAGTCT-3'
    MKV.B10.Mlu   5'-TACAGGATCC<u>ACGCGT</u>AGACATTCTGATGACCCAGTCT-3'
  Lambda chain
  constant domain
    MLC.F.Not   5'-TGACAAGCTT<u>GCGGCCGC</u>GGTGAGTGTGGGAGTGGACTTGGGCTG-3'
  Lambda chain
  variable domain
    MLV.B.Mlu   5'-TACAGGATCC<u>ACGCGT</u>ACAGGCTGTTGTGACTCAGGAA-3'

[a]Underlined, Recognition sequences of the restriction endonucleases used for cloning into pOPE101 or pSEX81plasmids.

11. Place the PCR tubes in a thermal cycler (see **Note 5**). Denature at 95°C for 5 min (see **Note 6**).
12. Perform 20 cycles with the following cycling program: 45 s of denaturation at 94°C, 45 s of annealing at the appropriate primer hybridization temperature (see **Note 9**), and 90 s of extension at 72°C. After 20 cycles, perform a final extension step for 5 min at 72°C.
13. Take out 1/10 volume of each PCR reaction for analytical agarose gel electrophoresis on a 2% agarose gel. Purify the PCR products using the kit. Elute DNA in H_2O (see **Note 8**).

3.2. Cloning into an Expression Vector (pOPE101)

Cloning of the two V-region gene fragments is done in two subsequent steps. First, the purified PCR product of the light chain is cloned into pOPE101:

1. Digest both the PCR product of the light chain and pOPE101-215(Yol) with *Mlu*I and *Not*I (see **Note 10**). To prevent self-ligation, the vector should be treated with CIP.
2. Run the digests on an agarose gel and gel-purify the vector backbone from the original V_L and gel- or PCR-purify the digested light chain. Elute plasmid DNA in H_2O.
3. Estimate DNA concentrations of vector and insert from an agarose gel using a calibrated marker. Prepare the ligation mix with an approximate molar ratio vector:insert of 1:3 in a reaction volume of 20 µL. The use of 100 ng of total DNA is recommended (see **Note 11**).
4. Incubate overnight at 16°C.
5. Add H_2O to a total volume of 50 µL and mix with 10 vol n-butanol, vortex, and centrifuge for 20 min at 15,000*g* at room temperature to precipitate the DNA. Let dry and redissolve the DNA in 20 µL H_2O.
6. Transform *E. coli* XL1-Blue cells (or any other suitable strain) and plate on LB agar plates containing 100 m*M* glucose and 100 µg/mL ampicillin (LB_{GA}). Incubate overnight at 37°C (see **Note 12**).
7. Pick five single colonies and grow in 5 mL LB medium containing 100 m*M* glucose and 100 µg/mL ampicillin. Shake overnight at 230 rpm, 37°C.
8. Miniprep the plasmid DNA and make glycerol stocks.
9. To confirm the presence of the appropriate insert, an analytical digest (run on agarose gel electrophoresis) with *Mlu*I and *Not*I as well as sequencing should be performed.

In a second step, the heavy chain is cloned:

10. Digest the PCR product of the heavy chain and the ligation product from **step 9** with *Nco*I and *Hin*dIII. The vector should be CIP-treated.
11. From here you can follow **steps 2–8** as given for the light chain DNA cloning.
12. Confirm the presence of the correct insert by digesting the construct with *Nco*I and *Hin*dIII as well as by sequencing.

3.3. Small-Scale Expression Test

To confirm the expression and correct size of the scFv fragment, a small-scale expression followed by Western blotting should be performed (see **Note 13**).

1. Prepare an overnight culture of *E. coli* cells transformed with the appropriate pOPE vector construct in 5 mL of LB_{GA} medium (see **Note 14**).
2. Dilute 300 µL of the overnight culture into 6 mL (1/20) of LB_{GA} and shake at 37°C and 230 rpm to an OD_{600} of 0.6–0.8 (see **Note 15**).
3. Separate the culture into two equal aliquots. To one of them add IPTG to a final concentration of 50 µ*M* (see **Note 16**).

4. Incubate the induced and the control culture for 3 h with vigorous shaking at 270–280 rpm at 25°C (see **Note 17**).
5. Incubate the cultures for 10 min on ice.
6. To check the production of the antibody fragment, take out 1 mL of each culture for immunoblotting. Centrifuge at 5000g for 5 min at 4°C and resuspend the pellet in 100 µL 2X SDS sample buffer. Boil at 95°C for 5 min and spin down for 3 min at 12,000g before loading the supernatant onto the SDS gel (see **Note 18**).
7. In addition, a periplasmic fraction can be prepared from the induced culture to test directly for antigen binding affinity in ELISA (see **Note 19**).

4. Notes

1. Use the RNeasy Mini Kit protocol for the isolation of total RNA from animal cells. Use QIAshredder column for the homogenization of the cells.
2. Use the Oligotex mRNA Spin-Column protocol for the isolation of poly A$^+$ mRNA from total RNA. Use up to 250 µg total RNA for the "miniprep" protocol. The concentration and purity of poly A$^+$ mRNA can be determined by measuring the absorbance at 260 nm and 280 nm in a spectrophotometer.
3. Use the protocol for First-Strand Synthesis Using Oligo(dT) for the isolated mRNA instead of total RNA, as given.
4. The PCR setup with the proofreading ProofStart DNA polymerase can be done at room temperature. For detailed information, see the ProofStart™ PCR Handbook. Alternatively, it is possible to use other proofreading polymerases, e.g., the Expand High Fidelity PCR System (Roche). Only a few primer combinations will give a PCR product. Please note that due to the conditions of the reaction, the same primary sequence may be amplified with several different primer combinations, depending on the individual sequence. The size of the resulting PCR product is approx 350 bp.
5. If using a thermal cycler with a heated lid, do not use mineral oil. Otherwise, overlay the reaction with approx 50 µL mineral oil to prevent evaporation.
6. The ProofStart DNA Polymerase is activated by this initial heating step.
7. For the oligonucleotide primers described in **Table 1**, an annealing temperature of 54°C should be tried initially. If no PCR products are found, decrease the annealing temperature by steps of 2 degrees.
8. Use a PCR Purification Kit. If multiple bands are found on the gel per reaction, use a Gel Extraction Kit for purification of the correct PCR product (size of approx 350 bp).
9. An annealing temperature of 56°C is recommended.
10. Per µg of vector, use 3 U *Not*I and 3 U *Mlu*I, 2 h, 37°C. Per µg of insert, apply 30 U *Not*I and 30 U *Mlu*I, 2 h, 37°C.
11. A critical factor for the proper function of the T4-DNA ligase is ATP. It is recommended to prepare aliquots of 10X reaction buffer, which has been supplemented with ATP to 100 mM and store frozen until used. **Caution:** This (high) ATP concentration inhibits blunt-end ligation reactions.
12. Electroporation is recommended because of its unsurpassed transformation efficiency. Use half of the ligation and 3–5 µL of electrocompetent cells (Stratagene) and bring to a volume of 50 µL with H$_2$O. For a 2-mm-diameter cuvet, use 2.5 kV/25 µF at 200 Ω. Alternatively, chemical transformation using a heat shock can be performed.
13. To save time, this test can be combined with the generation of cells for DNA "minipreps" and preparation of glycerol stocks by using the remainder of the overnight starter culture.

14. If possible, use glycerol stocks for the preparation of overnight cultures because clones on agar plates can mutate easily after prolonged storage even at 4°C. Glucose must always be present in the bacterial growth medium since it is necessary for the tight suppression of the synthetic promotor of the pOPE vector family and thus for the genetic stability of the insert.
15. Protein production from pOPE101 cannot be induced in bacteria grown to stationary phase.
16. With pOPE-vectors in *E. coli* XL1-Blue, we achieved optimal protein secretion with 20 μM IPTG at 25°C. This optimal IPTG concentration can vary between different Fv sequences by a factor of about 2. Higher IPTG concentrations lead to dramatically increased amounts of total recombinant protein, but in this case most of the scFv fragments still carry the bacterial leader sequence and form aggregates. However, for immunoblot analysis of total cellular SDS extracts, it is not necessary to discriminate between unprocessed and processed protein. Therefore, a higher IPTG concentration is used simply to increase the intensity of the protein band on the blot. To optimize expression conditions, IPTG concentrations between 10 and 100 μM should be tested for each individual fusion protein. Be aware that IPTG is light-sensitive and may decay during prolonged storage. Use stocks aliquoted and stored in brown 1.5-mL tubes at –20°C for not longer than 2 mo.
17. A maximum of functional scFv fragments was achieved at 25°C. Incubation times longer than 3 h lead to a slight increase in the amount of secreted protein, but a significantly higher contaminant concentration as well, possibly due to increased cell death. However, depending on the hybridoma antibody sequence, differences were found in solubility and the ability to be secreted. The expression of some antibodies even led to a strong growth inhibition during induction.
18. For detection of scFv produced from pOPE101, the monoclonal antibody Myc1-9E10 recognizing the c-myc tag or an antibody to the his tag is recommended. HRP-conjugated antibodies to mouse immunglobulins should be applied before TMB-stabilized substrate for HRP (Progen, Heidelberg, Germany) can be used for the detection of bound enzymatic activity.
19. Soluble antibody fragments can be isolated from induced *E. coli* cultures by osmotic shock. For immunodetection in ELISA, the same antibodies as in **Note 18** are suitable. Protocols for both methods can be found in Schmiedl et al. *(24)*.

Acknowledgment

We gratefully acknowledge the funding of O.B. by the Graduiertenkolleg 388 of the DFG.

References

1. Huston, J. S., Levinson, D., Mudgett-Hunter, M., et al. (1988) Protein engineering of antibody binding sites: recovery of specific activity in an anti-digoxin single-chain Fv analogue produced in *Escherichia coli*. *Proc. Natl. Acad. Sci. USA* **85,** 5879–5883.
2. Bird, R. E., Hardman, K. D., Jacobson, J. W., et al. (1988) Single-chain antigen-binding proteins. *Science* 242, 423–426.
3. Whitlow, M. and Filpula, D. (1991) Single-chain Fv proteins and their fusion proteins. *Methods Companion Methods Enzymol.* **2,** 97–105.
4. Orlandi, R., Güssow, D. H., Jones, P. T., and Winter, G. (1989), Cloning immunoglobulin variable domains for expression by the polymerase chain reaction. *Proc. Natl. Acad. Sci. USA* **86,** 3833–3837.

5. Ørum, H., Andersen, P. S., Øster, A., et al. (1993) Efficient method for constructing comprehensive murine Fab antibody libraries displayed on phage. *Nucleic Acids Res.* **21**, 4491–4498.
6. Dübel, S., Breitling, F., Fuchs, P., et al. (1994) Isolation of IgG antibody Fv-DNA from various mouse and rat bybridoma cell lines using the polymerase chain reaction with a simple set of primers. *J. Immunol. Methods.* **175**, 89–95.
7. Chen, Y. W., Word, C. J., Jones, S., Uhr, J. W., Tucker, P. W., and Vitetta, E. S. (1986) Double isotype production by a neoplastic B cell line. I. Cellular and biochemical characterization of a variant of BCL that expresses and secretes both IgM and IgG1. *J. Exp. Med.* **164**, 548.
8. Shimizu, A., Nussenzweig, M. C., Han, H., Sanchez, M., and Honjo, T. (1991) Trans-splicing as a possible molecular mechanism for the multiple isotype expression of the immunoglobulin gene. *J. Exp. Med.* **173**, 1385.
9. Pauza, M. E., Rehmann, J. A., and LeBien, T. W. (1993) Unusual patterns of immunoglobulin gene rearrangement and expression during human B cell ontogeny: human B cells can simultaneously express cell surface kappa and lambda light chains. *J. Exp. Med.* **178**, 139.
10. Welschof, M., Terness, P., Kolbinger, F., et al. (1995) Amino acid sequence based PCR primers for the amplification of human heavy and light chain immunoglobulin variable region genes. *J. Immunol. Methods* **179**, 203.
11. De Haard, H. and Kazemier, B. (1995) The effect of mutations in the primer encoded FR1 and FR4 regions of VH and VK on the reactivity of scFv-s. *Hum. Antibod. Hybridomas* **6**.
12. Krauss, J., Arndt, M. A. E., Martin, A. C. R., Liu, H., and Rybak, S. (2002) Specificity grafting of human antibody frameworks selected from a phage display library, in *Proceedings of the 13th International IBC Conference on Antibody Engineering*, IBC, Westborough, MA.
13. Marks, J. D., Hoogenboom, H. R., Bonnert, T. P., McCafferty, J., Griffiths, A. D., and Winter, G. (1991) By-passing immunization: human antibodies from V-gene libraries displayed on phage. *J. Mol. Biol.* **222**, 581–597.
14. Breitling, F., Dübel, S., Seehaus, T., Klewinghaus, I., and Little, M. (1991) A surface expression vector for antibody screening. *Gene* **104**, 147–153.
15. Dübel, S., Breitling, F., Fuchs, P., Braunagel, M., Klewinghaus, I., and Little, M. (1993) A family of vectors for surface display and production of antibodies. *Gene* **128**, 97–101.
16. Dörsam, H., Braunagel, M., Kleist, C., Moynet, D., and Welschof, M. (1997) Screening of phage displayed libraries, in *Molecular Diagnosis of Infectious Diseases, Methods in Molecular Medicine*, vol. 13 (Reischl, U., ed.), Humana, Totowa, NJ, pp. 595–610.
17. Li, J. Y., Sugimura, K., Boado, R. J., et al. (1999) Genetically engineered brain drug delivery vectors: cloning, expression, and in vivo application of an anti-transferrin receptor single chain antibody-streptavidin fusion gene and protein. *Protein Eng.* **12**, 787–796.
18. Plückthun, A. and Pack, P. (1997) New protein engineering approaches to multivalent and bispecific antibody fragments. *Immunotechnology* **3**, 83–105.
19. Kipriyanov, S. M., Moldenhauer, G., Martin, A. C., Kupriyanova, O. A., and Little, M. (1997) Two amino acid mutations in an anti-human CD3 single chain Fv antibody fragment that affect the yield on bacterial secretion but not the affinity. *Protein Eng.* **10**, 445–453.
20. Leung, S. O., Goldenberg, D. M., Dion, A. S., et al. (1995) Construction and characterization of a humanized, internalizing, B-cell (CD22)-specific, leukemia/lymphoma antibody, LL2. *Mol. Immunol.* **32**, 1413–1427.

21. Co, M. S., Scheinberg, D. A., Avdalovic, N. M., et al. (1993) Genetically engineered deglycosylation of the variable domain increases the affinity of an anti-CD33 monoclonal antibody. *Mol. Immunol.* **30,** 1361–1367.
22. Nishimura, E., Mochizuki, K., Kato, M., et al. (1999) Recombinant light chain of human monoclonal antibody HB4C5 as a potentially useful lung cancer-targeting vehicle. *Hum. Antibod.* **9,** 111–124.
23. Skerra, A. and Plückthun, A. (1988) Assembly of a functional immunoglobulin Fv fragment in *Escherichia coli. Science* **240,** 1038–1041.
24. Schmiedl, A., Breitling, F., Winter, C., Queitsch, I., and Dübel, S. (2000) Effects of unpaired cysteines on yield, solubility and activity of different recombinant antibody constructs expressed in *E. coli. J. Immunol. Methods* **242,** 101–114.
25. Kontermann, R. and Dübel, S. (eds.) (2001) *Antibody Engineering.* Springer-Verlag, New York (ISBN 3-540-41354-5).
26. Sambrook, J., Fritsch, E. F., and Maniatis, T. (1989) *Molecular Cloning. A Laboratory Manual*, 2nd ed. Cold Spring Harbor Laboratory Press, Cold Spring Harbor, NY.
27. Zhou, H., Fisher, R. J., and Papas, T. S. (1994) Optimization of primer sequences for mouse scFv repertoire display library construction. *Nucleic Acids Res.* **22,** 888–889.

Index

A

Adenovirus, 383
Affinity
  constants, 303–328
  maturation, 416
Algae, 191–196
  transformation, 193
Aminopeptidase M, 114
Antibodies
  avidity and affinities, 303–305, 416
  binding activity, 417
  determination of kinetic constants, 303–305
  hybridoma cells, 447–448
  monoclonal, 333–372
  phage-displayed antibody libraries, 449
  polyclonal, 333–372
  preparation of antibody-marker conjugates, 260–266
  purification, 236
  recombinant single chain fragments
    cloning, 451–454
    expression, 449–450, 454
    glycosylation, 450
    isolation of antibody DNA, 451
    oligonucleotides, 452–453
    small-scale induction, 454
  screening, 300

Antibody engineering, 447-454
Antibody columns, 112
Antigens
  B-cell
    identification, 8–9, 92
  T-cell
    cytotoxic T-cells, 127
    identification, 11, 92
    T-helper, 127
Aptamers, 321–330
Automated protein purification, 179–190

B

Baculovirus
  determination of virus titer, 146
  generation of recombinant baculoviruses, 45, 144–147
  infection of insect cells, 146
  propagation of insect cells, 144
Biosensor
  biomolecular interaction analysis (BIA)
    application examples, 303–305
    data preprocessing, 313
    determination of kinetic constants, 312–314
    immobilization of ligands, 310–311
    kinetic theory, 316–317
    principles, 300-303
    sensor chips, 307

quartz crystal microbalance (QCM)
  application examples, 321
  coating
    aptamers, 325
    proteins, 323-324
  determination of kinetic constants, 325–328
  fluid handling, 325
  principle, 322
Biotin, 41, 257–259
Borrelia burgdorferi, 393–428

C

Carboxypeptidase Y, 114
Cell fractionation, 7–8
Chaotropic agents, 226–227
Carrier proteins, 255
Chlamydomonas reinhardtii, 191–196
Chlorella vulgaris, 191–196
Chromium release assay, 149–151
CNBr-Sepharose, 112
Concanavalin A, 7, 354
Concentration steps, 7
Crosslinking, 112
Cytokines
  determination of production, 12, 148, 170–171

D

Depletion of cell subsets, 125
Diagnostic antigens
  conserved epitopes, 377–378
  designer antigens, 378
  expression
    strategies, 376-380
    systems, 380–384
Differential gene expression, 50–66, 67–90

DNA microarrays
  data mining
    clustering, 79
    literary data mining, 82-86
      MedMOLE, 84
    transcriptional change and functional meaning, 79–82
  data normalization, 70–75
  DNA spotted arrays, 68–69
  oligonucleotide arrays, 69–70
  statistical validation of differential gene expression, 75–78
DNA vaccination, 201
Drosophila DS-2
  generation of expression vectors, 167
  induction of protein expression, 168
  transfection, 168

E

Edman degradation, 280–282
ELIspot, 125–126, 148–149, 173
Endoproteinase LysC, 114
Enzyme linked immunoassay (ELISA), *see* Immunoassays
Epitope mapping
  by limited proteolysis, 110–120
  by ELIspot using peptide arrays, 121–132
  of linear epitopes, 110–120, 121–132
Expression cloning
  adsorption of antisera, 101
  colony screening, 101–103
  excision of plasmid, 103
  generation of expression libraries, 98–100
  principle, 91-97

F

Flow cytometry

determination of T-cell responses, 152
intracellular cytokine stain, 151–152, 171–172

G

Glutathione, 243

H

Helicobacter pylori
   cultivation, 169
   immunomodulation, 159–176
   infection of splenocytes, 170
   proteome, 21
High throughput protein expression, 179–190
His tag, 179–190
HIV, 109, 133–158, 197–210, 393–428

I

Immunoassays
   blocking, 358-361
   capture ELISA, 201
   detection of serum-IgG, 387–388
   guidelines, 364-367
   immobilization
      bacteria, viruses, cells, 356–357
      covalent bonding, 355–356
      direct adsorption, 347
      guidelines, 357–358, 361–362
      immunochemical, 354
      lectin, 354
      protein A/G, 354-355
      streptavidin-biotin, 354
   integrity of reactants, 349–354
   labels, 8
   matrices, 342–347
   PACE, 357
   principles
      diffusion dependence, 334–336
      hysteresis, 340–342
      interfacial interactions, 336–337
      solid phase vs solution, 337–340
   radioisotopes, 35
   rapid test, 222–223
   streptavidin, 42
   test-titration, 386–387
Immunoblotting
   blotting, 8–10, 27, 222
   digoxigenin labeling, 9–10, 41
   dot blots, 39
   electrophoresis, 8–10, 222
   membranes, 6, 23, 222, 272
   optimization of titers, 8, 27, 222
   quick protocol, 296–297
   sample preparation, 8–12, 222
   transfer buffers, 5, 23, 273
   visualization
      chromogenic, 10
      luminescent, 27
Immunoprecipitation
   direct immunoprecipitation, 36–37
   indirect immunoprecipitation, 37
   lysis of cells, 36
      buffers, 34
      detergents, 34, 36
      protease inhibitors, 34
   protein A, 37, 354–355
   protein G, 354–355
   quantization of antigens
      autoradiography, 37–38
      fluorography, 38
   separation of complexes, 37
Immunization of mice, 147
Insect cells expression systems, 382
Interferon-gamma
   determination, 12, 148, 170
   ELIspot, 125–126, 148–149, 173
   FACS, 151–152, 171–172

induction, 12
Internet
  compendia of resources, 432–433
  databases, 28, 84, 431, 439-440
  distributor based information, 438–439
  electronic classrooms, 440
  electronic journals, 438
  mailing lists, 433–438
  search engines, 433
  telemedicine, 440
In vitro protein expression, 384–385
In vitro transcription, 201
In vitro translation, 201

K

Keyhole limpet hemocyanin (KLH), 255, 260–262

L

Labeling of cells
  cell surface labeling, 36
  metabolic labeling, 35
Lamda ZAP, 98–100
Lectins, 39–42, 354
Lentivirus, 383
Liquid phase electrophoresis, 11

M

Magnetic beads, 125
MALDI, 29, 116, 282–284
Mass spectrometry
  ESI, 284–287
  liquid chromatography, 11
  MALDI/MS, 116, 282–284
  protein identification, 289–294
  sequencing, 287–289
Medline, 82
Molecular farming, 191
Mycobacterium tuberculosis, 3

N

Neisseria meningitides, 51

O

Overlapping peptides, 124

P

PBMC, 129
PCR, 50–66
Peptides
  purification by HPLC, 277–280
  sequencing, 25, 280
Periplasm
  polymyxin B, 167
  preparation, 167–168, 218
Posttranslational modifications, 23, 31, 111, 191–196, 214, 226, 250, 269, 375, 413, 449
pQE vector system, 180
Proteinases, 114, 277
  inhibitors, 5
Protein denaturation, 242–243
Protein expression
  cysteine rich proteins, 167–168
  determination of protein location, 217–218
  genetic stability, 376
  induction by temperature shift, 234
  native, 191–196
  rapid screening of expression, 216–216
  strategy, 376–377
  systems, 380–384
Protein–protein conjugation,
  antibody–antibody conjugates, 263
  antibody–enzymes conjugates, 262–263
  antibody–KLH conjugates, 260–262
Protein purification
  affinity

purification, 179–190
tags, 179–190, 248
basic proteins, 225–238
centrifugation, 385–386
continuous elution electrophoresis, 220–222
fusion proteins, 248
His tagged proteins
   denatured purification, 180–186
   immobilized metal ion affinity chromatography (IMAC), 180–186
   native purification, 180-186
   Ni-NTA magnetic agarose beads, 183
inclusion bodies
   isolation, 218–219, 242
   morphology, 240–241
   solubilization, 219–220, 227
insoluble proteins, 214-224, 227
ion exchange chromatography, 235
lysis of cells
   freeze/thaw cycles, 217
   lysis of E. coli cells, 182–190, 218–219, 235
   French press, 7
organic precipitation, 235
principles, 225–226
Protein refolding
   additives, 246
   chaperones, 249–250
   continuous refolding, 243
   L-arginine, 245–246
   matrix-assisted refolding, 248
   micelles, 248
   mixed disulfides, 243
   pH, 247
   protein disulfide isomerase, 249
   pulse renaturation, 243
   temperature, 245
Protein sequencing
   blotting, 272-273
   in-gel digestion, 10–11, 29, 276, 297
   sample preparation
      denaturation, 275
      reduction, 275–276
   staining using Coomassie G-250, 29
   strategies, 270–271
Protein–small molecule conjugation
   biotinylation, 257-259
   via amine residues, 260
   via carboxyl residues (NHS esters), 256–259
   via hydroxyl residues, 259-60
Proteomics
   immunoproteomics, 3–18, 19–32
   spot Identification, 28
Pubmed, 431–432

R

Representational differential analysis RDA, 49–66
   cDNA synthesis, 57
   cloning, 63–64
   Dpn II, 53
   driver, 52
   mung bean nuclease, 52
   oligonucleotides, 55
   PCR coupled substractive hybridization
      generation of representations, 59
      ligation of adapters, 59
      preparation of tester and driver populations, 60
      substractive hybridization, 61
      second difference product, 62–63

principle, 49-53
RNA isolation, 56-57
tester, 52
Retrovirus expression systems, 384

S

Serological laboratory diagnosis
  antigen definition, 410–412
  biological activities of antibodies, 417
  characteristics of antigens, 413–414
  confirmatory tests, 396
  expert systems and neural networks, 423
  external quality control, 421
  identification of antigens, 412–413
  internal quality control, 421
  nonlinear dilution effects, 417
  polyclonal antibodies, 415-416
  polyclonal B-cell activation, 410
  purity of antigens, 414-415
  receiver-operating characteristics, 399–401
  screening tests, 395–396
  S-distribution, 400
  sero-scars, 410
  serotypes and mutations, 414
  standardization, 418-421
  test of disease activity, 396-397
  test performance
    causal model, 404–408
    cut off, 397–399
    diagnostic model, 404
    efficiency, 401
    likelihood ratios, 402–404
    positive and negative predictive value, 401–402
    prognostic model, 408–409

  sensitivity, 398–399
  specificity, 398–399
Splenocytes
  induction, 12
  positive controls, 7
  preparation, 12, 124, 147–148, 169–170
Synthetic genes
  algae, 195
  cis-acting sequences, 202–205
  codon adaptation, 202–205
  GC content, 202–205
  increase of protein expression, 195, 197–210
  mRNA stability, 202–205

T

TA cloning, 64
Th1/Th2 type immune response, 160–163, 206–207
Two-dimensional gel electrophoresis
  data analysis, 24, 28
  isoelectric focusing
    preparation of samples, 8
    preparation of strips, 8
  SDS PAGE, 8
T4-ligase, 50–66

V

Virus-like particles
  generation via baculovirus expression system, 144–145
  induction of immune responses, 133–158
  purification, 147

Y

Yeast expression systems, 382